METHODS IN MOLECULAR BIOLOGY™

Series Editor
John M. Walker
School of Life Sciences
University of Hertfordshire
Hatfield, Hertfordshire, AL10 9AB, UK

For other titles published in this series, go to
www.springer.com/series/7651

The Blood-Brain and Other Neural Barriers

Reviews and Protocols

Edited by

Sukriti Nag

University of Toronto, Toronto, ON, Canada

☀ Humana Press

Editor
Sukriti Nag, MD, Ph.D.
Department of Laboratory Medicine and Pathobiology
University of Toronto
Toronto, ON
Canada
sukriti_nag@rush.edu

ISSN 1064-3745   e-ISSN 1940-6029
ISBN 978-1-60761-937-6   e-ISBN 978-1-60761-938-3
DOI 10.1007/978-1-60761-938-3
Springer New York Dordrecht Heidelberg London

Cover Illustration: Modified merged confocal image showing components of the neurovascular unit. Image provided
by Anish Kapadia and Sukriti Nag.

Printed on acid-free paper

Humana Press is part of Springer Science+Business Media (www.springer.com)

Preface

More than 100 years have elapsed since the discovery of the blood-brain barrier (BBB). Evolving technologies starting with tracer studies, and more recently with genomics and proteomics, have provided novel information about the molecular properties of cerebral endothelium and astrocytes. The concept of the neurovascular unit has provided an impetus for in vitro studies of the interaction of brain endothelial cells with other components of the neurovascular unit such as pericytes, astrocytes, and neurons in steady states. However, such studies have to be done in animal models of neurological diseases and in humans to get a clearer understanding of the pathogenesis of BBB breakdown in nervous system diseases. Determination of the temporal course of BBB breakdown and the parallel molecular alterations remain important goals to identify therapeutic windows and pertinent therapeutic agents which will modify the disease process and prevent irreversible brain damage. There is also the need to develop imaging techniques for early diagnosis of brain diseases before irreparable tissue damage results. Although, modest advances have been made in the area of the BBB, parallel advances have not been made in the other neural barriers. These need to be studied to obtain an overall picture of the disease process.

The Blood-Brain and Other Neural Barriers: Reviews and Protocols, a sequel to *The Blood-Brain Barrier: Biology and Research Protocols,* provides the reader with additional protocols to study the barriers. The first section consists of current reviews of the properties of some of the components of the neurovascular unit, namely the brain endothelium, pericytes, and astrocytes. In addition, current information about the blood- cerebrospinal fluid barrier, the blood-retinal and blood-nerve barriers is also provided. The second section of the book gives detailed protocols of specific techniques written by experts in the field. The protocols include applications as well as caveats of these techniques. The first part describes techniques to image the barriers in humans and experimental animals, followed by cutting-edge molecular techniques to study the BBB and novel models to study the barriers. The last part details some of the prevalent techniques for the delivery of therapeutic agents across the BBB. This is a rapidly growing and competitive field, and, in some cases, results are too preliminary for publication of detailed protocols. It is hoped that the detailed protocols given in this book will aid the research efforts of not only graduate students but also more experienced investigators and will enable more studies of the blood-cerebrospinal, blood-retinal, and blood-nerve barriers.

I would like to acknowledge Prof. John M. Walker for this opportunity and for his help and all the authors who have contributed their protocols without which this book would not be possible.

This book is dedicated to my parents, Mohit Kumar and Labonya Nag for their unconditional love and support and for giving me the opportunity to pursue my goals.

Sukriti Nag
Toronto, ON

Contents

Contributors

ABEDELNASSER ABULROB • *Cerebrovascular Research Group, Institute for Biological Sciences, National Research Council, Ottawa, ON, Canada*

DANIEL C. ANTHONY • *Departments of Pharmacology and Chemistry, University of Oxford, Oxford, UK*

DAVID A. ANTONETTI • *Department of Ophthalmology, Penn State College of Medicine, Hershey, PA, USA*

ROLAND J. BAINTON • *Department of Anesthesia and Perioperative Care, San Francisco General Hospital, University of California at San Francisco, San Francisco, CA, USA*

ANDREW BAIRD • *Department of Surgery, University of California San Diego, San Diego, CA, USA*

REINA BENDAYAN • *Department of Pharmaceutical Sciences, University of Toronto, Toronto, ON, Canada*

MARIE BLANCHETTE • *Departments of Neurosurgery and Neuro-oncology, University of Sherbrooke Hospital, Sherbrooke, QC, Canada*

ROMAIN CAYROL • *Neuroimmunology Research Laboratory, CHUM-Notre-Dame Hospital, Université de Montréal, Montréal, QC, Canada*

GARY N.Y. CHAN • *Department of Pharmaceutical Sciences, University of Toronto, Toronto, ON, Canada*

ROBIN P. CHOUDHURY • *Department of Cardiovascular Medicine, John Radcliffe Hospital, University of Oxford, Oxford, UK*

KRISTEN CLEARY • *Department of Neurology, Wayne State University School of Medicine, Detroit, MI, USA*

BEN DAVIS • *Departments of Pharmacology and Chemistry, University of Oxford, Oxford, UK*

AURORE DODELET-DEVILLERS • *Neuroimmunology Research Laboratory, CHUM-Notre-Dame Hospital, Université de Montréal, Montréal, QC, Canada*

PAULA DORE-DUFFY • *Department of Neurology, Wayne State University School of Medicine, Detroit, MI, USA*

KATARZYNA M. DZIEGIELEWSKA • *Department of Pharmacology, University of Melbourne, Parkville, Victoria, Australia*

C. JOAKIM EK • *Department of Pharmacology, University of Melbourne, Parkville, Victoria, Australia*

BRIAN P. ELICEIRI • *Department of Surgery, University of California San Diego, San Diego, CA, USA*

JEHAD EL-GINDI • *Institute of Biochemistry, Westfälische Wilhelms Universität Münster, Münster, Germany*

JAMES R. EWING • *Department of Neurology, Henry Ford Hospital, Detroit, MI, USA*

JOSEPH D. FENSTERMACHER • *Department of Anesthesiology, Henry Ford Hospital, Detroit, MI, USA*

DAVID FORTIN • *Departments of Neurosurgery and Neuro-oncology, University of Sherbrooke Hospital, Sherbrooke, QC, Canada*

HANS-JOACHIM GALLA • *Institute of Biochemistry, Westfälische Wilhelms Universität Münster, Münster, Germany*

ANA MARIA GONZALEZ • *Molecular Neuroscience Group, School of Medicine, University of Birmingham, Edgbaston, Birmingham, UK*

MARK D. HABGOOD • *Department of Pharmacology, University of Melbourne, Parkville, Victoria, Australia*

ROBIN D. HAMILTON • *School of Biomedical Sciences, University of Nottingham, Nottingham, UK*

WILLIAM S. HANCOCK • *Barnett Institute and Department of Chemistry, Northeastern University, Boston, MA, USA*

ARSALAN S. HAQQANI • *Proteomics Group, Institute of Biological Sciences, National Research Council, Ottawa, ON, Canada*

JENNIFER J. HILL • *Proteomics Group, Institute of Biological Sciences, National Research Council, Ottawa, ON, Canada*

IGAL IFERGAN • *Neuroimmunology Research Laboratory, CHUM-Notre-Dame Hospital, Université de Montréal, Montréal, QC, Canada*

UMAR IQBAL • *Cerebrovascular Research Group, Institute for Biological Sciences, National Research Council, Ottawa, ON, Canada*

CONRAD E. JOHANSON • *Department of Clinical Neuroscience, Alpert Medical School at Brown University, Providence, RI, USA*

PIA JOHANSSON • *Institute for Stem Cell Research, Helmholtz Zentrum München, Neuherberg, Germany*

TAKASHI KANDA • *Department of Neurology and Clinical Neuroscience, Yamaguchi University Graduate School of Medicine, Ube, Japan*

ANISH KAPADIA • *Department of Laboratory Medicine and Pathobiology, University of Toronto, Toronto, ON, Canada*

KISHOR KARKI • *Department of Neurology, Henry Ford Hospital, Detroit, MI, USA*

ANDREA KASSNER • *Department of Medical Imaging, University of Toronto, Toronto, ON, Canada*

ROBERT A. KNIGHT • *Department of Neurology, Henry Ford Hospital, Detroit, MI, USA*

LOPA LEACH • *School of Biomedical Sciences, University of Nottingham, Nottingham, UK*

WENDY LEADBEATER • *Molecular Neuroscience Group, School of Medicine, University of Birmingham, Edgbaston, Birmingham, UK*

SHANE LIDDELOW • *Department of Pharmacology, University of Melbourne, Parkville, Victoria, Australia*

MIRA LISCHPER • *Institute of Biochemistry, Westfälische Wilhelms Universität Münster, Münster, Germany*

QIAOZHAN LU • *Barnett Institute and Department of Chemistry, Northeastern University, Boston, MA, USA*

JENNIFER A. MACDONALD • *Department of Cell Biology, University of Connecticut Health Center, Farmington, CT, USA*

JANET L. MANIAS • *Department of Laboratory Medicine and Pathobiology, University of Toronto, Toronto, ON, Canada*

FAHIMA MAYER • *Department of Anesthesia and Perioperative Care, San Francisco General Hospital, University of California at San Francisco, San Francisco, CA, USA*

NASIMA MAYER • *Department of Anesthesia and Perioperative Care, San Francisco General Hospital, University of California at San Francisco, San Francisco, CA, USA*

MARTINA A. MCATEER • *Department of Cardiovascular Medicine, John Radcliffe Hospital, University of Oxford, Oxford, UK*

PAUL N. MCMILLAN • *Department of Pathology, Alpert Medical School at Brown University, Providence, RI, USA*

ANDREW P. MIZISIN • *Department of Pathology, School of Medicine, University of California San Diego, San Diego, CA, USA*

JAMES MULLEN • *Proteomics Group, Institute of Biological Sciences, National Research Council, Ottawa, ON, Canada*

NIVETHA MURUGESAN • *Department of Cell Biology, University of Connecticut Health Center, Farmington, CT, USA*

SUKRITI NAG • *Department of Laboratory Medicine and Pathobiology, University of Toronto, Toronto, ON, Canada*

TAVAREKERE N. NAGARAJA • *Department of Anesthesiology, Henry Ford Hospital, Detroit, MI, USA*

VIJAYA NAGESH • *Department of Radiation Oncology, University of Michigan, Ann Arbor, MI, USA*

JOEL S. PACHTER • *Department of Cell Biology, University of Connecticut Health Center, Farmington, CT, USA*

ROBERT L. PINSONNEAULT • *Department of Anesthesia and Perioperative Care, San Francisco General Hospital, University of California at San Francisco, San Francisco, CA, USA*

ALEXANDRE PRAT • *Department of Neurology, Neuroimmunology Research Laboratory, CHUM-Notre-Dame Hospital, Université de Montréal, Montréal, QC, Canada*

E. AARON RUNKLE • *Department of Cellular and Molecular Physiology, Penn State College of Medicine, Hershey, PA, USA*

YASUTERU SANO • *Department of Neurology and Clinical Neuroscience, Yamaguchi University Graduate School of Medicine, Ube, Japan*

NORMAN R. SAUNDERS • *Department of Pharmacology, University of Melbourne, Parkville, Victoria, Australia*

NICOLA R. SIBSON • *Experimental Neuroimaging Group, Department of Physiology, Anatomy and Genetics, University of Oxford, Oxford, UK*

DANICA B. STANIMIROVIC • *Proteomics Group, Institute of Biological Sciences, National Research Council of Canada, Ottawa, ON, Canada*

EDWARD G. STOPA • *Department of Pathology, Alpert Medical School at Brown University, Providence, RI, USA*

HUA SU • *Center for Cerebrovascular Research, Department of Anesthesia and Perioperative Care, University of California San Francisco, San Francisco, CA, USA*

NEBIYU TEGEGN • *Department of Anesthesia and Perioperative Care, San Francisco General Hospital, University of California at San Francisco, San Francisco, CA, USA*

GOKULAN THANABALASUNDARAM • *Institute of Biochemistry, Westfälische Wilhelms Universität Münster, Münster, Germany*

REBECCA THORNHILL • *Department of Medical Imaging, University of Toronto, Toronto, ON, Canada*

ANANDA WEERASURIYA • *Division of Basic Medical Sciences, School of Medicine, Mercer University, Macon, GA, USA*

CARSTEN WESSIG • *Department of Neurology, University of Würzburg, Würzburg, Germany*

SHIAW-LIN WU • *Barnett Institute and Department of Chemistry, Northeastern University, Boston, MA, USA*

GUO-YUAN YANG • *Med-X Research Institute, Sanghai JiaoTong University, Shanghai, China; Center for Cerebrovascular Research, Department of Anesthesia and Perioperative Care, University of California at San Francisco, San Francisco, CA, USA*

Part I

Biology of the Barriers

Chapter 1

Morphology and Properties of Brain Endothelial Cells

Sukriti Nag

Abstract

The molecular advances in various aspects of brain endothelial cell function in steady states are considerable and difficult to summarize in one chapter. Therefore, this chapter focuses on endothelial permeability mechanisms in steady states and disease namely vasogenic edema. The morphology and properties of caveolae and tight junctions that are involved in endothelial permeability to macromolecules are reviewed. Endothelial transport functions are briefly reviewed. Diseases with alterations of endothelial permeability are mentioned and details are provided of the molecular alterations in caveolae and tight junctions in vasogenic edema. Other factors involved in increased endothelial permeability such as the matrix metalloproteinases are briefly discussed. Of the modulators of endothelial permeability, angioneurins such as the vascular endothelial growth factors and angiopoietins are discussed. The chapter concludes with a brief discussion on delivery of therapeutic substances across endothelium.

Key words: Adherens junctions, Angioneurins, Angiopoietin 1, Angiopoietin 2, Blood-brain barrier, Brain endothelium, Carrier-mediated transport, Caveolae, Caveolin-1, Claudins, Cortical cold-injury model, Efflux transport, Junctional adhesion molecule-1, Matrix metalloproteinases, Occludin, Receptor-mediated transport, Tight junctions, Transcytosis, Vascular endothelial growth factor-A, Vascular endothelial growth factor-B, Vasogenic edema, Zonula occludens

1. Introduction

The recognition of a restrictive barrier that limits the passage of substances from the blood into the brain dates back to the work of Paul Ehrlich (1), who injected coerulean-S sulfate intravenously into rodents and observed that all the body organs were stained blue except for the brain. These findings were confirmed by other investigators and the concept of a blood-brain barrier (BBB) arose. Ongoing research led to the discovery of the transport properties of endothelium and a broader definition of the BBB emerged which included anatomical, physicochemical, and

Sukriti Nag (ed.), *The Blood-Brain and Other Neural Barriers: Reviews and Protocols*, Methods in Molecular Biology, vol. 686, DOI 10.1007/978-1-60761-938-3_1, © Springer Science+Business Media, LLC 2011

biochemical mechanisms at the level of the endothelium, which influence the exchange of materials between blood and brain and cerebrospinal fluid. There is growing recognition that integrated brain function and dysfunction arise from the complex interactions between a network of multiple cell types, such as neurons, glial cells including astrocytes, oligodendrocytes, microglia, and components of the brain vasculature including endothelial cells, smooth muscle cells, and pericytes. The interaction of these cells in steady states and their coordinated response to injury led to the concept that these cells constitute a functional unit, termed the neurovascular unit (2, 3) (Fig. 1).

Regulation of the BBB requires cross-talk between endothelial cells, pericytes, smooth muscle cells which invest the penetrating arterioles, the astrocytic foot processes that surround the entire intracerebral vascular network, and neuronal processes which either directly innervate the capillary endothelium or the astrocytic foot processes. This chapter will review the morphology and properties of cerebral endothelium, which is one of the best studied components of the BBB and the neurovascular unit. The biology of brain pericytes is reviewed in Chapter 2 and that of astrocytes in Chapter 3 and the interaction between these cell types as far as is known is discussed in these chapters.

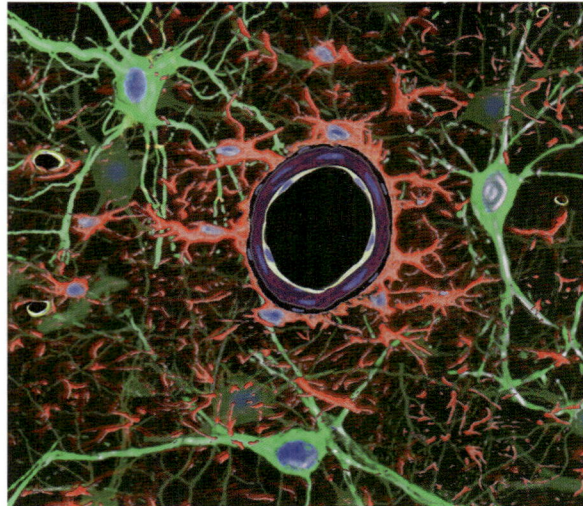

Fig. 1. A modified merged confocal image shows elements of the neurovascular unit in the cerebral cortex. An arteriole consisting of endothelial, smooth muscle, and modified leptomeningeal cell layers is surrounded by astrocytic processes (*red*). Some of the neuronal processes (*green*) in the surrounding neuropil are in close proximity to the perivascular astrocytes and the vessel wall.

2. Structural and Molecular Properties

Brain endothelial cells have many properties which are similar to those present in nonneural endothelium including the expression of glycoproteins (4, 5), adhesion molecules (6, 7), and integrin receptors (7) which will not be discussed. The principal morphological features which distinguish the endothelial cells of intracerebral vessels from those of nonneural vessels and form the structural basis of the BBB to proteins include reduced density of caveolae and the presence of circumferential tight junctions between endothelial cells. Brain endothelial cells also have increased density of mitochondria. Only these features will be discussed in this chapter.

2.1. Caveolae

Caveolae were identified by electron microscopy more than 50 years ago (8, 9) in many differentiated cell types including endothelial cells. They are flask-shaped membrane-bound vesicles having a mean diameter of ~70 nm which can open to both the luminal and abluminal plasma membrane through a neck 10–40 nm in diameter. Caveolae are also observed free in the cytoplasm of endothelial cells of most organs where they have a spherical shape (Fig. 2). These vesicles are distinct from clathrin-coated vesicles, which have an electron-dense coat and are involved in receptor-mediated endocytosis. Rapid-freeze followed by deep-etch electron microscopy shows that a striped or striated coat is present at the cytoplasmic surface of caveolae (10). Ultrastructural studies show thin protein barriers anchored in the neck of caveolae,which are named stomatal diaphragms (11–13). The function of these stomatal diaphragms, which are also associated with transendothelial channels

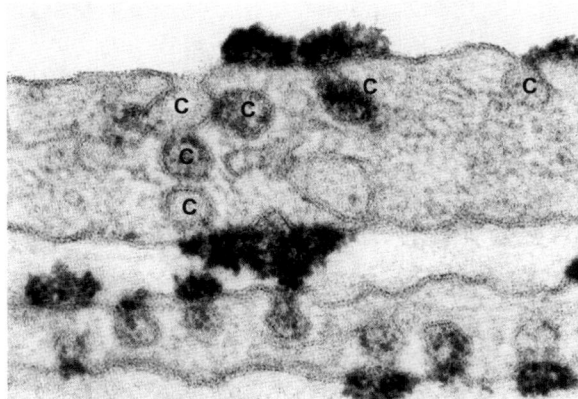

Fig. 2. A segment of arteriolar endothelium from a control rat shows caveolae © in the endothelial and smooth muscle cells. Many of the caveolae show electron-dense deposits representing Ca^{2+}-ATPase localization. X84, 000.

and vesiculo-vacuolar organelles is unknown. A major component of these diaphragms is the plasmalemmal vesicle protein (PV-1) (14, 15). An association between caveolae and both microtubules and microfilaments has also been reported in endothelial cells (13).

Intracerebral cortical vessels contain a mean of 5 caveolae/μm^2 in arteriolar (16) and capillary endothelium (17, 18). Cerebral endothelium contains 14-fold fewer vesicles as compared with endothelium of nonneural vessels such as myocardial capillaries (19). The decreased number of vesicles in cerebral endothelium implies limited transcellular traffic of solutes in steady states. In contrast, capillaries in areas where a BBB is absent such as the subfornicial organ and area postrema (20, 21) are highly permeable and have significantly higher numbers of endothelial caveolae.

2.1.1. Molecular Structure of Caveolae

Studies of nonneural endothelial cells and other cell types have provided information about the molecular structure of caveolae. It is generally accepted that caveolae are lipid rafts (22). The caveolae membrane is enriched in β-D-galactosyl and β-N-acetylglucosaminyl residues in palmitoleic and stearic acids (23) and in cholesterol and sphingolipids (sphingomyelin and glycosphingolipid). The sphingolipids are substrates for synthesis of a second intracellular messenger, the ceramides (24). Cholesterol provides a structural support for caveolae and creates the frame in which many caveolar molecules are inserted.

Located in the coat of caveolae are the caveolin (Cav) family of proteins, which comprise three members named Cav-1, 2, and 3. In brain, Cav-1 and 2 are primarily expressed in endothelial cells (25–28), while Cav-3 is expressed in astrocytes (26–28). Cav-2 is tightly coexpressed with Cav-1 in diverse cells including endothelial cells (26, 27), suggesting that both utilize identical transcription regulatory pathways (29). Cav-1, the specific marker and major component of caveolae, is an integral membrane protein (21–24 kDa) having both amino and carboxyl ends exposed on the cytoplasmic aspect of the membrane (30). The two major isoforms of Cav-1 are α and β and brain cells express predominantly the α-isoform (26).

There is a link between Cav-1 expression and caveolae formation, as Cav-1-null cells have no caveolae (31) and expression of Cav-1 in cells devoid of Cav-1 results in de novo caveolae formation (32). The precise mechanisms by which caveolins form caveolae are just starting to be unraveled (33, 34). Another molecule having a critical role in caveolae formation is cavin which is also termed polymerase I and transcript release factor. Cavin is abundant at the cytoplasmic face of caveolae (35) being roughly as abundant as Cav-1 in caveolae (36). Cavin expression parallels Cav-1 expression in various tissues and cell lines (36). Cavin is recruited to plasma membrane caveolae domains by Cav-1, and expression of full length cavin seems necessary for caveolae formation in the presence of

Cav-1 (36). When cavin expression decreases, Cav-1 diffuses in the plasma membrane and becomes internalized into the endolysosomal system where it is degraded, explaining why down-regulation of cavin results in lower expression of Cav-1 (36, 37).

Caveolin-1 expression is essential for transcytosis of macromolecules as described in the next section (38). Knockout of the Cav-1 gene results in defects in the uptake and transport of albumin in vivo (39). Cav-1 also acts as a multivalent docking site for recruiting and sequestering signaling molecules through the caveolin-scaffolding domain that recognizes a common sequence motif within caveolin-binding signaling molecules (40). Signaling molecules found in caveolar domains that form complexes with caveolin are: membrane proteins, G protein-coupled receptors, G proteins, nonreceptor tyrosine kinases, nonreceptor Ser/Thr kinases, GTPases, cellular proteins and adaptors, and structural proteins (41) (see Table 1). Cav-1 is also known to regulate endothelial nitric oxide synthase and angiogenesis (42).

Table 1
Proteins associated with endothelial caveolae and their function (41)

Protein	Function
Membrane proteins	
PDGF-R (324)	PDGF receptor
CD36 (63, 325)	Lipoprotein receptor
RAGE (63)	Advanced glycated end products receptor
Gp60 (326)	Albumin receptor
SR-BI (325, 327)	Lipoprotein receptor
Flk-1/KDR (328)	VEGF receptor
Tissue factor pathway inhibitor (329)	Down-regulates the procoagulant activity of tissue factor
Plasmalemmal vesicle protein-1(14)	Component of stomatal diaphragms of caveolae and transendothelial channels
P-glycoprotein (330)	ABC transporter
MMP-1(331)	Matrix metalloproteinase
MMP-2 (332)	Matrix metalloproteinase
Endothelial differentiation gene-1 (EDG-1) receptor (333)	EDG-1 product
uPAR (334)	Urokinase receptor
G protein-coupled receptors	
B2R (335)	Bradykinin receptor
ET_A (336)	Endothelin receptor

(continued)

Table 1
(continued)

Protein	Function
G Proteins	
Gα$_{S}$, Gα$_{11}$, Gα$_{i2}$, Gβγ, (63)	Regulate G protein-coupled receptor activity
Gq (63, 337)	•
Nonreceptor tyrosine kinases	
Src, Fyn, Yes, Lck, Lyn (63, 324)	Regulation of growth factor response
Tyk2, STAT3 (335)	Signal transduction and activator of transcription
Nonreceptor Ser/Thr kinases	
Raf (338)	Signal transduction of mitogenic signals
MEK (63)	Signal transduction of mitogenic signals
PI-3 kinase (63, 324)	Phosphorylation of phosphatidyl-inositol
PKC α, β (63, 324)	Ser/Thr kinase
Other enzymes	
eNOS (339, 340)	Production of NO
PLCγ (324)	Phospho-lipase
Prostacyclin synthase (341)	Production of prostacyclin (PGI2)
GTPases	GTPase
Ras, Rap1, Rap2 (63)	
Cellular proteins/adaptors	
Shc (63)	Regulates growth factor response
Grb2 (63)	Adaptor protein, associates growth factor receptors
Other proteins	
ER α and β (342)	Estrogen receptors
NCX (343)	Na+/Ca+ exchanger
Ca2+-ATPase (344, 345)	Calcium pump
IP3 receptor-like protein (344, 345)	Involved in calcium influx
Sprouty-1 and -2 (346)	Inhibitor of development-associated RTK signaling
Cationic arginine transporter-1 (347)	Arginine transporter
Structural proteins	
Actin (63)	Involved in cell motility
Annexin II and IV (63, 345)	Promotes membrane fusion and is involved in exocytosis
Dynamin (337, 348)	Involved in vesicular trafficking
NSF (345)	Involved in vesicle fusion
SNAP, SNARE (345)	Involved in vesicular transport
VAMP-2 (345)	Involved in the targeting/fusion of transported vesicles to their target membranes

2.1.2. Function of Caveolae

Several lines of research implicate caveolae in the process of vesicular trafficking in transcytosis of proteins (43–46), endocytosis (34, 43, 47), and potocytosis (43). Caveolae regulate a wide variety of signaling molecules (22, 48). In addition, caveolae function in the regulation of cell cholesterol and glycosylphosphatidylinositol (GPI)-linked proteins (49), in cell migration (34), as docking sites for glycolipids (50) and as flow sensors (34, 46).

Endothelial caveolae are involved in endocytosis, a process by which the permeant molecules are internalized within endothelial cells or they may be involved in transfer of molecules from blood across the cell to the interstitial fluid or in the reverse direction, a process termed transcytosis (51). Both endocytosis and transcytosis may be receptor-mediated or fluid phase and require ATP and can be inhibited by *N*-ethylmaleimide (NEM), an inhibitor of membrane fusion (52). Plasma proteins which are essential for many cellular functions are selectively taken up by endothelial cells in caveolae that actively carry cargo across the endothelial cell by receptor-mediated and receptor-independent transcytosis, generally bypassing lysosomes (45, 46, 53). Receptors present in caveolae membranes are involved in receptor-mediated transcytosis of low- and high-density lipoprotein, epidermal growth factor, tumor necrosis factor, albumin, transferrin, melanotransferrin, lactoferrin, ceruloplasmin, transcobalamin, advance glycation end products, leptin, and insulin, all of which are essential in maintaining cell and tissue homeostasis and are therefore referred to as the life receptors (46, 54). Also present are death receptors which are involved in cell apoptosis and include receptors for p75 and interleukin-1 (46).

Transcytosis is a multistep process that involves successive caveolae budding and fission from the plasma membrane, translocation across the cell, followed by docking and fusion with the opposite plasma membrane. Caveolae contain the molecular machinery for these processes. Isolated caveolae from lung capillaries demonstrate vesicle-associated membrane protein-2 (VAMP-2) (55), monomeric and trimeric GTPases, annexins II and IV, *N*-ethyl maleimide (NEF)-sensitive fusion factor (NSF), and its attachment protein – soluble NSF attachment protein (SNAP) and vesicle-associated SNAP receptor (v-SNARE) (56) (see Table 1). These molecules interact in the stages of transcytosis as follows: Caveolae form at the cell surface through ATP-, GTP-, and Mg^{2+}-polymerization of Cav-1 and 2, a process stabilized by cholesterol (30). Caveolin oligomers may also interact with glycosphingolipids (57); these protein–protein and protein–lipid interactions are thought to be the driving force for caveolae formation (58). A component of the caveolar fission machinery is the large GTPase, dynamin, which oligomerizes at the neck of caveolae and probably undergoes hydrolysis for fission and release of caveolae, so it becomes free in the cytoplasm (59). Localized at the caveolae neck is intersectin-2 which regulates the GTPase activity of dynamin and

controls dynamin "pinching" and caveolae-mediated endocytosis
(60), while endothelial nitric oxide synthase trafficking inducer
(NOSTRIN) functions as an adaptor recruiting dynamin-2 molecules required for vesicle membrane fission (61). Numerous proteins interact with intersectin including SNARE, 25 and 23 kDa
synaptosome-associated proteins (SNAP-25 and SNAP-23), actin-
binding proteins such as Wiskott-Aldrich syndrome protein, son of
sevenless, and a guanine nucleotide exchange factor (62).

The transcellular movement of caveolae is facilitated by the
association with the actin cytoskeleton-related proteins, such as
myosin HC, gelsolin, spectrin, and dystrophin (63). Fusion at the
abluminal membrane is aided by NSF, which interacts with SNAPs
that can associate with complementary SNAP receptors to form a
functional SNARE fusion complex. Prior to fusion of the target
and vesicle membrane, v-SNARE (VAMP), the targeting receptor
located on the vesicles recognizes and docks with its cognate
t-SNARE (syntaxin) on the target membrane (64, 65). Specific
docking at the opposite plasma membrane is aided by endothelial
VAMP-2 (55), which is localized in caveolae.

The evidence thus far favors the hypothesis that caveolae are
dynamic vesicular carriers budding off from the plasma membrane
to form free transport vesicles that traffic their cargo across cells
fusing with specific target molecules on the abluminal plasma
membrane as described previously (54, 66–69). Theoretical models of vesicular transport agree in predicting a transport time
across endothelium in the order of seconds (38, 70), even against
a concentration gradient (38).

Caveolae can also fuse to form transendothelial channels
extending from the luminal to the abluminal plasma membrane,
which allows passage of macromolecules from the blood to tissues
or in the reverse direction (53, 68). Such channels have been demonstrated in nonneural vessels in steady states (68), but not in normal cerebral endothelium either by freeze-fracture (71), transmission
(72), or high-voltage electron microscopy, which allows examination of 0.25–0.5 µm thick plastic sections (73). Transendothelial
channels have been observed in cerebral endothelium following
BBB breakdown as discussed in Subheading 5.2.1.

2.2. Endothelial Junctions

Tight junctions are present at the apical end of the interendothelial
space being intimately connected to and dependent on the cadherin-
based adherens junctions which are located near the basolateral
side of the interendothelial space. In brain injury, once breakdown of tight junctions occurs, adherens junctions do not impede
the passage of macromolecules.

2.2.1. Tight Junctions

Ultrastructural studies of the tight junctions of brain endothelial
cells demonstrate close approximation of the outer leaflets of
adjacent plasma membranes forming a pentalaminar structure
(74) (Fig. 3), which prevents the passage of tracers such as

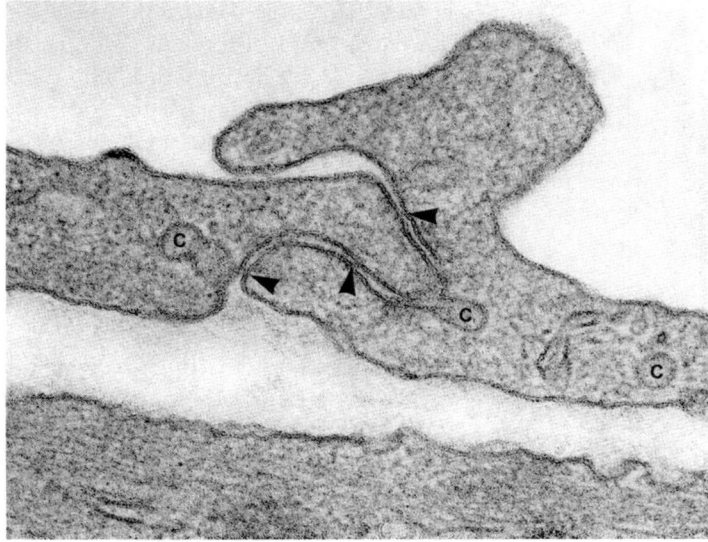

Fig. 3. A segment of arteriolar endothelium shows tight junctions (*arrowheads*) along the interendothelial space. Caveolae © are associated with the interendothelial space in two locations. ×162, 000.

horseradish peroxidase (HRP) into the brain (75). This suggests that these tight junctions extend circumferentially around the endothelial cells forming a barrier to paracellular passage of small hydrophilic molecules such as sodium, hydrogen, bicarbonate, and other ions, a property referred to as its "gate or barrier" function. Tight junctions also restrict the movement of membrane molecules between the functionally distinct apical and basolateral membrane surfaces, a property referred to as its "fence" function.

Freeze-fracture studies demonstrate that the tight junctions of cerebral endothelium consist of 8–12 parallel strands having no discontinuities, which run along the longitudinal axis of the vessel, with numerous lateral anastomotic strands (76). This pattern extends into the postcapillary venules, although in a less complex fashion. In cerebral arteries, tight junctions consist of simple networks of junctional strands, with occasional discontinuities, whereas collecting veins have tight junctional strands which are free-ending and widely discontinuous (76). Freeze-fracture studies also demonstrate that cerebral endothelial tight junctions have a high association with the protoplasmic (P)-face of the membrane leaflet, which is 55% as compared with endothelial cells of nonneural blood vessels, which have a P-face association of only 10% (77).

The physiological correlate of tightness in epithelial membranes is transepithelial resistance. Leaky epithelia generally exhibit electrical resistances between 100 and 200 Ω/cm^2, while the electrical resistance across the BBB in vivo is estimated to be approximately 4–8,000 Ω/cm^2 (78, 79). Cultured brain endothelial cells grown in the absence of astrocytes show a 100-fold decrease in electrical resistance to approximately 90 Ω/cm^2 (80), while coculture of brain microvascular endothelial with astrocytes increases transendothelial electrical resistance as described in Chapter 3, Subheading 3.2. The electrical resistance of cultured endothelial cells can also be increased to 400–1,000 Ω/cm² by using special substrata such as type IV collagen and fibronectin (81).

Tight junctions are localized at cholesterol-enriched regions along the plasma membrane associated with Cav-1 (82). Research using a variety of cell types including cerebral endothelial cells demonstrates that tight junctions are composed of an intricate combination of tetraspan and single-span transmembrane and cytoplasmic proteins linked to an actin-based cytoskeleton that allows these junctions to seal the paracellular space while remaining capable of rapid modulation and regulation (see Chapter 5, Fig. 3). The two tetraspan transmembrane proteins are the claudin family of proteins and occludin, while the single-span transmembrane protein is the junction adhesion molecule (JAM) family of proteins. The tetraspan proteins form the paracellular permeability barrier and determine the capacity and the selectivity of the paracellular diffusion pathway.

The claudin family consists of 24 members in mice and humans and exhibit distinct expression patterns in tissue and cells (83–86). Claudins are 18–27-kDa tetraspan proteins with a short cytoplasmic N-terminus, two extracellular loops, and a COOH-terminal cytoplasmic domain which ends in valine. The latter strongly attracts PDZ (PSD-95/DlgA/ZO-1)-containing proteins such as zonula occludens (ZO)-1-3, PATJ, PALS1, and MUPP1 (87). These interactions are thought to be important for junction assembly. Claudins are considered to be the main structural components of intramembrane strands (88) and recruit occludin to tight junctions (89). Occludin knockout mice are still capable of forming interendothelial tight junctions having normal morphology and barrier function in intestinal epithelial cells (90), while claudin knockout mice are nonviable (91). The extracellular loops of claudins are considered to create aqueous pores that have biophysical properties similar to those of traditional ion channels including ion charge selectivity, permeability dependence on ion concentration, and competition for movement of permeative molecules (92). These channels permit the passive diffusion of mostly cations, but anion passage has also been documented (86, 93). Claudin-5 regulates size-selective diffusion of small molecules since claudin-5 knockout mice show increased paracellular permeability to molecules <800 D (94).

Occludin is a regulatory protein, associated with the intramembrane strands at tight junctions, although it is not required for their assembly. Presence of occludin in brain endothelium is correlated with increased electrical resistance across the barrier and decreased paracellular permeability (95). Occludin is a ~60-kDa tetraspan membrane protein with two extracellular loops, a short intracellular turn, and N- and C-terminal cytoplasmic domains (96). The N-terminal domain regulates neutrophil transmigration and the C-terminal domain regulates the paracellular diffusion of small hydrophilic tracers as well as targeting of occludin to the basolateral membrane (97). The two extracellular loops of occludin also regulate its accumulation at junctions as well as paracellular permeability. Occludin has also been linked to the regulation of various subcellular signaling pathways, such as the MAP-kinase-dependent pathways, Rho, Raf1, and TGFβ signaling (98, 99), although the molecular mechanisms involved and the functional significance of these interactions remain to be determined.

The JAMs belong to the immunoglobulin superfamily and to the Cortical Thymocyte marker for xenopus family of molecules that lie at the crossroads between antigen-specific receptors and adhesion molecules. These proteins contain two immunoglobulin folds in their extracellular domain, one of the V_H- type and one of the C_2- type (100). The five JAMs identified designated JAM-A, JAM-B, JAM-C, JAM-4, and JAM-L can mediate homotypic cell–cell adhesion. Of these, only JAM-A is associated with tight junctions of brain endothelium, while other members of the JAM family are associated with nonneural endothelium of postcapillary high venules (101) and lymphatics (102) which have focal tight junctions. JAM-A, the first member of the family to be isolated is a 32-kDa glycoprotein, which is characterized by two immunoglobulin loops in the extracellular domain, a single transmembrane domain, and a cytoplasmic tail that ends in a classical type II PDZ domain-binding sequence. JAM-A has been implicated in a variety of physiologic and pathologic processes including tight junction assembly (103), leukocyte transmigration (104), angiogenesis (105), platelet activation (106, 107), and reovirus binding (108, 109). Overexpression of JAMs in cells that do not normally form tight junctions increases their resistance to the diffusion of soluble tracers, suggesting that JAM functionally contributes to permeability control (103). Several of the JAM proteins, as well as the related proteins, coxackievirus and adenovirus receptor and the endothelial-cell selective adhesion molecule, associate with tight junctions and interact with adaptor proteins of the cytoplasmic plaque (100, 110–112). JAMs are known to regulate cell polarity by binding to the complex between PAR3, PAR6, and atypical PKC (112–114).

Microfilaments, intermediate filaments, and microtubules form the cytoskeleton of brain endothelial cells (115, 116) and actin has also been localized in the endothelial plasma membrane by molecular techniques (117). Actin has known binding sites on claudin and occludin and on all ZO proteins (118). A cytoplasmic plaque is present between the cytoskeleton and the tight junction membrane where it functions in the regulation of adhesion and paracellular permeability, as well as in the transmission of signals from the junction to the cell interior to regulate cellular processes such as migration and gene expression. The cytoplasmic plaque in epithelial cells is formed by a complex network of adaptors, scaffolding, and cytoskeletal proteins that crosslink junctional membrane proteins and connect the tight junctions to the actin cytoskeleton (97) (Table 2). The cytoplasmic plaque also recruits

Table 2
The complex network of adaptors, scaffolding, and cytoskeletal proteins that form the cytoplasmic plaque that cross-links junctional membrane proteins to the actin cytoskeleton are listed

Signaling proteins	*Adaptors and complex forming proteins*
Atypical PKC (aPKC)	Zonula occludens (ZO)-1-3 (349–351)
GEF-H1/Lfc	Cingulin (352)
CDK4	Junction-associated coiled-coil protein (JACOP)
G proteins	Pals1 (protein associated with Lin Seven 1-associated TJ protein))
PTEN	PATJ (Pals 1-associated TJ protein)
RPTPβ	PAR3/ASIP
Yes	MAGI (membrane-associated guanylate kinase with an inverted orientation of protein–protein interaction domains) (87)
Tuba	MAGI-1
PP1	MAGI-2
PP2A	MUPP1 (multi-PDZ domain protein 1)
Rab3b	Shroom 2
Rab13	Junction-enriched and -associated protein (JEAP)
WNK-4	Sec 6/8
	Lysine-rich CEACAM-1 coisolated protein (LYRIC)
Transcriptional and Posttranscriptional regulators	PILT Barmotin
Zonula occludens-1 associated nucleic acid protein (ZONAB) (353)	
Symplekin	
AP1	
huASH1	

Adapted from (97)

an array of signaling components (Table 2) that includes classical signaling proteins, such as kinases and phosphatases, as well as dual-localization proteins that can reside at junctions as well as in the nucleus thus providing a mechanism by which tight junctions can regulate gene expression (119–121). Not all the molecules described in the cytoplasmic plaque of epithelial cells have been localized in cerebral endothelium thus far.

ZO-1, ZO-2, and ZO-3 are members of the MAGUK (membrane-associated guanylate kinase homologs) family with binding domains to adherens and tight junctions, in addition to the actin cytoskeleton (122). The MAGUK family is characterized by their PDZ domain, SH3 domain, and guanylate kinase homologous domain (123). These domains are important in signal transduction and in anchoring the transmembrane tight junction proteins to the cytoskeleton (124). In vitro assays demonstrate that ZO-1 (125, 126) and ZO-2 (127) bind directly to occludin at the C-terminal region. Upon tyrosine phosphorylation of occludin, interactions between all ZO proteins and occludin are reduced (128).

Occludin, claudins-3, 5, and 12, JAM-A and ZO-1 proteins have been localized in normal cerebral endothelium (77, 94, 129–131). Occludin, claudin-5, and JAM-A strands run parallel to the long axis of brain vessels in murine brain (129, 131). The ZO-1 protein occurs in approximately the same quantities (molecules per micron) as the intramembranous particles that constitute the junctional fibrils in freeze-fracture preparations (132, 133).

Tight junctions are also sites for vesicle targeting, cytoskeletal dynamics, and signals controlling proliferation and transcription. Details of these properties can be obtained from recent reviews (7, 88, 97, 120, 134–136).

2.2.2. Adherens Junctions

Adherens junctions are formed by the single-pass transmembrane adhesive protein named vascular endothelial (VE)-cadherin. Neural-cadherin is also expressed in large amounts in endothelial cells, but its distribution is diffuse in the cell membrane being absent in plasma membranes of cell–cell contacts (137). VE-cadherin is linked through its cytoplasmic tail to multiple intracellular proteins including β-catenin, p120, plakoglobin (γ-catenin), phosphatases such as endothelial-specific receptor-protein tyrosine phosphatase and density-enhanced phosphatase-1 (138). β-catenin and plakoglobin bind to α-catenin, which interacts with several actin-binding proteins, including α-actinin, ajuba, and ZO-1 (139). All these molecules are expressed at junctions in brain endothelial cells (140, 141). VE-cadherins and the associated catenins influence and are influenced by the actin cytoskeleton, although the molecular basis of this interaction is still uncertain. Adherens junctions promote cell-to-cell communication and transfer signals that mediate contact inhibition of cell growth, increase resistance to apoptosis, and regulate cell shape and polarity (142).

The VE-cadherin-catenin association is required for full cellular control of endothelial permeability and junction stabilization. Tyrosine phosphorylation of VE-cadherin and the other components of adherens junctions are associated with weak junctions and impaired permeability. The regulation of adherens junction proteins in the maintenance of junctional permeability has been reviewed previously (77, 122, 143–145).

2.3. Mitochondria

Most of the mitochondria in cerebral endothelium are located in the vicinity of the nucleus, but occasional mitochondria are present in the cytoplasm and these tend to be parallel to the cell surface. Murine cerebral endothelium has increased numbers of mitochondria (146), with 10% (147) and 13.7% (148) of the endothelial cell volume being occupied by mitochondria, which is greater than found in endothelia of other tissues. The increased mitochondria provide the metabolic work capacity for maintaining the ionic gradient across the BBB. Endothelial mitochondrial density is lower in capillaries of the subfornicial organ, which lies outside the BBB (21).

2.4. Gene Expression in Brain Endothelium

Suppression subtractive hybridization has generated a cDNA library containing clones for unique genes expressed in both human and rat brain microvasculatures (149–151). The serial analysis of gene expression (SAGE) method has identified mRNAs expressed in rat brain microvessels (152). These authors identified 864 genes enriched in microvessels by comparing their catalog to cortex and hippocampus SAGE catalogs. Sorting enriched genes based on function revealed groups that encode transporters (11%), receptors (5%), proteins involved in vesicle trafficking (4%), structural proteins (10%), and components of signal transduction pathways (17%). These results provide a useful resource, which can be accessed at the journal website (http://wwwnature. com/jcbfm).

3. Endothelial Transport Mechanisms

A wide range of lipid-soluble molecules can diffuse through the endothelium and enter brain passively (153). Bases, which carry a positive charge, have an advantage over acids in penetration of endothelial cells and it is probably the cationic nature of these molecules and an interaction with the negatively charged glycocalyx and phospholipid head groups of the outer leaflet of the cell membrane that facilitates their entry. Transcellular bidirectional transport across cerebral endothelium occurs by receptor-mediated transport, carrier-mediated transport, ion transport, peptide transport, and active efflux transport.

3.1. Receptor-Mediated Transport

This is described in Subheading 2.1.2.

3.2. Carrier-Mediated Transport

Carrier-mediated transport systems facilitate transport of nutrients into brain including hexoses (glucose, galactose); neutral, basic and acidic amino acids, and monocarboxylic acids (lactate, pyruvate, ketone bodies); nucleosides (adenosine, guanosine, uridine); purines (adenine, guanine), nucleotides, nucleobases, organic anion, and organic cations; amine and vitamins. The concentration gradients for nutrients are generally in the direction of blood to brain and are regulated by brain metabolic needs and by the concentration of substrates in plasma.

The glucose transporter (GLUT1), the L1 large neutral amino acid transporter, the CNT2 adenosine transporter, and the monocarboxylate transporter 1 were cloned from BBB-specific cDNA libraries (154). GLUT1, which is insulin insensitive, facilitates glucose entry down a concentration gradient from the plasma into the brain. In brain GLUT-1 is present as 2 distinct molecular forms with molecular weights of 55- and 45-kDa, which are encoded by the same gene and differ only in their extent of glycosylation (155). The 55-kDa isoform of GLUT1 is detected exclusively in brain endothelial cells, while the 45-kDa isoform is expressed in neurons and glial cells (156). An asymmetric distribution of GLUT1 is present in cerebral endothelium with a four-fold greater abundance on the abluminal compared with the luminal membrane (157, 158).

The amino acid transporters L1 and y+, on luminal and abluminal membranes, provide the brain with all essential amino acids (159). The sodium-dependent L1 facilitative transporter mediates transport of large neutral essential amino acids such as leucine, isoleucine, valine, tryptophan, tyrosine, phenylalanine, threonine, and methionine. The y+ system mediates transport of cationic amino acids, some of which are essential in the brain, for example lysine, and some of which are nonessential in the adult brain such as arginine and ornithine, but are essential during the juvenile period, for example l-arginine, a precursor of nitric oxide (136). Five sodium-dependent transporters for amino acids exist at the abluminal membrane that actively transfer every naturally occurring amino acid from the brain interstitial fluid to endothelial cells and then into the circulation. The sodium-dependent system for the excitatory acidic amino acids (EAAT), glutamate and aspartate, provides a mechanism for net removal of potentially neurotoxic amino acids from the brain. Glutamate transporters EAAT1, EAAT2, and EAAT3 determine the levels of extracellular glutamate and are essential in preventing excitotoxicity (160). The sodium-dependent system for nitrogen-rich amino acids removes glutamine and other nitrogen-rich amino acids such as histidine and asparagine from the brain. The facilitative transporters at the luminal side, x_G^- and n, mediate transport of acidic amino acids and nitrogen-rich amino acids from the endothelium to blood, respectively.

3.3. Ion Transport

Ion transporters in brain endothelial cells include the sodium pump (Na⁺, K⁺-ATPase), sodium–potassium-two chloride cotransporter (Na⁺-K⁺-2Cl⁻), sodium–hydrogen exchanger, chloride–bicarbonate exchanger, and the sodium–calcium exchanger (136). Localization of Na⁺, K⁺-ATPase is higher at the abluminal than luminal plasma membrane (161) and regulates sodium influx in exchange for potassium. Na⁺-K⁺-2Cl⁻ cotransporter present predominantly in the luminal plasma membrane transports sodium, potassium, and chloride from blood into the endothelium. The sodium–hydrogen exchanger is expressed on the luminal membrane, while the chloride–bicarbonate exchanger is expressed at each membrane (162). These two transporters play critical roles in regulating intracellular pH in the endothelium. The chloride–bicarbonate exchanger also regulates active secretion of bicarbonate across the endothelium. The sodium–calcium exchanger in the forward mode (Na⁺ entry/Ca²⁺ extrusion) mediates Ca²⁺ efflux from the endothelium. However, in the presence of altered Na⁺ gradients or in pathological conditions, it may allow calcium entry into the endothelium.

3.4. Peptide Transport

Endothelial cells at the BBB express several transport systems for neuroactive peptides such as enkephalins (163), tyrosine melanocyte-stimulating inhibitory factor 1 (tyr-MIF-1), delta-sleep inducing peptide (164), luteinizing-hormone releasing hormone (LHRH), and some cytokines and chemokines (165, 166). The V1-vasopressinergic receptor is required for transport of arginine-vasopressin (AVP) from blood to the brain (167). Peptide transport systems 1 and 2 at the abluminal membrane mediate efflux of encephalins/Tyr-MIF-1 and AVP, respectively, from the brain to the blood. PTS-3, on the luminal membrane, transports peptide T into the brain. PTS-4 transports LHRH bidirectionally.

3.5. Efflux Transport

Active efflux transport involves a transporter, which utilizes ATP to shuttle drugs and other solutes out of the brain and into the blood compartment. Efflux transporters minimize effective drug penetration into brain parenchyma, thus limiting the efficacy of drugs targeted at brain diseases. The efflux transporter proteins of the ABC family (ATP-binding cassette) transport a diverse range of lipid-soluble compounds out of brain endothelium. ABC transporters in the human are a superfamily of proteins containing 48 members, which on the basis of structural homology are grouped into seven subfamilies (168, 169). The ABC transporters of greatest significance are P-glycoprotein (Pg-P), the multidrug resistance-associated proteins (MRP), and the breast cancer resistance protein.

Localization of Pg-P in brain endothelium still appears to be controversial with some groups reporting abluminal localization of this protein, while others reporting luminal localization of this protein (170). Although a physiological role for P-gp has yet to be determined, it is believed that the main function of this efflux transporter is to protect the body against harmful xenobiotics from accumulating in cells. The mammalian MRP family (humans:MRP; rodents:Mrp), which includes MRP1/Mrp1, Mrp2, MRP4/Mrp4, Mrp5, and Mrp6, have been identified in mammalian cerebral endothelial cells (171–173). The breast cancer resistance protein (ABCG2) has been localized at the luminal plasma membrane of endothelial cells (174, 175) and in primary cultures of human brain microvessel endothelial cells (176, 177).

Further information about transport mechanism in brain endothelial cells can be obtained from reviews in the literature (136, 178–181).

4. Brain Diseases with BBB Dysfunction

Dysfunction of the BBB may be in the form of increased permeability or BBB breakdown to large and small molecules in brain diseases and/or may take the form of alterations in endothelial transport mechanisms. Well documented in the literature is the increased BBB permeability to plasma proteins, which occurs in conditions associated with vasogenic edema such as ischemic and hemorrhagic stroke, infections, inflammation, seizures, trauma, tumors, epilepsy, and hypertensive encephalopathy (182, 183). Increased permeability to C^{14} sucrose implying increased ionic permeability has been reported in peripheral inflammatory pain (184). Increased BBB permeability to ions and plasma proteins has been reported in human and experimental diabetes (184, 185). Global vascular changes and altered expression of Pg-p have been implicated in the pathogenesis of degenerative diseases such as Alzheimer's disease (186, 187) and Parkinson's disease (188) as reviewed previously (136, 189).

5. Endothelial Mechanisms in Vasogenic Edema

Vasogenic edema is the endpoint of many acute neurological disorders as described in the preceding section and is characterized by BBB breakdown to plasma proteins. Progressive brain swelling due to the edema leads to cerebral herniations and death, hence the importance of this condition.

5.1. Experimental Models

Permeability of the BBB to plasma proteins is best studied using an in vivo model since cultured brain endothelial cells do not develop junctions having the same degree of tightness as present in vivo as determined by measurement of transendothelial resistance, use of protein tracers, and morphology (190–192). Secondly, endothelial cells removed from their natural flow environment show decreased density of caveolae, which may be greater than tenfold (34, 193, 194); therefore increase in density of endothelial caveolae, which is a major route for protein passage into brain in various pathologies, is not observed in vitro. Thirdly, increased blood pressure and vasodilatation, which are important factors in BBB breakdown in vivo, are absent in vitro. Also, there is heterogeneity in endothelial cells in the different vessel types and only microvessel endothelial cells are present in the in vitro studies, while most in vivo studies have demonstrated that mainly small arterioles and veins show BBB breakdown rather than microvessels.

Novel models are being investigated to study the complex interactions of components of the neurovascular unit in vivo. One such model is the fruit fly, Drosophila melanogaster, which has septate junctions between subperineural glia that create a tight barrier to paracellular diffusion from the hemolymph as described in Chapter 17. Another model for in vivo BBB studies is the Zebrafish which is described in Chapter 18.

5.1.1. Cold- Injury Model

The cortical cold-injury model developed by Klatzo (195) to study the pathophysiology of vasogenic edema has been used extensively in the literature and in our studies (190, 196–198). There are variations in the method of producing the cold lesion which makes it difficult to compare the results obtained from different laboratories. In our studies, BBB breakdown starts at 6 h, and the BBB is restored to normal on day 6 postinjury (see Table 3) as determined using HRP as a tracer and by immunodetection of endogenous serum protein extravasation using antibodies to serum proteins or fibrinogen or fibronectin (196). Two peaks of active BBB breakdown occur in the cold-injury model (197, 198). An initial phase which extends from 6 h to day 2 affects mainly arterioles and large venules at the margin of the lesion and leads to extravasation of plasma proteins at the lesion site (Fig. 4a). There is spread of edema fluid through the ECS into the underlying white matter of the ipsilateral and contralateral side. The second phase of BBB breakdown accompanies angiogenesis and is maximal on day 4 (Fig. 4b). Arterioles, veins, and neovessels at the lesion site show extravasation of plasma proteins, which remain confined to the lesion site. There is florid angiogenesis in this model and a consistent inflammatory response. The time course of these processes has been described previously (196, 198).

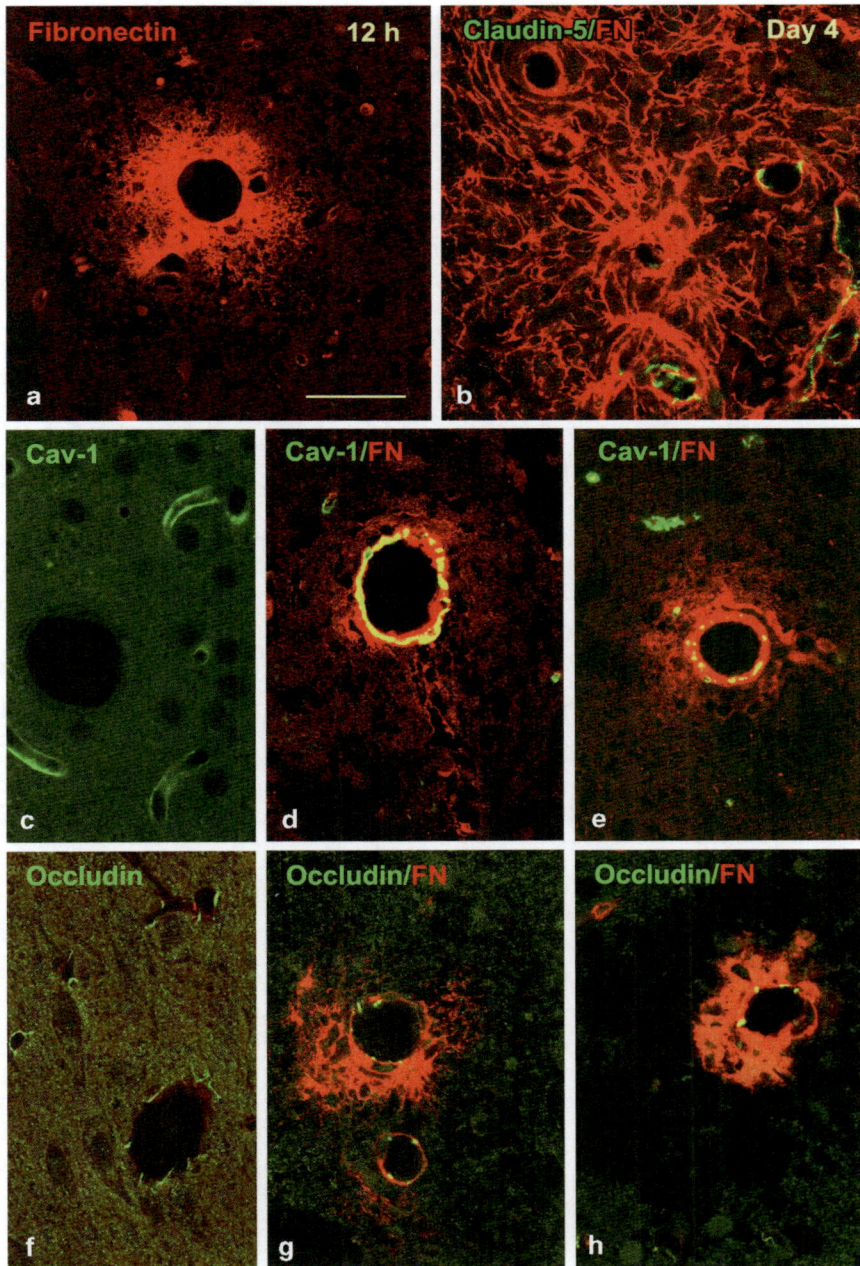

Fig. 4. Confocal images of the cold-injury model (**a**, **b**), control rat brains (**c**, **f**), and untreated (**d**, **g**) and Ang-1 treated (**e**, **h**) cold-injured rat brains are shown. (**a**) At 12 h, a vessel at the margin of the lesion shows BBB breakdown to fibronectin which has extravasated through the vessel wall and is present in the surrounding neuropil. (**b**) On day 4, BBB breakdown is diffuse involving many vessels, some of which show endothelial Claudin-5 localization (*green*). (**c**) Cortex of a control rat showing caveolin-1 (Cav-1) localization in the endothelium of microvessels but not in the endothelium of the arteriole. (**d**) Increased endothelial Cav-1 signal is present in an arteriole with BBB breakdown to fibronectin. It appears *yellow* due to the overlapping of the Cav-1 and fibronectin signals. (**e**) Following Ang-1 treatment, an arteriole leaking fibronectin shows decreased endothelial Cav-1 signal. (**f**) Cortex of a control rat showing endothelial localization of occludin in microvessels and in a large caliber vessel. (**g**) The occludin signal is decreased in a vessel with BBB breakdown on day 1 following the cold-injury. (**h**) Following Ang-1 treatment, there is restoration of the endothelial occludin signal in a vessel showing BBB breakdown. Scale bar = 50 μm.

In summary, time course studies of acute brain injury indicate that the main pathological processes are early BBB breakdown, neuronal and glial death, inflammation, and a period of repair characterized by angiogenesis and further BBB breakdown (Fig. 5). The end stage is restoration of the BBB to normal characteristics. The degree of changes in each of these stages varies depending on the clinical condition. For example, the inflammatory response is maximal in infections, while in hypertensive encephalopathy, BBB breakdown is massive and, without treatment, patients may die before the onset of an inflammatory response.

5.2. BBB Breakdown in Vasogenic Edema

This review will focus on the endothelial mechanisms leading to BBB breakdown to plasma proteins in vasogenic edema. Details of the inflammatory response can be obtained from current reviews (199–201).

5.2.1. Tracer Studies

Early studies utilized dyes such as trypan blue and Evans blue to detect BBB breakdown in experimental models of brain edema. These tracers can be detected on macroscopic examination of brains and provide information on the spatial localization of the areas of BBB breakdown. Evans blue was later combined with HRP, which has a molecular weight of 40 kDa to facilitate the sampling of areas for ultrastructural studies of BBB breakdown to HRP in experimental vasogenic edema. All these studies reported

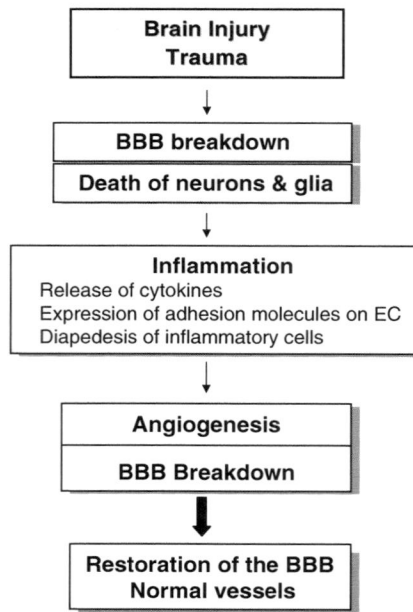

Fig. 5. Diagram showing the principal pathological processes following brain injury from the onset of BBB breakdown to the period when the BBB is restored.

an increased density of endothelial caveolae only in the vessels with BBB breakdown to HRP within minutes of the onset of pathological states such as hypertension, spinal cord injury, seizures, experimental autoimmune encephalomyelitis, excitotoxic brain damage, brain trauma, and BBB breakdown-induced by bradykinin, histamine, and leukotriene C4 (129, 182, 202) (Fig. 6). Transendothelial channels containing HRP were also demonstrated in these studies, while passage of HRP via tight junctions was not noted.

Thus, early BBB breakdown to HRP is associated with fluid-phase transcytosis of HRP and protein passage via transendothelial channels. Convincing demonstration of tight junction breakdown has only been reported 30 min after the intracarotid administration of hyperosmotic agents (203) using the tracer lanthanum, which is a marker of ionic permeability. Thus, junctional breakdown to proteins occurs late in the course of brain injury probably during end-stage disease and precedes endothelial cell breakdown.

The BBB is a dynamic system and recovery of BBB function can occur after transient injury as observed 15 min after an episode of acute hypertension when the BBB is no longer permeable to HRP. Permeability to smaller molecular weight tracers likely persists for a longer period as observed after spinal cord injury (204).

5.2.2. Surface Charge   The net negative charge on the luminal plasma membranes of brain endothelium is lost during BBB breakdown to HRP in acute hypertension (182) and cold-injury (205). This is due to loss of the terminal sialic acid groups on endothelium which exposes the β-D-gal-(1–3)-D-gal N-acetyl groups that normally are not

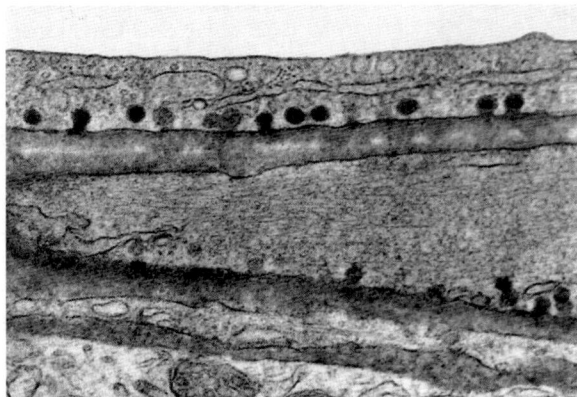

Fig. 6. Arteriolar segment from an acutely hypertensive rat shows BBB breakdown to HRP. Tracer is present in the endothelial and smooth muscle cell basement membranes and in the basement membrane between the attenuated leptomeningeal cell layer and the astrocytic foot process. The endothelium shows increased density of caveolae, many of which contain HRP. ×46,000.

accessible to peanut agglutinin and binding of this lectin occurs (206). Decrease in endothelial surface charge precedes the permeability changes since neutralization of the charge by intracarotid injections of polycationic protamine or poly-L-lysine results in BBB breakdown to tracers in the ipsilateral hemisphere (207, 208).

5.2.3. Caveolin-1

In parallel with the increase in endothelial caveolae, BBB breakdown in brain injury is associated with increased expression of Cav-1. Time course studies in the rat cortical cold-injury model demonstrate a significant increase in Cav-1α expression at the lesion site on days 0.5–4 postinjury by immunoblotting (27). At the cellular level, a marked increase in endothelial Cav-1 protein is present in vessels showing BBB breakdown to fibronectin (Fig. 4c, d). Further studies demonstrate that the endothelial Cav-1 in vessels with BBB breakdown is phosphorylated (209). It is well established that dilated vascular segments show enhanced permeability and leak protein (210, 211). Within dilated vascular segments, endothelial cells are subjected to hemodynamic shear stress, which is known to increase plasma membrane levels of Cav-1 due to redistribution of Cav-1 from the Golgi complex to the plasma membrane, and this is associated with an increase in the surface density of caveolae (193, 212). Shear and cellular stress can also result in phosphorylation of Cav-1 (193, 213). Phosphorylation of Cav-1 is known to be an essential step for fission of caveolae and the internalization of bound and fluid-phase macromolecules within caveolae, a step which precedes transcytosis of caveolae (42). Thus, phosphorylation of Cav-1 is essential for transcytosis of proteins across cerebral endothelium leading to BBB breakdown and brain edema following brain injury.

In summary, caveolae and Cav-1 have a significant role in early BBB breakdown occurring within 12 h after brain injury, therefore caveolae and Cav-1 could be potential therapeutic targets in the control of early brain edema.

5.2.4. Tight Junction Proteins

Decreased occludin immunoreactivity in lesion vessels with BBB breakdown was demonstrated in acquired immunodeficiency virus infection (214) and in multiple sclerosis plaques (215). Decreased expression of the tight junction proteins in vessels with BBB breakdown in the cold-injury model follows a specific sequence with transient decreases in expression of JAM-A on day 0.5 only (131), of claudin-5 on day 2 only, while occludin expression is attenuated from day 2 onwards and persists up to day 6 (27) (Fig. 4f, g). These studies support our previous observations that caveolae and Cav-1 changes precede significant tight junction changes during early BBB breakdown. The exact role of occludin in BBB permeability control is not entirely clear since, in our studies, lesion vessels without BBB breakdown to fibronectin also show decrease in occludin immunoreactivity.

Reduced expression of Cav-1, occludin, and ZO-1 has been reported in cultured brain microvascular cells exposed to the chemokine CCL2 (216). In this model, targeted knockdown of Cav-1 by the adenoviral-mediated small interfering RNA approach is associated with reduced expression of these tight junction proteins and increased paracellular permeability of the endothelial cells to fluorescein dextran (217). These results are opposed to the in vivo studies described above in Subheading 5.2.3 and can be attributed to the differences between the in vivo and in vitro systems as outlined in Subheading 5.1.

5.2.5. Cytoskeleton and Permeability Alterations

Integrity of the endothelial microfilaments is necessary for maintenance of the BBB to protein in steady states since increased cortical arteriolar permeability to HRP occurs in the presence of cytochalasin B, an actin-disrupting agent (115). Quantitative studies showed a significant increase in the density of endothelial caveolae in permeable vessels and lack of tracer passage via interendothelial junctions. As discussed previously in Subheading 2.2.1, actin microfilaments are necessary for maintenance of tight junction integrity as well since actin-disrupting substances, such as cytochalasin D, cytokines, and phalloidin, disrupt tight junction structure and function (218, 219). The importance of an intact cytoskeleton in the maintenance of the BBB in steady states is further supported by studies of the mdx mouse which lacks the actin-binding protein dystrophin (220). These mice show increased permeability of brain vessels due to disorganization of the α-actin cytoskeleton in endothelial cells and astrocytes, as well as altered localization of junctional proteins in the endothelium and aquaporin-4 in the astrocytic end feet.

Disruption of microtubules with colchicine has no effect on BBB permeability in steady states, suggesting that the microtubular network has no demonstrable role in cerebral homeostasis (115). However, pretreatment with colchicine attenuates BBB breakdown that is known to occur in acute hypertension (115). Microtubules are known to form intracellular pathways along which protein-bearing vesicles pass in neurons (221, 222) and nonneural endothelial cells (223). A possible explanation may be that disruption of microtubules impairs the transcytosis of endothelial vesicles from the luminal to the abluminal plasma membrane thus attenuating BBB breakdown. This is supported by the finding that colchicine reduces the transcytosis of fluorescein isothiocyanate dextran in bovine aortic endothelial cells (223).

5.2.6. Metalloproteinases

The MMPs are zinc- and calcium-dependent endopeptidases, which are known to cleave most components of the extracellular matrix including fibronectin, laminin, proteoglycans, and type IV collagen (224, 225). Activation of MMPs involves cleavage of the secreted proenzyme, while inhibition involves a group of four endogenous tissue inhibitors of metalloproteinases (TIMPs) (226) and α_2-macroglobulin. The balance between production, activation, and inhibition prevents excessive proteolysis or inhibition.

MMPs are found in all elements of the neurovascular unit, but different MMPs have a predilection for certain cell types. Endothelial cells express mainly MMP-9, pericytes express MMP-3 and -9, while astrocytic end feet express MMP-2 and its activator – membrane-type MMP (MT1-MMP) (227). MMP-2 is constitutively expressed, has a molecular weight of 72 kDa, and is normally present in a latent form tethered to the cell surface by MT1-MMP and requires the presence of TIMP-2 in order to undergo activation. This restricts the proteolytic action of MMP-2 to the immediate vicinity of the protease. MMP-9 is normally expressed at low levels but is markedly up-regulated in many brain diseases.

In human ischemic stroke, active MMP-2 is increased on days 2–5 compared with active MMP-9, which is elevated up to months after the ischemic episode (228). Molecular studies in experimental permanent and temporary ischemia have shown that MMPs contribute to disruption of the BBB leading to vasogenic edema and hemorrhage (227, 229, 230). Middle cerebral artery occlusion for 90 min with reperfusion in spontaneously hypertensive rats causes biphasic opening of the BBB in the piriform cortex with a transient, reversible opening at 3 h, which correlates with a transient increase in expression of MMP-2, and its activator MT1-MMP (231, 232). This is associated with a decrease in claudin-5 and occludin expression in cerebral vessels after 3 h of reperfusion. By 24 h, the tight junction proteins are no longer observed in lesion vessels, an alteration that is reversed by treatment with the MMP inhibitor, BB-1101 (231). The later BBB opening between 24 and 48 h is associated with a marked increase of MMP-9, which is released in the extracellular matrix where it degrades multiple proteins and produces more extensive blood vessel damage. The role of MMPs in BBB breakdown is further supported by the observation that treatment with MMP inhibitors or MMP neutralizing antibodies decreases infarct size and prevents BBB breakdown after focal ischemic stroke (233–235).

The MMP inhibitors used so far can restore early integrity of the BBB in rodent ischemia models, but are ineffective in the later opening at 48 h. Since these inhibitors block MMPs involved in angiogenesis and neurogenesis as well, they slow recovery. Therefore, identifying agents that will protect the BBB and block vasogenic edema without interfering with recovery remains a challenge.

5.2.7. Role of Astrocytes in Vasogenic Edema

This is summarized in Chapter 3, Subheading 4.1.

5.3. Modulators of Endothelial Permeability

Vasoactive agents known to increase BBB permeability have been discussed in previous reviews (183, 236, 237). There is increasing interest in the role of angiogenic factors such as the vascular endothelial growth factor (VEGF) and angiopoietin (Ang) families in BBB homeostasis. Initially, these agents were considered to be endothelial-specific; however, it is now evident that these growth factors are expressed in neurons as well and that they can affect both neural and vascular cell function, hence their recent designation as "angioneurins" (238, 239). Although many angioneurins are known to modulate the integrity of the BBB (238), this review will focus on VEGF-A and B and Ang1 and 2 expression in steady states and following brain injury.

5.3.1. Vascular Endothelial Growth Factors-A and B

VEGF, the first member of the six-member VEGF family to be discovered, is now designated as VEGF-A. VEGF-A has a significant role in vascular permeability and angiogenesis during embryonic vasculogenesis and in physiological and pathological angiogenesis (240–246). These effects are mediated by VEGF receptor-2 (VEGFR-2), which is present on endothelial cells (241).

Normal adult cortex shows basal expression of VEGF-A mRNA and protein, while high expression of VEGF-A mRNA and protein is present in normal choroid plexus epithelial cells and ependymal cells (198, 247, 248). Intracerebral injections of VEGF-A increases the permeability of brain vessels to HRP (249–251) by formation of interendothelial gaps and segmental fenestrae-like narrowings in endothelium (249). In contrast, VEGF-A-induced hyperpermeability of the blood–retinal barrier endothelium is associated predominantly with increased density of caveolae, while tight junction alterations or fenestrae in endothelium were not observed (252). These divergent results may be due to the different time points that permeability was assessed in these studies. Members of the Src family have been implicated in VEGF-dependent vascular hyperpermeability (253, 254) since Src–/– mice show reduced brain damage after induction of cortical ischemia, and a Src inhibitor has protective effects in wild-type mice in a similar brain injury model. VEGF-A can also increase permeability by altering the expression of tight junction proteins. Reduced occludin expression occurs in retinal (255) and brain (256) endothelial cells exposed to VEGF-A. In addition, there is disruption of ZO-1 and occludin organization leading to tight junction disassembly (256).

VEGF-B displays strong homology to VEGF-A (257, 258) and has two known isoforms, which bind to VEGFR-1 and neuropilin-1 (259, 260). Mice embryos (day 14) and adults show high expression of VEGF-B mRNA in most organs with very high levels in the heart and the nervous system (261, 262). Moderate down-regulation of VEGF-B occurs prior to birth (E17) and VEGF-B is the only member of the VEGF family that is expressed at detectable levels in the adult CNS. Constitutive expression of VEGF-B protein is present in the endothelium of all cerebral vessels including those of the choroid plexuses (198). Thus, VEGF-B may have a role in maintenance of the BBB in steady states and VEGF-B may be protective against BBB breakdown and edema formation.

VEGF-B also has a role in angiogenesis postinjury (198) and is implicated in various stages of adult neurogenesis (263).

5.3.2. Angiopoietins (Ang)1 and 2

Four members of this family have been isolated thus far and designated Ang1–4 (264–266). Best characterized are Ang1 and 2 which function as ligands for the Tie-2/Tek receptor with similar affinity (267–269) and do not bind to Tie-1 (270). Tie-2 (Tyrosine kinase with immunoglobulin and epidermal growth factor homology domain) is ubiquitously expressed in endothelial cells of all type of vessels in all organs, including brain (250, 271, 272). Ang1 induces autophosphorylation of Tie-2 and has a marked chemotactic effect on endothelial cells, whereas Ang2 competitively inhibits this effect, suggesting that it may be a naturally occurring inhibitor of Ang1/Tie-2 activity (268, 273). Endothelial Ang1 is expressed widely in normal adult tissues including brain (250, 271), consistent with its constitutive stabilization role by maintaining normal endothelial cell-to-cell and cell-to-matrix interactions (268). In steady states, the endothelium of all cerebral cortical vessels shows only weak Ang2 expression (250, 271).

Functional studies indicate that Ang1 and Ang2 have reciprocal effects on endothelial cells. Ang1 has an antiapoptotic effect on endothelial cells, which is mediated by the Akt/survivin/PI-3 kinase pathway (274, 275), while Ang2 is reported to be proapoptotic (250, 276, 277). Ang1 has a potent antileakage property that was established using transgenic mice (278) and nonneural endothelial cultures (279–282). Presence of Ang1 is also associated with fewer and smaller gaps in the endothelium of postcapillary venules during inflammation (283). Ang1 is reported to stabilize interendothelial junctions by increased expression of platelet endothelial cell adhesion molecule-1 (PECAM-1) and decreased phosphorylation of PECAM-1 and VE-cadherin (284) in nonneural endothelial cells and by up-regulating ZO-2 in brain endothelial cells (285).

5.3.3. Time Course of Growth Factor Expression Postinjury

The temporal and spatial alterations in expression of growth factors and their relation to BBB breakdown and angiogenesis have been studied in the cold-injury model (Table 3). In the early phase postinjury during the period of BBB breakdown up to day 2, there is loss of the constitutive factors normally expressed in endothelium, namely Ang1 and VEGF-B and increased expression of VEGF-A, VEGFR-2, and Ang2 proteins (198, 250, 271). Dual labeling shows colocalization of Ang2, but not Ang1, with activated caspase-3 in endothelium of lesion vessels on day 1 (250), implicating Ang2 as an effector of endothelial apoptosis and BBB breakdown in damaged vessels. This is supported by the finding that intracortical injections of Ang2 can produce BBB breakdown (250) and by the finding that Ang2 produces vascular

Table 3
Alterations in expression of Caveolin-1 (Cav-1),
phosphorylated Caveolin-1, tight junction proteins and
angiogenic proteins during the period of BBB breakdown,
and angiogenesis in the cortical cold-injury is shown

	12 h	Day 2	Day 4	Day 6
BBB Breakdown	+	+	+	−
Angiogenesis	-	+	+	+
Cav-1, PY14 Cav-1	↑	↑	↑	−
Tight junction proteins				
JAM-1	↓	−	−	−
Claudin-5	−	↓	−	−
Occludin	−	↓	↓	↓
Angiogenic factors				
VEGF-A	−	↑	↑	↑
VEGF-B	↓	↓	↑	−
VEGFR2	−	↑	↑	−
Ang1	↓	↓	↑	−
Ang2	−	↑	↑	−
Tie-2	−	−	−	−

"+" = present; "−" = no change

leakage in nonneural vessels (286). On days 4 and 6 postinjury, there is progressive increase in Ang1 and VEGF-B mRNA and protein and decrease in Ang2 and VEGF-A mRNA and protein coinciding with maturation of neovessels and restoration of the BBB.

The temporal expression of VEGF-A, Ang1, and Ang2 in focal cerebral ischemia models is very similar to that observed in the cold-injury model (287, 288). There is early up-regulation of VEGF-A and Ang2 mRNA and protein coinciding with the first opening of the BBB postischemia. At 48 h postischemia, there is increased expression of VEGF-A, Ang1, and Ang2 coinciding with the period of the second BBB opening postischemia and the onset of angiogenesis.

Increased expression of growth factors has been reported in gliomas. VEGF-A is overexpressed up to 50-fold in the perinecrotic tumor cells in glioblastomas, suggesting hypoxia-mediated transcriptional activation of the VEGF-A gene (289–291). This is associated with up-regulation of VEGF receptors in tumor endothelium, suggesting a paracrine mechanism of VEGF-driven tumor angiogenesis (290, 291). Increased expression of the angiopoietins has also been reported in glioblastomas. High expression of Ang1 has been reported in areas of high vascular density in all stages of glioblastoma progression (292–295), while high expression of Ang2 has been reported in endothelial cells in glioblastomas (295–299). In these studies a strong association is made between these growth factors and tumor angiogenesis. Since time course studies are difficult in human biopsy tissue, one assumes that these angiogenic factors play the same role in the genesis of vasogenic edema in brain tumors as observed in the cold-injury and the focal brain ischemia models. This is supported by reports of a strong correlation between VEGF-A expression and vascular permeability in gliomas (300) and between VEGF-A expression and the amount of peritumoral edema in meningiomas (301–303).

There is the potential of using inhibitors of VEGF-A or Ang1 or VEGF-B to treat early and massive edema associated with large hemispheric lesions which are lethal due to the effects of early edema. Pretreatment with VEGF-A receptor chimeric protein (Flt-(1-3)-IgG), which inactivates endogenous VEGF-A (304) or recombinant Ang1 (305) in the rodent ischemia model, attenuates BBB breakdown and edema associated with cerebral infarcts. Administration of Ang1 just after production of the cold lesion reduces extravasation of HRP at the lesion site on day 1 and further studies demonstrate that the antileakage effect of Ang1 is associated with attenuation of the Cav-1 increase (Fig. 4d, e) and occludin decrease (Fig. 4g, h) that occurs in untreated rats. The long-term effects of administering angiogenic agents on angiogenesis and repair are not known and must be assessed before these agents can be used to attenuate BBB breakdown and edema following brain injury.

6. Delivery of Therapeutic Substances Across the BBB

Multiple strategies, targeting both caveolae and tight junctions, have been designed to circumvent, manipulate, or disrupt the BBB for delivery of therapeutic substances across the BBB.

6.1. Transcytosis

Several laboratories have targeted transcytosis to deliver therapeutic agents into brain. Endogenous peptides such as insulin, insulin-like growth factor, and transferrin that cross the endothelium via receptor-mediated transcytosis have been used to ferry an attached therapeutic agent across the BBB, a process referred to as "molecular Trojan horse" technology (306). This technology has been used for BBB transport of peptides, recombinant proteins or nonviral plasmid DNA-encapsulated pegylated immunoliposomes, and RNA interference therapeutics in experimental models of CNS diseases (306–308). Nonantibody delivery systems have been developed, including histones (309), p97 (310), receptor-associated protein (311), the Tat transduction domain peptide (312), and other cationic peptides or polymers (313). Drugs conjugated with cationic peptide vectors cross the BBB by adsorptive-mediated transcytosis (314). Peptides designated as Angiopeps exhibit high transcytosis rates leading to subsequent accumulation in the brain parenchyma (315). The conjugation of low-density lipoprotein apoproteins to the surface of nanoparticles triggers receptor-mediated transcytosis across the BBB by the low-density lipoprotein receptor on brain endothelium (316). Thus, the manipulation of absorptive and receptor-mediated transcytosis holds promise for improved delivery of therapeutic agents across the BBB for treatment of CNS diseases.

Transcytosis has also been exploited for delivery of antibody or carbohydrates conjugated to microparticles of iron oxide, for detection of activated brain vessels with high magnetic resonance-detectable contrast in animal models of multiple sclerosis (317), brain metastases, stroke (318), and cerebral malaria (319). This technique allows early detection of activated brain endothelium, at a time when no structural abnormalities are detected with conventional magnetic resonance imaging. Details of this technology are given in Chapter 9.

6.2. Tight Junctions

Intracarotid infusion of hyperosmotic mannitol for reversible disruption of the BBB followed by intra-arterial chemotherapy began in 1979 for the treatment of human brain tumors (320). This invasive approach has been used to treat primary CNS lymphomas with favorable results and relatively little toxicity (321, 322). These authors have also investigated the use of BBB disruption, with or without intra-arterial chemotherapy, for the delivery of drugs which can be as large as the herpes virus (323). The technique to disrupt the BBB in experimental animals is given in Chapter 23.

7. Conclusions

Considerable advances have been made in understanding the reactivity of brain endothelium in health and disease in the last few decades. However, much remains to be done. The discovery of new molecules by genomics and proteomics presents an ongoing challenge to unravel their role in endothelial cell biology. Although information about the interaction of the endothelium with other components of the neurovascular unit in culture systems and brain slices is starting to emerge, studying these interactions in vivo in health and disease remains a challenge for future studies. Determination of the temporal course of BBB breakdown and the parallel molecular alterations remain important goals to identify therapeutic windows and pertinent therapeutic agents which will modify the disease process and prevent irreversible brain damage. Finally, parallel studies need to be done to find novel methods for delivery of the appropriate therapeutic agents to the brain to treat neurological diseases.

Acknowledgments

This work is supported by multiple grants from the Heart and Stroke Foundation of Ontario from 1978-2009.

References

1. Ehrlich P (1885) Das Sauerstoff-Bedürfnis des Organismus:Eine farbenanalytische studie

2. Lo EH, Dalkara T, Moskowitz MA (2003) Mechanisms, challenges and opportunities in stroke. Nat Rev Neurosci 4:399–415

3. Lok J, Gupta P, Guo S, Kim WJ, Whalen MJ, Van LK, Lo EH (2007) Cell-cell signaling in the neurovascular unit. Neurochem Res 32:2032–2045

4. Nag S (2003) Ultracytochemical studies of the compromised blood-brain barrier. Methods Mol Med 89:145–160

5. Vorbrodt AW, Dobrogowska DH, Lossinsky AS, Wisniewski HM (1986) Ultrastructural localization of lectin receptors on the luminal and abluminal aspects of brain micro-blood vessels. J Histochem Cytochem 34:251–261

6. Nag S (2003) Morphology and molecular properties of cellular components of normal cerebral vessels. Methods Mol Med 89:3–36

7. Wolburg H, Noell S, Mack A, Wolburg-Buchholz K, Fallier-Becker P (2009) Brain endothelial cells and the glio-vascular complex. Cell Tissue Res 335:75–96

8. Palade G (1953) Fine structure of blood capillaries. J Appl Physiol 24:1424–1436

9. Yamada E (1955) The fine structure of the gall bladder epithelium of the mouse. J Biophys Biochem Cytol 1:445–458

10. Rothberg KG, Heuser JE, Donzell WC, Ying YS, Glenney JR, Anderson RG (1992) Caveolin, a protein component of caveolae membrane coats. Cell 68:673–682

11. Stan RV (2005) Structure of caveolae. Biochim Biophys Acta 1746:334–348

12. Stan RV (2007) Endothelial stomatal and fenestral diaphragms in normal vessels and angiogenesis. J Cell Mol Med 11:621–643

13. Richter T, Floetenmeyer M, Ferguson C, Galea J, Goh J, Lindsay MR, Morgan GP, Marsh BJ, Parton RG (2008) High-resolution 3D quantitative analysis of caveolar ultra-structure and caveola-cytoskeleton interactions. Traffic 9:893–909

14. Stan RV, Ghitescu L, Jacobson BS, Palade GE (1999) Isolation, cloning, and localization of rat PV-1, a novel endothelial caveolar protein. J Cell Biol 145:1189–1198

15. Stan RV, Kubitza M, Palade GE (1999) PV-1 is a component of the fenestral and stomatal diaphragms in fenestrated endothelia. Proc Natl Acad Sci USA 96:13203–13207

16. Nag S, Robertson DM, Dinsdale HB (1979) Quantitative estimate of pinocytosis in experimental acute hypertension. Acta Neuropathol (Berl) 46:107–116

17. Connell CJ, Mercer KL (1974) Freeze-fracture appearance of the capillary endothe-

lium in the cerebral cortex of mouse brain. Am J Anat 140:595–599

18. Stewart PA, Magliocco M, Hayakawa K, Farrell CL, Del Maestro RF, Girvin J, Kaufmann JC, Vinters HV, Gilbert J (1987) A quantitative analysis of blood-brain barrier ultrastructure in the aging human. Microvasc Res 33:270–282

19. Simionescu M, Simionescu N, Palade GE (1974) Morphometric data on the endothelium of blood capillaries. J Cell Biol 60:128–152

20. Coomber BL, Stewart PA (1985) Morphometric analysis of CNS microvascular endothelium. Microvasc Res 30:99–115

21. Gross PM, Sposito NM, Pettersen SE, Fenstermacher JD (1986) Differences in function and structure of the capillary endothelium in gray matter, white matter and a circumventricular organ of rat brain. Blood Vessels 23:261–270

22. Thomas CM, Smart EJ (2008) Caveolae structure and function. J Cell Mol Med 12:796–809

23. Gafencu A, Stanescu M, Toderici AM, Heltianu C, Simionescu M (1998) Protein and fatty acid composition of caveolae from apical plasmalemma of aortic endothelial cells. Cell Tissue Res 293:101–110

24. Liu P, Anderson RG (1995) Compartmentalized production of ceramide at the cell surface. J Biol Chem 270:27179–27185

25. Virgintino D, Robertson D, Errede M, Benagiano V, Tauer U, Roncali L, Bertossi M (2002) Expression of caveolin-1 in human brain microvessels. Neuroscience 115:145–152

26. Ikezu T, Ueda H, Trapp BD, Nishiyama K, Sha JF, Volonte D, Galbiati F, Byrd AL, Bassell G, Serizawa H, Lane WS, Lisanti MP, Okamoto T (1998) Affinity-purification and characterization of caveolins from the brain: differential expression of caveolin-1, -2, and -3 in brain endothelial and astroglial cell types. Brain Res 804:177–192

27. Nag S, Venugopalan R, Stewart DJ (2007) Increased caveolin-1 expression precedes decreased expression of occludin and claudin-5 during blood-brain barrier breakdown. Acta Neuropathol (Berl) 114:459–469

28. Shin T, Kim H, Jin JK, Moon C, Ahn M, Tanuma N, Matsumoto Y (2005) Expression of caveolin-1, -2, and -3 in the spinal cords of Lewis rats with experimental autoimmune encephalomyelitis. J Neuroimmunol 165:11–20

29. Scherer PE, Okamoto T, Chun M, Nishimoto I, Lodish HF, Lisanti MP (1996) Identification,

sequence, and expression of caveolin-2 defines a caveolin gene family. Proc Natl Acad Sci USA 93:131–135

30. Monier S, Parton RG, Vogel F, Behlke J, Henske A, Kurzchalia TV (1995) VIP21-caveolin, a membrane protein constituent of the caveolar coat, oligomerizes in vivo and in vitro. Mol Biol Cell 6:911–927

31. Drab M, Verkade P, Elger M, Kasper M, Lohn M, Lauterbach B, Menne J, Lindschau C, Mende F, Luft FC, Schedl A, Haller H, Kurzchalia TV (2001) Loss of caveolae, vascular dysfunction, and pulmonary defects in caveolin-1 gene-disrupted mice. Science 293:2449–2452

32. Fra AM, Williamson E, Simons K, Parton RG (1995) De novo formation of caveolae in lymphocytes by expression of VIP21-caveolin. Proc Natl Acad Sci USA 92:8655–8659

33. Parton RG, Hanzal-Bayer M, Hancock JF (2006) Biogenesis of caveolae: a structural model for caveolin-induced domain formation. J Cell Sci 119:787–796

34. Parat MO (2009) The biology of caveolae: achievements and perspectives. Int Rev Cell Mol Biol 273:117–162

35. Vinten J, Johnsen AH, Roepstorff P, Harpoth J, Tranum-Jensen J (2005) Identification of a major protein on the cytosolic face of caveolae. Biochim Biophys Acta 1717:34–40

36. Hill MM, Bastiani M, Luetterforst R, Kirkham M, Kirkham A, Nixon SJ, Walser P, Abankwa D, Oorschot VM, Martin S, Hancock JF, Parton RG (2008) PTRF-Cavin, a conserved cytoplasmic protein required for caveola formation and function. Cell 132:113–124

37. Liu L, Pilch PF (2008) A critical role of cavin (polymerase I and transcript release factor) in caveolae formation and organization. J Biol Chem 283:4314–4322

38. Oh P, Borgstrom P, Witkiewicz H, Li Y, Borgstrom BJ, Chrastina A, Iwata K, Zinn KR, Baldwin R, Testa JE, Schnitzer JE (2007) Live dynamic imaging of caveolae pumping targeted antibody rapidly and specifically across endothelium in the lung. Nat Biotechnol 25:327–337

39. Schubert W, Frank PG, Razani B, Park DS, Chow CW, Lisanti MP (2001) Caveolae-deficient endothelial cells show defects in the uptake and transport of albumin in vivo. J Biol Chem 276:48619–48622

40. Li S, Couet J, Lisanti MP (1996) Src tyrosine kinases, Galpha subunits, and H-Ras share a common membrane-anchored scaffolding protein, caveolin. Caveolin binding negatively

regulates the auto-activation of Src tyrosine kinases. J Biol Chem 271:29182–29190

41. Frank PG, Woodman SE, Park DS, Lisanti MP (2003) Caveolin, caveolae, and endothelial cell function. Arterioscler Thromb Vasc Biol 23:1161–1168

42. Minshall RD, Sessa WC, Stan RV, Anderson RG, Malik AB (2003) Caveolin regulation of endothelial function. Am J Physiol Lung Cell Mol Physiol 285:L1179–L1183

43. Anderson RG, Kamen BA, Rothberg KG, Lacey SW (1992) Potocytosis: sequestration and transport of small molecules by caveolae. Science 255:410–411

44. Frank PG, Pavlides S, Lisanti MP (2009) Caveolae and transcytosis in endothelial cells: role in atherosclerosis. Cell Tissue Res 335:41–47

45. Predescu D, Vogel SM, Malik AB (2004) Functional and morphological studies of protein transcytosis in continuous endothelia. Am J Physiol Lung Cell Mol Physiol 287:L895–L901

46. Simionescu M, Popov D, Sima A (2009) Endothelial transcytosis in health and disease. Cell Tissue Res 335:27–40

47. Schnitzer JE, Oh P, McIntosh DP (1996) Role of GTP hydrolysis in fission of caveolae directly from plasma membranes. Science 274:239–242

48. Lisanti MP, Scherer PE, Tang Z, Sargiacomo M (1994) Caveolae, caveolin and caveolin-rich membrane domains: a signalling hypothesis. Trends Cell Biol 4:231–235

49. Rothberg KG, Ying YS, Kamen BA, Anderson RG (1990) Cholesterol controls the clustering of the glycophospholipid-anchored membrane receptor for 5-methyltetrahydrofolate. J Cell Biol 111:2931–2938

50. Parton RG (1994) Ultrastructural localization of gangliosides; GM1 is concentrated in caveolae. J Histochem Cytochem 42:155–166

51. Simionescu M, Simionescu N (1991) Endothelial transport of macromolecules: transcytosis and endocytosis. A look from cell biology. Cell Biol Rev 25:1–78

52. Schnitzer JE, Allard J, Oh P (1995) NEM inhibits transcytosis, endocytosis, and capillary permeability: implication of caveolae fusion in endothelia. Am J Physiol 268:H48–H55

53. Palade GE, Simionescu M, Simionescu N (1979) Structural aspects of the permeability of the microvascular endothelium. Acta Physiol Scand Suppl 463:11–32

54. Schnitzer JE (2001) Caveolae: from basic trafficking mechanisms to targeting transcytosis for tissue-specific drug and gene delivery in vivo. Adv Drug Deliv Rev 49:265–280

55. McIntosh DP, Schnitzer JE (1999) Caveolae require intact VAMP for targeted transport in vascular endothelium. Am J Physiol 277:H2222–H2232

56. Schnitzer JE, Liu J, Oh P (1995) Endothelial caveolae have the molecular transport machinery for vesicle budding, docking, and fusion including VAMP, NSF, SNAP, annexins, and GTPases. J Biol Chem 270:14399–14404

57. Fra AM, Masserini M, Palestini P, Sonnino S, Simons K (1995) A photo-reactive derivative of ganglioside GM1 specifically cross-links VIP21-caveolin on the cell surface. Febs Lett 375:11–14

58. Sargiacomo M, Scherer PE, Tang Z, Kubler E, Song KS, Sanders MC, Lisanti MP (1995) Oligomeric structure of caveolin: implications for caveolae membrane organization. Proc Natl Acad Sci USA 92:9407–9411

59. Oh P, McIntosh DP, Schnitzer JE (1998) Dynamin at the neck of caveolae mediates their budding to form transport vesicles by GTP-driven fission from the plasma membrane of endothelium. J Cell Biol 141:101–114

60. Hussain NK, Jenna S, Glogauer M, Quinn CC, Wasiak S, Guipponi M, Antonarakis SE, Kay BK, Stossel TP, Lamarche-Vane N, McPherson PS (2001) Endocytic protein intersectin-l regulates actin assembly via Cdc42 and N-WASP. Nat Cell Biol 3:927–932

61. Schilling K, Opitz N, Wiesenthal A, Oess S, Tikkanen R, Muller-Esterl W, Icking A (2006) Translocation of endothelial nitric-oxide synthase involves a ternary complex with caveolin-1 and NOSTRIN. Mol Biol Cell 17:3870–3880

62. Mehta D, Malik AB (2006) Signaling mechanisms regulating endothelial permeability. Physiol Rev 86:279–367

63. Lisanti MP, Scherer PE, Vidugiriene J, Tang Z, Hermanowski-Vosatka A, Tu YH, Cook RF, Sargiacomo M (1994) Characterization of caveolin-rich membrane domains isolated from an endothelial-rich source: implications for human disease. J Cell Biol 126:111–126

64. Hay JC, Scheller RH (1997) SNAREs and NSF in targeted membrane fusion. Curr Opin Cell Biol 9:505–512

65. Rothman JE (1994) Intracellular membrane fusion. Adv Second Messenger Phosphoprotein Res 29:81–96

66. Couet J, Belanger MM, Roussel E, Drolet MC (2001) Cell biology of caveolae and caveolin. Adv Drug Deliv Rev 49:223–235

67. McIntosh DP, Tan XY, Oh P, Schnitzer JE (2002) Targeting endothelium and its dynamic caveolae for tissue-specific transcytosis in vivo: a pathway to overcome cell barriers

to drug and gene delivery. Proc Natl Acad Sci USA 99:1996–2001

68. Simionescu M, Gafencu A, Antohe F (2002) Transcytosis of plasma macromolecules in endothelial cells: a cell biological survey. Microsc Res Tech 57:269–288

69. Stan RV (2002) Structure and function of endothelial caveolae. Microsc Res Tech 57:350–364

70. Shea SM, Raskova J (1983) Vesicular diffusion and thermal forces. Fed Proc 42: 2431–2434

71. Farrell CL, Shivers RR (1984) Capillary junctions of the rat are not affected by osmotic opening of the blood-brain barrier. Acta Neuropathol (Berl) 63:179–189

72. Nag S (1998) Blood-brain barrier permeability measured with histochemistry. In Pardridge WM (Ed) Introduction to the Blood-Brain Barrier. Methodology, biology and pathology, Cambridge University Press, Cambridge, pp 113–121

73. Shivers RR, Harris RJ (1984) Opening of the blood-brain barrier in Anolis carolinensis. A high voltage electron microscope protein tracer study. Neuropathol Appl Neurobiol 10:343–356

74. Muir AR, Peters A (1962) Quintuple-layered membrane junctions at terminal bars between endothelial cells. J Cell Biol 12:443–448

75. Reese TS, Karnovsky MJ (1967) Fine structural localization of a blood-brain barrier to exogenous peroxidase. J Cell Biol 34: 207–217

76. Nagy Z, Peters H, Huttner I (1984) Fracture faces of cell junctions in cerebral endothelium during normal and hyperosmotic conditions. Lab Invest 50:313–322

77. Kniesel U, Wolburg H (2000) Tight junctions of the blood-brain barrier. Cell Mol Neurobiol 20:57–76

78. Crone C, Olesen SP (1982) Electrical resistance of brain microvascular endothelium. Brain Res 241:49–55

79. Smith QR, Rapoport SI (1986) Cerebrovascular permeability coefficients to sodium, potassium, and chloride. J Neurochem 46: 1732–1742

80. Krause D, Mischeck U, Galla HJ, Dermietzel R (1991) Correlation of zonula occludens ZO-1 antigen expression and transendothelial resistance in porcine and rat cultured cerebral endothelial cells. Neurosci Lett 128: 301–304

81. Tilling T, Korte D, Hoheisel D, Galla HJ (1998) Basement membrane proteins influence brain capillary endothelial barrier function in vitro. J Neurochem 71:1151–1157

82. Nusrat A, Parkos CA, Verkade P, Foley CS, Liang TW, Innis-Whitehouse W, Eastburn KK, Madara JL (2000) Tight junctions are membrane microdomains. J Cell Sci 113 (Pt 10): 1771–1781

83. Morita K, Furuse M, Fujimoto K, Tsukita S (1999) Claudin multigene family encoding four-transmembrane domain protein components of tight junction strands. Proc Natl Acad Sci USA 96:511–516

84. Tsukita S, Furuse M, Itoh M (2001) Multifunctional strands in tight junctions. Nat Rev Mol Cell Biol 2:285–293

85. Turksen K, Troy TC (2004) Barriers built on claudins. J Cell Sci 117:2435–2447

86. Van Itallie CM, Anderson JM (2006) Claudins and epithelial paracellular transport. Annu Rev Physiol 68:403–429

87. Hamazaki Y, Itoh M, Sasaki H, Furuse M, Tsukita S (2002) Multi-PDZ domain protein 1 (MUPP1) is concentrated at tight junctions through its possible interaction with claudin-1 and junctional adhesion molecule. J Biol Chem 277:455–461

88. Furuse M, Tsukita S (2006) Claudins in occluding junctions of humans and flies. Trends Cell Biol 16:181–188

89. Furuse M, Sasaki H, Fujimoto K, Tsukita S (1998) A single gene product, claudin-1 or -2, reconstitutes tight junction strands and recruits occludin in fibroblasts. J Cell Biol 143:391–401

90. Saitou M, Furuse M, Sasaki H, Schulzke JD, Fromm M, Takano H, Noda T, Tsukita S (2000) Complex phenotype of mice lacking occludin, a component of tight junction strands. Mol Biol Cell 11:4131–4142

91. Gow A, Southwood CM, Li JS, Pariali M, Riordan GP, Brodie SE, Danias J, Bronstein JM, Kachar B, Lazzarini RA (1999) CNS myelin and sertoli cell tight junction strands are absent in Osp/claudin-11 null mice. Cell 99:649–659

92. Tang VW, Goodenough DA (2003) Paracellular ion channel at the tight junction. Biophys J 84:1660–1673

93. Krause G, Winkler L, Piehl C, Blasig I, Piontek J, Muller SL (2009) Structure and function of extracellular claudin domains. Ann N Y Acad Sci 1165:34–43

94. Nitta T, Hata M, Gotoh S, Seo Y, Sasaki H, Hashimoto N, Furuse M, Tsukita S (2003) Size-selective loosening of the blood-brain barrier in claudin-5-deficient mice. J Cell Biol 161:653–660

95. Hirase T, Staddon JM, Saitou M, Ando-Akatsuka Y, Itoh M, Furuse M, Fujimoto K, Tsukita S, Rubin LL (1997) Occludin as a

possible determinant of tight junction permeability in endothelial cells. J Cell Sci 110 (Pt 14): 1603–1613

96. Furuse M, Hirase T, Itoh M, Nagafuchi A, Yonemura S, Tsukita S, Tsukita S (1993) Occludin: a novel integral membrane protein localizing at tight junctions. J Cell Biol 123:1777–1788

97. Balda MS, Matter K (2008) Tight junctions at a glance. J Cell Sci 121:3677–3682

98. Barrios-Rodiles M, Brown KR, Ozdamar B, Bose R, Liu Z, Donovan RS, Shinjo F, Liu Y, Dembowy J, Taylor IW, Luga V, Przulj N, Robinson M, Suzuki H, Hayashizaki Y, Jurisica I, Wrana JL (2005) High-throughput mapping of a dynamic signaling network in mammalian cells. Science 307:1621–1625

99. Wang Z, Mandell KJ, Parkos CA, Mrsny RJ, Nusrat A (2005) The second loop of occludin is required for suppression of Raf1-induced tumor growth. Oncogene 24:4412–4420

100. Bazzoni G (2003) The JAM family of junctional adhesion molecules. Curr Opin Cell Biol 15:525–530

101. Palmeri D, van ZA, Huang CC, Hemmerich S, Rosen SD (2000) Vascular endothelial junction-associated molecule, a novel member of the immunoglobulin superfamily, is localized to intercellular boundaries of endothelial cells. J Biol Chem 275:19139–19145

102. Arrate MP, Rodriguez JM, Tran TM, Brock TA, Cunningham SA (2001) Cloning of human junctional adhesion molecule 3 (JAM3) and its identification as the JAM2 counter-receptor. J Biol Chem 276:45826–45832

103. Martin-Padura I, Lostaglio S, Schneemann M, Williams L, Romano M, Fruscella P, Panzeri C, Stoppacciaro A, Ruco L, Villa A, Simmons D, Dejana E (1998) Junctional adhesion molecule, a novel member of the immunoglobulin superfamily that distributes at intercellular junctions and modulates monocyte transmigration. J Cell Biol 142: 117–127

104. Ostermann G, Weber KS, Zernecke A, Schroder A, Weber C (2002) JAM-1 is a ligand of the beta(2) integrin LFA-1 involved in transendothelial migration of leukocytes. Nat Immunol 3:151–158

105. Naik MU, Mousa SA, Parkos CA, Naik UP (2003) Signaling through JAM-1 and alphav-beta3 is required for the angiogenic action of bFGF: dissociation of the JAM-1 and alphav-beta3 complex. Blood 102:2108–2114

106. Kornecki E, Walkowiak B, Naik UP, Ehrlich YH (1990) Activation of human platelets by a stimulatory monoclonal antibody. J Biol Chem 265:10042–10048

107. Ozaki H, Ishii K, Arai H, Horiuchi H, Kawamoto T, Suzuki H, Kita T (2000) Junctional adhesion molecule (JAM) is phosphorylated by protein kinase C upon platelet activation. Biochem Biophys Res Commun 276:873–878

108. Barton ES, Forrest JC, Connolly JL, Chappell JD, Liu Y, Schnell FJ, Nusrat A, Parkos CA, Dermody TS (2001) Junction adhesion molecule is a receptor for reovirus. Cell 104:441–451

109. Forrest JC, Campbell JA, Schelling P, Stehle T, Dermody TS (2003) Structure-function analysis of reovirus binding to junctional adhesion molecule 1. Implications for the mechanism of reovirus attachment. J Biol Chem 278:48434–48444

110. Bazzoni G, MartinezEstrada OM, Orsenigo F, Cordenonsi M, Citi S, Dejana E (2000) Interaction of junctional adhesion molecule with the tight junction components ZO-1, cingulin, and occludin. J Biol Chem 275:20520–20526

111. Ebnet K, Schulz CU, Meyer zu Brickwedde MK, Pendl GG, Vestweber D (2000) Junctional adhesion molecule interacts with the PDZ domain-containing proteins AF-6 and ZO-1. J Biol Chem 275:27979-27988

112. Ebnet K, Aurrand-Lions M, Kuhn A, Kiefer F, Butz S, Zander K, Meyer zu Brickwedde MK, Suzuki A, Imhof BA, Vestweber D (2003) The junctional adhesion molecule (JAM) family members JAM-2 and JAM-3 associate with the cell polarity protein PAR-3: a possible role for JAMs in endothelial cell polarity. J Cell Sci 116:3879–3891

113. Bradfield PF, Nourshargh S, Aurrand-Lions M, Imhof BA (2007) JAM family and related proteins in leukocyte migration (Vestweber series). Arterioscler Thromb Vasc Biol 27:2104–2112

114. Weber C, Fraemohs L, Dejana E (2007) The role of junctional adhesion molecules in vascular inflammation. Nat Rev Immunol 7:467–477

115. Nag S (1995) Role of the endothelial cytoskeleton in blood-brain-barrier permeability to protein. Acta Neuropathol (Berl) 90:454–460

116. Nag S, Robertson DM, Dinsdale HB (1978) Cytoplasmic filaments in intracerebral cortical vessels. Ann Neurol 3:555–559

117. Pardridge WM, Nowlin DM, Choi TB, Yang J, Calaycay J, Shively JE (1989) Brain capillary 46,000 dalton protein is cytoplasmic

actin and is localized to endothelial plasma membrane. J Cereb Blood Flow Metab 9:675–680

118. Itoh M, Furuse M, Morita K, Kubota K, Saitou M, Tsukita S (1999) Direct binding of three tight junction-associated MAGUKs, ZO-1, ZO-2, and ZO-3, with the COOH termini of claudins. J Cell Biol 147:1351–1363

119. Guillemot L, Paschoud S, Pulimeno P, Foglia A, Citi S (2008) The cytoplasmic plaque of tight junctions: a scaffolding and signalling center. Biochim Biophys Acta 1778:601–613

120. Matter K, Balda MS (2007) Epithelial tight junctions, gene expression and nucleo-junctional interplay. J Cell Sci 120:1505–1511

121. Paris L, Tonutti L, Vannini C, Bazzoni G (2008) Structural organization of the tight junctions. Biochim Biophys Acta 1778:646–659

122. Hartsock A, Nelson WJ (2008) Adherens and tight junctions: structure, function and connections to the actin cytoskeleton. Biochim Biophys Acta 1778:660–669

123. Funke L, Dakoji S, Bredt DS (2005) Membrane-associated guanylate kinases regulate adhesion and plasticity at cell junctions. Annu Rev Biochem 74:219–245

124. Staddon JM, Rubin LL (1996) Cell adhesion, cell junctions and the blood-brain barrier. Curr Opin Neurobiol 6:622–627

125. Fanning AS, Jameson BJ, Jesaitis LA, Anderson JM (1998) The tight junction protein ZO-1 establishes a link between the transmembrane protein occludin and the actin cytoskeleton. J Biol Chem 273:29745–29753

126. Furuse M, Itoh M, Hirase T, Nagafuchi A, Yonemura S, Tsukita S, Tsukita S (1994) Direct association of occludin with ZO-1 and its possible involvement in the localization of occludin at tight junctions. J Cell Biol 127:1617–1626

127. Itoh M, Morita K, Tsukita S (1999) Characterization of ZO-2 as a MAGUK family member associated with tight as well as adherens junctions with a binding affinity to occludin and alpha catenin. J Biol Chem 274:5981–5986

128. Kale G, Naren AP, Sheth P, Rao RK (2003) Tyrosine phosphorylation of occludin attenuates its interactions with ZO-1, ZO-2, and ZO-3. Biochem Biophys Res Commun 302:324–329

129. Nag S (2007) Structure and pathology of the blood-brain barrier. In Lathja A, (Ed) Handbook of Neurochemistry and Molecular Neurobiology, Springer, New York, pp 58–78

130. Vorbrodt AW, Dobrogowska DH (2003) Molecular anatomy of intercellular junctions in brain endothelial and epithelial barriers: electron microscopist's view. Brain Res Brain Res Rev 42:221–242

131. Yeung D, Manias JL, Stewart DJ, Nag S (2008) Decreased junctional adhesion molecule-A expression during blood-brain barrier breakdown. Acta Neuropathol 115:635–642

132. Anderson JM, Stevenson BR, Jesaitis LA, Goodenough DA, Mooseker MS (1988) Characterization of ZO-1, a protein component of the tight junction from mouse liver and Madin-Darby canine kidney cells. J Cell Biol 106:1141–1149

133. Stevenson BR, Anderson JM, Goodenough DA, Mooseker MS (1988) Tight junction structure and ZO-1 content are identical in two strains of Madin-Darby canine kidney cells which differ in transepithelial resistance. J Cell Biol 107:2401–2408

134. Chiba H, Osanai M, Murata M, Kojima T, Sawada N (2008) Transmembrane proteins of tight junctions. Biochim Biophys Acta 1778:588–600

135. Forster C (2008) Tight junctions and the modulation of barrier function in disease. Histochem Cell Biol 130:55–70

136. Zlokovic BV (2008) The blood-brain barrier in health and chronic neurodegenerative disorders. Neuron 57:178–201

137. Navarro P, Ruco L, Dejana E (1998) Differential localization of VE- and N-cadherins in human endothelial cells: VE-cadherin competes with N-cadherin for junctional localization. J Cell Biol 140:1475–1484

138. Bazzoni G, Dejana E (2004) Endothelial cell-to-cell junctions: molecular organization and role in vascular homeostasis. Physiol Rev 84:869–901

139. Weis WI, Nelson WJ (2006) Re-solving the cadherin-catenin-actin conundrum. J Biol Chem 281:35593–35597

140. Staddon JM, Herrenknecht K, Smales C, Rubin LL (1995) Evidence that tyrosine phosphorylation may increase tight junction permeability. J Cell Sci 108 (Pt 2):609–619

141. Schulze C, Smales C, Rubin LL, Staddon JM (1997) Lysophosphatidic acid increases tight junction permeability in cultured brain endothelial cells. J Neurochem 68:991–1000

142. Rudini N, Dejana E (2008) Adherens junctions. Curr Biol 18:R1080–R1082

143. Dejana E, Orsenigo F, Lampugnani MG (2008) The role of adherens junctions and

VE-cadherin in the control of vascular permeability. J Cell Sci 121:2115–2122

144. Dejana E, Orsenigo F, Molendini C, Baluk P, McDonald DM (2009) Organization and signaling of endothelial cell-to-cell junctions in various regions of the blood and lymphatic vascular trees. Cell Tissue Res 335:17–25

145. Rubin LL, Staddon JM (1999) The cell biology of the blood-brain barrier. Annu Rev Neurosci 22:11–28

146. Oldendorf WH, Brown WJ (1975) Greater number of capillary endothelial cell mitochondria in brain than in muscle. Proc Soc Exp Biol Med 149:736–738

147. Oldendorf WH, Cornford ME, Brown WJ (1977) The large apparent work capability of the blood-brain barrier: a study of the mitochondrial content of capillary endothelial cells in brain and other tissues of the rat. Ann Neurol 1:409–417

148. Claudio L, Kress Y, Norton WT, Brosnan CF (1989) Increased vesicular transport and decreased mitochondrial content in blood-brain barrier endothelial cells during experimental autoimmune encephalomyelitis. Am J Pathol 135:1157–1168

149. Li JY, Boado RJ, Pardridge WM (2001) Blood-brain barrier genomics. J Cereb Blood Flow Metabol 21:61–68

150. Li JY, Boado RJ, Pardridge WM (2002) Rat blood-brain barrier genomics. II. J Cereb Blood Flow Metab 22:1319-1326

151. Shusta EV, Boado RJ, Mathern GW, Pardridge WM (2002) Vascular genomics of the human brain. J Cereb Blood Flow Metab 22:245–252

152. Enerson BE, Drewes LR (2006) The rat blood-brain barrier transcriptome. J Cereb Blood Flow Metab 26:959–973

153. Liu X, Tu M, Kelly RS, Chen C, Smith BJ (2004) Development of a computational approach to predict blood-brain barrier permeability. Drug Metab Dispos 32:132–139

154. Pardridge WM (2005) Molecular biology of the blood-brain barrier. Mol Biotechnol 30:57–70

155. Birnbaum MJ, Haspel HC, Rosen OM (1986) Cloning and characterization of a cDNA encoding the rat brain glucose-transporter protein. Proc Natl Acad Sci USA 83:5784–5788

156. Maher F, Vannucci SJ, Simpson IA (1994) Glucose transporter proteins in brain. FASEB J 8:1003–1011

157. Cornford EM, Hyman S, Swartz BE (1994) The human brain GLUT1 glucose transporter: ultrastructural localization to the blood-brain barrier endothelia. J Cereb Blood Flow Metab 14:106–112

158. Farrell CL, Pardridge WM (1991) Blood-brain barrier glucose transporter is asymmetrically distributed on brain capillary endothelial lumenal and ablumenal membranes: an electron microscopic immunogold study. Proc Natl Acad Sci USA 88:5779–5783

159. Hawkins RA, O'Kane RL, Simpson IA, Vina JR (2006) Structure of the blood-brain barrier and its role in the transport of amino acids. J Nutr 136:218S–226S

160. Lipton SA (2005) The molecular basis of memantine action in Alzheimer's disease and other neurologic disorders: low-affinity, uncompetitive antagonism. Curr Alzheimer Res 2:155–165

161. Nag S (1990) Ultracytochemical localisation of Na+, K(+)-ATPase in cerebral endothelium in acute hypertension. Acta Neuropathol (Berl) 80:7–11

162. Taylor CJ, Nicola PA, Wang S, Barrand MA, Hladky SB (2006) Transporters involved in regulation of intracellular pH in primary cultured rat brain endothelial cells. J Physiol 576:769–785

163. Zlokovic BV, Mackic JB, Djuricic B, Davson H (1989) Kinetic analysis of leucine-enkephalin cellular uptake at the luminal side of the blood-brain barrier of an in situ perfused guinea-pig brain. J Neurochem 53:1333–1340

164. Zlokovic BV, Susic VT, Davson H, Begley DJ, Jankov RM, Mitrovic DM, Lipovac MN (1989) Saturable mechanism for delta sleep-inducing peptide (DSIP) at the blood-brain barrier of the vascularly perfused guinea pig brain. Peptides 10:249–254

165. Banks WA (2006) The CNS as a target for peptides and peptide-based drugs. Expert Opin Drug Deliv 3:707–712

166. Zlokovic BV (1995) Cerebrovascular permeability to peptides: manipulations of transport systems at the blood-brain barrier. Pharm Res 12:1395–1406

167. Zlokovic BV, Hyman S, McComb JG, Lipovac MN, Tang G, Davson H (1990) Kinetics of arginine-vasopressin uptake at the blood-brain barrier. Biochim Biophys Acta 1025:191–198

168. Borst P, Elferink RO (2002) Mammalian ABC transporters in health and disease. Annu Rev Biochem 71:537–592

169. Dean M, Allikmets R (2001) Complete characterization of the human ABC gene family. J Bioenerg Biomembr 33:475–479

170. Ronaldson PT, Persidsky Y, Bendayan R (2008) Regulation of ABC membrane transporters in

glial cells: relevance to the pharmacotherapy of brain HIV-1 infection. Glia 56:1711–1735

171. Leggas M, Adachi M, Scheffer GL, Sun D, Wielinga P, Du G, Mercer KE, Zhuang Y, Panetta JC, Johnston B, Scheper RJ, Stewart CF, Schuetz JD (2004) Mrp4 confers resistance to topotecan and protects the brain from chemotherapy. Mol Cell Biol 24: 7612–7621

172. Miller DS, Nobmann SN, Gutmann H, Toeroek M, Drewe J, Fricker G (2000) Xenobiotic transport across isolated brain microvessels studied by confocal microscopy. Mol Pharmacol 58:1357–1367

173. Zhang Y, Schuetz JD, Elmquist WF, Miller DW (2004) Plasma membrane localization of multidrug resistance-associated protein homologs in brain capillary endothelial cells. J Pharmacol Exp Ther 311:449–455

174. Hori S, Ohtsuki S, Tachikawa M, Kimura N, Kondo T, Watanabe M, Nakashima E, Terasaki T (2004) Functional expression of rat ABCG2 on the luminal side of brain capillaries and its enhancement by astrocyte-derived soluble factor(s). J Neurochem 90: 526–536

175. Lee YJ, Kusuhara H, Jonker JW, Schinkel AH, Sugiyama Y (2005) Investigation of efflux transport of dehydroepiandrosterone sulfate and mitoxantrone at the mouse blood-brain barrier: a minor role of breast cancer resistance protein. J Pharmacol Exp Ther 312:44–52

176. Lee G, Babakhanian K, Ramaswamy M, Prat A, Wosik K, Bendayan R (2007) Expression of the ATP-binding cassette membrane transporter, ABCG2, in human and rodent brain microvessel endothelial and glial cell culture systems. Pharm Res 24:1262–1274

177. Zhang W, Mojsilovic-Petrovic J, Andrade MF, Zhang H, Ball M, Stanimirovic DB (2003) The expression and functional characterization of ABCG2 in brain endothelial cells and vessels. FASEB J 17:2085–2087

178. Abbott NJ, Patabendige AA, Dolman DE, Yusof SR, Begley DJ (2009) Structure and function of the blood-brain barrier. Neurobiol Dis 37:13–25

179. Begley DJ (2004) ABC transporters and the blood-brain barrier. Curr Pharm Des 10: 1295–1312

180. Carvey PM, Hendey B, Monahan AJ (2009) The blood-brain barrier in neurodegenerative disease: a rhetorical perspective. J Neurochem 111:291–314

181. Nag S and Begley DJ (2005) Blood-brain barrier, exchange of metabolites and gases. In Kalimo H (Ed) Pathology and Genetics. Cerebrovascular Diseases, ISN Neuropath Press, Basel, pp 22–29

182. Nag S (2003) Pathophysiology of blood-brain barrier breakdown. Methods Mol Med 89:97–119

183. Nag S, Manias JL, Stewart DJ (2009) Pathology and new players in the pathogenesis of brain edema. Acta Neuropathol 118: 197–217

184. Hawkins BT, Egleton RD (2008) Pathophysiology of the blood-brain barrier: animal models and methods. Curr Top Dev Biol 80:277–309

185. Starr JM, Wardlaw J, Ferguson K, MacLullich A, Deary IJ, Marshall I (2003) Increased blood-brain barrier permeability in type II diabetes demonstrated by gadolinium magnetic resonance imaging. J Neurol Neurosurg Psychiatry 74:70–76

186. Bell RD, Zlokovic BV (2009) Neurovascular mechanisms and blood-brain barrier disorder in Alzheimer's disease. Acta Neuropathol 118:103–113

187. Vogelgesang S, Cascorbi I, Schroeder E, Pahnke J, Kroemer HK, Siegmund W, Kunert-Keil C, Walker LC, Warzok RW (2002) Deposition of Alzheimer's beta-amyloid is inversely correlated with P-glycoprotein expression in the brains of elderly non-demented humans. Pharmacogenetics 12: 535–541

188. Kortekaas R, Leenders KL, van Oostrom JC, Vaalburg W, Bart J, Willemsen AT, Hendrikse NH (2005) Blood-brain barrier dysfunction in parkinsonian midbrain in vivo. Ann Neurol 57:176–179

189. Lee G, Bendayan R (2004) Functional expression and localization of P-glycoprotein in the central nervous system: relevance to the pathogenesis and treatment of neurological disorders. Pharm Res 21:1313–1330

190. Nag S (2002) The blood-brain barrier and cerebral angiogenesis: lessons from the cold-injury model. Trends Mol Med 8:38–44

191. Wolburg H, Neuhaus J, Kniesel U, Krauss B, Schmid EM, Ocalan M, Farrell C, Risau W (1994) Modulation of tight junction structure in blood-brain barrier endothelial cells. Effects of tissue culture, second messengers and cocultured astrocytes. J Cell Sci 107 (Pt 5): 1347–1357

192. Wolburg H, Wolburg-Buchholz K, Kraus J, Rascher-Eggstein G, Liebner S, Hamm S, Duffner F, Grote EH, Risau W, Engelhardt B (2003) Localization of claudin-3 in tight junctions of the blood-brain barrier is selectively lost during experimental autoimmune encephalomyelitis and human glioblastoma

multiforme. Acta Neuropathol (Berl) 105: 586–592

193. Rizzo V, Morton C, DePaola N, Schnitzer JE, Davies PF (2003) Recruitment of endothelial caveolae into mechanotransduction pathways by flow conditioning in vitro. Am J Physiol Heart Circ Physiol 285:H1720–H1729

194. Schnitzer JE, Carley WW, Palade GE (1988) Specific albumin binding to microvascular endothelium in culture. Am J Physiol 254:H425–H437

195. Klatzo I, Piraux A, Laskowski EJ (1958) The relationship between edema, blood-brain-barrier and tissue elements in a local brain injury. J Neuropathol Exp Neurol 17:548–564

196. Nag S (1996) Cold-injury of the cerebral cortex: immunolocalization of cellular proteins and blood-brain barrier permeability studies. J Neuropathol Exp Neurol 55: 880–888

197. Nag S, Picard P, Stewart DJ (2001) Expression of nitric oxide synthases and nitrotyrosine during blood-brain barrier breakdown and repair after cold injury. Lab Invest 81:41–49

198. Nag S, Eskandarian MR, Davis J, Eubanks JH (2002) Differential expression of vascular endothelial growth factor-A (VEGF-A) and VEGF-B after brain injury. J Neuropathol Exp Neurol 61:778–788

199. Banks WA, Erickson MA (2010) The blood-brain barrier and immune function and dysfunction. Neurobiol Dis 37:26–32

200. Engelhardt B, Sorokin L (2009) The blood-brain and the blood-cerebrospinal fluid barriers: function and dysfunction. Semin Immunopathol 31:497–511

201. Pachter JS, de Vries HE, Fabry Z (2003) The blood-brain barrier and its role in immune privilege in the central nervous system. J Neuropathol Exp Neurol 62: 593–604

202. Lossinsky AS, Shivers RR (2004) Structural pathways for macromolecular and cellular transport across the blood-brain barrier during inflammatory conditions. Review. Histol Histopathol 19:535–564

203. Brightman MW, Zis K, Anders J (1983) Morphology of cerebral endothelium and astrocytes as determinants of the neuronal microenvironment. Acta Neuropathol Suppl (Berl) 8:21–33

204. Habgood MD, Bye N, Dziegielewska KM, Ek CJ, Lane MA, Potter A, Morganti-Kossmann C, Saunders NR (2007) Changes in blood-brain barrier permeability to large and small molecules following traumatic brain injury in mice. Eur J Neurosci 25:231–238

205. Vorbrodt AW (1993) Morphological evidence of the functional polarization of brain microvascular endothelium. In Pardridge WM (Ed) The Blood-Brain Barrier. Cellular and Molecular Biology, Raven, New York, pp 137–164

206. Nag S (1986) Cerebral endothelial plasma membrane alterations in acute hypertension. Acta Neuropathol (Berl) 70:38–43

207. Hardebo JE, Kahrstrom J (1985) Endothelial negative surface charge areas and blood-brain barrier function. Acta Physiol Scand 125:495–499

208. Nagy Z, Peters H, Huttner I (1983) Charge-related alterations of the cerebral endothelium. Lab Invest 49: 662–671

209. Nag S, Manias JL, Stewart DJ (2009) Expression of endothelial phosphorylated caveolin-1 is increased in brain injury. Neuropathol Appl Neurobiol 35:417–426

210. Mayhan WG (2000) Nitric oxide donor-induced increase in permeability of the blood-brain barrier. Brain Res 866:101–108

211. Unterberg A, Wahl M, Baethmann A (1984) Effects of bradykinin on permeability and diameter of pial vessels in vivo. J Cereb Blood Flow Metab 4:574–585

212. Boyd NL, Park H, Yi H, Boo YC, Sorescu GP, Sykes M, Jo H (2003) Chronic shear induces caveolae formation and alters ERK and Akt responses in endothelial cells. Am J Physiol Heart Circ Physiol 285:H1113–H1122

213. Volonte D, Galbiati F, Pestell RG, Lisanti MP (2001) Cellular stress induces the tyrosine phosphorylation of caveolin-1 (Tyr14) via activation of p38 mitogen-activated protein kinase and c-Src kinase. J Biol Chem 276:8094–8103

214. Dallasta LM, Pisarov LA, Esplen JE, Werley JV, Moses AV, Nelson JA, Achim C L (1999) Blood-brain barrier tight junction disruption in human immunodeficiency virus-1 encephalitis. Am J Pathol 155:1915–1927

215. Plumb J, McQuaid S, Mirakhur M, Kirk J (2002) Abnormal endothelial tight junctions in active lesions and normal-appearing white matter in multiple sclerosis. Brain Pathol 12:154–169

216. Song L, Pachter JS (2004) Monocyte chemoattractant protein-1 alters expression of tight junction-associated proteins in brain microvascular endothelial cells. Microvasc Res 67:78–89

217. Song L, Ge S, Pachter JS (2007) Caveolin-1 regulates expression of junction-associated proteins in brain microvascular endothelial cells. Blood 109:1515–1523

218. Bentzel CJ, Hainau B, Edelman A, Anagnostopoulos T, Benedetti EL (1976) Effect of plant cytokinins on microfilaments and tight junction permeability. Nature 264:666–668

219. Stevenson BR, Begg DA (1994) Concentration-dependent effects of cytochalasin D on tight junctions and actin filaments in MDCK epithelial cells. J Cell Sci 107 (Pt 3):367–375

220. Nico B, Frigeri A, Nicchia GP, Corsi P, Ribatti D, Quondamatteo F, Herken R, Girolamo F, Marzullo A, Svelto M, Roncali L (2003) Severe alterations of endothelial and glial cells in the blood-brain barrier of dystrophic mdx mice. Glia 42:235–251

221. Allen RD, Weiss DG, Hayden JH, Brown DT, Fujiwake H, Simpson M (1985) Gliding movement of and bidirectional transport along single native microtubules from squid axoplasm: evidence for an active role of microtubules in cytoplasmic transport. J Cell Biol 100:1736–1752

222. Schnapp BJ, Vale RD, Sheetz MP, Reese TS (1985) Single microtubules from squid axoplasm support bidirectional movement of organelles. Cell 40:455–462

223. Liu SM, Magnusson KE, Sundqvist T (1993) Microtubules are involved in transport of macromolecules by vesicles in cultured bovine aortic endothelial cells. J Cell Physiol 156:311–316

224. Sternlicht MD, Werb Z (2001) How matrix metalloproteinases regulate cell behavior. Annu Rev Cell Dev Biol 17:463–516

225. Rosenberg GA (2002) Matrix metalloproteinases in neuroinflammation. Glia 39:279–291

226. Cunningham LA, Wetzel M, Rosenberg GA (2005) Multiple roles for MMPs and TIMPs in cerebral ischemia. Glia 50:329–339

227. Candelario-Jalil E, Yang Y, Rosenberg GA (2008) Diverse roles of matrix metalloproteinases and tissue inhibitors of metalloproteinases in neuroinflammation and cerebral ischemia. Neuroscience 158:983–994

228. Clark AW, Krekoski CA, Bou SS, Chapman KR, Edwards DR (1997) Increased gelatinase A (MMP-2) and gelatinase B (MMP-9) activities in human brain after focal ischemia. Neurosci Lett 238:53–56

229. Mun-Bryce S, Rosenberg GA (1998) Matrix metalloproteinases in cerebrovascular disease. J Cereb Blood Flow Metab 18:1163–1172

230. Rosenberg GA, Yang Y (2007) Vasogenic edema due to tight junction disruption by matrix metalloproteinases in cerebral ischemia. Neurosurg Focus 22:E4

231. Yang Y, Estrada EY, Thompson JF, Liu W, Rosenberg GA (2007) Matrix metalloproteinase-mediated disruption of tight junction proteins in cerebral vessels is reversed by synthetic matrix metalloproteinase inhibitor in focal ischemia in rat. J Cereb Blood Flow Metab 27:697–709

232. Chang DI, Hosomi N, Lucero J, Heo JH, Abumiya T, Mazar AP, del Zoppo GJ (2003) Activation systems for latent matrix metalloproteinase-2 are upregulated immediately after focal cerebral ischemia. J Cereb Blood Flow Metab 23:1408–1419

233. Asahi M, Asahi K, Jung JC, del Zoppo GJ, Fini ME, Lo EH (2000) Role for matrix metalloproteinase 9 after focal cerebral ischemia: effects of gene knockout and enzyme inhibition with BB-94. J Cereb Blood Flow Metab 20:1681–1689

234. Asahi M, Wang X, Mori T, Sumii T, Jung JC, Moskowitz MA, Fini ME, Lo EH (2001) Effects of matrix metalloproteinase-9 gene knock-out on the proteolysis of blood-brain barrier and white matter components after cerebral ischemia. J Neurosci 21:7724–7732

235. Rosenberg GA, Estrada EY, Dencoff JE (1998) Matrix metalloproteinases and TIMPs are associated with blood-brain barrier opening after reperfusion in rat brain. Stroke 29:2189–2195

236. Abbott NJ (2000) Inflammatory mediators and modulation of blood-brain barrier permeability. Cell Mol Neurobiol 20:131–147

237. Schilling L, Wahl M (1997) Brain edema: pathogenesis and therapy. Kidney Int Suppl 59:S69–S75

238. Segura I, De SF, Hohensinner PJ, Almodovar CR, Carmeliet P (2009) The neurovascular link in health and disease: an update. Trends Mol Med 15:439–451

239. Zacchigna S, Lambrechts D, Carmeliet P (2008) Neurovascular signalling defects in neurodegeneration. Nat Rev Neurosci 9:169–181

240. Adams RH, Alitalo K (2007) Molecular regulation of angiogenesis and lymphangiogenesis. Nat Rev Mol Cell Biol 8:464–478

241. Ferrara N, Gerber HP, LeCouter J (2003) The biology of VEGF and its receptors. Nat Med 9:669–676

242. Otrock ZK, Makarem JA, Shamseddine AI (2007) Vascular endothelial growth factor family of ligands and receptors: review. Blood Cells Mol Dis 38:258–268

243. Raab S, Plate KH (2007) Different networks, common growth factors: shared growth factors and receptors of the vascular and the nervous system. Acta Neuropathol 113: 607–626

244. Roy H, Bhardwaj S, Yla-Herttuala S (2006) Biology of vascular endothelial growth factors. Febs Lett 580:2879–2887

245. Yla-Herttuala S, Rissanen TT, Vajanto I, Hartikainen J (2007) Vascular endothelial growth factors: biology and current status of clinical applications in cardiovascular medicine. J Am Coll Cardiol 49:1015–1026

246. Segura AM, Luna RE, Horiba K, Stetler Stevenson WG, McAllister HA, Willerson JT, Ferrans VJ (1998) Immunohistochemistry of matrix metalloproteinases and their inhibitors in thoracic aortic aneurysms and aortic valves of patients with Marfan's syndrome. Circulation 98:II331-II337

247. Breier G, Albrecht U, Sterrer S, Risau W (1992) Expression of vascular endothelial growth factor during embryonic angiogenesis and endothelial cell differentiation. Development 114:521–532

248. Nag S, Takahashi JL, Kilty DW (1997) Role of vascular endothelial growth factor in blood-brain barrier breakdown and angiogenesis in brain trauma. J Neuropathol Exp Neurol 56:912–921

249. Dobrogowska DH, Lossinsky AS, Tarnawski M, Vorbrodt AW (1998) Increased blood-brain barrier permeability and endothelial abnormalities induced by vascular endothelial growth factor. J Neurocytol 27:163–173

250. Nag S, Papneja T, Venugopalan R, Stewart DJ (2005) Increased angiopoietin2 expression is associated with endothelial apoptosis and blood-brain barrier breakdown. Lab Invest 85:1189–1198

251. Proescholdt MA, Heiss JD, Walbridge S, Muhlhauser J, Capogrossi MC, Oldfield EH, Merrill MJ (1999) Vascular endothelial growth factor (VEGF) modulates vascular permeability and inflammation in rat brain. J Neuropathol Exp Neurol 58:613–627

252. Hofman P, Blaauwgeers HGT, Tolentino MJ, Adamis AP, Cardozo BJN, Vrensen GFJ M, Schlingemann RO (2000) VEGF-A induced hyperpermeability of blood-retinal barrier endothelium in vivo is predominantly associated with pinocytotic vesicular transport and not with formation of fenestrations. Curr Eye Res 21:637–645

253. Eliceiri BP, Paul R, Schwartzberg PL, Hood JD, Leng J, Cheresh DA (1999) Selective requirement for Src kinases during VEGF-induced angiogenesis and vascular permeability. Mol Cell 4: 915–924

254. Paul R, Zhang ZG, Eliceiri BP, Jiang Q, Boccia AD, Zhang RL, Chopp M, Cheresh DA (2001) Src deficiency or blockade of Src activity in mice provides cerebral protection following stroke. Nat Med 7:222–227

255. Antonetti DA, Barber AJ, Khin S, Lieth E, Tarbell JM, Gardner TW (1998) Vascular permeability in experimental diabetes is associated with reduced endothelial occludin content: Vascular endothelial growth factor decreases occludin in retinal endothelial cells. Diabetes 47:1953–1959

256. Wang W, Dentler WL, Borchardt RT (2001) VEGF increases BMEC monolayer permeability by affecting occludin expression and tight junction assembly. Am J Physiol Heart Circ Physiol 280:H434–H440

257. Olofsson B, Pajusola K, Kaipainen A, von Euler G, Joukov V, Saksela O, Orpana A, Pettersson RF, Alitalo K, Eriksson U (1996) Vascular endothelial growth factor B, a novel growth factor for endothelial cells. Proc Natl Acad Sci USA 93:2576–2581

258. Olofsson B, Pajusola K, von Euler G, Chilov D, Alitalo K, Eriksson U (1996) Genomic organization of the mouse and human genes for vascular endothelial growth factor B (VEGF-B) and characterization of a second splice isoform. J Biol Chem 271:19310–19317

259. Makinen T, Olofsson B, Karpanen T, Hellman U, Soker S, Klagsbrun M, Eriksson U, Alitalo K (1999) Differential binding of vascular endothelial growth factor B splice and proteolytic isoforms to neuropilin-1. J Biol Chem 274:21217–21222

260. Olofsson B, Korpelainen E, Pepper MS, Mandriota SJ, Aase K, Kumar V, Gunji Y, Jeltsch MM, Shibuya M, Alitalo K, Eriksson U (1998) Vascular endothelial growth factor B (VEGF-B) binds to VEGF receptor- 1 and regulates plasminogen activator activity in endothelial cells. Proc Natl Acad Sci USA 95:11709–11714

261. Aase K, Lymboussaki A, Kaipainen A, Olofsson B, Alitalo K, Eriksson U (1999) Localization of VEGF-B in the mouse embryo suggests a paracrine role of the growth factor in the developing vasculature. Develop Dynam 215:12–25

262. Lagercrantz J, Farnebo F, Larsson C, Tvrdik T, Weber G, Piehl, F (1998) A comparative study of the expression patterns for vegf, vegf-b/vrf and vegf-c in the developing and adult mouse. Bba Gene Struct Express 1398:157–163

263. Sun Y, Jin K, Childs JT, Xie L, Mao XO, Greenberg DA (2006) Vascular endothelial growth factor-B (VEGFB) stimulates

neurogenesis: evidence from knockout mice and growth factor administration. Dev Biol 289:329–335

264. Davis S, Aldrich TH, Jones PF, Acheson A, Compton DL, Jain V, Ryan TE, Bruno J, Radziejewski C, Maisonpierre PC, Yancopoulos GD (1996) Isolation of angiopoietin-1, a ligand for the TIE2 receptor, by secretion-trap expression cloning. Cell 87:1161–1169

265. Jones N, Iljin K, Dumont DJ, Alitalo K (2001) Tie receptors: new modulators of angiogenic and lymphangiogenic responses. Nat Rev Mol Cell Biol 2:257–267

266. Valenzuela DM, Griffiths JA, Rojas J, Aldrich TH, Jones PF, Zhou H, McClain J, Copeland NG, Gilbert DJ, Jenkins NA, Huang T, Papadopoulos N, Maisonpierre PC, Davis S, Yancopoulos GD (1999) Angiopoietins 3 and 4: diverging gene counterparts in mice and humans. Proc Natl Acad Sci USA 96:1904–1909

267. Dumont DJ, Gradwohl GJ, Fong GH, Auerbach R, Breitman ML (1993) The endothelial-specific receptor tyrosine kinase, tek, is a member of a new subfamily of receptors. Oncogene 8:1293–1301

268. Maisonpierre PC, Suri C, Jones PF, Bartunkova S, Wiegand SJ, Radziejewski C, Compton D, McClain J, Aldrich TH, Papadopoulos N, Daly TJ, Davis S, Sato TN, Yancopoulos GD (1997) Angiopoietin-2, a natural antagonist for Tie2 that disrupts in vivo angiogenesis. Science 277:55–60

269. Runting AS, Stacker SA, Wilks AF (1993) tie2, a putative protein tyrosine kinase from a new class of cell surface receptor. Growth Factors 9:99–105

270. Puri MC, Rossant J, Alitalo K, Bernstein A, Partanen J (1995) The receptor tyrosine kinase TIE is required for integrity and survival of vascular endothelial cells. EMBO J 14:5884–5891

271. Nourhaghighi N, Teichert-Kuliszewska K, Davis J, Stewart DJ, Nag S (2003) Altered expression of angiopoietins during blood-brain barrier breakdown and angiogenesis. Lab Invest 83:1211–1222

272. Wong AL, Haroon ZA, Werner S, Dewhirst MW, Greenberg CS, Peters KG (1997) Tie2 expression and phosphorylation in angiogenic and quiescent adult tissues. Circ Res 81:567–574

273. Witzenbichler B, Asahara T, Murohara T, Silver M, Spyridopoulos I, Magner M, Principe N, Kearney M, Hu JS, Isner JM (1998) Vascular endothelial growth factor-c (VEGF-C/VEGF-2) promotes angiogenesis in the setting of tissue ischemia. Am J Pathol 153: 381–394

274. Fujikawa K, Scherpenseel ID, Jain SK, Presman E, Varticovski L (1999) Role of PI 3-kinase in angiopoietin-1-mediated migration and attachment-dependent survival of endothelial cells. Exp Cell Res 253:663–672

275. Papapetropoulos A, Fulton D, Mahboubi K, Kalb RG, OConnor DS, Li FZ, Altieri DC, Sessa WC (2000) Angiopoietin-1 inhibits endothelial cell apoptosis via the Akt/ survivin pathway. J Biol Chem 275:9102–9105

276. Cohen B, Barkan D, Levy Y, Goldberg I, Fridman E, Kopolovic J, Rubinstein M (2001) Leptin induces angiopoietin-2 expression in adipose tissues. J Biol Chem 276:7697–7700

277. Zagzag D, Amirnovin R, Greco MA, Yee H, Holash J, Wiegand SJ, Zabski S, Yancopoulos GD, Grumet M (2000) Vascular apoptosis and involution in gliomas precede neovascularization: A novel concept for glioma growth and angiogenesis. Lab Invest 80:837–849

278. Thurston G, Suri C, Smith K, McClain J, Sato TN, Yancopoulos GD, McDonald DM (1999) Leakage-resistant blood vessels in mice transgenically overexpressing angiopoietin-1. Science 286:2511–2514

279. Gamble JR, Drew J, Trezise L, Underwood A, Parsons M, Kasminkas L, Rudge J, Yancopoulos G, Vadas MA (2000) Angiopoietin-1 is an antipermeability and anti-inflammatory agent in vitro and targets cell junctions. Circ Res 87:603–607

280. Jho D, Mehta D, Ahmmed G, Gao XP, Tiruppathi C, Broman M, Malik AB (2005) Angiopoietin-1 opposes VEGF-induced increase in endothelial permeability by inhibiting TRPC1-dependent Ca2 influx. Circ Res 96:1282–1290

281. Li X, Hahn CN, Parsons M, Drew J, Vadas MA, Gamble JR (2004) Role of protein kinase Czeta in thrombin-induced endothelial permeability changes: inhibition by angiopoietin-1. Blood 104:1716–1724

282. Pizurki L, Zhou Z, Glynos K, Roussos C, Papapetropoulos A (2003) Angiopoietin-1 inhibits endothelial permeability, neutrophil adherence and IL-8 production. Br J Pharmacol 139:329–336

283. Baffert F, Le T, Thurston G, McDonald DM (2006) Angiopoietin-1 decreases plasma leakage by reducing number and size of endothelial gaps in venules. Am J Physiol Heart Circ Physiol 290:H107–H118

284. Thurston G, Rudge JS, Ioffe E, Zhou H, Ross L, Croll SD, Glazer N, Holash J, McDonald DM, Yancopoulos GD (2000) Angiopoietin-1 protects the adult vasculature against plasma leakage. Nature Med 6:460–463

285. Lee SW, Kim WJ, Jun HO, Choi YK, Kim KW (2009) Angiopoietin-1 reduces vascular endothelial growth factor-induced brain endothelial permeability via upregulation of ZO-2. Int J Mol Med 23:279–284

286. Roviezzo F, Tsigkos S, Kotanidou A, Bucci M, Brancaleone V, Cirino G, Papapetropoulos A (2005) Angiopoietin-2 causes inflammation in vivo by promoting vascular leakage. J Pharmacol Exp Ther 314:738–744

287. Croll SD, Wiegand SJ (2001) Vascular growth factors in cerebral ischemia. Mol Neurobiol 23:121–135

288. Hansen TM, Moss AJ, Brindle NP (2008) Vascular endothelial growth factor and angiopoietins in neurovascular regeneration and protection following stroke. Curr Neurovasc Res 5:236–245

289. Machein MR, Plate KH (2000) VEGF in brain tumors. J Neurooncol 50:109–120

290. Plate KH, Breier G, Weich HA, Risau W (1992) Vascular endothelial growth factor is a potential tumour angiogenesis factor in human gliomas in vivo. Nature 359: 845–848

291. Plate KH (1999) Mechanisms of angiogenesis in the brain. J Neuropathol Exp Neurol 58:313–320

292. Audero E, Cascone I, Zanon I, Previtali SC, Piva R, Schiffer D, Bussolino F (2001) Expression of angiopoietin-1 in human glioblastomas regulates tumor-induced angiogenesis: in vivo and in vitro studies. Arterioscler Thromb Vasc Biol 21:536–541

293. Holash J, Maisonpierre PC, Compton D, Boland P, Alexander CR, Zagzag D, Yancopoulos G D, Wiegand SJ (1999) Vessel cooption, regression, and growth in tumors mediated by angiopoietins and VEGF. Science 284:1994–1998

294. Holash J, Wiegand SJ, Yancopoulos GD (1999) New model of tumor angiogenesis: dynamic balance between vessel regression and growth mediated by angiopoietins and VEGF. Oncogene 18:5356–5362

295. Reiss Y, Machein MR, Plate KH (2005) The role of angiopoietins during angiogenesis in gliomas. Brain Pathol 15:311–317

296. Ding H, Roncari L, Wu X, Lau N, Shannon P, Nagy A, Guha A (2001) Expression and hypoxic regulation of angiopoietins in human astrocytomas. Neuro oncol 3:1–10

297. Stratmann A, Risau W, Plate KH (1998) Cell type-specific expression of angiopoietin-1 and angiopoietin-2 suggests a role in glioblastoma angiogenesis. Am J Pathol 153: 1459–1466

298. Zadeh G, Guha A (2003) Neoangiogenesis in human astrocytomas: expression and functional role of angiopoietins and their cognate receptors. Front Biosci 8:e128–e137

299. Zagzag D, Hooper A, Friedlander DR, Chan W, Holash J, Wiegand SJ, Yancopoulos G D, Grumet M (1999) In situ expression of angiopoietins in astrocytomas identifies angiopoietin-2 as an early marker of tumor angiogenesis. Exp Neurol 159:391–400

300. Machein MR, Kullmer J, Fiebich BL, Plate KH, Warnke PC (1999) Vascular endothelial growth factor expression, vascular volume, and capillary permeability in human brain tumors. Neurosurgery 44:732–740

301. Goldman CK, Bharara S, Palmer CA, Vitek J, Tsai JC, Weiss HL, Gillespie GY (1997) Brain edema in meningiomas is associated with increased vascular endothelial growth factor expression. Neurosurgery 40:1269–1277

302. Kalkanis SN, Carroll RS, Zhang J, Zamani AA, Black PM (1996) Correlation of vascular endothelial growth factor messenger RNA expression with peritumoral vasogenic cerebral edema in meningiomas. J Neurosurg 85:1095–1101

303. Provias J, Claffey K, delAguila L, Lau N, Feldkamp M, Guha A (1997) Meningiomas: role of vascular endothelial growth factor/ vascular permeability factor in angiogenesis and peritumoral edema. Neurosurgery 40:1016–1026

304. vanBruggen N, Thibodeaux H, Palmer JT, Lee WP, Fu L, Cairns B, Tumas D, Gerlai R, Williams S P, Campagne M V, Ferrara N (1999) VEGF antagonism reduces edema formation and tissue damage after ischemia/ reperfusion injury in the mouse brain. J Clin Invest 104:1613–1620

305. Zhang ZG, Zhang L, Croll SD, Chopp M (2002) Angiopoietin-1 reduces cerebral blood vessel leakage and ischemic lesion volume after focal cerebral embolic ischemia in mice. Neuroscience 113:683–687

306. Pardridge WM (2006) Molecular Trojan horses for blood-brain barrier drug delivery. Curr Opin Pharmacol 6:494–500

307. Boado RJ (2007) Blood-brain barrier transport of non-viral gene and RNAi therapeutics. Pharm Res 24:1772–1787

308. Kumar P, Wu H, McBride JL, Jung KE, Kim MH, Davidson BL, Lee SK, Shankar P, Manjunath N (2007) Transvascular delivery of small interfering RNA to the central nervous system. Nature 448:39–43

309. Pardridge WM, Triguero D, Buciak J (1989) Transport of histone through the blood-brain barrier. J Pharmacol Exp Ther 251:821–826

310. Demeule M, Poirier J, Jodoin J, Bertrand Y, Desrosiers RR, Dagenais C, Nguyen T, Lanthier J, Gabathuler R, Kennard M, Jefferies WA, Karkan D, Tsai S, Fenart L, Cecchelli R, Beliveau R (2002) High transcytosis of melanotransferrin (P97) across the blood-brain barrier. J Neurochem 83:924–933

311. Prince WS, McCormick LM, Wendt DJ, Fitzpatrick PA, Schwartz KL, Aguilera AI, Koppaka V, Christianson TM, Vellard MC, Pavloff N, Lemontt JF, Qin M, Starr CM, Bu G, Zankel TC (2004) Lipoprotein receptor binding, cellular uptake, and lysosomal delivery of fusions between the receptor-associated protein (RAP) and alpha-L-iduronidase or acid alpha-glucosidase. J Biol Chem 279:35037–35046

312. Schwarze SR, Ho A, Vocero-Akbani A, Dowdy SF (1999) In vivo protein transduction: delivery of a biologically active protein into the mouse. Science 285:1569–1572

313. Broadwell RD, Balin BJ, Salcman M (1988) Transcytotic pathway for blood-borne protein through the blood-brain barrier. Proc Natl Acad Sci USA 85:632–636

314. Adenot M, Merida P, Lahana R (2007) Applications of a blood-brain barrier technology platform to predict CNS penetration of various chemotherapeutic agents. 2. Cationic peptide vectors for brain delivery. Chemotherapy 53:73–76

315. Demeule M, Currie JC, Bertrand Y, Che C, Nguyen T, Regina A, Gabathuler R, Castaigne JP, Beliveau R (2008) Involvement of the low-density lipoprotein receptor-related protein in the transcytosis of the brain delivery vector angiopep-2. J Neurochem 106:1534–1544

316. Kreuter J, Hekmatara T, Dreis S, Vogel T, Gelperina S, Langer K (2007) Covalent attachment of apolipoprotein A-I and apolipoprotein B-100 to albumin nanoparticles enables drug transport into the brain. J Control Release 118:54–58

317. Serres S, Anthony DC, Jiang Y, Broom KA, Campbell SJ, Tyler DJ, van Kasteren SI, Davis BG, Sibson NR (2009) Systemic inflammatory response reactivates immune-mediated lesions in rat brain. J Neurosci 29:4820–4828

318. van Kasteren SI, Campbell SJ, Serres S, Anthony DC, Sibson NR, Davis BG (2009) Glyconanoparticles allow pre-symptomatic in vivo imaging of brain disease. Proc Natl Acad Sci USA 106:18–23

319. von Zur MC, Sibson NR, Peter K, Campbell SJ, Wilainam P, Grau GE, Bode C, Choudhury RP, Anthony DC (2008) A contrast agent recognizing activated platelets reveals murine cerebral malaria pathology undetectable by conventional MRI. J Clin Invest 118: 1198–1207

320. Kroll RA, Pagel MA, Muldoon LL, RomanGoldstein S, Fiamengo, SA, Neuwelt EA (1998) Improving drug delivery to intracerebral tumor and surrounding brain in a rodent model: A comparison of osmotic versus bradykinin modification of the blood-brain and/or blood-tumor barriers. Neurosurgery 43:879–886

321. Jahnke K, Doolittle ND, Muldoon LL, Neuwelt EA (2006) Implications of the blood-brain barrier in primary central nervous system lymphoma. Neurosurg Focus 21:E11

322. Neuwelt EA (2004) Mechanisms of disease: the blood-brain barrier. Neurosurgery 54: 131–140

323. Muldoon LL, Nilaver G, Kroll RA, Pagel MA, Breakefield XO, Chiocca EA, Davidson BL, Weissleder R, Neuwelt EA (1995) Comparison of intracerebral inoculation and osmotic blood-brain barrier disruption for delivery of adenovirus, herpesvirus, and iron oxide particles to normal rat brain. Am J Pathol 147:1840–1851

324. Liu J, Oh P, Horner T, Rogers RA, Schnitzer JE (1997) Organized endothelial cell surface signal transduction in caveolae distinct from glycosylphosphatidylinositol-anchored protein microdomains. J Biol Chem 272: 7211–7222

325. Uittenbogaard A, Shaul PW, Yuhanna IS, Blair A, Smart EJ (2000) High density lipoprotein prevents oxidized low density lipoprotein-induced inhibition of endothelial nitric-oxide synthase localization and activation in caveolae. J Biol Chem 275: 11278–11283

326. Tiruppathi C, Song W, Bergenfeldt M, Sass P, Malik AB (1997) Gp60 activation mediates albumin transcytosis in endothelial cells by tyrosine kinase-dependent pathway. J Biol Chem 272:25968–25975

327. Yuhanna IS, Zhu Y, Cox BE, Hahner LD, Osborne-Lawrence S, Lu P, Marcel YL, Anderson RG, Mendelsohn ME, Hobbs HH, Shaul PW (2001) High-density lipoprotein binding to scavenger receptor-BI activates endothelial nitric oxide synthase. Nat Med 7:853–857

328. Feng Y, Venema VJ, Venema RC, Tsai N, Behzadian MA, Caldwell RB (1999) VEGF-induced permeability increase is mediated by caveolae. Invest Ophthalmol Vis Sci 40: 157–167

329. Lupu C, Goodwin CA, Westmuckett AD, Emeis JJ, Scully MF, Kakkar VV, Lupu F (1997) Tissue factor pathway inhibitor in endothelial cells colocalizes with glycolipid microdomains/caveolae. Regulatory mechanism(s) of the anticoagulant properties of the endothelium. Arterioscler Thromb Vasc Biol 17:2964–2974

330. Demeule M, Jodoin J, Gingras D, Beliveau R (2000) P-glycoprotein is localized in caveolae in resistant cells and in brain capillaries. Febs Lett 466:219–224

331. Annabi B, Lachambre M, Bousquet-Gagnon N, Page M, Gingras D, Beliveau R (2001) Localization of membrane-type 1 matrix metalloproteinase in caveolae membrane domains. Biochem J 353:547–553

332. Puyraimond A, Fridman R, Lemesle M, Arbeille B, Menashi S (2001) MMP-2 colocalizes with caveolae on the surface of endothelial cells. Exp Cell Res 262:28–36

333. Igarashi J, Michel T (2000) Agonist-modulated targeting of the EDG-1 receptor to plasmalemmal caveolae. eNOS activation by sphingosine 1-phosphate and the role of caveolin-1 in sphingolipid signal transduction. J Biol Chem 275:32363–32370

334. Wei Y, Yang X, Liu Q, Wilkins JA, Chapman HA (1999) A role for caveolin and the urokinase receptor in integrin-mediated adhesion and signaling. J Cell Biol 144:1285–1294

335. Ju H, Venema VJ, Liang H, Harris MB, Zou R, Venema RC (2000) Bradykinin activates the Janus-activated kinase/signal transducers and activators of transcription (JAK/STAT) pathway in vascular endothelial cells: localization of JAK/STAT signalling proteins in plasmalemmal caveolae. Biochem J 351:257–264

336. Chun M, Liyanage UK, Lisanti MP, Lodish HF (1994) Signal transduction of a G protein-coupled receptor in caveolae: colocalization of endothelin and its receptor with caveolin. Proc Natl Acad Sci USA 91:11728–11732

337. Oh P, Schnitzer JE (2001) Segregation of heterotrimeric G proteins in cell surface microdomains. G(q) binds caveolin to concentrate in caveolae, whereas G(i) and G(s) target lipid rafts by default. Mol Biol Cell 12:685–698

338. Rizzo V, Sung A, Oh P, Schnitzer JE (1998) Rapid mechanotransduction in situ at the luminal cell surface of vascular endothelium and its caveolae. J Biol Chem 273:26323–26329

339. Shaul PW, Smart EJ, Robinson LJ, German Z, Yuhanna IS, Ying Y, Anderson RG, Michel T (1996) Acylation targets emdothelial nitric-oxide synthase to plasmalemmal caveolae. J Biol Chem 271:6518–6522

340. Garcia-Cardena G, Oh P, Liu J, Schnitzer JE, Sessa WC (1996) Targeting of nitric oxide synthase to endothelial cell caveolae via palmitoylation: implications for nitric oxide signaling. Proc Natl Acad Sci USA 93: 6448–6453

341. Spisni E, Griffoni C, Santi S, Riccio M, Marulli R, Bartolini G, Toni M, Ullrich V, Tomasi V (2001) Colocalization prostacyclin (PGI2) synthase–caveolin-1 in endothelial cells and new roles for PGI2 in angiogenesis. Exp Cell Res 266:31–43

342. Teubl M, Groschner K, Kohlwein SD, Mayer B, Schmidt K (1999) Na(+)/Ca(2+) exchange facilitates Ca(2+)-dependent activation of endothelial nitric-oxide synthase. J Biol Chem 274:29529–29535

343. Impagnatiello MA, Weitzer S, Gannon G, Compagni A, Cotten M, Christofori G (2001) Mammalian sprouty-1 and -2 are membrane-anchored phosphoprotein inhibitors of growth factor signaling in endothelial cells. J Cell Biol 152:1087–1098

344. Fujimoto T, Nakade S, Miyawaki A, Mikoshiba K, Ogawa K (1992) Localization of inositol 1,4,5-trisphosphate receptor-like protein in plasmalemmal caveolae. J Cell Biol 119:1507–1513

345. Schnitzer JE, Oh P, Jacobson BS, Dvorak AM (1995) Caveolae from luminal plasmalemma of rat lung endothelium: microdomains enriched in caveolin, Ca(2+)-ATPase, and inositol trisphosphate receptor. Proc Natl Acad Sci USA 92:1759–1763

346. Chambliss KL, Yuhanna IS, Mineo C, Liu P, German Z, Sherman TS, Mendelsohn ME, Anderson RG, Shaul PW (2000) Estrogen receptor alpha and endothelial nitric oxide synthase are organized into a functional signaling module in caveolae. Circ Res 87:E44–E52

347. McDonald KK, Zharikov S, Block ER, Kilberg MS (1997) A caveolar complex between the cationic amino acid transporter 1 and endothelial nitric-oxide synthase may explain the "arginine paradox". J Biol Chem 272:31213–31216

348. Henley JR, Krueger EW, Oswald BJ, McNiven MA (1998) Dynamin-mediated internalization of caveolae. J Cell Biol 141:85–99

349. Gumbiner B, Lowenkopf T, Apatira D (1991) Identification of a 160-kDa polypeptide that binds to the tight junction protein ZO-1. Proc Natl Acad Sci USA 88:3460–3464

350. Haskins J, Gu L, Wittchen ES, Hibbard J, Stevenson BR (1998) ZO-3, a novel member of the MAGUK protein family found at

the tight junction, interacts with ZO-1 and occludin. J Cell Biol 141:199–208

351. Stevenson BR, Siliciano JD, Mooseker MS, Goodenough DA (1986) Identification of ZO-1: a high molecular weight polypeptide associated with the tight junction (zonula occludens) in a variety of epithelia. J Cell Biol 103:755–766

352. Citi S, Sabanay H, Jakes R, Geiger B, Kendrick-Jones J (1988) Cingulin, a new peripheral component of tight junctions. Nature 333:272–276

353. Balda MS, Garrett MD, Matter K (2003) The ZO-1-associated Y-box factor ZONAB regulates epithelial cell proliferation and cell density. J Cell Biol 160:423–432

Chapter 2

Morphology and Properties of Pericytes

Paula Dore-Duffy and Kristen Cleary

Abstract

Pericytes were described in 1873 by the French scientist Charles-Marie Benjamin Rouget and were originally called Rouget cells. The Rouget cell was renamed some years later due to its anatomical location abluminal to the endothelial cell (EC) and luminal to parenchymal cells. In the brain, pericytes are located in precapillary arterioles, capillaries and postcapillary venules. They deposit elements of the basal lamina and are totally surrounded by this vascular component. Pericytes are important cellular constituents of the blood–brain barrier (BBB) and actively communicate with other cells of the neurovascular unit such as ECs, astrocytes, and neurons. Pericytes are local regulatory cells that are important for the maintenance of homeostasis and hemostasis, and are a source of adult pluripotent stem cells. Further understanding of the role played by this intriguing cell may lead to novel targeted therapies for neurovascular diseases.

Key words: Angiogenesis, Blood–brain barrier, Capillaries, Contractility, DNA repair, Endothelial cells, Gap junction, Homeostasis, Migration, Neurovascular unit, Pericyte, Stem cells, Stress response, Vascular injury

1. Introduction

Pericytes were described in the late 1800s by the French scientist Charles-Marie Benjamin Rouget (1) and were referred to as the Rouget cell. It was not until the early 1900s that Rouget's work was confirmed as reviewed by Doré (2) and the Rouget cell was renamed as the pericyte. Since its discovery there has been considerable confusion and controversy reflected in the numerous conflicting definitions of the pericyte found in the literature. The pericyte has been referred to as: (a) A contractile, motile cell that surrounds the capillary in a tunic-like fashion (1). (b) A branching contractile cell on the external wall of a capillary and peculiar elongated, contractile cell wrapped around precapillary arterioles "outside" the basement membrane (3). (c) A slender, relatively undifferentiated connective tissue cell in the capillaries or other

Sukriti Nag (ed.), *The Blood-Brain and Other Neural Barriers: Reviews and Protocols*, Methods in Molecular Biology, vol. 686, DOI 10.1007/978-1-60761-938-3_2, © Springer Science+Business Media, LLC 2011

small blood vessels also called the adventitial cell (4). (d) A smooth muscle/pericyte or smooth muscle cell of the capillaries (5, 6). A broad flat cell with slender projections that wrap around the capillaries (6). (e) A stem or mesenchymal-like cell, associated with the walls of small blood vessels. As a relatively undifferentiated cell, it serves to support these vessels, but it can differentiate into a fibroblast, smooth muscle cell, or macrophage as well if required (6–14).

Despite nearly 130 years of investigation, the role of the pericyte is still somewhat of a mystery. This is due, in part, to the relatively low numbers of pericytes in most tissues. The ratio of pericytes to EC varies from species to species and organ to organ and varies even within the capillary bed. In the brain the average ratio of pericytes to EC in the rat capillary is 1:5. In the mouse the ratio is 1:4 and in humans 1:3–4. This low number is further augmented by the difficulty of isolating pure primary pericytes (15). Once isolated, pericytes rapidly differentiate along multiple lineages depending on the regulatory signals present in the microenvironment. It is this pluripotentiality and the ability to migrate as well as the lack of a pericyte specific marker that has lead to the enormous confusion about this cell (14). In this chapter, the role of the pericyte as an adaptive regulatory cell of the neurovascular unit that is important in the maintenance of tissue homeostasis will be discussed, as well as the role of the pericytes as a source of adult stem cells and the potential role of pericytes in development of disease pathology.

2. Morphology

2.1. Central Nervous System (CNS) Pericyte Morphology

In the mature CNS capillary, the pericyte is located between the EC and parenchymal astrocytes and neurons (Fig. 1). Pericytes have a prominent round nucleus that clearly differs from the elongated cigar shaped nucleus of the EC. The pericyte extends long

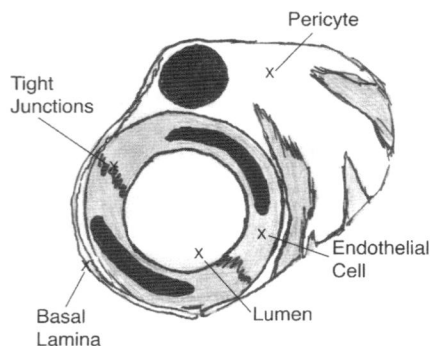

Fig. 1. Cross section of a CNS capillary is shown in cartoon form. The pericyte is located between the endothelial cells (EC) and astrocytes (not shown). Pericyte projections (*white*) wrap around the capillary (*gray*). The pericyte is totally surrounded by basal lamina.

processes that extend over the vessel wall. The morphological pattern of projections appears to be somewhat heterogeneous (Fig. 2). Pericyte projections can extend around the capillary (Fig. 2a–c) as originally described by Rouget (1). The classic wrapping pattern is also somewhat heterogeneous. The most common association of the pericyte with the capillary is one in which the pericyte processes are broad and span a large somewhat continuous surface of the vessel (Fig. 2a, d). Alternatively these

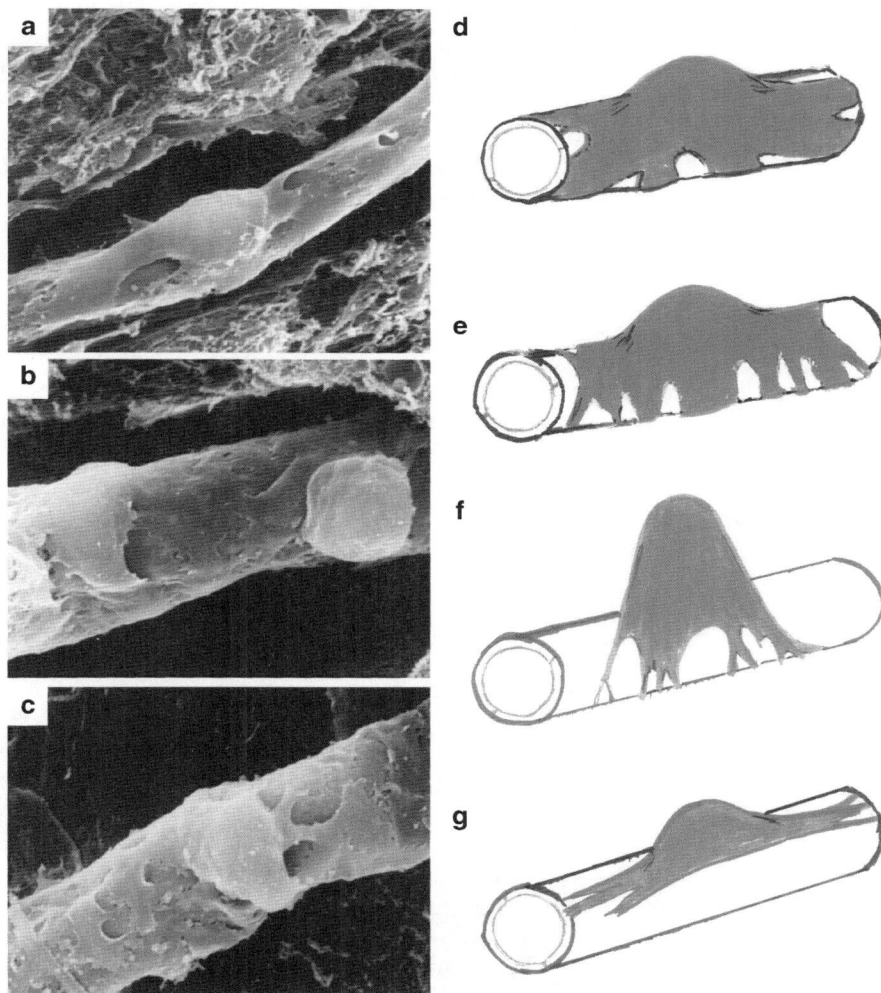

Fig. 2. Three scanning electron micrographs of segments of rat CNS capillaries are shown (a–c) along with four cartoons depicting the structural association of pericytes with microvessels (d–g). The common pattern is the pericyte encircling the capillary with broad, virtually continuous projections that cover a large surface area of the microvessel (a, d). The second pattern shows pericytes wrapping around the capillary, but the area is more defined and smaller and the pericyte projections are finger-like in shape (b, c, e). The third pattern is that of migrating pericytes (b, f). This pattern is predominantly seen following injury and during the early stages of angiogenesis. The fourth pattern shows that the pericyte is positioned longitudinally in a polar fashion along the microvessel (g). This pattern may reflect pericytes migrating along the vessel or may reflect transition pericytes. This pattern is seen at arteriolar/capillary junctions and during angiogenesis.

processes may form finger-like projections that are more confined and surround a more finite portion of vessels (Fig. 2b, c, e). A third pattern of pericyte orientation in the microvessel involves a retraction of projections and this represents a migrating pericyte (Fig. 2b, f) (16). Pericytes may also extend along the long axis of the capillary which represents longitudinal migration (Fig. 2g). This pattern is more commonly seen during angiogenesis. In normal capillaries, the wrapping pattern predominates but under pathological conditions the migrating pattern increases. It is likely that these patterns represent functional differentiation of pericytes rather than heterologous subsets.

The CNS pericyte is surrounded by the basal lamina on all sides. During development and during angiogenesis the pericyte deposits basal lamina components (14). Even pericyte projections, observed using electron microscopy, have a thin layer of basal lamina. The basal lamina has been shown to thicken or thin in response to stress stimuli (17–20). Changes in the basal lamina can be directly associated with pericyte expression of proteases (16) and ultimate migration from its vascular location (17, 21).

The intact basal lamina may not only provide anchoring and structural integrity to the capillary but it may also be involved in regulation of pericyte function and differentiation. It seems intuitive that there must be a reason why the pericyte is surrounded by laminal proteins. Avβ8 integrin is important in neurovascular cell adhesion (22, 23). Pericytes encased in the basal lamina or exposed to laminal proteins do not usually differentiate (Dore-Duffy, unpublished observations). Thus migration through the basal lamina is necessary before cells can function in their stem cell capacity. Regulation at the level of the basal lamina may also be integral to vascular adaptability to an ever-changing environment and to pericyte signaling mechanisms (24).

Within its capillary location, the pericyte may signal nearby EC (25), astrocytes (26), neurons, smooth muscle cells, and perhaps other pericytes (14). Pericyte-EC contacts include peg and socket arrangements (27, 28) and gap junctions (29, 30). Gap junctions allow pericytes to communicate through the exchange of ions and small molecules. Peg-and-socket contacts enable pericytes to penetrate through the basal lamina and make contact with other cells and nearby vessels (27, 28). Junctional complexes including adhesion plaques also support transmission of contractile forces from pericytes to other cells. Pericyte gap junctions contain N-cadherin, a variety of adhesion molecules, β-catenin, extracellular matrix (ECM) molecules such as fibronectin, and a number of integrins (30, 31). Thus, pericytes are involved in highly complex signaling cascades that enable this cell to respond to changes in the microenvironment. However, it is unclear whether gap junctions and peg and socket contacts are naturally present or whether they are initiated during changes in functional activity.

For example, it is known that pericytes interdigitate with ECs during the early phases of angiogenesis and with neurons during the maturation of newly formed vessels (32). These sites of communication are altered under pathological conditions. During cerebral edema or diabetes, gap junctions are substantially decreased or disrupted in retinal pericytes (33–35). Diabetes-induced changes in gap junctions may be regulated by high glucose (36). Pericyte-EC communication via gap junctions is fundamental to the adaptive responses to compromised bioenergetic homeostasis (37). Crosstalk between ECs, pericytes, as well as astrocytes is involved in regulation of insulin transport (26). Pericyte/endothelial cross talk is also integral to physiological angiogenesis (38), and is likely to be important in adaptation to hypoxic injury and focal capillary contractility.

2.2. Pericyte Markers   Identification of the pericyte in culture or in situ is difficult. They can be definitively identified at the electron microscopic (EM) level or in semithin sections where their location relative to the basal lamina can be seen (Fig. 1). EM morphology is detailed and discussed in a number of excellent articles (21, 39–41). Pericytes can also be identified in capillary isolates by the shape of their nucleus which is round while the EC nucleus is elongated and cigar-shaped and can be easily delineated using a nuclear dye (14). Many investigators have used antibody directed against alpha smooth muscle actin (αSMA) to identify pericytes (Table 1). While pericytes are capable of expressing αSMA, the expression of this protein in vivo may be associated with functional heterogeneity within the capillary and in vitro may be a marker of dedifferentiation. In their capillary location, most pericytes are αSMA negative (8, 14, 42, 43). In brain, only those pericytes that are located near arterioles are routinely immunoreactive for αSMA (14). Expression of this protein can be induced within the capillary and may be related to the role of the pericyte in focal regulation of capillary blood flow (44–47) and in the acute stress response (14). In primary cultures, less than 5% of freshly isolated capillary pericytes express αSMA (8, 43) but nearly 100% express this marker by day 7 (Fig. 3) (48).

Bovine retinal pericytes express potassium (K+) channels (49). In vasa recta pericytes, elevation of extracellular K+ hyperpolarizes pericytes and this is reversed by barium (Ba2+), confirming the presence of strong inward rectifier K+ channels (Kir) (50). Kir, however, is also strongly expressed in EC and arteriolar smooth muscle cells (51). Transcription profiling of pericyte-deficient brain microvessels isolated from platelet-derived growth factor beta (PDGFβ) –/– and PDGFβ receptor (PDGFβR) –/– mouse mutants has identified new candidates for pericyte markers. The ATP-sensitive potassium-channel Kir6.1 (also known as Kcnj8) and sulfonylurea receptor 2, (SUR2, also known as Abcc9), as well as delta homologue 1 (DLK1), have been proposed as specific markers for brain pericytes. The three

Table 1
This table outlines pericyte markers

Pericyte marker	Reference
3G5-defined ganglioside	(62)
140 kDa Aminopeptidase N	(60)
Angiopoietin 1	(61)
DLK1	(52)
ICAM-1	(42)
Kcnj8	(52)
K+ channels/Kir	(49–51)
Nestin	(9)
NG2	(8, 9, 53, 59)
OX-42/αM	(54)
PDFGαR	(14)
PDGFβR	(8, 14, 43, 58)
RGS-5	(55)
αSMA	(8, 14, 42, 43, 145, 146)
SUR2/Abcc9	(52)
VCAM-1	(42, 147)
Vimentin	(8, 13, 56, 148)

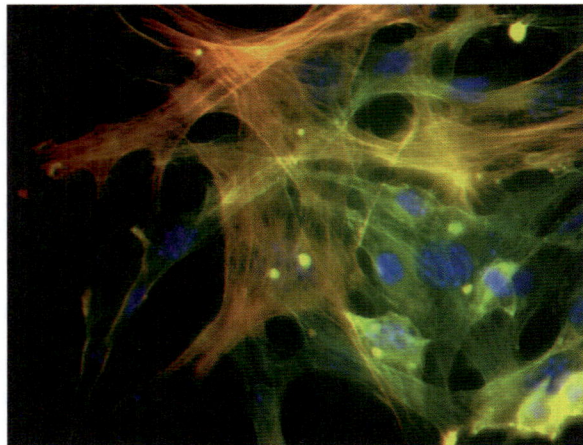

Fig. 3. Primary rat pericytes (4 days old) grown in DMEM plus 20% fetal calf serum. Cells were fixed in 4 % glutaraldehyde, permeabilized with Triton X-100 and dual stained for expression of beta actin (*green*) and alpha smooth muscle actin (*red*). In this culture 30% of the cells expressed the smooth muscle phenotype. One hundred percent of the cells expressed the receptor for platelet derived growth factor beta (not shown).

proposed brain pericyte markers are signaling molecules implicated in ion transport and intercellular signaling (52). The selectivity of expression has yet to be confirmed and it is possible that these markers are not expressed in adult pericytes.

Pericytes are positive for the chondroitin sulfate proteoglycan NG2, formerly known as the high molecular weight melanoma associated antigen (53) and nestin (9). They also express vascular cell adhesion molecule-1 (VCAM-1) and intercellular adhesion molecule-1 (ICAM-1) (42). We have reported that pericytes express the OX-42 marker (αM) in vivo and in isolated capillaries (54). However, in primary culture, this marker associates at focal adhesion sites and is down-regulated with time. The adult CNS pericyte expresses a number of proteins that are useful in their identification (8, 48, 55). They express vimentin but not desmin (8, 56). Developmental markers such as the regulator of G-protein signaling (RGS-5) protein have been identified in knockout mice (57). This protein is expressed during embryonic development and lost after birth. In our hands it is not expressed in normal adult CNS pericytes *(Dore-Duffy unpublished observations)*. However, expression of RGS5 has been reported in tumor pericytes during angiogenesis (55). Adult pericytes express PDGFβ, PDGFβR (58), and NG2 chondroitin sulfate proteoglycan (8, 53, 59). Pericytes are also known to express PDGFαR (14) but we have not yet tested this in brain sections. Other markers expressed by pericytes include the 140 kDa aminopeptidase N (60), angiopoietin 1 (61), and a 3G5-defined ganglioside (62). The ganglioside 3G5 is not expressed by all pericytes (63) and is expressed by a large number of other cells including islet cells, follicular cells, melanocytes, and pancreatic and adrenal cells.

The lack of a definitive pan-marker for pericytes may be due to the fact that these cells are multipotent self-renewing cells (9, 14). This will be discussed in more detail under Subheading 3. When pericytes are subcultured from freshly isolated capillaries they undergo a period of quiescence that is followed by development of the αSMA phenotype. This may reflect either dedifferentiation, if one assumes that pericytes are derived from mesenchymal cells, or may reflect differentiation of a quiescent stem cell along the mesenchymal lineage (64). Thus αSMA cannot be used to definitively identify pericytes nor can the lack of expression be used to exclude pericytes.

3. Properties

3.1. Pericytes and the Blood–Brain Barrier (BBB)

The BBB regulates the passage of various nutrients and essential components, proteins, chemical substances, and microscopic organisms between the bloodstream and the parenchymal tissue.

The anatomical constituents of the BBB are the EC, pericytes, and basal lamina (matrix proteins) that together with the astrocytes, neurons, and possibly other glial cells comprise the neurovascular unit. Coordinated cell-to-cell interactions between cells of the neurovascular unit regulate a wide variety of functions that include development, BBB permeability, cerebral blood flow, and the stress response.

Dysregulation at the neurovascular level is linked to many common human CNS pathologies, making the neurovascular unit a potential target for therapeutic intervention. Together the cells of the neurovascular unit adapt to environmental changes and make fine-tuned regulatory decisions that maintain homeostasis and promote tissue survival. Nowhere is such tight regulation more important than in the brain where bioenergetic and metabolic homeostasis is integral for neuronal survival. The role of the CNS pericyte in the neurovascular unit is still largely unknown. Pericytes were once thought to function as a support or scaffolding structure. It is known that pericytes are highly complex regulatory cells that communicate with ECs and other cells of the neurovascular unit such as neurons and astrocytes by direct physical contact and through autocrine and paracrine signaling pathways (8, 11–13, 27, 65). It seems intuitive that loss or dysfunction of the pericyte or of any of the cells comprising the neurovascular unit has highly deleterious effects.

3.2. Contractility

The concept that pericytes regulate blood flow at the capillary level was originally proposed by Steinach and Kahn in 1903 (45) and Ni in 1922 (66). Both scientific groups studied the effects of toxic and electrical stimulation on capillary diameter. Doré reviewed this area in 1923 (2). The concepts put forward in this review are on target with what is known today. As stated by

Doré, (2): *Until a few years ago the capillaries were regarded as elastic tubes undergoing passive distension in accordance with the general blood pressure, the state of contraction or dilatation of the supplying arterioles, and the nutrition of the vascular walls. There is now, however, conclusive evidence that the capillaries play an independent part in the peripheral circulation, that they possess the intrinsic property of contraction and relaxation, and are under the direct influence of the nervous system.*

Pericytes have receptors for a large number of vasoactive signaling molecules (8, 14) suggesting that they have the capacity to be involved in cerebrovascular autoregulation. Nonneural pericyte expression of αSMA and desmin, two proteins found in smooth muscle cells, as well as their adherence to the endovascular cells make them potential candidates in regulation of capillary diameter and focal capillary blood flow (8, 27, 67–70). Electrical stimulation of retinal and cerebellar pericytes is reported to evoke a localized capillary constriction (47). ATP in the retina or noradrenaline in the cerebellum also results in constriction of capillaries by pericytes. Glutamate reverses the constriction produced by noradrenaline. Following simulated ischemia and traumatic brain injury (TBI), capillary pericytes are induced to express αSMA. Thus, it is likely that pericytes modulate blood flow in response to acute changes in neural activity and/or metabolic need. For example, other investigators have shown that capillary contraction can be directly linked to metabolic need (69, 71). Exposure to lactate increases pericyte calcium, contraction, and capillary lumina become constricted. The contractile response appears to involve a cascade of events resulting in the inhibition of Na^+/Ca^{2+} exchangers on the EC (71). Hypoxia, which closes gap junctions, switches the effect of lactate from contraction to relaxation. This further suggests that when energy supplies are ample, lactate may stimulate vasoconstriction, and under hypoxic conditions, induce vasodilation. Thus, pericyte function may be linked with local vascular adaptation to changes in local bioenergetic requirements.

3.3. Pericytes Are Multipotential Stem Cells in the Adult Brain

Adaptations to stress at the vascular level include functional and phenotypic changes involving differentiation along mesenchymal and neural lineages, and lend credence to the idea that pericytes are multipotential stems cells in the adult brain and in other tissues. That pericytes are stem cells is supported by a host of information from historical work and from more recent literature. We will speculate on the importance of pericyte stem cell activity in survival and DNA repair and how dysregulation of pericyte function may lead to disease.

The pluripotentiality of pericytes has been proposed for many years and has been reviewed (8, 10–12, 14, 65). As early as 1970, it was proposed that there is a similarity between neuroglial cells and pericytes (72). Katenkamp and Stiller (73) proposed that myofibroblasts were derived from pericytes and pericytes were multipotent stem cells. They further proposed that these cells are not only functional in dermatofibroma but are integral to connective tissue regeneration (74) and involved in interferon gamma (IFN-γ) (75) release. Nestin is induced in liver stellate cells during transition from the quiescent to the activated phenotype in culture (76). These cells also express glial fibrillary acidic protein (GFAP) and neural cell adhesion molecule (NCAM). They proposed a potential embryonic origin of these cells. In the adult liver, the replicating cells including endothelial, Kupffer, stellate cells (Ito or pericytes), bile duct epithelium, and granular lymphocytes (pit cells), were found to be stem cells (77). The ability of pericytes to form bone nodules in vitro (78) provided the basis for a number of elegant studies predominantly by Anne Canfield's laboratory showing that pericytes are a source of osteogenic progenitor cells (79–83). Pericytes have also been shown to produce chondrocytes and adipocytes (84). As early as 1993, it was shown that bone marrow stromal cells are mesenchymal, express vimentin, and can be induced to express αSMA in culture (85). These cells are multipotent and share many pericyte characteristics. An alternative hypothesis is that these cells are derived from a common precursor. Multipotent stem cells isolated from human reaming debris collected during surgical treatment of long bone diaphyseal fractures differentiate along the osteogenic pathway and can be redirected to a neuronal phenotype (86). These cells also resemble pericytes. It has been reported that growth factors such as bovine fibroblast growth factor (bFGF) and epidermal growth factor (EGF) stimulate pericyte proliferation and angiogenesis (87). Both EGF and bFGF responsive vascular stem cells have been reported in the rat and avian microvasculature (88–90). Additional reports suggest that pericytes differentiate into fibroblasts (17, 83, 91), endothelial cells (90), adipocytes (92), chondrocytes (82), and macrophages/dendritic cells (54, 93).

We have investigated the neural potential of primary pericytes subcultured from isolated rat CNS capillaries (9). Using fluorescence-activated cell sorting (FACs) analysis, our study demonstrates that adult CNS capillaries contain NG2 and nestin-positive pericytes, markers not expressed in EC populations. Pericyte BRDU/nestin-positive, bFGF-induced spheres ultimately differentiate and are composed of cells of neural cell lineage. Pericytes undergo self-renewal and increase in number after subculturing. By clonal analysis, multipotent pericytes differentiate along multiple lineages that include astrocytes, neurons, oligodendrocytes, and αSMA-positive cells that are NG2/nestin-positive and resemble primary pericytes. There is no evidence of cell fusion in these studies. When spheres are disrupted, cells coexpressing oligodendrocyte and astrocyte markers are noted. Pericytes also generate neurospheres directly from cultured rat capillaries at a faster rate than seen with primary pericytes (14) suggesting that ECs can enhance this process. ECs secrete a substance that enhances neurosphere formation (94) while smooth muscle cells do not enhance neurosphere formation. While capillary pericytes are a source of adult stem cells, ECs within the vascular niche provide trophic support.

Subsequent studies have confirmed these findings and identified pericytes from other organs as adult stem cells. Liver pericytes (Ito cells, stellate cells) are liver cell progenitors (95, 96). Skin pericytes are the source of regenerating skin tissue in adults (97). Rajkumar et al. (98) examined mechanisms by which microvascular injury leads to dermal fibrosis in diffuse cutaneous systemic sclerosis. They hypothesized that microvascular pericytes or fibroblasts transdifferentiate into myofibroblasts. Purified pericytes also demonstrate high myogenic potential in culture and in vivo (99–101). Cells of testicular blood vessels (vascular smooth muscle cells or pericytes) are the progenitors of Leydig cells (102). Resembling stem cells of the nervous system, the Leydig cell progenitors are characterized by the expression of nestin. The pulp of human teeth contains a population of cells with stem cell properties and it has been suggested that these cells originate from pericytes (103). Pulp stem cells express molecules of the Notch signaling pathway (Notch3). Notch3 was coexpressed with RGS5. RGS5 induction may be coregulated with stem cell activity. Thus, in six separate tissues namely, teeth, brain, skin, muscle, prostate and liver, pericytes are a source of tissue progenitor cells in adult tissue.

The pericyte has a very broad stem cell potential that goes beyond organ specific production of progenitor cells. In serum-containing culture medium, primary CNS pericytes (2–4 days old) take on a macrophage/dendritic cell-like phenotype (14, 54, 93). During this period, pericytes express macrophage markers and can present antigen (54). Upon exposure to interferon, pericytes express MHC class II antigen and can present antigen to sensitized splenic T-cells (93). Pericytes continue to differentiate becoming 100% αSMA positive on days 7–10 and express other markers such as the integrin β1 characteristic of mesenchymal stem cells (MSC). With prolonged culture, pericytes form nodules that produce mineralized bone by 21 days in culture (14, 81). These nodules are alizaren red positive. In the same culture, there are other cell types known to be derived from mesenchymal lineage. Thus, in vitro data supports the concept that with the correct environmental cues CNS pericytes may form MSC and then differentiate to bone, adipocytes, smooth muscle cells, and endothelial cells. With different cues, pericytes differentiate along the neural lineage.

In the adult, proliferating stem cell activity is usually found in a perivascular location in response to stress or injury (104, 105). However, the exact mechanism that regulates induction, proliferation then reprogramming, and differentiation of adult stem cells is not known. Pericytes migrate from their vascular location in response to stress injury (21) and remain in a perivascular location where they may encounter local signaling molecules that dictate their subsequent activities such as migration, proliferation or differentiation. It is likely that they may also migrate back to the vascular location. Perivascular pericytes proliferate during the developmental angiogenic response (106) and during physiological angiogenesis (48). Pericyte signaling molecules that are involved in regulation of angiogenesis are also involved in neurogenesis (107, 108). Pericytes synthesize the proangiogenic cytokine, vascular endothelial growth factor (VEGF) (109). VEGF augments pericyte proliferation in an autocrine fashion (109), promotes differentiation of multipotent chondrocytic stem cells (110), and promotes migration and vascular instability (111). Pericytes are also responsive to growth factors and other signaling molecules important in regulation of neurogenesis (8, 109, 112). Thus, it is possible that the response to injury and stress at the tissue level is coregulated with stem cell differentiation in the adult.

4. Pericytes and Disease

As discussed above, pericytes play an important role in the maintenance of vascular and tissue homeostasis and are integral to injury responses. Under normal conditions the pericyte is relatively quiescent and is essential for vascular stability. Under conditions of stress or injury, the pericyte undergoes phenotypic and functional changes that may include migration, proliferation or differentiation. How these events that include pericyte reprogramming are coordinated at the molecular level needs to be determined. However, it is clear that pericyte dysfunction or the loss of pericytes is likely to play an important role in the pathogenesis of disease.

Pericyte loss or a reduced pericyte to EC ratio may be achieved through migration of pericytes from their microvascular location under pathological or physiological conditions, selective pericyte death or from reduced pericyte turnover or maintenance. Pericytes migrate naturally during the early phases of physiological angiogenesis to make way for growing sprouts (113, 114), or in response to stress or injury (21). Migration following TBI is thought to promote survival as pericytes remaining in their vascular location show signs of degenerative activity (21). However, migration is also thought to play a pathogenic role in diabetic retinopathy (115). Decreased pericyte to EC ratios have been observed following TBI (21) and stroke (116), multiple sclerosis (117–121), brain tumor (122, 123), diabetic retinopathy (124), aging (125, 126), and in a variety of angiopathies (127). Pericyte loss may also play a role in Alzheimer's disease, however; enhanced pericyte coverage of some vessels may suggest that increased proliferation of pericytes is an adaptation to focal loss of bioenergetic homeostasis (121, 126, 128–130). Pericyte loss due to cellular degeneration/apoptosis has been shown in hypertrophic scars, keloids (131), early diabetic retinopathy (132, 133), cancer (134–137), hyperglycemia (138), and during development (139). Premature infants have decreased pericyte coverage (140). Increased pericyte coverage may also be an indicator of vascular dysfunction. Pericyte proliferation has been associated with development of muscularization during pulmonary hypertension and is thought to be due to platelet activating factor (141).

On a more subtle level, loss of pericyte stem cell activity may also lead to disease. Stem cells must maintain a functional genome. Under continuous exogenous and endogenous stress, uncontrolled self-renewal cells can accumulate DNA errors, drive proliferative expansion, and ultimately transform into cancer stem cells. Tumor stem cells are thought to be involved in hematological malignancies, such as hemangiomas and pericytomas, as well as in solid tumors. The complex cellular mechanisms including cell cycle arrest, transcription induction and DNA repair are activated but may be dysregulated with an absence of repair machinery. Mismatch repair gene defects have been recently identified in hematopoietic malignancies, leukemia, and lymphoma cell lines (142, 143). Pericyte differentiation within the vascular wall may be considered dysfunctional. For example, differentiation along the mesenchymal lineage with bone formation may result in microcalcifications within small vessels (144) or even fatty deposition in the vascular walls. With continued knowledge of pericyte biology, it is likely that their role in disease pathology may expand.

5. Conclusion

After its identification by Rouget in the late 1800s, relatively little was published about the pericyte until 1902 when the presence of this intriguing cell was confirmed. The development of tissue culture techniques has generated considerable interest in pericyte biology. The ability to isolate pure primary pericytes has enabled scientists to study these cells in vitro. The development of sophisticated molecular biological techniques has enabled us to begin to clearly delineate the complex mechanisms by which pericytes communicate with other cells of the neurovascular unit. A better understanding of the mechanisms by which pericytes communicate with other cells and how altered communication may result in disease pathology is likely to yield exciting new insights as well as help in the development of a new therapeutic target in CNS disorders.

Acknowledgments

The work discussed in this manuscript was supported by grants from the National Institutes of Health and the National Multiple Sclerosis Society.

References

1. Rouget, C. (1874) Note sur le developpement de la tunique contractile des vaisseaux *Compt Rend Acad Sci* **59**, 559–562.

2. Doré, SE. (1923) On the contractility and nervous supply of the capillaries *Brit J Derma* **35**, 398–404.

3. Blood, D. C., and Studdert, V.P. (1988) *Baillière's Comprehensive Veterinary Dictionary*, Baillère Tindall, London.

4. Stedman, T. L. (1995) *Stedman's Medical Dictionary*, Williams & Wilkins, Baltimore.

5. Fabry, Z., Fitzsimmons, K. M., Herlein, J. A., Moninger, T. O., Dobbs, M. B., and Hart, M. N. (1993) Production of the cytokines interleukin 1 and 6 by murine brain microvessel endothelium and smooth muscle pericytes *J Neuroimmunol* **47**, 23–34.

6. Ding, R., Darland, D.C., Parmacek, M.S., and D'Amore, P.A. (2004) Endothelial-mesenchymal interactions in vitro reveal molecular mechanisms of smooth muscle/pericyte differentiation *Stem Cells Dev* **13**, 509–520.

7. Pericytes [online]. [last accessed 2009 Jan 09]. Available from: http://en.wikipedia.org/wiki/Pericytes.

8. Balabanov, R. and Dore-Duffy, P. (1998) Role of the CNS microvascular pericyte in the blood-brain barrier *J Neurosci Res* **53**, 637–644.

9. Dore-Duffy, P., Katychev, A., Wang, X., and Van Buren, E. (2006) CNS microvascular pericytes exhibit multipotential stem cell activity *J Cereb Blood Flow Metab* **26**, 613–624.

10. Krüger, M., and Bechmann, I. (2008) Pericytes, in *Central Nervous System Diseases and Inflammation* (Lane, T. E., Bergmann, C., Carson, M., Wyss-Coray, T., Ed.), pp 33–43, Springer, Berlin.

11. Sims, D.E. (1991) Recent advances in pericyte biology–implications for health and disease *Can J Cardiol* **7**, 431–443.

12. Tilton, R.G. (1991) Capillary pericytes: perspectives and future trends *J Electron Microsc Tech* **19**, 327-344.

13. Shepro, D., and Morel, N.M. (1993) Pericyte physiology *FASEB J* **7**, 1031–1038.

14. Dore-Duffy, P. (2008) Pericytes: Pluripotent cells of the blood brain barrier *Curr Pharm Des* **14**, 1581–1593.

15. Dore-Duffy, P. (2003) Isolation and characterization of cerebral microvascular pericytes *Methods Mol Med* **89**, 375-82.

16. Du, R., Petritsch, C., Lu, K., Liu, P., Haller, A., Ganss, R., Song, H., Vandenberg, S., and Bergers, G. (2008) Matrix metalloproteinase-2 regulates vascular patterning and growth affecting tumor cell survival and invasion in GBM *Neuro Oncol* **10**, 254–264.

17. Gonul, E., Duz, B., Kahraman, S., Kayali, H., Kubar, A., and Timurkaynak, E. (2002) Early pericyte response to brain hypoxia in cats: an ultrastructural study *Microvasc Res* **64**, 116–119.

18. Arismendi-Morillo, G., and Castellano, A. (2005) Tumoral micro-blood vessels and vascular microenvironment in human astrocytic tumors. A transmission electron microscopy study *J Neurooncol* **73**, 211–217.

19. Wiley, L.A., Rupp, G.R., and Steinle, J.J. (2005) Sympathetic innervation regulates basement membrane thickening and pericyte number in rat retina *Invest Ophthalmol Vis Sci* **46**, 744–748.

20. Hughes, S. J., Wall, N., Scholfield, C. N., McGeown, J. G., Gardiner, T. A., Stitt, A. W., and Curtis, T. M. (2004) Advanced glycation endproduct modified basement membrane attenuates endothelin-1 induced [Ca2+]i signalling and contraction in retinal microvascular pericytes, *Mol Vis* **10**, 996–1004.

21. Dore-Duffy, P., Owen, C., Balabanov, R., Murphy, S., Beaumont, T., and Rafols, J.A. (2000) Pericyte migration from the vascular wall in response to traumatic brain injury *Microvasc Res* **60**, 55–69.

22. McCarty, J.H., Monahan-Earley, R.A., Brown, L.F., Keller, M., Gerhardt,H., Rubin, K., Shani, M., Dvorak, H.F., Wolburg, H., Bader, B.L., Dvorak, A.M., and Hynes, R.O. (2002) Defective associations between blood vessels and brain parenchyma lead to cerebral hemorrhage in mice lacking alphav integrins *Mol Cell Biol* **22**, 7667–7677.

23. McCarty, J.H. (2005) Cell biology of the neurovascular unit: implications for drug delivery across the blood-brain barrier *Assay Drug Dev Technol* **3**, 89–95.

24. Hayden, M.R., Sowers, J.R., and Tyagi, S.C. (2005) The central role of vascular extracellular matrix and basement membrane remodeling in metabolic syndrome and type 2 diabetes: the matrix preloaded *Cardiovasc Diabetol* **4**, 9.

25. Ryan, U.S., Ryan, J.W., and Whitaker, C. (1979) How do kinins affect vascular tone? *Adv Exp Med Biol* **120A**, 375–391.

26. Nakaoke, R., Verma, S., Niwa, M., Doghu, S., and Banks, W.A. (2007) Glucose regulated blood-brain barrier transport of insulin: Pericyte-astrocyte-endothelial cell cross talk *IJNN* **3**, 195–200.

27. Rucker, H.K., Wynder, H.J., and Thomas, W.E. (2000) Cellular mechanisms of CNS pericytes *Brain Res Bull* **51**, 363–369.

28. Carlson, E.C. (1989) Fenestrated subendothelial basement membranes in human retinal capillaries *Invest Ophthalmol Vis Sci* **30**, 1923–1932.

29. Larson, D.M., Haudenschild, C.C., and Beyer, E.C. (1990) Gap junction messenger RNA expression by vascular wall cells *Circ Res* **66**, 1074–1080.

30. Cuevas, P., Gutierrez-Diaz, J. A., Reimers, D., Dujovny, M., Diaz, F. G., and Ausman, J. I. (1984) Pericyte endothelial gap junctions in human cerebral capillaries *Anat Embryol (Berl)* **170**, 155–159.

31. Nakamura, K., Kamouchi, M., Kitazono, T., Kuroda, J., Matsuo, R., Hagiwara, N., Ishikawa, E., Ooboshi, H., Ibayashi, S., and Iida, M. (2008) Role of NHE1 in calcium signaling and cell proliferation in human CNS pericytes *Am J Physiol Heart Circ Physiol* **294**, H1700–H1707.

32. Jójárt, I., Joó, F., Siklós, L., and László, F.A. (1984) Immunoelectronhistochemical evidence for innervation of brain microvessels by vasopressin-immunoreactive neurons in the rat *Neurosci Lett* **51**, 259–264.

33. Castejón, O.J. (1980) Electron microscopic study of capillary wall in human cerebral edema *J Neuropathol Exp Neurol* **39**, 296–328.

34. Oku, H., Kodama, T., Sakagami, K., and Puro, D.G. (2001) Diabetes-induced disruption of gap junction pathways within the retinal microvasculature *Invest Ophthalmol Vis Sci* **42**, 1915–1920.

35. Kawamura, H., Oku, H., Li, Q., Sakagami, K., and Puro, D.G. (2002) Endothelin-induced changes in the physiology of retinal pericytes *Invest Ophthalmol Vis Sci* **43**, 882–888.

36. Li, A.F., Sato, T., Haimovici, R., Okamoto, T., and Roy, S. (2003) High glucose alters connexin 43 expression and gap junction intercellular communication activity in retinal pericytes. *Invest Ophthalmol Vis Sci* **44**, 5376–5382.

37. Allt, G., and Lawrenson, J.G. (2001) Pericytes: cell biology and pathology *Cells Tissues Organs* **169**, 1–11

38. Virgintino, D., Girolamo, F., Errede, M., Capobianco, C., Robertson, D., Stallcup, W. B., Perris, R., and Roncali, L. (2007) An intimate interplay between precocious, migrating pericytes and endothelial cells governs human fetal brain angiogenesis. *Angiogenesis* **10**, 35–45.

39. Rauch, S., and Reale, E. (1968) The pericytes (Rouget cells) of the stria vascularis vessels [German]. *Arch Klin Exp Ohren Nasen Kehlkopfheilkd* **192**, 82–90.

40. Rodriguez-Baeza, A., Reina-De La Torre, F., Ortega-Sanchez, M., and Sahuquillo-Barris, J. (1998) Perivascular structures in corrosion casts of the human central nervous system: a confocal laser and scanning electron microscope study *Anat Rec* **252**, 176–184.

41. Mottow-Lippa, L., Tso, M.O., Peyman, G.A., and Chejfec, G. (1983) Von Hippel angiomatosis. A light, electron microscopic, and immunoperoxidase characterization *Ophthalmology* **90**, 848–855.

42. Verbeek, M.M., Otte-Höller, I., Wesseling, P., Ruiter, D.J., and de Waal, R.M. (1994) Induction of alpha-smooth muscle actin expression in cultured human brain pericytes by transforming growth factor-beta 1 *Am J Pathol* **144**, 372–382.

43. Liebner, S., Fischmann, A., Rascher, G., Duffner, F., Grote, E. H., Kalbacher, H., and Wolburg, H. (2000) Claudin-1 and claudin-5 expression and tight junction morphology are altered in blood vessels of human glioblastoma multiforme. *Acta Neuropathol* **100**, 323–331.

44. Kutcher, M.E., Kolyada, A.Y., Surks, H.K., and Herman, I.M. (2007) Pericyte Rho GTPase mediates both pericyte contractile phenotype and capillary endothelial growth state *Am J Pathol* **171**, 693–701.

45. Steinach, E., and Kahn, B.H. (1903) Echte Contractilität und motorische Innervation der Blutcapillaren [German] *Pflügers Archiv* **97**, 195.

46. Anderson, D.R. (1996) Glaucoma, capillaries and pericytes. 1. Blood flow regulation *Ophthalmologica* **210**, 257–262.

47. Peppiatt, C.M., Howarth, C., Mobbs, P., and Attwell, D. (2006) Bidirectional control of CNS capillary diameter by pericytes *Nature* **443**, 700–704.

48. Dore-Duffy, P., and LaManna, J. C. (2007) Physiologic angiodynamics in the brain *Antioxid Redox Signal* **9**, 1363–1371.

49. Quignard, J.F., Harley, E.A., Duhault, J., Vanhoutte, P.M., and Félétou, M. (2003) K+ channels in cultured bovine retinal pericytes:

effects of beta-adrenergic stimulation. *J Cardiovasc Pharmacol* **42**, 379–388.

50. Cao, C., Goo, J.H., Lee-Kwon, W., and Pallone, T.L. (2006) Vasa recta pericytes express a strong inward rectifier K+ conductance *Am J Physiol Regul Integr Comp Physiol* **290**, R1601–R1607.

51. Jackson, W.F. (2005) Potassium channels in the peripheral microcirculation *Microcirculation* **12**, 113–127.

52. Bondjers, C., He, L., Takemoto, M., Norlin, J., Asker, N., Hellström, M., Lindahl, P., and Betsholtz, C. (2006) Microarray analysis of blood microvessels from PDGF-B and PDGF-Rbeta mutant mice identifies novel markers for brain pericytes *FASEB J* **20**, 1703–1705.

53. Stallcup, W.B. (2002) The NG2 proteoglycan: past insights and future prospects. *J Neurocytol* **31**, 423–435.

54. Balabanov, R., Washington, R., Wagnerova, J., and Dore-Duffy, P. (1996) CNS microvascular pericytes express macrophage-like function, cell surface integrin alpha M, and macrophage marker ED-2. *Microvasc Res* **52**, 127–142.

55. Bergers, G., and Song, S. (2005) The role of pericytes in blood-vessel formation and maintenance *Neuro Oncol* **7**, 452–464.

56. Bandopadhyay, R., Orte, C., Lawrenson, J.G., Reid, A.R., De Silva, S., and Allt, G. (2001) Contractile proteins in pericytes at the blood-brain and blood-retinal barriers *J Neurocytol* **30**, 35–44.

57. Cho, H., Kozasa, T., Bondjers, C., Betsholtz, C., and Kehrl, J.H. (2003) Pericyte-specific expression of Rgs5: implications for PDGF and EDG receptor signaling during vascular maturation *FASEB J* **17**, 440–442.

58. Hellström, M., Kalén, M., Lindahl, P., Abramsson, A., and Betsholtz, C. (1999) Role of PDGF-B and PDGFR-beta in recruitment of vascular smooth muscle cells and pericytes during embryonic blood vessel formation in the mouse *Development* **126**, 3047–3055.

59. Schlingemann, R.O., Rietveld, F.J., de Waal, R.M., Ferrone, S., and Ruiter, D.J. (1990) Expression of the high molecular weight melanoma-associated antigen by pericytes during angiogenesis in tumors and in healing wounds. *Am J Pathol* **136**, 1393–1405.

60. Kunz, J., Krause, D., Kremer, M., and Dermietzel, R. (1994) The 140-kDa protein of blood-brain barrier-associated pericytes is identical to aminopeptidase N. *J Neurochem* **62**, 2375–2386.

61. Sundberg, C., Kowanetz, M., Brown, L.F., Detmar, M., and Dvorak, H.F. (2002) Stable expression of angiopoietin-1 and other markers by cultured pericytes: phenotypic similarities to a subpopulation of cells in maturing vessels during later stages of angiogenesis in vivo. *Lab Invest* **82**, 387–401.

62. Nayak, R.C., Berman, A.B., George, K.L., Eisenbarth, G.S., and King, G.L. (1988) A monoclonal antibody (3G5)-defined ganglioside antigen is expressed on the cell surface of microvascular pericytes *J Exp Med* **167**, 1003–1015.

63. Gushi, A., Tanaka, M., Tsuyama, S., Nagai, T., Kanzaki, T., Kanekura, T., and Matsuyama, T. (2008) The 3G5 antigen is expressed in dermal mast cells but not pericytes. *J Cutan Pathol* **35**, 278–284.

64. Charbord, P., Oostendorp, R., Pang, W., Herault, O., Noel, F., Tsuji, T., Dzierzak, E., and Peault, B. (2002) Comparative study of stromal cell lines derived from embryonic, fetal, and postnatal mouse blood-forming tissues *Exp Hematol* **30**, 1202–1210.

65. Fisher, M. (2009) Pericyte signaling in the neurovascular unit *Stroke* **40**, S13–S15.

66. Ni, T.G. (1922) The active response of capillaries of frogs, tadpoles, fish, bats, and men to various forms of excitation. *Am J Phys* **62**, 282–309.

67. Kelley, C., D'Amore, P., Hechtman, H.B., and Shepro, D. (1988) Vasoactive hormones and cAMP affect pericyte contraction and stress fibres in vitro *J Muscle Res Cell Motil* **9**, 184–194.

68. Das, A., Frank, R.N., Weber, M.L., Kennedy, A., Reidy, C.A., and Mancini, M.A. (1988) ATP causes retinal pericytes to contract in vitro *Exp Eye Res* **46**, 349–362.

69. Edelman, D.A., Jiang, Y., Tyburski, J., Wilson, R.F., and Steffes, C. (2006) Pericytes and their role in microvasculature homeostasis. *J Surg Res* **135**, 305–311.

70. Hughes, S., Gardiner, T., Hu, P., Baxter, L., Rosinova, E., and Chan-Ling, T. (2006) Altered pericyte-endothelial relations in the rat retina during aging: implications for vessel stability *Neurobiol Aging* **27**, 1838–1847.

71. Yamanishi, S., Katsumura, K., Kobayashi, T., and Puro, D.G. (2006) Extracellular lactate as a dynamic vasoactive signal in the rat retinal microvasculature *Am J Physiol Heart Circ Physiol* **290**, H925–H934.

72. King, J.S., and Schwyn, R.C. (1970) The fine structure of neuroglial cells and pericytes in the primate red nucleus and substantia nigra *Z Zellforsch Mikrosk Anat* **106**, 309–321.

73. Katenkamp, D., and Stiller, D. (1975) Cellular composition of the so-called dermatofibroma (histiocytoma cutis) *Virchows Arch A Pathol Anat Histol* **367**, 325–336.

74. Katenkamp, D., Stiller, D., and Schulze, E. (1976) Ultrastructural cytology of regenerating tendon–an experimental study *Exp Pathol (Jena)* **12**, 25–37.

75. Desmoulière, A., Rubbia-Brandt, L., Abdiu, A., Walz, T., Macieira-Coelho, A., and Gabbiani, G. (1992) Alpha-smooth muscle actin is expressed in a subpopulation of cultured and cloned fibroblasts and is modulated by gamma-interferon *Exp Cell Res* **201**, 64–73.

76. Niki, T., Pekny, M., Hellemans, K., Bleser, P. D., Berg, K. V., Vaeyens, F., Quartier, E., Schuit, F., and Geerts, A. (1999) Class VI intermediate filament protein nestin is induced during activation of rat hepatic stellate cells *Hepatology* **29**, 520–527.

77. Williams, G.M., and Iatropoulos, M.J. (2002) Alteration of liver cell function and proliferation: differentiation between adaptation and toxicity *Toxicol Pathol* **30**, 41–53.

78. Schor, A.M., Canfield, A.E., Sutton, A.B., Allen, T.D., Sloan, P., and Schor, S.L. (1992) The behaviour of pericytes in vitro: relevance to angiogenesis and differentiation *EXS* **61**, 167–178.

79. Diaz-Flores, L., Gutierrez, R., Lopez-Alonso, A., Gonzalez, R., and Varela, H. (1992) Pericytes as a supplementary source of osteoblasts in periosteal osteogenesis *Clin Orthop Relat Res* **275**, 280–286.

80. Brighton, C.T., Lorich, D.G., Kupcha, R., Reilly, T.M., Jones, A.R., and Woodbury, R.A. II. (1992) The pericyte as a possible osteoblast progenitor cell. *Clin Orthop Relat Res* **275**, 287–299.

81. Canfield, A.E., Sutton, A.B., Hoyland, J.A., and Schor, A.M. (1996) Association of thrombospondin-1 with osteogenic differentiation of retinal pericytes in vitro. *J Cell Sci* **109**, 343–353.

82. Reilly, T.M., Seldes, R., Luchetti, W., and Brighton, C.T. (1998) Similarities in the phenotypic expression of pericytes and bone cells *Transpl Immunol* **346**, 95–103.

83. Doherty, M.J., and Canfield, A.E. (1999) Gene expression during vascular pericyte differentiation *Crit Rev Eukaryot Gene Expr* **9**, 1–17.

84. Farrington-Rock, C., Crofts, N.J., Doherty, M.J., Ashton, B.A., Griffin-Jones, C., and Canfield, A.E. (2004) Chondrogenic and adipogenic potential of microvascular pericytes *Circulation* **110**, 2226–2232.

85. Galmiche, M. C., Koteliansky, V. E., Briere, J., Herve, P., and Charbord, P. (1993) Stromal cells from human long-term marrow cultures are mesenchymal cells that differentiate following a vascular smooth muscle differentiation pathway *Blood* **82**, 66–76.

86. Wenisch, S., Trinkaus, K., Hild, A., Hose, D., Herde, K., Heiss, C., Kilian, O., Alt, V., and Schnettler, R. (2005) Human reaming debris: a source of multipotent stem cells *Bone* **36**, 74–83.

87. Nico, B., Mangieri, D., Corsi, P., De Giorgis, M., Vacca, A., Roncali, L., and Ribatti, D. (2004) Vascular endothelial growth factor-A, vascular endothelial growth factor receptor-2 and angiopoietin-2 expression in the mouse choroid plexuses. *Brain Res* **1013**, 256–259.

88. Palmer, T.D., Willhoite, A.R., and Gage, F.H. (2000) Vascular niche for adult hippocampal neurogenesis *J Comp Neurol* **425**, 479–494.

89. Jin, K., Mao, X. O., Sun, Y., Xie, L., Jin, L., Nishi, E., Klagsbrun, M., and Greenberg, D. A. (2002) Heparin-binding epidermal growth factor-like growth factor: hypoxia-inducible expression in vitro and stimulation of neurogenesis in vitro and in vivo *J Neurosci* **22**, 5365–5373.

90. Louissaint, A. Jr, Rao, S., Leventhal, C., and Goldman, S.A. (2002) Coordinated interaction of neurogenesis and angiogenesis in the adult songbird brain *Neuron* **34**, 945–960.

91. Chaudhry, A.P., Montes, M., and Cohn, G.A. (1978) Ultrastructure of cerebellar hemangioblastoma *Cancer* **42**, 1834–1850.

92. Cinti, S., Cigolini, M., Bosello, O., and Björntorp, P. (1984) A morphological study of the adipocyte precursor *J Submicrosc Cytol* **16**, 243–251.

93. Balabanov, R., Beaumont, T., and Dore-Duffy, P. (1999) Role of central nervous system microvascular pericytes in activation of antigen-primed splenic T-lymphocytes *J Neurosci Res* **55**, 578–587.

94. Shen, P. J., Yuan, C. G., Ma, J., Cheng, S., Yao, M., Turnley, A. M., and Gundlach, A. L. (2005) Galanin in neuro(glio)genesis: expression of galanin and receptors by progenitor cells in vivo and in vitro and effects of galanin on neurosphere proliferation. *Neuropeptides* **39**, 201–205.

95. Suematsu, M, and Aiso, S. (2001) Professor Toshio Ito: a clairvoyant in pericyte biology *Keio J Med* **50**, 66–71.

96. Lardon, J., Rooman, I., and Bouwens, L. (2002) Nestin expression in pancreatic stellate cells and angiogenic endothelial cells *Histochem Cell Biol* **117**, 535–540.

97. Kadoya, K., Fukushi, J., Matsumoto, Y., Yamaguchi, Y., and Stallcup, W. B. (2008) NG2 proteoglycan expression in mouse skin: altered postnatal skin development in the NG2 null mouse *J Histochem Cytochem* **56**, 295–303.

98. Rajkumar, V.S., Howell, K., Csiszar, K., Denton, C.P., Black, C.M., and Abraham, D.J. (2005) Shared expression of phenotypic markers in systemic sclerosis indicates a convergence of pericytes and fibroblasts to a myofibroblast lineage in fibrosis *Arthritis Res Ther* **7**, R1113–R1123.

99. Crisan, M., Zheng, B., Zambidis, E.T., Yap, S., Tavian, M., Sun, B., Giacobino, J-P., Casteilla, L., Huard, J., and Péault, B. (2007) Blood Vessels as a Source of Progenitor Cells in Human Embryonic and Adult Life In: Stem Cells and Their Potential for Clinical Application, in *NATO Science for Peace and Security* (Bilko, N. M., Fehse, B., Ostertag, W., Stocking, C., Zander, A.R., Ed.), pp 137–147, Springer, Netherlands.

100. Dellavalle, A., Sampaolesi, M., Tonlorenzi, R., Tagliafico, E., Sacchetti, B., Perani, L., Innocenzi, A., Galvez, B. G., Messina, G., Morosetti, R., Li, S., Belicchi, M., Peretti, G., Chamberlain, J. S., Wright, W. E., Torrente, Y., Ferrari, S., Bianco, P., and Cossu, G. (2007) Pericytes of human skeletal muscle are myogenic precursors distinct from satellite cells *Nat Cell Biol* **9**, 255–267.

101. Péault, B., Rudnicki, M., Torrente, Y., Cossu, G., Tremblay, J. P., Partridge, T., Gussoni, E., Kunkel, L. M., and Huard, J. (2007) Stem and progenitor cells in skeletal muscle development, maintenance, and therapy *Mol Ther* **15**, 867–877.

102. Davidoff, M.S., Middendorff, R., Enikolopov, G., Riethmacher, D., Holstein, A.F., and Müller, D. (2004) Progenitor cells of the testosterone-producing Leydig cells revealed *J Cell Biol* **167**, 935–944.

103. Lovschall, H., Mitsiadis, T.A., Poulsen, K., Jensen, K.H., and Kjeldsen, A.L. (2007) Coexpression of Notch3 and Rgs5 in the pericyte-vascular smooth muscle cell axis in response to pulp injury *Int J Dev Biol* **51**, 715–721.

104. Puri, P. L., Bhakta, K., Wood, L. D., Costanzo, A., Zhu, J., and Wang, J. Y. (2002) A myogenic differentiation checkpoint activated by genotoxic stress *Nat Genet* **32**, 585–593.

105. Song, S., Ewald, A. J., Stallcup, W., Werb, Z., and Bergers, G. (2005) PDGFRbeta+ perivascular progenitor cells in tumours regulate pericyte differentiation and vascular survival *Nat Cell Biol* **7**, 870–879.

106. Gerhardt, H., and Betsholtz, C. (2003) Endothelial-pericyte interactions in angiogenesis *Cell Tissue Res* **314**, 15–23.

107. Mancuso MR, Kuhnert F, Kuo C.J. (2008) Developmental angiogenesis of the central nervous system *Lymphat Res Biol* **6**, 173–180.

108. Wang, L., Chopp, M., Gregg, S.R., Zhang, R.L., Teng, H., Jiang, A., Feng, Y., and Zhang, Z.G. (2008) Neural progenitor cells treated with EPO induce angiogenesis through the production of VEGF *J Cereb Blood Flow Metab* **28**, 1361–1318.

109. Yonekura, H., Sakurai, S., Liu, X., Migita, H., Wang, H., Yamagishi, S., Nomura, M., Abedin, M. J., Unoki, H., Yamamoto, Y., and Yamamoto, H. (1999) Placenta growth factor and vascular endothelial growth factor B and C expression in microvascular endothelial cells and pericytes. Implication in autocrine and paracrine regulation of angiogenesis *J Biol Chem* **274**, 35172–35178.

110. de la Fuente, R., Abad, J. L., Garcia-Castro, J., Fernandez-Miguel, G., Petriz, J., Rubio, D., Vicario-Abejon, C., Guillen, P., Gonzalez, M. A., and Bernad, A. (2004) Dedifferentiated adult articular chondrocytes: a population of human multipotent primitive cells *Exp Cell Res* **297**, 313–328.

111. Greenberg, J.I., Shields, D.J., Barillas, S.G., Acevedo, L.M., Murphy, E., Huang, J., Scheppke, L., Stockmann, C., Johnson, R.S., Angle, N., and Cheresh, D.A. (2008) A role for VEGF as a negative regulator of pericyte function and vessel maturation *Nature* **456**, 809–813.

112. Dore-Duffy, P., Wang, X., Mehedi, A., Kreipke, C.W., and Rafols, J.A. (2007) Differential expression of capillary VEGF isoforms following traumatic brain injury *Neurol Res* **29**: 395–403.

113. Diaz-Flores, L., Gutierrez, R., Valladares, F., Varela, H., and Perez, M. (1994) Intense vascular sprouting from rat femoral vein induced by prostaglandins E1 and E2 *Anat Rec* **238**, 68–76.

114. Nehls, V., Denzer, K., and Drenckhahn, D. (1992) Pericyte involvement in capillary sprouting during angiogenesis in situ *Cell Tissue Res* **270**, 469-474.

115. Pfister, F., Feng, Y., vom Hagen, F., Hoffmann, S., Molema, G., Hillebrands, J. L., Shani, M., Deutsch, U., and Hammes, H. P. (2008) Pericyte migration: a novel mechanism of pericyte loss in experimental diabetic retinopathy *Diabetes* **57**, 2495–2502.

116. Duz, B., Oztas, E., Erginay, T., Erdogan, E., and Gonul, E. (2007) The effect of moderate

hypothermia in acute ischemic stroke on pericyte migration: an ultrastructural study *Cryobiology* **55**, 279–284.

117. Claudio, L., and Brosnan, C. F. (1992) Effects of prazosin on the blood-brain barrier during experimental autoimmune encephalomyelitis *Brain Res* **594**, 233–243.

118. Kunz, J., Krause, D., Gehrmann, J., and Dermietzel, R. (1995) Changes in the expression pattern of blood-brain barrier-associated pericytic aminopeptidase N (pAP N) in the course of acute experimental autoimmune encephalomyelitis *J Neuroimmunol* **59**, 41–55.

119. Dore-Duffy, P., Balabanov, R., Rafols, J., and Swanborg, R. H. (1996) Recovery phase of acute experimental autoimmune encephalomyelitis in rats corresponds to development of endothelial cell unresponsiveness to interferon gamma activation *J Neurosci Res* **44**, 223–234.

120. Bolton, C. (1997) Neurovascular damage in experimental allergic encephalomyelitis: a target for pharmacological control *Mediators Inflamm* **6**, 295–302.

121. Zlokovic, B. V. (2008) The blood-brain barrier in health and chronic neurodegenerative disorders *Neuron* **57**, 178–201.

122. Ho, K. L. (1985) Ultrastructure of cerebellar capillary hemangioblastoma. IV. Pericytes and their relationship to endothelial cells *Acta Neuropathol* **67**, 254–264.

123. Winkler, F., Kozin, S. V., Tong, R. T., Chae, S. S., Booth, M. F., Garkavtsev, I., Xu, L., Hicklin, D. J., Fukumura, D., di Tomaso, E., Munn, L. L., and Jain, R. K. (2004) Kinetics of vascular normalization by VEGFR2 blockade governs brain tumor response to radiation: role of oxygenation, angiopoietin-1, and matrix metalloproteinases *Cancer Cell* **6**, 553–563.

124. Hammes, H. P., Lin, J., Renner, O., Shani, M., Lundqvist, A., Betsholtz, C., Brownlee, M., and Deutsch, U. (2002) Pericytes and the pathogenesis of diabetic retinopathy *Diabetes* **51**, 3107–3112.

125. Frank, R. N., Turczyn, T. J., and Das, A. (1990) Pericyte coverage of retinal and cerebral capillaries *Invest Ophthalmol Vis Sci* **31**, 999–1007.

126. Feng, Y., Pfister, F., Schreiter, K., Wang, Y., Stock, O., Vom Hagen, F., Wolburg, H., Hoffmann, S., Deutsch, U., and Hammes, H. P. (2008) Angiopoietin-2 deficiency decelerates age-dependent vascular changes in the mouse retina *Cell Physiol Biochem* **21**, 129–136.

127. Szpak, G. M., Lewandowska, E., Wierzba-Bobrowicz, T., Bertrand, E., Pasennik, E.,

Mendel, T., Stepien, T., Leszczynska, A., and Rafalowska, J. (2007) Small cerebral vessel disease in familial amyloid and non-amyloid angiopathies: FAD-PS-1 (P117L) mutation and CADASIL. Immunohistochemical and ultrastructural studies *Folia Neuropathol* **45**, 192–204.

128. Wegiel, J., and Wisniewski, H. M. (1992) Tubuloreticular structures in microglial cells, pericytes and endothelial cells in Alzheimer's disease *Acta Neuropathol* **83**, 653–658.

129. Stewart, P. A., Hayakawa, K., Akers, M. A., and Vinters, H. V. (1992) A morphometric study of the blood-brain barrier in Alzheimer's disease *Lab Invest* **67**, 734–742.

130. Yamagishi, S., and Imaizumi, T. (2005) Pericyte biology and diseases *Int J Tissue React* **27**, 125–135.

131. Kischer, C. W. (1992) The microvessels in hypertrophic scars, keloids and related lesions: a review *J Submicrosc Cytol Pathol* **24**, 281–296.

132. Li, W., Yanoff, M., Liu, X., and Ye, X. (1997) Retinal capillary pericyte apoptosis in early human diabetic retinopathy *Chin Med J (Engl)* **110**, 659–663.

133. Shojaee, N., Patton, W. F., Hechtman, H. B., and Shepro, D. (1999) Myosin translocation in retinal pericytes during free-radical induced apoptosis *J Cell Biochem* **75**, 118–129.

134. Zagzag, D., Amirnovin, R., Greco, M. A., Yee, H., Holash, J., Wiegand, S. J., Zabski, S., Yancopoulos, G. D., and Grumet, M. (2000) Vascular apoptosis and involution in gliomas precede neovascularization: a novel concept for glioma growth and angiogenesis *Lab Invest* **80**, 837–849.

135. Machein, M. R., Knedla, A., Knoth, R., Wagner, S., Neuschl, E., and Plate, K. H. (2004) Angiopoietin-1 promotes tumor angiogenesis in a rat glioma model. *Am J Pathol* **165**, 1557–1570.

136. Vermeulen, P. B., Colpaert, C., Salgado, R., Royers, R., Hellemans, H., Van Den Heuvel, E., Goovaerts, G., Dirix, L. Y., and Van Marck, E. (2001) Liver metastases from colorectal adenocarcinomas grow in three patterns with different angiogenesis and desmoplasia *J Pathol* **195**, 336–342.

137. Bexell, D., Gunnarsson, S., Tormin, A., Darabi, A., Gisselsson, D., Roybon, L., Scheding, S., and Bengzon, J. (2009) Bone marrow multipotent mesenchymal stroma cells act as pericyte-like migratory vehicles in experimental gliomas *Mol Ther* **17**, 183–190.

138. Li, W., Liu, X., Yanoff, M., Cohen, S., and Ye, X. (1996) Cultured retinal capillary pericytes

die by apoptosis after an abrupt fluctuation from high to low glucose levels: a comparative study with retinal capillary endothelial cells *Diabetologia* **39**, 537–547.

139. Zhu, M., Madigan, M. C., van Driel, D., Maslim, J., Billson, F. A., Provis, J. M., and Penfold, P. L. (2000) The human hyaloid system: cell death and vascular regression *Exp Eye Res* **70**, 767–776.

140. Braun, A., Xu, H., Hu, F., Kocherlakota, P., Siegel, D., Chander, P., Ungvari, Z., Csiszar, A., Nedergaard, M., and Ballabh, P. (2007) Paucity of pericytes in germinal matrix vasculature of premature infants *J Neurosci* **27**, 12012–12024.

141. Khoury, J., and Langleben, D. (1996) Platelet-activating factor stimulates lung pericyte growth in vitro *Am J Physiol* **270**, L298–L304.

142. Seifert, M., and Reichrath, J. (2006) The role of the human DNA mismatch repair gene hMSH2 in DNA repair, cell cycle control and apoptosis: implications for pathogenesis, progression and therapy of cancer *J Mol Histol* **37**, 301–307.

143. Heyer, J., Yang, K., Lipkin, M., Edelmann, W., and Kucherlapati, R. (1999) Mouse models for colorectal cancer *Oncogene* **18**, 5325–5333.

144. Canfield, A.E., Farrington, C., Dziobon, M.D., Boot-Handford, R.P., Heagerty, A.M., Kumar, S.N., and Roberts, I.S. (2002) The involvement of matrix glycoproteins in vascular calcification and fibrosis: an immunohistochemical study *J Pathol* **196**, 228–234.

145. Nehls V, and Drenckhahn D. (1991) Heterogeneity of microvascular pericytes for smooth muscle type alpha-actin *J Cell Biol* **113**, 147–154.

146. Newcomb, P.M., and Herman, IM. (1993) Pericyte growth and contractile phenotype: modulation by endothelial-synthesized matrix and comparison with aortic smooth muscle *J Cell Physiol* **155**, 385–393.

147. Verbeek, M.M., Westphal, J.R., Ruiter, D.J., and de Waal, R.M. (1995) T lymphocyte adhesion to human brain pericytes is mediated via very late antigen-4/vascular cell adhesion molecule-1 interactions. *J Immunol* **154**, 5876–5884.

148. Fujimoto, T., and Singer, S.J. (1987) Immunocytochemical studies of desmin and vimentin in pericapillary cells of chicken *J Histochem Cytochem* **35**, 1105–1115.

Chapter 3

Morphology and Properties of Astrocytes

Sukriti Nag

Abstract

Astrocytes were identified about 150 years ago, and, for the longest time, were considered to be supporting cells in the brain providing trophic, metabolic, and structural support for neural networks. Research in the last 2 decades has uncovered many novel molecules in astrocytes and the finding that astrocytes communicate with neurons via Ca^{2+} signaling, which leads to release of chemical transmitters, termed gliotransmitters, has led to renewed interest in their biology. This chapter will briefly review the unique morphology and molecular properties of astrocytes. The reader will be introduced to the role of astrocytes in blood-brain barrier (BBB) maintenance, in Ca^{2+} signaling, in synaptic transmission, in CNS synaptogenesis, and as neural progenitor cells. Mention is also made of the diseases in which astrocyte dysfunction has a role.

Key words: Astrocyte, Brain edema, Calcium, Cerebral blood flow, Gap junctions, Gliotransmitters, GFAP, Aquaporin-4, Connexin43, Brain edema, Glutamate, Neural progenitor cells, Synaptic transmission, Synaptogenesis, Tripartite synapse

1. Introduction

The neuron is well recognized as the principal signaling unit of neurotransmission and key to the nervous system function while astrocytes were for the longest time considered to be supporting cells in the brain providing trophic, metabolic, and structural support for neurons. They were bypassed by neurophysiologists since they were found to be electrically nonexcitable cells. However, the finding that they can communicate with neurons via Ca^{2+} signaling has led to renewed interest in their biology. The last 2 decades have seen an explosion of research on astrocytes and it is now recognized that astrocytes have a much greater role in the nervous system than just a supportive one. This chapter will briefly review the unique morphology and molecular properties

Sukriti Nag (ed.), *The Blood-Brain and Other Neural Barriers: Reviews and Protocols*, Methods in Molecular Biology, vol. 686, DOI 10.1007/978-1-60761-938-3_3, © Springer Science+Business Media, LLC 2011

of astrocytes. The reader will be introduced to the role of astrocytes in blood-brain barrier (BBB) maintenance, in Ca^{2+} signaling, in synaptic transmission, in CNS synaptogenesis, and as neural progenitor cells. Mention is also made of the diseases in which astrocyte dysfunction has a role. This chapter is designed to whet your appetite for astrocyte biology which is reviewed in detail in a recent monograph (1).

2. Morphology and Anatomic Distribution of Astrocytes

The interstitial tissue between neurons and blood vessels which contain stellate and spindle-shaped cells was named as "neuroglia" or "nerve glue" by Virchow in 1860 (2). About 100 years ago, use of the gold sublimate staining method showed well-developed processes to emerge from many sides of astrocytes giving them their stellate shape and allowing the distinction of astrocytes from other glia (3). Cajal also noted that the tips of astrocytic processes having bulbous dilatations or end-feet terminated on vessel walls and that astrocytes could form a physical bridge between neurons and vessels. The Weigert technique (4) demonstrated the cytoplasmic fibrils in astrocytes and allowed the distinction between fibrillary astrocytes in the white matter, which have abundant fibrils, and protoplasmic astrocytes in the cerebral cortex which have fewer fibrils. Astrocytic processes also combine at the surface of the brain to form the glia limitans (Fig. 1a). The Bergmann astrocytes of the cerebellum have processes predominantly oriented in one direction and extending all the way from their cell bodies in the Purkinje layer to the surface of the molecular layer, where their end-feet form the glia limitans.

Ultrastructural studies show that astrocytes have relatively "clear" cytoplasm containing small, highly electron-dense granules that are glycogen, and all the usual organelles and lipid droplets. Microtubules are rarely seen in mature astrocytes. Their nuclei are oval and contain evenly distributed moderately abundant DNA components. A feature of the cytoplasm is the presence of 9-nm intermediate filaments which may also occur in parallel bundles in their processes (Figs. 2 and 3). Thin short extensions appear to interconnect some of these filaments.

Adjacent astrocytes are separated by a 15–20 nm extracellular space along which two types of junctions are present. The first type of junction is the puncta adhaerentia where adjacent astrocyte membranes are parallel being separated by a wider space of 25–30 nm (5). Slightly increased electron density of the gap and adjacent cytoplasm is also present. The second type of junction is the gap junction where adjacent membranes are separated by a 2- to 3-nm wide gap (Fig. 3). Both junctions allow the penetration of tracers such as horseradish peroxidase and lanthanum (6).

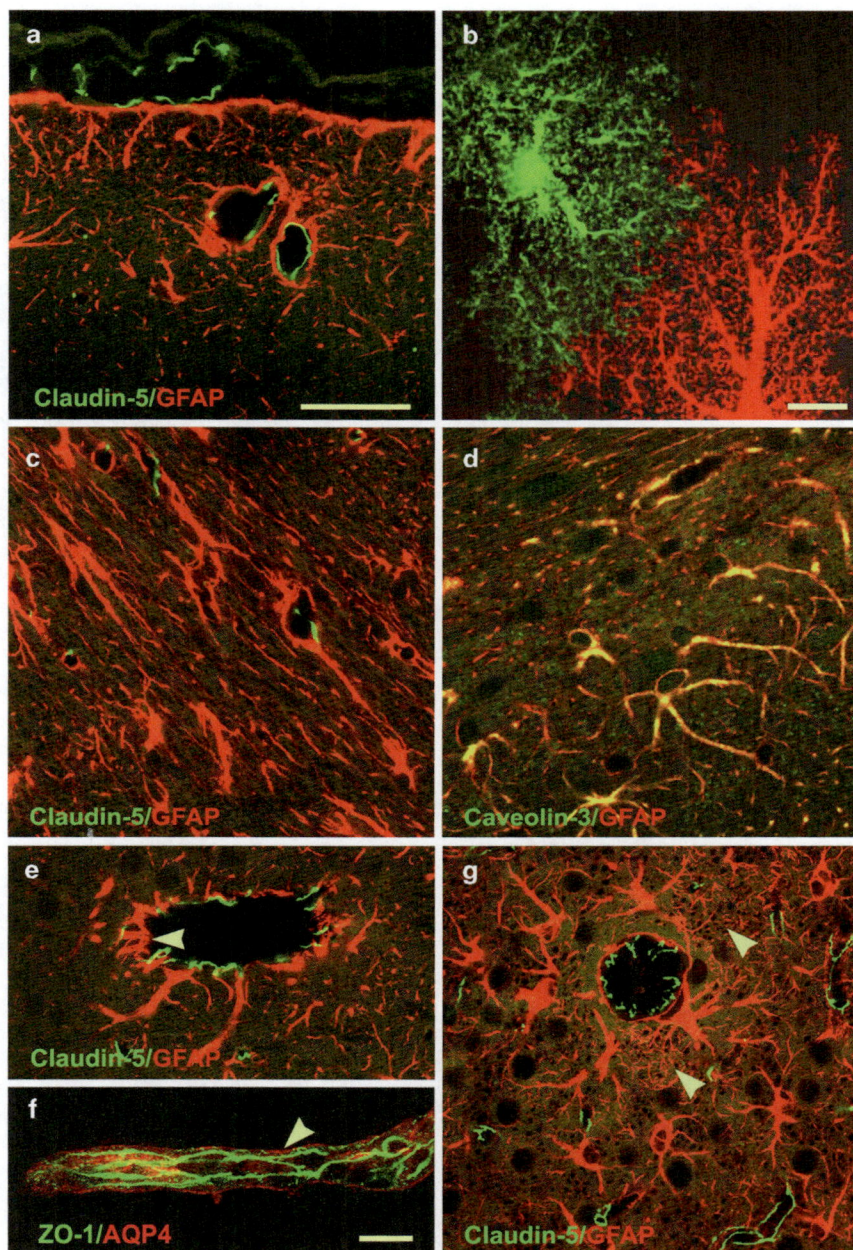

Fig. 1. Merged confocal images from rat brain dual-labeled for GFAP (*red*) and Claudin-5 (*green*) (**a**, **c**, **e**, and **g**), for caveolin-3 and GFAP (**d**), and zonula occludens-1 (ZO-1) and aquaporin4 (AQP4) (**f**) are shown. (**a**) The glia limitans formed by astrocytic processes is present at the brain surface and astrocytic end-feet surround the entire circumference of two intracerebral arterioles. (**b**) One-month-old postnatal rats showing two protoplasmic astrocytes, one filled with *Lucifer Yellow* (*green*) and the other with Alexa 568. Astrocytic domains are well established and the fine spongiform processes of adjacent astrocytes intermingle in an area only a couple of microns wide. (**c**) Fibrillary astrocytes of the white matter have few processes and the long axes of these cells are parallel to the white matter axons. (**d**) Normal rat brain shows colocalization (*yellow*) of caveolin-3 and GFAP in the white matter and in hippocampal astrocytes. (**e**) A large cortical vessel shows the termination of end-feet in the form of looped processes (*arrowhead*). (**f**) Vessel segment shows endothelial ZO-1 fibrils (*green*) surrounded by astrocytic foot processes which are immunoreactive for AQP4 (*red, arrowhead*). (**g**) Reactive astrocytosis adjacent to a cortical cold lesion is shown. The density and size of astrocytes are increased. Note also that their processes terminate into multiple mini processes which form a network (*arrowheads*). (**f**) Scale bar = 50 μm (**a**, **c–e**, **g**); 10 μm (**b**) and 20 μm (**f**). (**b**) Reproduced with permission (13); (**f**) reproduced with permission (64).

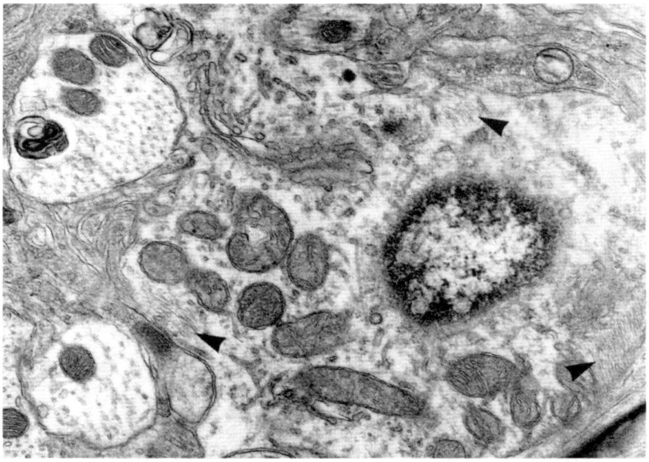

Fig. 2. An electron micrograph of an astrocyte in layer 3 of the cerebral cortex shows relatively electron-lucent cytoplasm in which the normal complement of organelles is present. Groups of 9-nm intermediate filaments are present in the cytoplasm (*arrowheads*). ×18,500.

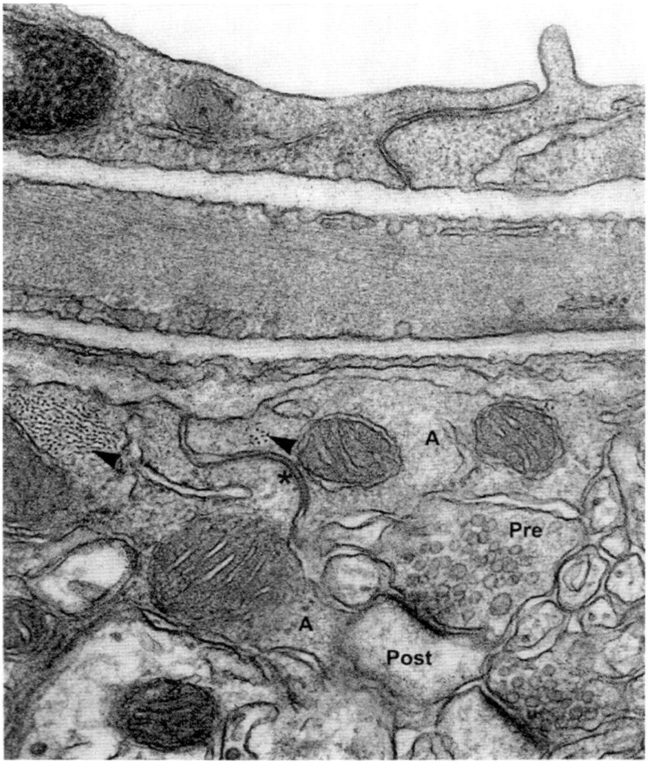

Fig. 3. An electron micrograph taken at the level of layer 3 of the cerebral cortex shows an arteriolar segment which consists of endothelial, smooth muscle, and attenuated leptomeningeal cell layers. The latter is separated from two astrocytic end-feet (*A*) by a basement membrane. Note that in well-fixed vessels, a perivascular space is not present. The end-feet show mitochondria and the cross-sections of intermediate filaments (*arrowheads*) in the cytoplasm and are separated by a gap junction (*). Note the proximity of the end-feet to pre- and postsynaptic terminals which together form a tripartite synapse. ×50,000.

Gap junctions are evenly distributed along the astrocytic processes, often interconnecting adjacent astrocytic processes derived from the same cell referred to as autocellular junctions (7, 8). Only in the narrow interface of adjacent cells do gap junctions couple processes from different cells (9). At the gap junction, each of the joined cells contributes a hemi-channel or connexon to each cell–cell channel. Each hemi-channel comprises a hexamer of connexins arranged around a central pore, and the cell–cell channels are gated by several stimuli, including transjunctional voltage, low pH, and various pharmacological agents. Connexin30 and connexin43 colocalize at gap junctions (10). Zonula occludens-1 (ZO-1) has also been localized at astrocytic gap junctions where it is found to colocalize with connexin30 and 43 and ZONAB (ZO-1-associated nucleic acid-binding protein) (11). Connexins are known to have adhesive properties and the autocellular junctions may stabilize the complex network of astrocytic processes and may also facilitate intracellular diffusion of energy metabolites and possible signaling molecules, such as Ca^{2+} and inositol (1,4,5)-triphosphate (IP_3), between fine astrocytic processes (8).

2.1. Astrocyte Domains

Astrocytes typically extend between five to eight major processes, each of which is highly ramified into innumerable delicate leaflet-like processes, which are insinuated between and around the various components of the nervous tissue (12). Microinjection of single hippocampal astrocytes with fluorescent dyes demonstrates that each astrocyte occupies a discrete area that is free of processes from any adjacent astrocytes thus defining its own anatomical domain (Fig. 1b). Only the most peripheral processes interdigitate with one another in a narrow interface within which <5% of the volumes of adjacent astrocytes overlap (13, 14). Glial fibrillary acidic protein (GFAP) labels only the major processes of astrocytes, many of the smaller processes being nonreactive with GFAP. These smaller processes fill a volume that is best defined as a polyhedron (12–14). These fine processes have a significantly higher density of mitochondria as compared with the surrounding neuronal processes, synapses, other glial processes, and endothelial cells, supporting the concept that oxidative metabolism is a major part of the energy metabolism in protoplasmic astrocytes (15).

Within a single astrocyte domain, 300–600 neuronal dendrites (16) and 10^5 synapses are present in the rodent cortex and hippocampus. In contrast, in the human cortex, a single astrocyte might sense the activity and regulate the function of more than one million synapses within its domain (17). The distribution of astrocytes throughout the brain and spinal cord is highly organized being evenly distributed, such that their cell bodies and larger processes are not in contact with each other (18).

The functional significance of these nonoverlapping astrocytic domains is unknown, although all synapses lying within a given volumetrically defined compartment may be under the sole influence of a single astrocyte (19).

Neurons are dispersed among the astrocytic domains, with the innumerable fine neurites penetrating each astrocytic domain and being surrounded by its processes. The ratio of glia to neurons is higher in humans than most other species (19–21).

2.2. Types of Astrocytes in the Human Cerebral Cortex

Based on GFAP immunostaining, human cortical astrocytes are reported to have four distinct morphologies being named protoplasmic, interlaminar, polarized, and fibrous or fibrillary astrocytes (17).

2.2.1. Protoplasmic Astrocytes

Protoplasmic astrocytes are the most abundant type in human cortex, being present in cortical layers 2–6. Human protoplasmic astrocytes are larger and more elaborate than their rodent counterparts (22). Although the cell body of human astrocytes is only ~10 μm in diameter, their processes span 100–200 μm, giving them a 27-fold greater volume than their rodent counterparts (17). The synaptic density in the rat cortex has been estimated to be 1,397 million synapses/mm^3, while that of human cortex is ~1,100 million synapses/mm^3 (23). This suggests that synaptic density alone does not account for the increased capacity of human brain. The majority of the GFAP-positive processes of protoplasmic astrocytes do not overlap indicating a domain organization.

2.2.2. Interlaminar Astrocytes

Interlaminar astrocytes were first described in cortical layer 1 of primate cortex (24, 25) where they extend striking long, frequently unbranched, processes extending through the cortical layers, terminating in layers 3 or 4 (26). The cell bodies of these astrocytes are ~10 μm in diameter and extend two types of processes: three to six fibers that contribute to the astrocytic network near the pial surface, and another one or two that penetrate deeper layers of the cortex. The latter have a constant diameter and can extend up to 1 mm in length (26). These processes are tortuous and, although largely unbranched, occasionally send collaterals to the vasculature (26). The endings of these interlaminar fibers deep in the cortex might be in the form of a "terminal mass" or end bulb containing a multilaminar structure and mitochondria (27). The function of these interlaminar astrocytes is unknown. Their interlaminar fibers clearly violate the domain organization and might serve as a nonsynaptic pathway for long-distance signaling and integration of activity within cortical columns (17).

2.2.3. Polarized Astrocytes

These unipolar cells are relatively uncommon and are present in layers 5 and 6 of the cortex, near the white matter, and extend one or two long GFAP-positive processes, which are up to 1 mm in length, away from the white matter (17, 22). These long processes are straight, frequently unbranched or branch once, have a constant diameter of ~2–3 μm, and have numerous "beads" or varicosities. Occasionally, polarized astrocytes extend processes to the vasculature, but most terminate in the neuropil (22). Polarized astrocytes do not respect the domain boundaries of their neighbors, because the long processes from these cells travel directly through other protoplasmic astrocytic domains. The function of these cells has not been investigated, but these cells might serve as an alternative pathway for long-distance communication across cortical layers, perhaps forming links between functionally related domains in different laminae, or between gray and white matter (17).

2.2.4. Fibrous Astrocytes

These white matter astrocytes have fewer primary GFAP-positive processes and their fibers are straighter and less branched than those of other glia (22, 28) (Fig. 1c). These cells are roughly equidistant from one another. In contrast to protoplasmic astrocytes, the processes of adjacent fibrous astrocytes intermingle and overlap (22, 28). The simple morphology of these cells and their relative uniformity suggest that their function might be limited to metabolic support and not extend to information processing and modulation of neural activity.

3. Properties of Astrocytes

3.1. Molecular Properties of Astrocytes

3.1.1. Genes Expressed by Astrocytes

Transcriptome databases are available for genes expressed in acutely isolated mouse astrocytes from postnatal days 1–30 (29) and genes expressed in protoplasmic astrocytes in adult mouse cortex (15). Postnatal mouse astrocytes have high expression of many transcription factors, signaling transmembrane receptors, and secreted proteins (see supplemental material in www.jneurosci.org). Astrocytes were found to be enriched in specific metabolic and lipid synthetic pathways, including draper/Megf10 and Mertk/integrin $\alpha_v\beta_5$ phagocytic pathways, suggesting that astrocytes are professional phagocytes. Since the gene profiles of astrocytes and oligodendrocytes were found to be very dissimilar, it was suggested that they should no longer be classed together as "glia."

Genomic expression profiling was used to identify metabolic pathways in protoplasmic astrocytes and neurons from adult mouse cortex (15). The analysis showed that both astrocytes and neurons express transcripts for oxidative metabolism of glucose; however, the expression of the majority of enzymes in the tricarboxylic acid cycle was higher in astrocytes than neurons. These findings support the presence of robust oxidative metabolism in astrocytes.

3.1.2. Molecules in Astrocyte Cell Bodies

GFAP and S100β are well recognized and widely used specific markers of astrocytes (30). Dye injection in adult rodents shows that GFAP expression occupies only 15% of the total volume of the cell (12, 14). Other markers of astrocytes are the gap junction proteins connexin30 (10) and connexin43 (31). Immunohistochemistry shows diffuse connexin43 localization in brain, while localization of connexin30 is more heterogeneous being present in gap junctions of gray, but not white, matter astrocytes (10). Caveolin-3 is expressed in both cortical and white matter astrocytes (Fig. 1d). Basic fibroblast growth factor (FGF)-2 (32) is present in adult astrocytes and expression of FGF receptor -2 and 3 is reported in astrocytes and oligodendrocytes (33).

Aldehyde dehydrogenase 1 family member L1 (Aldh1L1) has been identified as a new astrocyte-specific marker (29) in mouse brain. The mRNA signal for this gene is present throughout the CNS in a pattern consistent with pan-astrocyte expression. Immunohistochemistry shows Aldh1L1 localization in both the gray and white matter astrocytes with labeling of the cell body and its extensive processes, while GFAP only labels the thick main processes of some astrocytes (29, 34). Dual labeling demonstrates that Aldh1L1 does not label other cell types in the brain making it a useful astrocyte-specific marker (29).

High-affinity astrocyte-specific glutamate transporters, excitatory amino acid transporter (EAAT)1 (GLAST) and EAAT2 (GLT1), and subtypes are enriched in astrocytic processes and play a major role in glutamate clearance in the adult CNS (35, 36). Localization of P-gp is described in Subheading 3.1.3. Astrocytes also express urea transporter 3 (37), nucleoside transporters (38), and ryanodine receptors (39). Other efflux transport proteins including Multidrug Resistance Proteins 1–6 and the Breast Cancer Resistance Protein (ABCG2) have been reported in astrocytes, although expression varies depending on the species examined and the age of the rat, whether embryonic or adult (40).

3.1.3. Molecules in Astrocytic End-Feet

Cerebral capillaries are typically positioned along the interfaces between adjacent astrocytic domains and the end-feet of adjacent astrocytes provide a contiguous but nonoverlapping sheath around the capillaries bordering its domain (41). The end-feet of astrocytes cover >99% of the vascular surface facing endothelial cells or pericytes, but are not always GFAP positive, giving the false impression that astrocytic coverage of the vasculature is incomplete (41, 42). Penetrating arterioles are normally surrounded by GFAP-positive end-feet (Fig. 1a), whereas end-feet around capillaries are GFAP negative. The significance of GFAP expression in end-feet is not clear, but GFAP expression is upregulated by mechanical stress and may be induced by arterial pulsation (43). Confocal microscopy shows that astrocyte end-feet terminate into multiple looped processes lying on the vessel wall forming a rosette-like structure (42) (Fig. 1e). This morphology greatly increases the surface area of the end-feet in contact with the blood vessel. Each astrocyte has at least one process with end-feet surrounding a blood vessel (41), although individual astrocytes can contact several endothelial cells via multiple end-feet, even those lying some distance away. Thus, vessels can be covered with many end-feet derived from distinct astrocytes (Fig. 1, Chapter 1).

Astrocyte polarity refers to the molecular and structural heterogeneity of specific membrane domains on the astroglial surface. Vascular end-feet are highly polarized expressing several specialized proteins at their luminal surface, including glucose transporter-1 (45-kDa form) which facilitates rapid transfer of glucose to metabolically demanding dendrites (42), P-glycoprotein (44) which is involved in the removal of lipophilic molecules and BBB differentiation, the purinergic receptors P2Y(2) and P2Y(4) which are mediators of astrocytic Ca^{2+} signaling and colocalize with GFAP around larger vessels in the cortex (41), and functional α_1 and β adrenergic receptors which, when activated, cause prominent elevations in end-foot Ca^{2+} (45). The gap junction protein connexin43 is also highly expressed in astrocytic end-feet (41).

AQP4, the principal AQP in mammalian brain, is expressed in astrocytes at the borders between major water compartments and the brain parenchyma (46, 47). Expression of AQP4 in astrocytic foot processes brings it in close proximity to intracerebral vessels and thus the blood-brain interface (Fig. 1f). Water molecules moving from the blood pass through the luminal and abluminal endothelial membranes by diffusion and across the astrocytic foot processes through the AQP4 channels. AQP4 is also expressed in the basolateral membrane of the ependymal cells lining the cerebral ventricles, in subependymal astrocytes which are located at the ventricular cerebrospinal fluid (CSF)–brain interface, and in the dense astrocytic processes that form the glia limitans at the subarachnoid–CSF fluid interface.

Freeze-fracture studies have shown orthogonal arrays of intramembranous particles (OAPs) in astrocytic end-feet (48–52) (Fig. 4). A positive relationship between the OAP-based polarity and an intact BBB has been postulated (50, 53). When the BBB becomes leaky as occurs in brain tumors, the OAP-related polarity of astrocytes decreases (49).

The Kir 4.1 K^+ channel and AQP4 are located in the OAPs anchored to α-syntrophin, an adaptor protein associated with the dystrophin–dystroglycan complex (DDC) (54, 55). Restriction of AQP4 immunoreactivity to the astrocytic end-feet membrane is dependent on the extracellular heparin sulfate proteoglycan, agrin, which is deposited by both astrocytes and endothelial cells

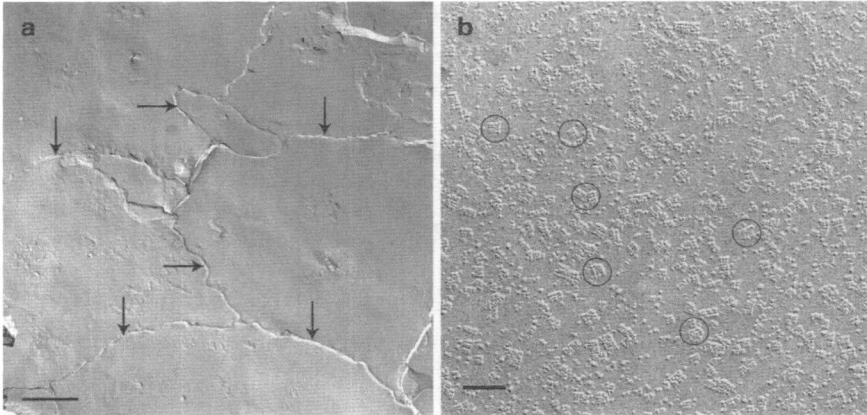

Fig. 4. Freeze-fracture images from mouse brain showing astrocytic end-feet at the glia limitans in low (**a**) and high magnification (**b**). (**a**) The borders between astrocytic end-feet are marked by *arrows*. (**b**) High magnification of (**a**) shows orthogonal arrays of intramembranous particles, some of which are circled. Scale bar = (**a**) 1 μm, (**b**) 100 nm. Contributed by Dr. H.Wolburg, University of Tuebingen, Germany.

(56). Agrin binding to α-dystroglycan couples to AQP4 through α-syntrophin. If agrin is absent from the basal lamina, AQP4 immunoreactivity becomes diffuse, being present across the entire cell surface (57).

Altered expression of AQP4 in brain injury has been reported in experimental autoimmune encephalomyelitis (EAE) (58). Areas showing perivascular accumulation of inflammatory cells show loss of the polarized localization of AQP4 to end-feet and localization becomes diffuse over the entire astrocytic cell surface associated with loss of OAPs as observed in freeze-fracture replicas (58). Loss of β-dystroglycan immunoreactivity is also present in these areas suggesting that loss of β-dystroglycan-mediated astrocyte foot process anchoring to the basement membrane leads to loss of polarized AQP4 localization in astrocytic end-feet, and thus to edema formation in EAE.

3.1.4. K⁺ Channels

At least four K⁺ channels have been identified in cortical astrocytes, namely Kir 2.1, 2.2, 2.3, and 4.1 (59). Based on studies in transgenic mice, it appears that one member in particular, Kir 4.1, is the predominant K⁺ channel in mature astrocytes and almost solely responsible for establishing the astrocyte negative resting membrane potential (60, 61). The distribution of the potassium channel Kir 4.1 and K⁺ conductivity is similar to that of the DDC and AQP4 (62–64). This codistribution seen in astrocytes and retinal Müller cells may enable these cells to respond to the potassium uptake with water influx (46, 60).

3.1.5. Transmitter Release from Astrocytes

One of the principal functions of astrocytes is uptake of neurotransmitters released from nerve terminals. However, astrocytes can also release neuroactive agents, including transmitters, eicosanoids, steroids, neuropeptides, and growth factors (65).

Neuropeptides expressed at the mRNA or protein level in astrocytes include angiotensin (66–68), atrial natriuretic peptide (69–71), enkephalin (72), neuropeptide Y (73, 74), nociceptin (75), somatostatin (76, 77), substance P (78), and vasoactive intestinal peptide (79, 80). The regional distribution of these peptides in basal conditions and postinjury and their possible role in neuron–astrocyte interactions have been reviewed previously (81). For many of these agents, it is still uncertain whether the expressed transcript translates into a peptide, whether all these agents are expressed in vivo, and whether the target of the released neuropeptides are receptors on glia or neurons, or both.

3.2. Astrocytes and the Blood-Brain Barrier

The proximity of astrocytes to brain capillaries suggested that these cells may have a role in maintaining the barrier properties of cerebral endothelium. The first study to demonstrate the inductive influence of astrocytes and the neural microenvironment on barrier features in brain vessels was performed by grafting neural tissue in the coelomic cavity of different bird species (82). The newly formed vessels originating from the host displayed BBB characteristics. Immediately after isolation, cerebral microvessels and rodent and human brain endothelial cells lose their functional BBB properties as indicated by their low electrical resistance (83). This suggests that the milieu surrounding the vasculature determines the characteristics of the blood vessels. Brain endothelial cells cocultured with astrocytes display a significant increase in tight junction formation, together with an increase in enzymatic systems such as γ-glutamyl transpeptidase, $Na^+ K^+$ ATPase, alkaline phosphatase, and transporters for neutral amino acids (84–89). Up-regulation of low-density lipoprotein receptors and P-glycoprotein was also reported (90, 91). Endothelial cells cultured with astrocyte-conditioned media also demonstrate an increase in barrier features including tight junction formation and increased electrical resistance, decrease in permeability and expression of γ–GT, ATPase, HT7, and neurothelin, suggesting that soluble factors released by astrocytes are responsible for this effect (88, 92–95). It was further proposed that the basal lamina between capillary endothelial cells and astrocytic foot processes promotes interaction between the astroglia and endothelium, by increasing the local concentration of soluble factors secreted by astrocytes (89).

Over the years, several agents derived from astrocytes have been shown to increase the barrier properties in cultured endothelium. Treatment of cultured endothelial cells with src-suppressed C-kinase substrate-conditioned medium, a factor which stimulates the expression of angiopoietin-1 in astrocytes, results in an increase in tight junction proteins and decreased permeability to ^{3}H-sucrose (96). Various molecules released by astrocytes such as transforming growth factor β_1 (97), glial-derived neurotrophic factor (98), FGF-2 (99), and interleukin-6 (100) are able to induce some of the barrier properties in cultured brain endothelial cells. Human brain endothelial cells cultured in astrocyte-conditioned media show increased activity of protein kinase C, implicating a receptor-mediated action of an astrocyte-derived factor (101). Using angiotensinogen knockout mice with BBB disruption, it was shown that angiotensinogen production by reactive astrocytes was required to reinduce BBB function (102).

3.3. Astrocyte Signaling Mechanisms

3.3.1. Ca²⁺ Signaling

Astrocytes are electrically nonexcitable cells. Advances in Ca^{2+} imaging techniques led to the finding that astrocytes can communicate by Ca^{2+} signaling in two major ways. Firstly, signaling is expressed as repetitive monophasic oscillations in cytosolic Ca^{2+} concentrations ($[Ca^{2+}]_i$) limited to a single cell when activated by different transmitters, including glutamate, GABA, and ATP (adenosine 5'-triphosphate) (103, 104). They can also be evoked by changes in the extracellular environment including lowering of extracellular Ca^{2+}, hypo-osmotic conditions, local application of potassium, or mechanical stress (105). Astrocytic Ca^{2+} oscillations are known to involve activation of phospholipase C, IP_3 production, and release of Ca^{2+} from intracellular stores, rather than influx through membrane channels (106, 107).

The second type of Ca^{2+} signaling is in the form of propagating Ca^{2+} waves which can be stimulated by focal electric stimulation, mechanical stimulation, lowering extracellular Ca^{2+} levels, or by local application of transmitters such as glutamate or ATP (103). High-frequency neuronal spiking has been shown to induce astrocytic Ca^{2+} waves in organotypic slices and in anesthetized mice following sensory stimulation (108, 109). In general, Ca^{2+} waves propagate with a velocity of about 8–20 μm/s and engage as many as 50 neighboring astrocytes per wave (110). These signals are transmitted to other cell types in the brain including neurons, microglial cells, and oligodendrocytes (111, 112). Initially, these intracellular waves were thought to be propagated by diffusion of IP_3 or calcium through intercellular gap junctions (103). Pharmacological approaches demonstrate that ATP is the diffusible messenger (113) and that connexin hemi-channels are the most significant mechanism of ATP release from astrocytes (110). Wave propagation is mediated by P2Y receptor subtypes including P2Y1, P2Y2, and P2Y4 (114). ATP, once released, can be broken down by ecto-ATPase and ecto-5-nucleosidase into adenosine, which has a dilating effect on cerebral vessels during functional hyperemia (115) and also has presynaptic and postsynaptic effects. Presynaptically, adenosine A1 receptors inhibit Ca^{2+} channel opening resulting in inhibition of excitatory synaptic transmission (116). Postsynaptically, A1 receptors open K^+ channels resulting in hyperpolarization and decreased neuronal activity (117).

Both modalities of astrocytic signaling, Ca^{2+} oscillations and Ca^{2+} waves, are readily transmitted to surrounding neurons, which display prolonged increases in intracellular $[Ca^{2+}]$ (111, 118). Further details of astrocyte signaling can be obtained from reviews in the literature (41, 110, 119–121).

3.3.2. Astrocyte Signaling and Cerebral Blood Flow

Neural activity is known to increase cerebral blood flow (CBF) within seconds in the activated region, a process referred to as functional hyperemia (122–124). A variety of agents, including H^+, K^+, lactate, neurotransmitters, adenosine, arachidonic acid metabolites, and glutamate-induced activation of neuronal nitric oxide (NO) synthase are implicated in this increase in CBF (125, 126). The cellular source of these agents was not known and it was uncertain whether neurons were the source of the signals mediating vasodilatation or whether other cells were involved. The proximity of astrocytic processes with synapses and blood vessels suggested that astrocytes may have a role in CBF regulation. The first demonstration that astrocytic activity can influence vascular tone was observed in brain slices where electrical stimulation of neuronal processes increased the amount of intracellular Ca^{2+} in astrocytic end-feet leading to slow dilatation of local cerebral arterioles (127). Inhibitors of metabotropic glutamate (mGluR) receptors block the vascular response, while activation of these receptors by agonists increases the amount of intracellular Ca^{2+} and reproduces the vasodilatation observed by neuronal stimulation (127). Direct electrical and mechanical stimulation of individual astrocytes increases the amount of intracellular Ca^{2+} and induces vasodilatation (127). Several mediators are implicated in these vascular changes, including vasoactive metabolites of the cyclooxygenase (COX) or cytochrome P450 ω-hydroxylase pathways as reviewed previously (126).

The availability of two-photon laser scanning microscopy allowed the investigation of the relationships between neural activity, astrocytic Ca^{2+}, and CBF in vivo in anesthetized mice (128). These authors reported that increase in astrocytic Ca^{2+} by photolysis of caged Ca^{2+} caused vasodilatation of cortical arterioles in less than a second. An 18% increase in arterial cross-sectional area corresponding to an almost 40% increase in local perfusion was observed. Pharmacological studies indicated that the vascular responses evoked from astrocytes are mediated by metabolites of the COX-1 pathway. The nonselective COX inhibitor indomethacin and the selective COX-1 inhibitor SC-560 attenuated the vascular response, whereas inhibitors of NO synthase, COX-2, p450 hepoxygenases, or adenosine receptors had no effect. Stimulation of neural activity by electrical stimulation resulted in an increase in astrocytic Ca^{2+} and vasodilatation of adjacent arterioles, which was attenuated by COX-1 inhibition suggesting that astrocytes induce vasodilatation predominantly through COX-1 reaction products. This is supported by a recent study demonstrating that COX-1 is the primary mediator of astrocyte-induced vasodilatation (129).

These studies provide strong evidence that during neural activity, changes in intracellular Ca^{2+} in astrocytic end-feet modulate vascular tone in adjacent arterioles. This is an exciting development in the field and reinforces the concept of a neurovascular unit in which neurons, astrocytes, and endothelial cells work together to maintain homeostasis of the brain microenvironment.

Ca^{2+} increases in astrocytic end-feet activates soluble PLA_2, leading to the production of multiple vasoactive substances. PLA_2 generates diffusible arachidonic acid (AA) from the plasma membrane, which can be converted into a number of compounds, some of which induce vasodilatation while others induce vasoconstriction. Dilating products include PGE_2 from the action of COX enzymes (127, 128, 130) and several epoxyeicosatrienoic acids (EETs) (5,6-EET;8,9-EET; 11,12-EET and 14,15-EET) (131, 132) from the activity of a subtype of cytochrome P450 (CYP450) enzyme. Constricting molecules consist of PGF_2 (133) and thromboxane A_2 (134–136) from COX activity, endothelin peptide (137, 138), as well as 20-HETE (45, 139). The enzymes governing the production of these vasoactive products are sensitive to NO, suggesting that NO levels may dictate the direction of the vessel response (140).

Decrease in CBF occurs in vivo in response to norepinephrine (NE) (141), an effect that may help maintain CBF at a constant rate at higher blood pressures. In vitro work shows that NE triggers robust intracellular Ca^{2+} increases in astrocytes via activation of α_1 and β adrenergic receptors (142). This precedes prominent vasoconstriction (45). When astrocytes are loaded with BAPTA-AM to chelate rises in intracellular Ca^{2+}, NE-induced vascular constrictions are drastically reduced, suggesting Ca^{2+} is critical for the astrocyte-mediated effect. Thus, NE is another vasoactive transmitter affecting blood vessels besides glutamate.

Pial arteriolar dilatation is known to accompany increased neuronal activity in the cerebral cortex, despite the absence of direct neuronal connections from the cortex to these vessels. A recent study demonstrates that the vasodilating signal arising in the parenchyma is transmitted to pial arterioles via an astrocytic, rather than a vascular route (143). Selective injury to the glia limitans, but not endothelium, prevents neural activation-induced pial arteriolar dilatation, suggesting a key role for astrocytes in this process. This astrocytic signaling pathway is sensitive to connexin43 blockade, suggesting an important contribution from gap junctions and/or hemi-channels.

Certain findings question whether Ca^{2+} signaling is indeed a key factor in mediating functional hyperemia in vivo. Astrocyte Ca^{2+} increase often occurs independent of neuronal activity (144) and spontaneous Ca^{2+} oscillations have been reported both in vivo (145) and in vitro (146). Secondly, neuronal activity does not consistently increase astrocyte Ca^{2+} with a time course that follows the time of activation (147). Also, the astrocyte Ca^{2+} waves that are considered to be important for relaying neuronal information through the astrocyte syncytium towards vessels are not always observed in vivo, suggesting that this signaling process may actually be attributable to aspects of the slice or culture condition, or that it manifests more easily in patho-physiological conditions such as epilepsy (148, 149), rather than being the "normal" method of signaling for neurovascular coupling. Further studies are required to understand the significance of Ca^{2+} signals in the control of cerebral vessels.

In summary, astrocytes are capable of eliciting changes in vessel diameter in both directions. Additional details of the role of astrocytes in modulation of CBF can be obtained from reviews in the literature (126, 140, 150, 151).

3.3.3. Signaling in Relation to Cell Volume

Astrocytes undergo swelling in response to a decrease in extracellular osmolarity and this is followed by a corrective process leading to restoration of normal volume (152, 153). The latter is an active process resulting from extrusion of mainly K^+, Cl^- and organic molecules such as pyroles, and organic amines (153). Efflux of taurine, GABA, glutamate, and glycine has been documented in astrocyte cultures in response to hypo-osmotic stimuli. An increase in the cellular volume triggers swell-activated channels (153, 154). Volume-regulated anion channels (VRAC) tightly regulate cell volume homeostasis and act as release routes for transmitters including excitatory amino acids and ATP and chloride currents (155). VRACs contribute to neuronal damage via excitatory amino acid release in pathological conditions. For example, persistent swelling of astrocytes induced by lactacidosis appears to be secondary to inhibition of VRACs (156). For further details about ionic homeostasis in astrocytes, the reader is referred to reviews on this subject (119, 155, 157).

3.4. Astrocytes and Synaptic Transmission

3.4.1. The Tripartite Synapse

The presynaptic terminal and the postsynaptic neuron are well-known functionally important elements of the synapse. However, a third cellular component consisting of astrocytic processes is often associated with synapses in the cerebral (158, 159) (Fig. 3) and cerebellar cortex (160, 161), while synapses in the retina are contacted by Müller cells (astrocyte-like radial glia) (162). Thus, synapses should be considered a tripartite, rather than a bipartite, structure (163).

3.4.2. Neuron/Astrocyte Interactions

Astrocytes express a wide variety of neurotransmitter receptors including metabotropic glutamate receptors (164, 165), $GABA_B$ receptors (166), and muscarinic acetyl choline (Ach) receptors

(167). In cell culture systems and in brain slices, following electrical stimulation of neurons, a variety of neurotransmitters, including glutamate, GABA, adrenaline, ATP, serotonin, Ach, and several peptides, can activate astrocyte receptors leading to increases in intracellular [Ca^{2+}] concentrations (164, 168). This in turn initiates gliotransmission and the release of gliotransmitters (to distinguish them from neurotransmitters released from neurons), including glutamate, ATP, adenosine, D-serine, eicosanoids such as prostaglandin and 20-hydroxyeicosatetraenoic acid, cytokines such as tumor necrosis factor-α, and proteins and peptides such as acetylcholine-binding protein and atrial natriuretic peptide (121, 169). The most significant of these agents is glutamate which plays a central bidirectional role in astrocyte–neuronal interactions. Release of glutamate from astrocytes in response to increases in [Ca^{2+}) has been demonstrated in both culture (118, 170–172) and brain slice preparations (165, 166, 173, 174). Glutamate release from astrocytes may occur by numerous mechanisms including exocytosis (175–177), VRAC (178), hemi-channels (179), purinergic P2X receptors (180), and pannexins (181). The state of the brain dictates which mechanism is utilized to release transmitters

3.4.3. Synaptic Modulation

In vitro studies demonstrate that activated astrocytes can regulate synaptic transmission by release of glutamate and ATP. Glutamate can have presynaptic effects that are mediated either by metabotropic glutamate receptors (182) or by kainate receptors that induce an enhancement of transmitter release (183). ATP can act through postsynaptic P2X receptors to induce an elevation of postsynaptic Ca^{2+} level, which is thought to drive α-amino-3-hydroxy-5-methyl-4-isoxazolepropionic acid (AMPA) receptors to mediate an increase of synaptic transmission (184). Extracellular hydrolysis of released ATP can cause reduction in synaptic transmission that is mediated by presynaptic adenosine A1 receptors (116, 185, 186) and modify neuronal excitability by activating K+ conductance, which hyperpolarizes the neuronal membrane potential (187). Experiments in brain slices of the hippocampus (165, 173) and thalamus (188) demonstrate that application of glutamate agonists and PGE$_2$ in addition to neuronal stimulation evokes a Ca^{2+}-dependent glutamate release from astrocytes which activates neighboring neurons and results in increases in neuronal Ca^{2+} levels (162). This astrocytic modulation of neurotransmitter release from the presynaptic terminal and stimulation of postsynaptic neurons constitutes direct modulation of synaptic regulation.

Astrocytes can also regulate synaptic transmission by "indirect" mechanisms, one of which is the uptake of glutamate via the high-affinity glutamate transporters, EAAT1 and EAAT2 subtypes, which are enriched in astrocytic processes (189, 190). Transmission at glutamatergic synapses is terminated by removal of glutamate from the synaptic cleft. Brain astrocytes and Müller cells in the retina account for the bulk of the glutamate uptake at synapses (36, 191, 192). Within the astrocyte, glutamate is converted to glutamine through an ATP-requiring reaction catalyzed by the astrocyte-specific enzyme glutamine synthetase (193–195). Glutamine is subsequently released to the extracellular space for uptake by neurons and recycled into glutamate for glutamatergic neurotransmission.

Astrocytes can also modulate synaptic transmission by releasing chemical cofactors. The best documented example of such modulation is activation of the NMDA receptor, which requires the presence of glutamate as well as the cofactor D-serine that binds to the glycine-binding site of the receptor (196, 197). Serine racemase is an enzyme highly expressed in astrocytes and is responsible for conversion of L- to D-serine (196, 198). Brain astrocytes and Müller cells are the sole source of D-serine in the CNS and most likely modulate synaptic transmission at the NMDA synapses by releasing D-serine (198, 199).

A third indirect mechanism of modulation is via glial regulation of extracellular ion levels. Neuronal activity leads to substantial variations in the concentrations of K^+ and H^+ in the extracellular space (200–202). These variations can alter synaptic transmission since increases in K^+ levels depolarize synaptic terminals (203), while H^+ blocks presynaptic Ca^{2+} channels (204, 205) and NMDA receptors (206). K^+ and H^+ taken up by astrocytes are dissipated through many cells via gap junction coupling, thus regulating the extracellular concentrations of these ions (207, 208).

Astrocytes can also modulate synaptic transmission by directly controlling synaptogenesis as discussed in the next section.

In summary, release of neurotransmitter from the presynaptic terminal not only stimulates the postsynaptic neuron, but also activates the perisynaptic astrocytic processes. The activated astrocyte, in turn, releases gliotransmitters that can directly stimulate the postsynaptic neuron and can feed back onto the presynaptic terminal either to enhance or to depress further release of neurotransmitter. Thus, the perisynaptic astrocyte is an active partner in synaptic transmission. Further details of astrocytic modulation of synaptic transmission can be obtained from reviews in the literature (121, 162, 169, 209–211).

3.5. Astrocytes and CNS Synaptogenesis

Astrocytes are known to enhance the formation of functional synapses in the CNS. Retinal ganglion cells (RGCs) cultured in the absence of astrocytes, even after many weeks in culture, exhibit very little spontaneous synaptic activity when excitatory postsynaptic currents are measured by patch clamp (212). However, when RGCs are cultured in the presence of a feeder layer of astrocytes or in astrocyte-conditioned medium, they exhibit high levels of synaptic activity (212). Two subsequent studies show that a glial factor (or factors) enhances the number of synapses between RGCs sevenfold without changing neuronal survival or neurite growth (213, 214). Astrocytes are also required for synapse stability and maintenance. Coculture of purified RGCs with a feeder layer of astrocytes cultured on a removable insert results in synapse formation in 1 week. When the astrocyte insert is removed and the neurons are examined after 1 week by immunostaining and patch-clamp recording, the majority of synapses are no longer present (214).

The synaptogenic factor contained in the glia-conditioned medium was identified as cholesterol (215, 216). Other factors derived from astrocytes which are important for synapse development include tumor necrosis factor-α (217) and the thrombospondins (218). Astrocytes may also play a role in the sculpting of synaptic structure and function within the developing or even adult brain as reviewed previously (219, 220).

3.6. Astrocytes as Neural Progenitor Cells

Several studies have proposed that astrocytes might not only regulate neurogenesis but also themselves be neuronal progenitor cells (221–223). Studies show that radial glia in development and specific subpopulations of astrocytes located in the subventricular zone (SVZ) (224), along the walls of the lateral ventricles and the subgranular zone within the dentate gyrus of the hippocampus (225) of adults mammals, function as primary progenitors or neural stem cells (NSC) that give rise to differentiated neurons and glial cells during development and in the postnatal brain (226). Antimitotic ablation of the SVZ kills off the neuroblasts, but the SVZ-astrocytes survive and can repopulate the entire SVZ including the neural precursors (227). This is a novel finding which contradicts the classic teaching that neurons and glia are derived from distinct pools of progenitor cells. During development and in the adult brain, neural progenitors are capable of giving rise to transit-amplifying, or intermediate, progenitor cells (IPCs) which can, in the subventicular zone, divide rapidly to expand the available pool of neural precursors (228). Within NSCs and IPCs, genetic programs unfold for generating the extraordinary diversity of cell types in the CNS. The timing in development and location of NSCs, a property tightly linked to their neuroepithelial origin, appear to be key determinants of the types of neurons generated. Further discussion of the role of astrocytes as neural progenitor cells can be obtained from published reviews (226, 228, 229).

4. Astrocytes in Brain Diseases

The occurrence of reactive astrocytosis (Fig. 3.1g) and up-regulation of several molecules including GFAP, S100β, iNOS, and NFκB in response to neuronal injury in various brain pathologies are well known. It is now recognized that astrocyte excitation and gliotransmission are affected in various pathological conditions and astrocyte dysfunction plays a major role in the pathogenesis of many diseases (Table 1). A discussion of all these conditions is beyond the scope of this chapter and readers are referred to reviews in the literature (Table 1). Brain edema is common to many neurological disorders and a brief review of the role of astrocytes in brain edema follows.

4.1. Brain Edema

Brain edema is defined as an increase in brain volume resulting from a localized or diffuse abnormal accumulation of fluid within the brain parenchyma. The realization that brain edema is associated with either extra- or intracellular accumulation of abnormal fluid led to its classification into vasogenic and cytotoxic edema (230, 231). Vasogenic edema is associated with dysfunction of the BBB, which allows increased passage of plasma proteins and water into the extracellular compartment and occurs following brain hemorrhage, infections, seizures, trauma, tumors, radiation injury, hypertensive encephalopathy, and the late stages of cerebral ischemia. Cytotoxic edema results from abnormal water uptake

Table 1
Diseases in which astrocyte dysfunction plays a role are listed

Brain diseases

AIDS-related neuropathology	(121)
Degenerative diseases	(121, 251–253)
Alzheimer's disease	(254, 255)
Amyotrophic lateral sclerosis	(209, 256, 257)
Huntington's disease	(258)
Parkinsonism	(169, 253)
Epilepsy	(169, 257, 259–261)
Hepatic encephalopathy	(262)
Leukodystrophy	(253, 263, 264)
Alexander's disease	
Metabolic disease	(252)
Niemann-Pick disease type C	
Tumors	(121, 265–267)
Gliomas (primary brain tumors)	
Schizophrenia	(169)
Traumatic brain injury	(251)
Vascular disease	(110, 119, 155, 251, 257, 268)
Ischemic stroke	

by injured brain cells and is most commonly encountered in cerebral ischemia. The third type of edema is hydrocephalic or interstitial edema in which a rise in the intraventricular pressure causes CSF to migrate through the ependyma into the periventricular white matter, thus increasing the extracellular fluid volume (232, 233). Further information about brain edema can be obtained from reviews in the literature (234, 235).

Morphological studies of the white matter in areas of vasogenic edema show an astrocytosis and increased expression of GFAP mRNA occurs as early as 6 h after a cortical cold injury (236). Astrocytes have a role in the clearance of edema fluid since they are known to phagocytose and digest the extravasated serum proteins in the extracellular space (237). In cytotoxic edema, cytoplasmic swelling of astrocytes is noted in the affected area. Astrocytes are known to swell up to five times their normal size and since astrocytes outnumber neurons 20:1 in humans, glial swelling is the main finding in cytotoxic edema (238).

More recently, the discovery of the aquaporins has provided insight into the pathogenesis and resolution of brain edema. AQP4, a key component of astrocytic foot processes at the brain–fluid interfaces (Subheading 3.1.3) has a key role in brain water balance as evident from studies of AQP4-null mice (239–244).

Data derived from AQP4-null mice suggest that AQP4 is involved in the clearance of extracellular fluid from the brain parenchyma in vasogenic edema. Models in which vasogenic edema is the predominant form of edema, including the cortical cold injury, tumor implantation, and brain abscess models, demonstrate that the AQP4-null mice have a significantly greater increase in brain water content and ICP than the wild-type mice, suggesting that brain water elimination is defective after AQP4 deletion (240, 245).

Swelling of astrocytic foot processes is a major finding in cytotoxic edema, and since AQP4 channels are located in the astrocytic foot processes, it was hypothesized that they may have a role in formation of cell swelling. This was found to be the case since AQP4-null mice subjected to ischemic stroke (246) and bacterial meningitis (247) show decreased cerebral edema and improved outcome. As discussed previously, AQP4 protein in the plasma membranes is bound to an aggregate of intracellular proteins including α-syntrophin (54). Reduced brain swelling after cerebral ischemia and water intoxication is also observed in α-syntrophin-null mice which have reduced AQP4 expression in astrocyte foot processes (248).

Obstructive hydrocephalus produced by injecting kaolin in the cistern magna of AQP4-null mice shows accelerated ventricular enlargement compared with wild-type mice (249). Reduced water permeability of the ependymal layer, subependymal astrocytes, astrocytic foot processes, and glia limitans produced by

AQP4 deletion reduces the elimination rate of CSF across these routes. Thus, AQP4 induction can be evaluated as a nonsurgical treatment for hydrocephalus.

In summary, AQP4 has opposing roles in the pathogenesis of vasogenic and hydrocephalic edema when compared to cytotoxic edema. Therefore, AQP4 activators or upregulators have the potential to facilitate the clearance of vasogenic and hydrocephalic edema, while AQP4 inhibitors have the potential to protect the brain in cytotoxic edema. This is an area of ongoing research since none of the AQP4 activators or inhibitors investigated thus far are suitable for development for clinical use (241). However, drugs designed to treat a specific type of edema may have limited clinical value since in most clinical situations there is a combination of different types of edema depending on the time course of the primary disease.

5. Conclusion

In conclusion, in vitro studies in the last 2 decades have provided novel information about the biology of astrocytes using cultured cells and brain slices. Many astrocyte functions need to be confirmed in vivo and the effect of astrocyte dysfunction in brain pathology is just starting to emerge. Clearly, much work remains to be done in understanding the function of astrocytes in health and disease; however, the astrocyte has evolved from its role as a supportive cell in the brain, a role which it held for approximately 150 years, to being an essential partner of neurons in the control of brain function. As mentioned very aptly, astrocytes are "stars at last" (250).

Acknowledgments

Thanks are expressed to Drs. Cynthia Hawkins and Dittkavi S.R. Sarma for reviewing this manuscript and for their helpful suggestions.

References

1. Parpura V, Haydon PG (2009) Astrocytes in (Patho) Physiology of the Nervous System. Springer Science + Business Media, New York

2. Virchow R (1860) Cellular Pathology. Churchill, London

3. Cajal SR (1913) Sobre un neuvo proceder de impregnacion de la neuriglia y sus resultados en los centros nerviosos del hombre y animales. Trab lab Invest Biol Univ Madr 11:219–237

4. Weigert F (1895) Beitrage zur Kermtnis der normalen menschlichen Neuroglia Weisbrod, Frankfurt am Main

5. Farquhar MG, Palade GE (1963) Junctional complexes in various epithelia. J Cell Biol 17:375–412

6. Brightman MW, Reese TS (1969) Junctions between intimately apposed cell membranes in the vertebrate brain. J Cell Biol 40:648–677

7. Rouach N, Avignone E, Meme W, Koulakoff A, Venance L, Blomstrand F, Giaume C (2002) Gap junctions and connexin expression in the normal and pathological central nervous system. Biol Cell 94:457–475

8. Wolff JR, Stuke K, Missler M, Tytko H, Schwarz P, Rohlmann A, Chao TI (1998) Autocellular coupling by gap junctions in cultured astrocytes: a new view on cellular autoregulation during process formation. Glia 24:121–140

9. Rohlmann, A, Wolff, J (1996) Subcellular Topography and Plasticity of Gap Junction Distribution in Astrocytes. RG Landes, Austin

10. Nagy JI, Patel D, Ochalski PA, Stelmack GL (1999) Connexin30 in rodent, cat and human brain: selective expression in gray matter astrocytes, co-localization with connexin43 at gap junctions and late developmental appearance. Neuroscience 88: 447–468

11. Penes MC, Li X, Nagy JI (2005) Expression of zonula occludens-1 (ZO-1) and the transcription factor ZO-1-associated nucleic acid-binding protein (ZONAB)-MsY3 in glial cells and colocalization at oligodendrocyte and astrocyte gap junctions in mouse brain. Eur J Neurosci 22:404–418

12. Bushong EA, Martone ME, Jones YZ, Ellisman MH (2002) Protoplasmic astrocytes in CA1 stratum radiatum occupy separate anatomical domains. J Neurosci 22: 183–192

13. Bushong EA, Martone ME, Ellisman MH (2004) Maturation of astrocyte morphology and the establishment of astrocyte domains during postnatal hippocampal development. Int J Dev Neurosci 22:73–86

14. Ogata K, Kosaka T (2002) Structural and quantitative analysis of astrocytes in the mouse hippocampus. Neuroscience 113:221–233

15. Lovatt D, Sonnewald U, Waagepetersen HS, Schousboe A, He W, Lin JH, Han X, Takano T, Wang S, Sim FJ, Goldman SA, Nedergaard M (2007) The transcriptome and metabolic gene signature of protoplasmic astrocytes in the adult murine cortex. J Neurosci 27:12255–12266

16. Halassa MM, Fellin T, Takano H, Dong JH, Haydon PG (2007) Synaptic islands defined by the territory of a single astrocyte. J Neurosci 27:6473–6477

17. Oberheim NA, Wang X, Goldman S, Nedergaard M (2006) Astrocytic complexity distinguishes the human brain. Trends Neurosci 29:547–553

18. Chan-Ling T, Stone J (1991) Factors determining the migration of astrocytes into the developing retina: migration does not depend on intact axons or patent vessels. J Comp Neurol 303:375–386

19. Nedergaard M, Ransom B, Goldman SA (2003) New roles for astrocytes: redefining the functional architecture of the brain. Trends Neurosci 26:523–530

20. Bass NH, Hess HH, Pope A, Thalheimer C (1971) Quantitative cytoarchitectonic distribution of neurons, glia, and DNa in rat cerebral cortex. J Comp Neurol 143:481–490

21. Leuba G, Garey LJ (1989) Comparison of neuronal and glial numerical density in primary and secondary visual cortex of man. Exp Brain Res 77:31–38

22. Cajal, RSY (1995) Histology of the Nervous System of Man and Vertebrates. Oxford University Press, New York

23. Defelipe J, Alonso-Nanclares L, Arellano JI (2002) Microstructure of the neocortex: comparative aspects. J Neurocytol 31:299–316

24. Andriezen L (1893) The neuroglia elements in the human brain. BMJ 29:227–230

25. Retzius G (1894) Die neuroglia des Gehirns beim Menschen und bei Saeugethieren. Biol Untersuchungen 6:1–28

26. Colombo JA, Reisin HD (2004) Interlaminar astroglia of the cerebral cortex: a marker of the primate brain. Brain Res 1006:126–131

27. Colombo JA, Gayol S, Yanez A, Marco P (1997) Immunocytochemical and electron microscope observations on astroglial interlaminar processes in the primate neocortex. J Neurosci Res 48:352–357

28. Butt AM, Colquhoun K, Berry M (1994) Confocal imaging of glial cells in the intact rat optic nerve. Glia 10:315–322

29. Cahoy JD, Emery B, Kaushal A, Foo LC, Zamanian JL, Christopherson KS, Xing Y, Lubischer JL, Krieg PA, Krupenko SA, Thompson WJ, Barres BA (2008) A transcriptome database for astrocytes, neurons, and oligodendrocytes: a new resource for understanding brain development and function. J Neurosci 28:264–278

30. Ludwin SK, Kosek JC, Eng LF (1976) The topographical distribution of S-100 and GFA proteins in the adult rat brain: an immunohistochemical study using horseradish peroxidase-labelled antibodies. J Comp Neurol 165:197–207

31. Dermietzel R, Hertberg EL, Kessler JA, Spray DC (1991) Gap junctions between

cultured astrocytes: immunocytochemical, molecular, and electrophysiological analysis. J Neurosci 11:1421–1432

32. Kuzis K, Reed S, Cherry NJ, Woodward WR, Eckenstein FP (1995) Developmental time course of acidic and basic fibroblast growth factors' expression in distinct cellular populations of the rat central nervous system. J Comp Neurol 358:142–153

33. Yazaki N, Hosoi Y, Kawabata K, Miyake A, Minami M, Satoh M, Ohta M, Kawasaki T, Itoh N (1994) Differential expression patterns of mRNAs for members of the fibroblast growth factor receptor family, FGFR-1-FGFR-4, in rat brain. J Neurosci Res 37:445–452

34. Neymeyer V, Tephly TR, Miller MW (1997) Folate and 10-formyltetrahydrofolate dehydrogenase (FDH) expression in the central nervous system of the mature rat. Brain Res 766:195–204

35. Rauen T, Taylor WR, Kuhlbrodt K, Wiessner M (1998) High-affinity glutamate transporters in the rat retina: a major role of the glial glutamate transporter GLAST-1 in transmitter clearance. Cell Tissue Res 291:19–31

36. Rothstein JD, Dykes-Hoberg M, Pardo CA, Bristol LA, Jin L, Kuncl RW, Kanai Y, Hediger MA, Wang Y, Schielke JP, Welty DF (1996) Knockout of glutamate transporters reveals a major role for astroglial transport in excitotoxicity and clearance of glutamate. Neuron 16:675–686

37. Berger UV, Tsukaguchi H, Hediger MA (1998) Distribution of mRNA for the facilitated urea transporter UT3 in the rat nervous system. Anat Embryol (Berl) 197:405–414

38. Sinclair CJ, LaRiviere CG, Young JD, Cass CE, Baldwin SA, Parkinson FE (2000) Purine uptake and release in rat C6 glioma cells: nucleoside transport and purine metabolism under ATP-depleting conditions. J Neurochem 75:1528–1538

39. Simpson PB, Holtzclaw LA, Langley DB, Russell JT (1998) Characterization of ryanodine receptors in oligodendrocytes, type 2 astrocytes, and O-2A progenitors. J Neurosci Res 52:468–482

40. Ronaldson PT, Persidsky Y, Bendayan R (2008) Regulation of ABC membrane transporters in glial cells: relevance to the pharmacotherapy of brain HIV-1 infection. Glia 56:1711–1735

41. Simard M, Arcuino G, Takano T, Liu QS, Nedergaard M (2003) Signaling at the gliovascular interface. J Neurosci 23:9254–9262

42. Kacem K, Lacombe P, Seylaz J, Bonvento G (1998) Structural organization of the perivascular astrocyte endfeet and their relationship with the endothelial glucose transporter: a confocal microscopy study. Glia 23:1–10

43. Pekny M, Pekna M (2004) Astrocyte intermediate filaments in CNS pathologies and regeneration. J Pathol 204:428–437

44. Schinkel AH, Smit JJ, van Tellingen O, Beijnen JH, Wagenaar E, van Deemter L, Mol CA, van der Valk MA, Robanus-Maandag EC, te Riele HP (1994) Disruption of the mouse mdr1a P-glycoprotein gene leads to a deficiency in the blood-brain barrier and to increased sensitivity to drugs. Cell 77:491–502

45. Mulligan SJ, MacVicar BA (2004) Calcium transients in astrocyte endfeet cause cerebrovascular constrictions. Nature 431:195–199

46. Nielsen S, Nagelhus EA, miry-Moghaddam M, Bourque C, Agre P, Ottersen OP (1997) Specialized membrane domains for water transport in glial cells: high-resolution immunogold cytochemistry of aquaporin-4 in rat brain. J Neurosci 17:171–180

47. Rash JE, Yasumura T, Hudson CS, Agre P, Nielsen S (1998) Direct immunogold labeling of aquaporin-4 in square arrays of astrocyte and ependymocyte plasma membranes in rat brain and spinal cord. Proc Natl Acad Sci U S A 95:11981–11986

48. Dermietzel R, Leibstein AG (1978) The microvascular pattern and perivascular linings of the area postrema. A combined freeze-etching and ultrathin section study. Cell Tissue Res 186:97–110

49. Neuhaus J (1990) Orthogonal arrays of particles in astroglial cells: quantitative analysis of their density, size, and correlation with intramembranous particles. Glia 3:241–251

50. Wolburg H (1995) Orthogonal arrays of intramembranous particles: a review with special reference to astrocytes. J Hirnforsch 36:239–258

51. Furman CS, Gorelick-Feldman DA, Davidson KG, Yasumura T, Neely JD, Agre P, Rash JE (2003) Aquaporin-4 square array assembly: opposing actions of M1 and M23 isoforms. Proc Natl Acad Sci U S A 100: 13609–13614

52. Rash JE, Davidson KG, Yasumura T, Furman CS (2004) Freeze-fracture and immunogold analysis of aquaporin-4 (AQP4) square arrays, with models of AQP4 lattice assembly. Neuroscience 129:915–934

53. Nico B, Frigeri A, Nicchia GP, Quondamatteo F, Herken R, Errede M, Ribatti D, Svelto M, Roncali L (2001) Role of aquaporin-4 water channel in the development and integrity of

the blood-brain barrier. J Cell Sci 114: 1297–1307

54. Amiry-Moghaddam M, Frydenlund DS, Ottersen OP (2004) Anchoring of aquaporin-4 in brain: molecular mechanisms and implications for the physiology and pathophysiology of water transport. Neuroscience 129:999–1010

55. Nicchia GP, Cogotzi L, Rossi A, Basco D, Brancaccio A, Svelto M, Frigeri A (2008) Expression of multiple AQP4 pools in the plasma membrane and their association with the dystrophin complex. J Neurochem 105(6):2156–2165

56. Warth A, Kroger S, Wolburg H (2004) Redistribution of aquaporin-4 in human glioblastoma correlates with loss of agrin immunoreactivity from brain capillary basal laminae. Acta Neuropathol 107:311–318

57. Rascher G, Fischmann A, Kroger S, Duffner F, Grote EH, Wolburg H (2002) Extracellular matrix and the blood-brain barrier in glioblastoma multiforme: spatial segregation of tenascin and agrin. Acta Neuropathol 104:85–91

58. Wolburg-Buchholz K, Mack AF, Steiner E, Pfeiffer F, Engelhardt B, Wolburg H (2009) Loss of astrocyte polarity marks blood-brain barrier impairment during experimental autoimmune encephalomyelitis. Acta Neuropathol 118:219–233

59. Schroder W, Seifert G, Huttmann K, Hinterkeuser S, Steinhauser C (2002) AMPA receptor-mediated modulation of inward rectifier K+ channels in astrocytes of mouse hippocampus. Mol Cell Neurosci 19: 447–458

60. Kofuji P, Ceelen P, Zahs KR, Surbeck LW, Lester HA, Newman EA (2000) Genetic inactivation of an inwardly rectifying potassium channel (Kir4.1 subunit) in mice: phenotypic impact in retina. J Neurosci 20: 5733–5740

61. Olsen ML, Campbell SL, Sontheimer H (2007) Differential distribution of Kir4.1 in spinal cord astrocytes suggests regional differences in K+ homeostasis. J Neurophysiol 98:786–793

62. Amiry-Moghaddam M, Ottersen OP (2003) The molecular basis of water transport in the brain. Nat Rev Neurosci 4:991–1001

63. Blake DJ, Kroger S (2000) The neurobiology of duchenne muscular dystrophy: learning lessons from muscle? Trends Neurosci 23:92–99

64. Wolburg H, Noell S, Mack A, Wolburg-Buchholz K, Fallier-Becker P (2009) Brain endothelial cells and the glio-vascular complex. Cell Tissue Res 335:75–96

65. Martin DL (1992) Synthesis and release of neuroactive substances by glial cells. Glia 5:81–94

66. Bunnemann B, Fuxe K, Metzger R, Bjelke B, Ganten D (1992) The semi-quantitative distribution and cellular localization of angiotensinogen mRNA in the rat brain. J Chem Neuroanat 5:245–262

67. Intebi AD, Flaxman MS, Ganong WF, Deschepper CF (1990) Angiotensinogen production by rat astroglial cells in vitro and in vivo. Neuroscience 34:545–554

68. Stornetta RL, Hawelu-Johnson CL, Guyenet PG, Lynch KR (1988) Astrocytes synthesize angiotensinogen in brain. Science 242: 1444–1446

69. McKenzie JC (1992) Atrial natriuretic peptide-like immunoreactivity in astrocytes of parenchyma and glia limitans of the canine brain. J Histochem Cytochem 40:1211–1222

70. McKenzie JC, Berman NE, Thomas CR, Young JK, Compton LY, Cothran LN, Liu WL, Klein RM (1994) Atrial natriuretic peptide-like (ANP-LIR) and ANP prohormone immunoreactive astrocytes and neurons of human cerebral cortex. Glia 12: 228–243

71. McKenzie JC, Juan YW, Thomas CR, Berman NE, Klein RM (2001) Atrial natriuretic peptide-like immunoreactivity in neurons and astrocytes of human cerebellum and inferior olivary complex. J Histochem Cytochem 49:1453–1467

72. Vilijn MH, Vaysse PJ, Zukin RS, Kessler JA (1988) Expression of preproenkephalin mRNA by cultured astrocytes and neurons. Proc Natl Acad Sci U S A 85:6551–6555

73. Barnea A, guila-Mansilla N, Bigio EH, Worby C, Roberts J (1998) Evidence for regulated expression of neuropeptide Y gene by rat and human cultured astrocytes. Regul Pept 75–76:293–300

74. Barnea A, Roberts J, Keller P, Word RA (2001) Interleukin-1beta induces expression of neuropeptide Y in primary astrocyte cultures in a cytokine-specific manner: induction in human but not rat astrocytes. Brain Res 896:137–145

75. Buzas B (2002) Regulation of nociceptin/orphanin FQ gene expression in astrocytes by ceramide. Neuroreport 13:1707–1710

76. Shinoda H, Marini AM, Cosi C, Schwartz JP (1989) Brain region and gene specificity of neuropeptide gene expression in cultured astrocytes. Science 245:415–417

77. Shinoda H, Marini AM, Schwartz JP (1992) Developmental expression of the proenkephalin and prosomatostatin genes in cultured

cortical and cerebellar astrocytes. Brain Res Dev Brain Res 67:205–210

78. Too HP, Marriott DR, Wilkin GP (1994) Preprotachykinin-A and substance P receptor (NK1) gene expression in rat astrocytes in vitro. Neurosci Lett 182:185–187

79. Paspalas CD, Halasy K, Gerics B, Papadopoulos GC, Hajos F (2001) Vasoactive intestinal polypeptide in neuroglia? Immunoelectron microscopic localization in astrocytes of the rat mesencephalon. Glia 34:229–233

80. Virgintino D, Benagiano V, Maiorano E, Rizzi A, Errede M, Bertossi M, Roncali L, Ambrosi G (1996) Vasoactive intestinal polypeptide-like immunoreactivity in astrocytes of the human brain. Neuroreport 7: 1577–1581

81. Ubink R, Calza L, Hokfelt T (2003) 'Neuro'-peptides in glia: focus on NPY and galanin. Trends Neurosci 26:604–609

82. Stewart PA, Wiley MJ (1981) Developing nervous tissue induces formation of blood-brain barrier characteristics in invading endothelial cells: a study using quail--chick transplantation chimeras. Dev Biol 84: 183–192

83. Rubin LL, Barbu K, Bard F, Cannon C, Hall DE, Horner H, Janatpour M, Liaw C, Manning K, Morales J (1991) Differentiation of brain endothelial cells in cell culture. Ann N Y Acad Sci 633:420–425

84. Beck DW, Vinters HV, Hart MN, Cancilla PA (1984) Glial cells influence polarity of the blood-brain barrier. J Neuropathol Exp Neurol 43:219–224

85. Cancilla PA, DeBault LE (1983) Neutral amino acid transport properties of cerebral endothelial cells in vitro. J Neuropathol Exp Neurol 42:191–199

86. Meresse S, Dehouck MP, Delorme P, Bensaid M, Tauber JP, Delbart C, Fruchart JC, Cecchelli R (1989) Bovine brain endothelial cells express tight junctions and monoamine oxidase activity in long-term culture. J Neurochem 53:1363–1371

87. Raub TJ, Kuentzel SL, Sawada GA (1992) Permeability of bovine brain microvessel endothelial cells in vitro: barrier tightening by a factor released from astroglioma cells. Exp Cell Res 199:330–340

88. Rubin LL, Hall DE, Porter S, Barbu K, Cannon C, Horner HC, Janatpour M, Liaw CW, Manning K, Morales J (1991) A cell culture model of the blood-brain barrier. J Cell Biol 115:1725–1735

89. Tao-Cheng JH, Nagy Z, Brightman MW (1987) Tight junctions of brain endothelium in vitro are enhanced by astroglia. J Neurosci 7:3293–3299

90. Dehouck B, Dehouck MP, Fruchart JC, Cecchelli R (1994) Upregulation of the low density lipoprotein receptor at the blood-brain barrier: intercommunications between brain capillary endothelial cells and astrocytes. J Cell Biol 126:465–473

91. Gaillard PJ, Voorwinden LH, Nielsen JL, Ivanov A, Atsumi R, Engman H, Ringbom C, De Boer AG, Breimer DD (2001) Establishment and functional characterization of an in vitro model of the blood-brain barrier, comprising a co-culture of brain capillary endothelial cells and astrocytes. Eur J Pharm Sci 12:215–222

92. Arthur FE, Shivers RR, Bowman PD (1987) Astrocyte-mediated induction of tight junctions in brain capillary endothelium: an efficient in vitro model. Brain Res 433:155–159

93. Lobrinus JA, Juillerat-Jeanneret L, Darekar P, Schlosshauer B, Janzer RC (1992) Induction of the blood-brain barrier specific HT7 and neurothelin epitopes in endothelial cells of the chick chorioallantoic vessels by a soluble factor derived from astrocytes. Brain Res Dev Brain Res 70:207–211

94. Maxwell K, Berliner JA, Cancilla PA (1987) Induction of gamma-glutamyl transpeptidase in cultured cerebral endothelial cells by a product released by astrocytes. Brain Res 410:309–314

95. Prat A, Biernacki K, Wosik K, Antel JP (2001) Glial cell influence on the human blood-brain barrier. Glia 36:145–155

96. Lee SW, Kim WJ, Choi YK, Song HS, Son MJ, Gelman IH, Kim YJ, Kim KW (2003) SSeCKS regulates angiogenesis and tight junction formation in blood-brain barrier. Nat Med 9:900–906

97. Tran ND, Correale J, Schreiber SS, Fisher M (1999) Transforming growth factor-beta mediates astrocyte-specific regulation of brain endothelial anticoagulant factors. Stroke 30:1671–1678

98. Igarashi Y, Utsumi H, Chiba H, Yamada-Sasamori Y, Tobioka H, Kamimura Y, Furuuchi K, Kokai Y, Nakagawa T, Mori M, Sawada N (1999) Glial cell line-derived neurotrophic factor induces barrier function of endothelial cells forming the blood-brain barrier. Biochem Biophys Res Commun 261: 108–112

99. Sobue K, Yamamoto N, Yoneda K, Hodgson M E, Yamashiro K, Tsuruoka N, Tsuda T, Katsuya H, Miura Y, Asai K, Kato T (1999) Induction of blood-brain barrier properties

in immortalized bovine brain endothelial cells by astrocytic factors. Neurosci Res:35:155–164

100. Sun D, Lytle C, O'Donnell ME (1997) IL-6 secreted by astroglial cells regulates Na-K-Cl cotransport in brain microvessel endothelial cells. Am J Physiol 272:C1829–C1835

101. Stanimirovic DB, Ball R, Durkin JP (1995) Evidence for the role of protein kinase C in astrocyte-induced proliferation of rat cere-bromicrovascular endothelial cells. Neurosci Lett 197:219–222

102. Kakinuma Y, Hama H, Sugiyama F, Yagami K, Goto K, Murakami K, Fukamizu A (1998) Impaired blood-brain barrier function in angiotensinogen-deficient mice. Nat Med 4:1078–1080

103. Charles AC, Merrill JE, Dirksen ER, Sanderson MJ (1991) Intercellular signaling in glial cells: calcium waves and oscillations in response to mechanical stimulation and glu-tamate. Neuron 6:983–992

104. Cornell-Bell AH, Finkbeiner SM, Cooper MS, Smith SJ (1990) Glutamate induces cal-cium waves in cultured astrocytes: long-range glial signaling. Science 247:470–473

105. Cotrina ML, Lin JH, Nedergaard M (1998) Cytoskeletal assembly and ATP release regu-late astrocytic calcium signaling. J Neurosci 18:8794–8804

106. Berridge MJ, Lipp P, Bootman MD (2000) The versatility and universality of calcium signalling. Nat Rev Mol Cell Biol 1:11–21

107. Carmignoto G, Pasti L, Pozzan T (1998) On the role of voltage-dependent calcium chan-nels in calcium signaling of astrocytes in situ. J Neurosci 18:4637–4645

108. Dani JW, Chernjavsky A, Smith SJ (1992) Neuronal activity triggers calcium waves in hippocampal astrocyte networks. Neuron 8:429–440

109. Wang X, Lou N, Xu Q, Tian GF, Peng WG, Han X, Kang J, Takano T, Nedergaard M (2006) Astrocytic Ca2+ signaling evoked by sensory stimulation in vivo. Nat Neurosci 9:816–823

110. Takano T, Oberheim N, Cotrina ML, Nedergaard M (2009) Astrocytes and isch-emic injury. Stroke 40:S8–12

111. Nedergaard M (1994) Direct signaling from astrocytes to neurons in cultures of mamma-lian brain cells. Science 263:1768–1771

112. Smith SJ (1994) Neural signalling. Neuromodulatory astrocytes. Curr Biol 4:807–810

113. Cotrina ML, Lin JH, Alves-Rodrigues A, Liu S, Li J, zmi-Ghadimi H, Kang J, Naus CC, Nedergaard M (1998) Connexins regulate

calcium signaling by controlling ATP release. Proc Natl Acad Sci U S A 95:15735–15740

114. Cotrina ML, Lin JH, Lopez-Garcia JC, Naus CC, Nedergaard M (2000) ATP-mediated glia signaling. J Neurosci 20:2835–2844

115. Shi Y, Liu X, Gebremedhin D, Falck JR, Harder DR, Koehler RC (2008) Interaction of mechanisms involving epoxyeicosa-trienoic acids, adenosine receptors, and metabotropic glutamate receptors in neu-rovascular coupling in rat whisker barrel cortex. J Cereb Blood Flow Metab 28:111–125

116. Zhang JM, Wang HK, Ye CQ, Ge W, Chen Y, Jiang ZL, Wu CP, Poo MM, Duan S (2003) ATP released by astrocytes mediates glutamatergic activity-dependent heterosyn-aptic suppression. Neuron 40:971–982

117. Dunwiddie TV, Masino SA (2001) The role and regulation of adenosine in the central nervous system. Annu Rev Neurosci 24:31–55

118. Parpura V, Basarsky TA, Liu F, Jeftinija K, Jeftinija S, Haydon PG (1994) Glutamate-mediated astrocyte-neuron signalling. Nature 369:744–747

119. Simard M, Nedergaard M (2004) The neu-robiology of glia in the context of water and ion homeostasis. Neuroscience 129:877–896

120. Tian GF, Takano T, Lin JH, Wang X, Bekar L, Nedergaard M (2006) Imaging of cortical astrocytes using 2-photon laser scanning microscopy in the intact mouse brain. Adv Drug Deliv Rev 58:773–787

121. Volterra A, Meldolesi J (2005) Astrocytes, from brain glue to communication elements: the revolution continues. Nat Rev Neurosci 6:626–640

122. Cox SB, Woolsey TA, Rovainen CM (1993) Localized dynamic changes in cortical blood flow with whisker stimulation corresponds to matched vascular and neuronal architecture of rat barrels. J Cereb Blood Flow Metab 13:899–913

123. Ngai AC, Ko KR, Morii S, Winn HR (1988) Effect of sciatic nerve stimulation on pial arterioles in rats. Am J Physiol 254: H133–H139

124. Silva AC, Lee SP, Iadecola C, Kim SG (2000) Early temporal characteristics of cerebral blood flow and deoxyhemoglobin changes during somatosensory stimulation. J Cereb Blood Flow Metab 20:201–206

125. Iadecola C (1993) Regulation of the cerebral microcirculation during neural activity: is nitric oxide the missing link? Trends Neurosci 16:206–214

126. Iadecola C, Nedergaard M (2007) Glial regulation of the cerebral microvasculature. Nat Neurosci 10:1369–1376

127. Zonta M, Angulo MC, Gobbo S, Rosengarten B, Hossmann KA, Pozzan T, Carmignoto G (2003) Neuron-to-astrocyte signaling is central to the dynamic control of brain microcirculature. Nat Neurosci 6:43–50

128. Takano T, Tian GF, Peng W, Lou N, Libionka W, Han X, Nedergaard M (2006) Astrocyte-mediated control of cerebral blood flow. Nat Neurosci 9:260–267

129. Petzold GC, Albeanu DF, Sato TF, Murthy VN (2008) Coupling of neural activity to blood flow in olfactory glomeruli is mediated by astrocytic pathways. Neuron 58:897–910

130. Niwa K, Haensel C, Ross ME, Iadecola C (2001) Cyclooxygenase-1 participates in selected vasodilator responses of the cerebral circulation. Circ Res 88:600–608

131. Ellis EF, Police RJ, Yancey L, McKinney JS, Amruthesh SC (1990) Dilation of cerebral arterioles by cytochrome P-450 metabolites of arachidonic acid. Am J Physiol 259: H1171–H1177

132. Gebremedhin D, Ma YH, Falck JR, Roman RJ, VanRollins M, Harder DR (1992) Mechanism of action of cerebral epoxyeicosatrienoic acids on cerebral arterial smooth muscle. Am J Physiol 263:H519–H525

133. Ellis EF, Wei EP, Cockrell CS, Choi S, Kontos HA (1983) The effect of PGF2 alpha on in vivo cerebral arteriolar diameter in cats and rats. Prostaglandins 26:917–923

134. Benyo Z, Gorlach C, Wahl M (1998) Involvement of thromboxane A2 in the mediation of the contractile effect induced by inhibition of nitric oxide synthesis in isolated rat middle cerebral arteries. J Cereb Blood Flow Metab 18:616–618

135. Filosa JA, Bonev AD, Nelson MT (2004) Calcium dynamics in cortical astrocytes and arterioles during neurovascular coupling. Circ Res 95:e73–e81

136. Ishimoto H, Matsuoka I, Nakanishi H, Nakahata N (1996) A comparative study of arachidonic acid metabolism in rabbit cultured astrocytes and human astrocytoma cells (1321N1). Gen Pharmacol 27:313–317

137. Faraci FM (1989) Effects of endothelin and vasopressin on cerebral blood vessels. Am J Physiol 257:H799–H803

138. MacCumber MW, Ross CA, Snyder SH (1990) Endothelin in brain: receptors, mitogenesis, and biosynthesis in glial cells. Proc Natl Acad Sci U S A 87:2359–2363

139. Lange A, Gebremedhin D, Narayanan J, Harder D (1997) 20-Hydroxyeicosatetraenoic acid-induced vasoconstriction and inhibition of potassium current in cerebral vascular smooth muscle is dependent on activation of protein kinase C. J Biol Chem 272:27345–27352

140. Gordon GR, Mulligan SJ, MacVicar BA (2007) Astrocyte control of the cerebrovasculature. Glia 55:1214–1221

141. Raichle ME, Hartman BK, Eichling JO, Sharpe LG (1975) Central noradrenergic regulation of cerebral blood flow and vascular permeability. Proc Natl Acad Sci U S A 72:3726–3730

142. Duffy S, MacVicar BA (1995) Adrenergic calcium signaling in astrocyte networks within the hippocampal slice. J Neurosci 15:5535–5550

143. Xu HL, Mao L, Ye S, Paisansathan C, Vetri F, Pelligrino DA (2008) Astrocytes are a key conduit for upstream signaling of vasodilation during cerebral cortical neuronal activation in vivo. Am J Physiol Heart Circ Physiol 294:H622–H632

144. Nett WJ, Oloff SH, McCarthy KD (2002) Hippocampal astrocytes in situ exhibit calcium oscillations that occur independent of neuronal activity. J Neurophysiol 87:528–537

145. Hirase H, Qian L, Bartho P, Buzsaki G (2004) Calcium dynamics of cortical astrocytic networks in vivo. PLoS Biol 2:E96

146. Parri HR, Crunelli V (2001) Pacemaker calcium oscillations in thalamic astrocytes in situ. Neuroreport 12:3897–3900

147. Gobel W, Kampa BM, Helmchen F (2007) Imaging cellular network dynamics in three dimensions using fast 3D laser scanning. Nat Methods 4:73–79

148. Tashiro A, Goldberg J, Yuste R (2002) Calcium oscillations in neocortical astrocytes under epileptiform conditions. J Neurobiol 50:45–55

149. Balazsi G, Cornell-Bell AH, Moss F (2003) Increased phase synchronization of spontaneous calcium oscillations in epileptic human versus normal rat astrocyte cultures. Chaos 13:515–518

150. Gordon GRJ, Mulligan SJ, MacVicar BA (2009) Astrocyte control of blood flow. In: Parpura V, Haydon P G (eds) Astrocytes in (Patho)Physiology of the Nervous System, Springer Science + Business Media, New York, pp 461–486

151. Koehler RC, Roman RJ, Harder DR (2009) Astrocytes and the regulation of cerebral blood flow. Trends Neurosci 32:160–169

152. Pasantes-Morales H (1996) Volume regulation in brain cells: cellular and molecular mechanisms. Metab Brain Dis 11:187–204

153. Pasantes-Morales H, Franco R, Ochoa L, Ordaz B (2002) Osmosensitive release of neurotransmitter amino acids: relevance and mechanisms. Neurochem Res 27:59–65

154. Mizuno A, Matsumoto N, Imai M, Suzuki M (2003) Impaired osmotic sensation in mice lacking TRPV4. Am J Physiol Cell Physiol 285:C96–101

155. Kimelberg HK, MacVicar BA, Sontheimer H (2006) Anion channels in astrocytes: biophysics, pharmacology, and function. Glia 54:747–757

156. Duffy S, MacVicar BA (1996) In vitro ischemia promotes calcium influx and intracellular calcium release in hippocampal astrocytes. J Neurosci 16:71–81

157. Abbott N, Ronnback L, Hansson E (2006) Astrocyte-endothelial interactions at the blood-brain barrier. Nat Rev Neurosci 7:41–53

158. Schikorski T, Stevens CF (1999) Quantitative fine-structural analysis of olfactory cortical synapses. Proc Natl Acad Sci U S A 96:4107–4112

159. Ventura R, Harris KM (1999) Three-dimensional relationships between hippocampal synapses and astrocytes. J Neurosci 19:6897–6906

160. Grosche J, Matyash V, Moller T, Verkhratsky A, Reichenbach A, Kettenmann H (1999) Microdomains for neuron-glia interaction: parallel fiber signaling to Bergmann glial cells. Nat Neurosci 2:139–143

161. Grosche J, Kettenmann H, Reichenbach A (2002) Bergmann glial cells form distinct morphological structures to interact with cerebellar neurons. J Neurosci Res 68:138–149

162. Newman EA (2003) New roles for astrocytes: regulation of synaptic transmission. Trends Neurosci 26:536–542

163. Araque A, Parpura V, Sanzgiri RP, Haydon PG (1999) Tripartite synapses: glia, the unacknowledged partner. Trends Neurosci 22:208–215

164. Porter JT, McCarthy KD (1997) Astrocytic neurotransmitter receptors in situ and in vivo. Prog Neurobiol 51:439–455

165. Pasti L, Volterra A, Pozzan T, Carmignoto G (1997) Intracellular calcium oscillations in astrocytes: a highly plastic, bidirectional form of communication between neurons and astrocytes in situ. J Neurosci 17:7817–7830

166. Kang J, Jiang L, Goldman SA, Nedergaard M (1998) Astrocyte-mediated potentiation of inhibitory synaptic transmission. Nat Neurosci 1:683–692

167. Araque A, Martin ED, Perea G, Arellano JI, Buno W (2002) Synaptically released acetylcholine evokes Ca2+ elevations in astrocytes in hippocampal slices. J Neurosci 22:2443–2450

168. Finkbeiner SM (1993) Glial calcium. Glia 9:83–104

169. Halassa MM, Fellin T, Haydon PG (2007) The tripartite synapse: roles for gliotransmission in health and disease. Trends Mol Med 13:54–63

170. Araque A, Parpura V, Sanzgiri RP, Haydon PG (1998) Glutamate-dependent astrocyte modulation of synaptic transmission between cultured hippocampal neurons. Eur J Neurosci 10:2129–2142

171. Araque A, Sanzgiri RP, Parpura V, Haydon PG (1998) Calcium elevation in astrocytes causes an NMDA receptor-dependent increase in the frequency of miniature synaptic currents in cultured hippocampal neurons. J Neurosci 18:6822–6829

172. Innocenti B, Parpura V, Haydon PG (2000) Imaging extracellular waves of glutamate during calcium signaling in cultured astrocytes. J Neurosci 20:1800–1808

173. Bezzi P, Carmignoto G, Pasti L, Vesce S, Rossi D, Rizzini BL, Pozzan T, Volterra A (1998) Prostaglandins stimulate calcium-dependent glutamate release in astrocytes. Nature 391:281–285

174. Pasti L, Zonta M, Pozzan T, Vicini S, Carmignoto G (2001) Cytosolic calcium oscillations in astrocytes may regulate exocytotic release of glutamate. J Neurosci 21:477–484

175. Bezzi P, Gundersen V, Galbete JL, Seifert G, Steinhauser C, Pilati E, Volterra A (2004) Astrocytes contain a vesicular compartment that is competent for regulated exocytosis of glutamate. Nat Neurosci 7:613–620

176. Montana V, Ni Y, Sunjara V, Hua X, Parpura V (2004) Vesicular glutamate transporter-dependent glutamate release from astrocytes. J Neurosci 24:2633–2642

177. Zhang Q, Fukuda M, Van BE, Pascual O, Haydon PG (2004) Synaptotagmin IV regulates glial glutamate release. Proc Natl Acad Sci U S A 101:9441–9446

178. Takano T, Kang J, Jaiswal JK, Simon SM, Lin JH, Yu Y, Li Y, Yang J, Dienel G, Zielke HR, Nedergaard M (2005) Receptor-mediated glutamate release from volume sensitive channels in astrocytes. Proc Natl Acad Sci U S A 102:16466–16471

179. Ye ZC, Wyeth MS, Baltan-Tekkok S, Ransom BR (2003) Functional hemichannels in

astrocytes: a novel mechanism of glutamate release. J Neurosci 23:3588–3596

180. Fellin T, Pozzan T, Carmignoto G (2006) Purinergic receptors mediate two distinct glutamate release pathways in hippocampal astrocytes. J Biol Chem 281:4274–4284

181. Barbe MT, Monyer H, Bruzzone R (2006) Cell-cell communication beyond connexins: the pannexin channels. Physiology (Bethesda) 21:103–114

182. Fiacco TA, McCarthy KD (2004) Intracellular astrocyte calcium waves in situ increase the frequency of spontaneous AMPA receptor currents in CA1 pyramidal neurons. J Neurosci 24:722–732

183. Liu QS, Xu Q, Arcuino G, Kang J, Nedergaard M (2004) Astrocyte-mediated activation of neuronal kainate receptors. Proc Natl Acad Sci U S A 101:3172–3177

184. Gordon GR, Baimoukhametova DV, Hewitt SA, Rajapaksha WR, Fisher TE, Bains JS (2005) Norepinephrine triggers release of glial ATP to increase postsynaptic efficacy. Nat Neurosci 8:1078–1086

185. Pascual O, Casper KB, Kubera C, Zhang J, Revilla-Sanchez R, Sul JY, Takano H, Moss SJ, McCarthy K, Haydon PG (2005) Astrocytic purinergic signaling coordinates synaptic networks. Science 310:113–116

186. Serrano A, Haddjeri N, Lacaille JC, Robitaille R (2006) GABAergic network activation of glial cells underlies hippocampal heterosynaptic depression. J Neurosci 26:5370–5382

187. Newman EA (2003) Glial cell inhibition of neurons by release of ATP. J Neurosci 23:1659–1666

188. Parri HR, Gould TM, Crunelli V (2001) Spontaneous astrocytic Ca2+ oscillations in situ drive NMDAR-mediated neuronal excitation. Nat Neurosci 4:803–812

189. Anderson CM, Swanson RA (2000) Astrocyte glutamate transport: review of properties, regulation, and physiological functions. Glia 32:1–14

190. Bergles DE, Diamond JS, Jahr CE (1999) Clearance of glutamate inside the synapse and beyond. Curr Opin Neurobiol 9:293–298

191. Bergles DE, Jahr CE (1998) Glial contribution to glutamate uptake at Schaffer collateral-commissural synapses in the hippocampus. J Neurosci 18:7709–7716

192. Oliet SH, Piet R, Poulain DA (2001) Control of glutamate clearance and synaptic efficacy by glial coverage of neurons. Science 292:923–926

193. Fonnum F (1984) Glutamate: a neurotransmitter in mammalian brain. J Neurochem 42:1–11

194. Hallermayer K, Harmening C, Hamprecht B (1981) Cellular localization and regulation of glutamine synthetase in primary cultures of brain cells from newborn mice. J Neurochem 37:43–52

195. Loo DT, Althoen MC, Cotman CW (1995) Differentiation of serum-free mouse embryo cells into astrocytes is accompanied by induction of glutamine synthetase activity. J Neurosci Res 42:184–191

196. Mothet JP, Parent AT, Wolosker H, Brady RO, Jr., Linden DJ, Ferris CD, Rogawski MA, Snyder SH (2000) D-serine is an endogenous ligand for the glycine site of the N-methyl-D-aspartate receptor. Proc Natl Acad Sci U S A 97:4926–4931

197. Wolosker H, Blackshaw S, Snyder SH (1999) Serine racemase: a glial enzyme synthesizing D-serine to regulate glutamate-N-methyl-D-aspartate neurotransmission. Proc Natl Acad Sci U S A 96:13409–13414

198. Schell MJ, Molliver ME, Snyder SH (1995) D-serine, an endogenous synaptic modulator: localization to astrocytes and glutamate-stimulated release. Proc Natl Acad Sci U S A 92:3948–3952

199. Stevens ER, Esguerra M, Kim PM, Newman EA, Snyder SH, Zahs KR, Miller RF (2003) D-serine and serine racemase are present in the vertebrate retina and contribute to the physiological activation of NMDA receptors. Proc Natl Acad Sci U S A 100:6789–6794

200. Chesler M, Kaila K (1992) Modulation of pH by neuronal activity. Trends Neurosci 15:396–402

201. Kelly JP, Van E (1974) Cell structure and function in the visual cortex of the cat. J Physiol 238:515–547

202. Karwoski CJ, Newman EA, Shimazaki H, Proenza LM (1985) Light-evoked increases in extracellular K+ in the plexiform layers of amphibian retinas. J Gen Physiol 86:189–213

203. Rausche G, Igelmund P, Heinemann U (1990) Effects of changes in extracellular potassium, magnesium and calcium concentration on synaptic transmission in area CA1 and the dentate gyrus of rat hippocampal slices. Pflugers Arch 415:588–593

204. Barnes S, Bui Q (1991) Modulation of calcium-activated chloride current via pH-induced changes of calcium channel properties in cone photoreceptors. J Neurosci 11:4015–4023

205. Prod'hom B, Pietrobon D, Hess P (1989) Interactions of protons with single open L-type calcium channels. Location of protonation site and dependence of proton-induced current fluctuations on concentration and species of permeant ion. J Gen Physiol 94:23–42

206. Traynelis SF, Cull-Candy SG (1990) Proton inhibition of N-methyl-D-aspartate receptors in cerebellar neurons. Nature 345:347–350

207. Chen JC, Chesler M (1992) pH transients evoked by excitatory synaptic transmission are increased by inhibition of extracellular carbonic anhydrase. Proc Natl Acad Sci U S A 89:7786–7790

208. Newman EA (1996) Acid efflux from retinal glial cells generated by sodium bicarbonate cotransport. J Neurosci 16:159–168

209. Blackburn D, Sargsyan S, Monk PN, Shaw PJ (2009) Astrocyte function and role in motor neuron disease: a future therapeutic target? Glia 57:1251–1264

210. Haydon PG, Carmignoto G (2006) Astrocyte control of synaptic transmission and neurovascular coupling. Physiol Rev 86:1009–1031

211. Haydon PG, Blendy J, Moss SJ, Rob JF (2009) Astrocytic control of synaptic transmission and plasticity: a target for drugs of abuse? Neuropharmacology 56:83–90

212. Pfrieger FW, Barres BA (1997) Synaptic efficacy enhanced by glial cells in vitro. Science 277:1684–1687

213. Nagler K, Mauch DH, Pfrieger FW (2001) Glia-derived signals induce synapse formation in neurones of the rat central nervous system. J Physiol 533:665–679

214. Ullian EM, Sapperstein SK, Christopherson KS, Barres BA (2001) Control of synapse number by glia. Science 291:657–661

215. Mauch DH, Nagler K, Schumacher S, Goritz C, Muller EC, Otto A, Pfrieger FW (2001) CNS synaptogenesis promoted by glia-derived cholesterol. Science 294:1354–1357

216. Pfrieger FW (2003) Outsourcing in the brain: do neurons depend on cholesterol delivery by astrocytes? Bioessays 25:72–78

217. Beattie EC, Stellwagen D, Morishita W, Bresnahan JC, Ha BK, Von ZM, Beattie MS, Malenka RC (2002) Control of synaptic strength by glial TNFalpha. Science 295:2282–2285

218. Christopherson KS, Ullian EM, Stokes CC, Mullowney CE, Hell JW, Agah A, Lawler J, Mosher DF, Bornstein P, Barres BA (2005) Thrombospondins are astrocyte-secreted proteins that promote CNS synaptogenesis. Cell 120:421–433

219. Slezak M, Pfrieger FW (2003) New roles for astrocytes: regulation of CNS synaptogenesis. Trends Neurosci 26:531–535

220. Ullian EM, Christopherson KS, Barres BA (2004) Role for glia in synaptogenesis. Glia 47:209–216

221. Malatesta P, Hartfuss E, Gotz M (2000) Isolation of radial glial cells by fluorescent-activated cell sorting reveals a neuronal lineage. Development 127:5253–5263

222. Miyata T, Kawaguchi A, Okano H, Ogawa M (2001) Asymmetric inheritance of radial glial fibers by cortical neurons. Neuron 31:727–741

223. Noctor SC, Flint AC, Weissman TA, Dammerman RS, Kriegstein AR (2001) Neurons derived from radial glial cells establish radial units in neocortex. Nature 409:714–720

224. Doetsch F, Caille I, Lim DA, Garcia-Verdugo JM, Alvarez-Buylla A (1999) Subventricular zone astrocytes are neural stem cells in the adult mammalian brain. Cell 97:703–716

225. Seri B, Garcia-Verdugo JM, McEwen BS, varez-Buylla A (2001) Astrocytes give rise to new neurons in the adult mammalian hippocampus. J Neurosci 21:7153–7160

226. Kriegstein A, varez-Buylla A (2009) The glial nature of embryonic and adult neural stem cells. Annu Rev Neurosci 32:149–184

227. Doetsch F, Garcia-Verdugo JM, varez-Buylla A (1999) Regeneration of a germinal layer in the adult mammalian brain. Proc Natl Acad Sci U S A 96:11619–11624

228. Ihrie RA, Alvarez-Buylla A (2009) Neural stem cells disguised as astrocytes. In: Parpara V, Haydon PG (eds) Astrocytes in (Patho) physiology of the nervous system, Springer Science + Business Media, New York, pp 27–47

229. Goldman S (2003) Glia as neural progenitor cells. Trends Neurosci 26:590–596

230. Klatzo I (1967) Presidental address. Neuropathological aspects of brain edema. J Neuropathol Exp Neurol 26:1–14

231. Klatzo I (1994) Evolution of brain edema concepts. Acta Neurochir Suppl (Wien) 60:3–6

232. Milhorat TH, Clark RG, Hammock MK (1970) Experimental hydrocephalus. 2. Gross pathological findings in acute and subacute obstructive hydrocephalus in the dog and monkey. J Neurosurg 32:390–399

233. Milhorat TH, Clark RG, Hammock MK, McGrath PP (1970) Structural, ultrastructural, and permeability changes in the ependyma and surrounding brain favoring

equilibration in progressive hydrocephalus. Arch Neurol 22:397–407

234. Marmarou A (2007) A review of progress in understanding the pathophysiology and treatment of brain edema. Neurosurg Focus 22:E1

235. Nag S, Manias JL, Stewart DJ (2009) Pathology and new players in the pathogenesis of brain edema. Acta Neuropathol 118:197–217

236. Cancilla PA, Bready J, Berliner J, Sharifi-Nia H, Toga AW, Santori EM, Scully S, deVellis J (1992) Expression of mRNA for glial fibrillary acidic protein after experimental cerebral injury. J Neuropathol Exp Neurol 51:560–565

237. Klatzo I, Chui E, Fujiwara K, Spatz M (1980) Resolution of vasogenic brain edema. Adv Neurol 28:359–373

238. Kimelberg HK (1995) Current concepts of brain edema. Review of laboratory investigations. J Neurosurg 83:1051–1059

239. Bloch O, Manley GT (2007) The role of aquaporin-4 in cerebral water transport and edema. Neurosurg Focus 22:E3

240. Papadopoulos MC, Verkman AS (2007) Aquaporin-4 and brain edema. Pediatr Nephrol 22:778–784

241. Papadopoulos MC, Verkman AS (2008) Potential utility of aquaporin modulators for therapy of brain disorders. Prog Brain Res 170:589–601

242. Tait MJ, Saadoun S, Bell BA, Papadopoulos MC (2008) Water movements in the brain: role of aquaporins. Trends Neurosci 31:37–43

243. Verkman AS (2005) More than just water channels: unexpected cellular roles of aquaporins. J Cell Sci 118:3225–3232

244. Zador Z, Bloch O, Yao X, Manley GT (2007) Aquaporins: role in cerebral edema and brain water balance. Prog Brain Res 161:185–194

245. Papadopoulos MC, Manley GT, Krishna S, Verkman AS (2004) Aquaporin-4 facilitates reabsorption of excess fluid in vasogenic brain edema. FASEB J 18:1291–1293

246. Manley GT, Fujimura M, Ma T, Noshita N, Filiz F, Bollen AW, Chan P, Verkman AS (2000) Aquaporin-4 deletion in mice reduces brain edema after acute water intoxication and ischemic stroke. Nat Med 6:159–163

247. Papadopoulos MC, Verkman AS (2005) Aquaporin-4 gene disruption in mice reduces brain swelling and mortality in pneumococcal meningitis. J Biol Chem 280:13906–13912

248. Amiry-Moghaddam M, Otsuka T, Hurn PD, Traystman RJ, Haug FM, Froehner SC, Adams ME, Neely JD, Agre P, Ottersen OP, Bhardwaj A (2003) An alpha-syntrophin-dependent pool of AQP4 in astroglial end-feet confers bidirectional water flow between blood and brain. Proc Natl Acad Sci U S A 100:2106–2111

249. Bloch O, Auguste KI, Manley GT, Verkman AS (2006) Accelerated progression of kaolin-induced hydrocephalus in aquaporin-4-deficient mice. J Cereb Blood Flow Metab 26:1527–1537

250. Ransom B, Behar T, Nedergaard M (2003) New roles for astrocytes (stars at last). Trends Neurosci 26:520–522

251. Chvatal A, Anderova M, Neprasova H, Prajerova I, Benesova J, Butenko O, Verkhratsky A (2008) Pathological potential of astroglia. Physiol Res 57 Suppl 3:S101–S110

252. Guo S, Lo EH (2009) Dysfunctional cell-cell signaling in the neurovascular unit as a paradigm for central nervous system disease. Stroke 40:S4–S7

253. Maragakis NJ, Rothstein JD (2006) Mechanisms of disease: astrocytes in neurodegenerative disease. Nat Clin Pract Neurol 2:679–689

254. Rodriguez JJ, Olabarria M, Chvatal A, Verkhratsky A (2009) Astroglia in dementia and Alzheimer's disease. Cell Death Differ 16:378–385

255. Wyss-Coray T, Loike JD, Brionne TC, Lu E, Anankov R, Yan F, Silverstein SC, Husemann J (2003) Adult mouse astrocytes degrade amyloid-beta in vitro and in situ. Nat Med 9:453–457

256. Rothstein JD, Van KM, Levey AI, Martin LJ, Kuncl RW (1995) Selective loss of glial glutamate transporter GLT-1 in amyotrophic lateral sclerosis. Ann Neurol 38:73–84

257. Seifert G, Schilling K, Steinhauser C (2006) Astrocyte dysfunction in neurological disorders: a molecular perspective. Nat Rev Neurosci 7:194–206

258. Gu X, Andre VM, Cepeda C, Li SH, Li XJ, Levine MS, Yang XW (2007) Pathological cell-cell interactions are necessary for striatal pathogenesis in a conditional mouse model of Huntington's disease. Mol Neurodegener 2:8

259. Binder DK, Steinhauser C (2009) Role of astrocytes in epilepsy. In: Parpura V, Haydon PG (eds) Astrocytes in (Patho)Physiology of the Nervous System, Springer Science + Business Media, New York, pp 649–672

260. Jabs R, Seifert G, Steinhauser C (2008) Astrocytic function and its alteration in the epileptic brain. Epilepsia 49 Suppl 2: 3–12

261. Tian GF, Azmi H, Takano T, Xu Q, Peng W, Lin J, Oberheim N, Lou N, Wang X, Zielke HR, Kang J, Nedergaard M (2005) An astrocytic basis of epilepsy. Nat Med 11: 973–981

262. Butterworth RF (2009) Hepatic encephalopathy: a primary astrocytopathy. In: Parpura V, Haydon PG (eds) Astrocytes in (Patho) Physiology of the Nervous System, Springer Science + Business Media, New York, pp 673–692

263. Brenner M, Goldman JE, Quinlan RA, Messing A (2009) Alexander disease: a genetic disorder of astrocytes. In: Parpura V, Haydon PG (eds) Astrocytes in (Patho)Physiology of the Nervous System, Springer Science + Business Media, New York, pp 592–648

264. Quinlan RA, Brenner M, Goldman JE, Messing A (2007) GFAP and its role in Alexander disease. Exp Cell Res 313:2077–2087

265. Higashimori H, Sontheimer H (2007) Role of Kir4.1 channels in growth control of glia. Glia 55:1668–1679

266. McCoy E, Sontheimer H (2007) Expression and function of water channels (aquaporins) in migrating malignant astrocytes. Glia 55:1034–1043

267. Sontheimer H (2003) Malignant gliomas: perverting glutamate and ion homeostasis for selective advantage. Trends Neurosci 26:543–549

268. Nedergaard M, Dirnagl U (2005) Role of glial cells in cerebral ischemia. Glia 50:281–286

Chapter 4

The Blood–Cerebrospinal Fluid Barrier: Structure and Functional Significance

Conrad E. Johanson, Edward G. Stopa, and Paul N. McMillan

Abstract

The choroid plexus (CP) of the blood–CSF barrier (BCSFB) displays fundamentally different properties than blood-brain barrier (BBB). With brisk blood flow (10 × brain) and highly permeable capillaries, the human CP provides the CNS with a high turnover rate of fluid (~400,000 µL/day) containing micronutrients, peptides, and hormones for neuronal networks. Renal-like basement membranes in microvessel walls and underneath the epithelium filter large proteins such as ferritin and immunoglobulins. Type IV collagen (α3, α4, and α5) in the subepithelial basement membrane confers kidney-like permselectivity. As in the glomerulus, so also in CP, the basolateral membrane utrophin A and colocalized dystrophin impart structural stability, transmembrane signaling, and ion/water homeostasis. Extensive infoldings of the plasma-facing basal labyrinth together with lush microvilli at the CSF-facing membrane afford surface area, as great as that at BBB, for epithelial solute and water exchange. CSF formation occurs by basolateral carrier-mediated uptake of Na^+, Cl^-, and HCO_3^-, followed by apical release via ion channel conductance and osmotic flow of water through AQP1 channels. Transcellular epithelial active transport and secretion are energized and channeled via a highly dense organelle network of mitochondria, endoplasmic reticulum, and Golgi; bleb formation occurs at the CSF surface. Claudin-2 in tight junctions helps to modulate the lower electrical resistance and greater permeability in CP than at BBB. Still, ratio analyses of influx coefficients (K_{in}) for radiolabeled solutes indicate that paracellular diffusion of small nonelectrolytes (e.g., urea and mannitol) through tight junctions is restricted; molecular sieving is proportional to solute size. Protein/peptide movement across BCSFB is greatly limited, occurring by paracellular leaks through incomplete tight junctions and low-capacity transcellular pinocytosis/exocytosis. Steady-state concentration ratios, CSF/plasma, ranging from 0.003 for IgG to 0.80 for urea, provide insight on plasma solute penetrability, barrier permeability, and CSF sink action to clear substances from CNS.

Key words: Adherens junctions, Aging choroid plexus, Angiopoietin-1, Arginine vasopressin, Ascorbate, Atrial natriuretic peptide, Choroid plexus blood flow, Choroid plexus cell lines, Cilia, Circumventricular organs, Collagen, CSF turnover rate, Diffusion coefficient, Ecto-5'-nucleotidase, Endothelium, Ependyma, Epithelial monolayer, Folate, Horseradish peroxidase, Hyponatremia, Kolmer cells, Metalloproteinase-3, Microperoxidase, Na^+, K^+-ATPase, Na^+-K^+-2 Cl^- cotransport, Periventricular regions, Permeability coefficient, Syntrophin, Thyroid hormone transcription factor-1, Transthyretin, Vascular endothelial growth factor, Zonulae occludentes

Sukriti Nag (ed.), *The Blood-Brain and Other Neural Barriers: Reviews and Protocols*, Methods in Molecular Biology, vol. 686, DOI 10.1007/978-1-60761-938-3_4, © Springer Science+Business Media, LLC 2011

1. Introduction

There are multiple loci of the blood–cerebrospinal barrier (BCSFB). Structurally and functionally, the BCSFB displays considerable heterogeneity. CSF abuts the vascular system at multiple epithelial sites in both interior and exterior regions of the central nervous system (CNS). Two prominent areas where the blood is near CSF are the arachnoid membrane blanket over the subarachnoid space and the choroid plexus (CP) invagination of the ventricles. Figure 1 shows the BCSFB relative to other CNS transport interfaces. Comparative features of the arachnoid vs. choroid epithelial systems have been presented elsewhere (1, 2). While all BCSFB regions concurrently engage influx/efflux of solutes and water (Fig. 1, arrows 11–14), the arachnoid is best known for preponderant fluid reabsorption (Fig. 1, 14), whereas the CP is specialized for high-capacity secretion

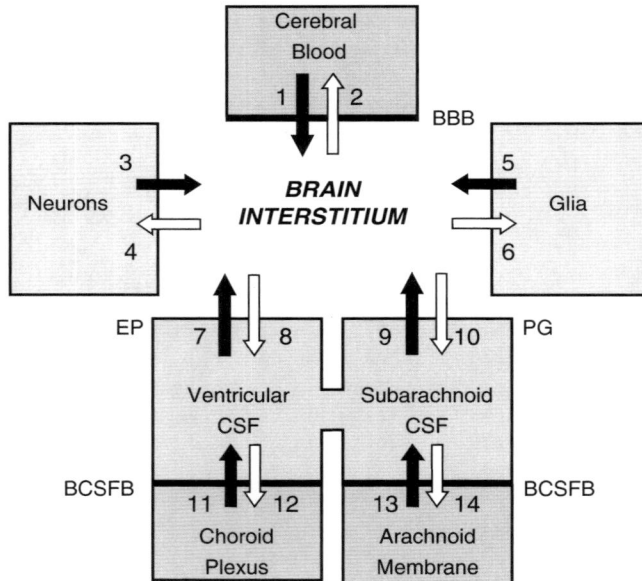

Fig. 1. The major transport interfaces in the central nervous system (CNS) are shown. Several sites where the blood and CSF interface with brain impact interstitial fluid (ISF) composition. Endothelial cells at BBB and epithelial cells of blood–CSF barrier (BCSFB) regulate molecular fluxes by carrier-mediated mechanisms and permeability barriers (tight junctions). Two regions of BCSFB are the single and multiple epithelial cell layers, respectively, in choroid plexus (CP) and arachnoid membrane. At all barrier sites and parenchymal cell membranes of neurons and glia (3–6), numerous secretory (*outward filled arrows*) and reabsorptive (*inward unfilled arrows*) fluxes of water, ions, and molecules occur simultaneously. The neuronal interstitial environment is set by the summation of these diverse exchange processes. Where CSF interfaces the ependyma (EP) and pia-glia (PG) (7–10), the gap junctions offer less restriction to diffusion between brain, and the ventricular fluid and subarachnoid space CSF, respectively. Hydrostatic pressure gradient-driven bulk flow facilitates fluid convection between ISF and CSF compartments, thus promoting distribution. Reabsorptive transport, for example, of organic acid anions and amyloid peptide fragments takes place at *2, 12,* and *14.* Prominent secretory transport occurs at *1,* very extensively at *11,* but to a much lesser extent at *13. Odd* and *even-numbered arrows*, respectively, represent transport into CNS extracellular fluid and uptake from it.

(Fig. 1, 11). Because this review emphasizes CNS-inward transport at the BCSFB, it mainly treats CP morphology and blood-to-ventricle permeation of molecules (Fig. 1, 11). The CPs of the lateral, 3rd, and 4th ventricles exhibit embryological, anatomical, and functional differences (3–5). However, in adulthood the choroidal tissues in these ventricular regions are more similar than different. Information presented herein is mainly for the most commonly studied rodent lateral ventricle tissue. Throughout the lifespan, there are substantial changes in BCSFB transport and permeability, from fetal development (6) to senescence (7). This chapter stresses water and solute permeation into the CP-CSF system of healthy adult mammals.

2. Overview of Choroid Plexus Structure and Function

Long known for its pivotal role in fostering brain development, the CP continually gains appreciation as a stabilizer of fluids in the adult and disease-challenged CNS. Operating in concert with the blood-brain barrier (BBB) (Fig. 1, 1), the choroid epithelial interface of the BCSFB has distinctive structural features that facilitate regulatory functions for ultimately promoting neuronal well-being. Brain extracellular fluid stability is vitally linked to an array of CP-CSF homeostatic transport mechanisms. In supplying nutritional and trophic substances for neurons (Fig. 1, 4), the choroid epithelial cells furnish the CSF-brain nexus (Fig. 1, 7 and 9) with vitamins (8, 9), growth factors, peptides, neurotrophins, hormones, and protein stabilizers such as transthyretin (10).

Suspended in all four ventricles, the CPs impart endocrine, micronutritional, and excretory benefits for areas of brain adjacent to CSF (11). Periventricular regions impacted by CSF include the hippocampus, subventricular zone, hypothalamus, aqueductal gray matter, brain stem, and cerebellum. As the main generator of CSF, the CP effectively furnishes a buffering and distributional medium for the brain interior. Lacking true lymphatic capillaries, the inner surface of the brain requires help to clear harmful proteins and anionic catabolites. Fluid throughput via the CP-CSF circulation fills this need in young adulthood, but less efficiently so at late stages of life (12).

3. Cell Types and Structures in CP

Transport-relevant cellular elements and biomarkers of CP are presented in Table 1. Choroid endothelial cells, staining for factor VIII (13), lack the tight junctions, or *zonulae occludentes*, characteristic of brain microvessel counterparts. The basement membrane

Table 1
Biomarkers for cell types and structures in mammalian choroid plexus

Cell or structure	Morphological features	Biomarker	Species
Endothelial cell	Thin wall, lacking tight junctions	Factor VIII(IHC)	Rat (cell line) (13)
Subendothelial basement membrane	Thin filter, composed of Type IV collagen	α1 and α2 chains (monoclonal)	Mouse (intact tissue) (61)
Epithelial cell	Cuboidal parenchyma 10-micrometer cubes, single layer	Transthyretin IHC (polyclonal) and mRNA	Rat (intact tissue and culture); fetal and adult (21, 135)
Subepithelial basement membrane	Kidney-like filter, composed of Type IV collagen	α3, α4, and α5 chains (monoclonal)	Mouse (intact tissue) (61)
Basolateral membrane	Complex infolding, and intertwining with neighboring cell base	Glucose transporter-1 (18); thrombospondin (136)	Cow and rat tissues (18) pig (cultured cells) (136)
Adherens junctions	Located between CP epithelial cells, under the tight junctions	The long isoform of p120ctn stabilizes BCSFB structure	Mouse (137)
Apical membrane	Lush filamentous microvilli, with great surface area	3E6 (IHC, monoclonal antibody)	Rat culture (4VCP) (20)
Tight junction (zonulae occludentes)	Tight junctions confirmed by electron microscopy	Express occludin, claudin-1, and ZO1	Rat (Z310 line) (138)
Ciliary process	Apical flagellar-like appendage	Type III adenylyl cyclase (IHC)	Mouse tissue (intact) (27)
Nerve fiber to parenchyma	Sympathetic fibers running between epithelial cells	Dopamine D-2 like receptors (Western blot and IHC)	Rat CNS tissue (15)
Kolmer cell	Macrophage-like, fingers interdigitating with apical microvilli	Complement receptor CRI (mRNA, in situ hybridization)	Human tissue (postmortem and surgery) (28)

underlying the plexus capillary wall has a different collagen composition than the basement membrane subjacent to each epithelial cell (Table 1). Unmyelinated, efferent nerve fibers penetrating into the stroma (14), and between epithelial cells (15), likely modulate CSF formation. Histologically, the main parenchymal cell type at the BCSFB is the choroidal epithelium; these cells have complete-belt tight junctions (16) along the apical surface (Fig. 2). Other less restrictive junctions occur between choroid epithelial cells (17). As integrin-mediated anchors for the epithelium, the adherens junctions are near tight junctions but

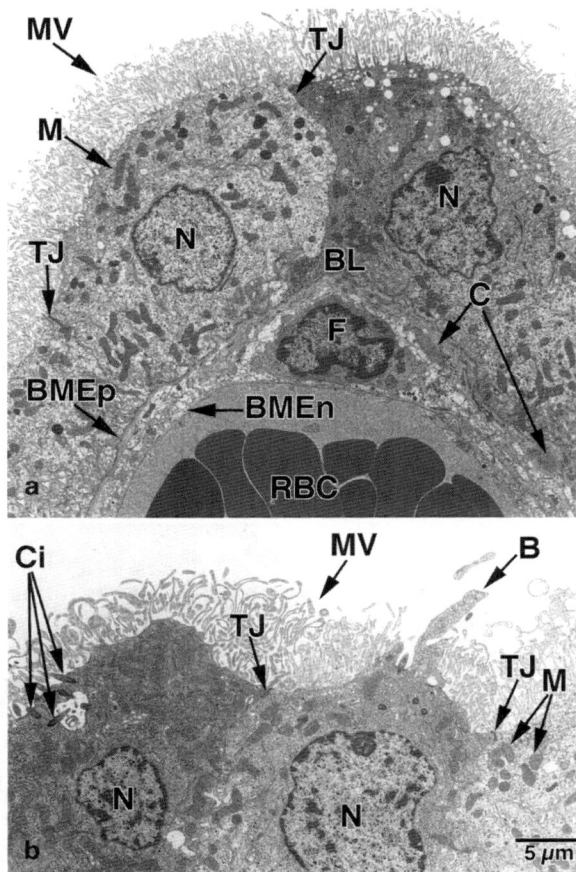

Fig. 2. Components of adult rat lateral ventricle CP villus: (**a**) Three epithelial cells comprise the apex of a villus projecting into CSF. Surfaces of epithelial cells are replete with microvilli (MV). Tight junctions (TJ) adjoin apical membranes of neighboring CP epithelial cells. Subjacent to epithelial cells is a core containing a fibroblast (F) and a sinusoidal blood vessel with several red cells (RBC). Clusters of collagen fibers (C) are seen in cross-section. BL is the basal labyrinth. Basement membranes (BMEp) lining the basal surfaces of CP epithelial cells are apparent as is the basemembrane (BMEn) of an endothelial cell lining the blood vessel. Numerous cytoplasmic organelles are visible in CP epithelial cells such as nuclei (N) and mitochondria (M). (**b**) A sequence of three CP epithelial cells of varying cytoplasmic electron density is displayed. Epithelial surfaces contain lush microvilli (MV) and barrier-forming tight junctions (TJ). A cluster of 7 cilia (Ci) is in the upper left-hand quadrant of the "dark" CP epithelial cell. B represents bleb formation. Scale bar = 5 μm.

more basal; they connect the cytoskeletal actin filaments to the extracellular matrix (ECM). Villi within the fronds contain the epithelium as a single concentric layer in adult CP (Fig. 2). Epithelial cells immunostain at the basolateral (plasma-facing) surface for glucose transporter-1 (18) and at the apical (CSF-facing) membrane for both Na^+, K^+-ATPase, and the Na^+-K^+-$2Cl^-$ cotransporter (19). Microvilli stain specifically with the 3E6 antibody (20). Of the entire cellular complement in the CNS, just the CP epithelium synthesizes transthyretin (TTR) protein (21); thus, at the earliest stage of formation in fetal life, TTR is the most reliable biomarker to identify emerging epithelial cells. Light (normal) and dark (electron-dense) epithelial cells are noted by electron microscopy; there are more dark cells when CSF formation is decreased (22). Dark cells, their volume shrunken by a third, are neuroendocrine-like epithelial cells induced by basic fibroblast growth factor (23) and arginine vasopressin (24). Atrial natriuretic peptide (factor) also induces dark cells (25) and increases BCSFB permeability (26). Ciliary processes, emanating from some epithelial cells, display type III adenylyl cyclase as a localizing marker (27). Sending out finger-like processes that intertwine with apical microvilli, a few macrophage-like Kolmer cells staining specifically for complement receptor 1 (CR1) (28) adhere tightly to the free CSF border of the CP epithelium. However, volume-wise and in number, the choroid epithelial cells are the main anatomical substrate of the BCSFB.

4. Flow vs. Transport and Permeability Properties

To engender the CSF circulation, a great flux of water and endogenous solutes continually streams across the CPs into the ventricles (Fig. 1, 11). A substantial rate of nascent CSF formation arises from the brisk blood flow to the plexuses. How does choroidal hemodynamics functionally affect CSF hydrodynamics? Arterial systolic pulsations in CP transmitted to ventricular CSF promote circulation of CSF. Most mammalian species normally form CSF at a rate of 0.4 mL/min/g (29). Reduced net choroidal perfusion pressure, resulting from arterial hypotension and/or elevated intracranial pressure (ICP), can lower CSF formation rate. CSF production is curtailed when augmented ICP opposes the hydrostatic pressure within the choroidal microvessels, thus reducing filtration of plasma out of capillaries. However, with a typical mean arterial pressure of 100 mmHg and a CSF pressure of 10 mmHg, a sufficient volume of plasma ultrafiltrate (30) is delivered to CP epithelium for adequately supporting CSF formation.

Upon delivery of plasma ultrafiltrate to the interstitium, the movement of solutes through the BCSFB into the ventricles depends upon the state of active transporters in the CP epithelial membranes as well as the barriers to molecular diffusion. At the plasma-face of the CP epithelium, there is restriction to diffusion of even low molecular weight water-soluble molecules (31); therefore, active or facilitated-transport carriers are essential for loading up the epithelial cells with ions and organic substrates such as vitamin C (8) for subsequent regulated release across the CSF face. Consequently, carrier-mediated transport of ions, vitamins, nucleotides, peptides, and hormones keeps CSF levels tightly maintained over narrow concentration ranges (7–9, 32–34). Integrated transport of inorganic (Na^+, K^+, Cl^-) and organic (bicarbonate) ions is necessary for the formation of CSF with a stable composition. Whereas a detailed analysis of ion transport (11), serially, across basolateral and apical membranes is beyond the scope of this chapter, the overview of these complex processes in Fig. 3 nevertheless provides a condensed perspective. The BCSFB permeability discussion that follows centers on the distribution, mainly via diffusion, of water, nonelectrolytes, and proteins through the various compartments of CPs and from blood to the ventricles (Fig. 1, 11).

5. Compartments Within the Choroidal Tissue

Truly a multifaceted functional organ, the CP is structurally designed to impart a wide spectrum of physiological actions on brain. Several unique transport features, compared to BBB, enable the choroidal tissue to carry out diverse tasks. Having renal- (32) and hepatic-like roles (35) and mediating immune (36) as well as endocrine phenomena (37), the BCSFB interface controls secretory and reabsorptive fluxes between blood and ventricular fluid. Like the kidney handling of systemic fluid, the CP transfers a large volume of fluid (water plus ions) across its permselective basolateral and apical membranes. In addition to transporting plasma-derived solutes, the choroidal epithelium *synthesizes* and secretes a wide spectrum of peptides (38, 39) and proteins into CSF. To energize and channel this high degree of molecular trafficking across the BCSFB, the epithelial compartment utilizes a great density of organelles such as mitochondria, endoplasmic reticulum, and Golgi apparatus (Fig. 2). At the ventricular side of the epithelium, a complex array of microvilli and cilia interact to culminate CP transport and CSF volume flow into and through the ventricles.

In transit from the choroidal blood to interstitium and then to parenchyma and CSF (Fig. 1, 11), molecules must thus traverse a series of tissue compartments that offer variable degrees of resistance to diffusion. Each step in the permeation process that assures optimal formation and regulation of CSF will be discussed (Fig. 4).

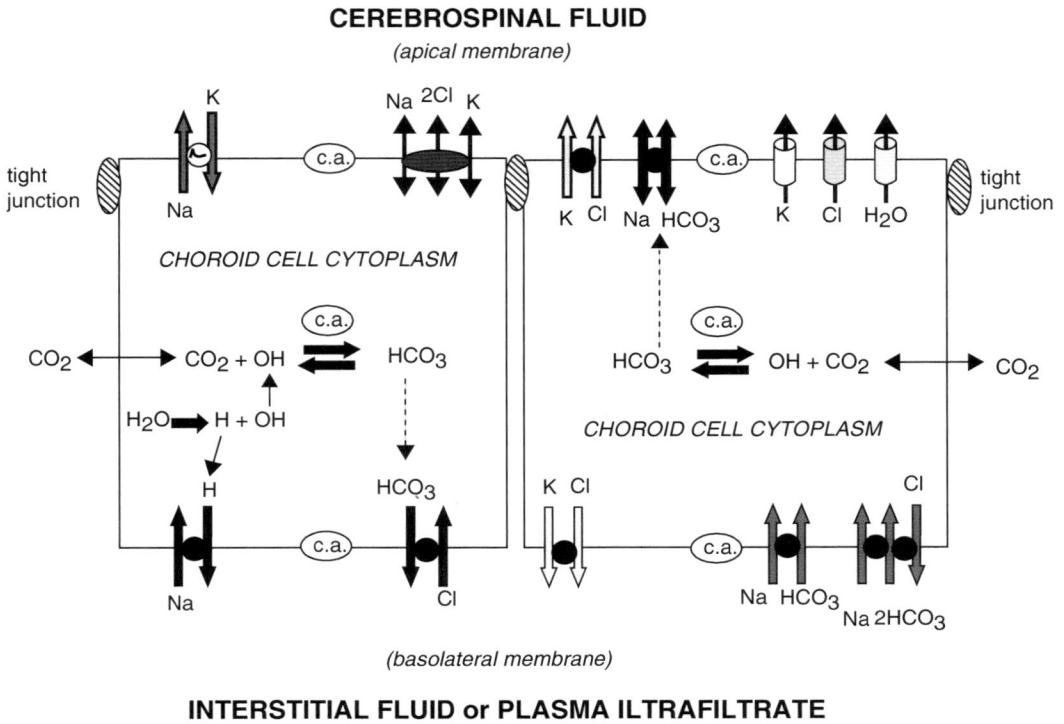

Fig. 3. Ion transport in CP to generate CSF formation: Numerous active transport systems, along with ion and water channels, participate in epithelial elaboration of CSF. CSF production is basically a two-step process: (1) basolateral plasma-facing membrane loading of the choroid cell with ions and molecules from plasma ultrafiltrate in the interstitial space, and (2) apical CSF-facing membrane release of ions and molecules into the ventricles. CSF formation is essentially the net blood-to-CSF transport of Na^+, Cl^-, K^+, HCO_3^-, and water (11). Basolateral loading processes: Plasma-derived Na^+ and Cl^-, as fixed ions incapable of diffusing through epithelial membranes, undergo facilitated or carrier-mediated transport into the choroid cell. The following transporters load up Na^+, Cl^-, and HCO_3^-: the Na^+-H^+, Cl^--HCO_3^-, and Sodium-dependent-Cl^-, HCO_3^- exchangers (73, 140). Relatively low cell (Na^+), from active Na^+ extrusion into CSF (40), sets up the inward transmembrane Na^+ gradient driving force for the Na^+-coupled cotransporters. Cl^- uptake is facilitated by exchange (141) with cellular HCO_3^- generated by carbonic anhydrase (142). Evidence for loading of basolateral K^+ (32) and water awaits more experimentation. Apical release processes: While the Na^+ pump actively discharges Na^+ into CSF (40), there are concurrently a number of processes to release Na^+, K^+, Cl^-, HCO_3^-, and water into ventricles. Because steady-state K^+, Cl^-, and HCO_3^- levels in CP epithelial cells are above electrochemical equilibrium (143–145), these ions diffuse down gradients through specific anion and cation channels on the CSF-facing membrane (93). Additional mechanisms for ion extrusion into CSF include facilitated efflux via the Na-K-2Cl, KCl, and NaHCO$_3$ cotransporters (91, 146, 147). Translocated ions set up an osmotic gradient, from nascent CSF to cytoplasm. To complete CSF secretion, water then diffuses down its chemical potential gradient, via Aquaporin 1 channels (148), onto the ventricular surface of CP. The rodent distribution of transporters is similar to human CP (149). Modified Fig. 4.5 (11) reproduced with permission.

Accordingly, here we discuss the several anatomical compartments of CP involved in CSF secretion: (1) a voluminous choroidal vasculature to expedite substantial blood flow, (2) highly permeable capillaries in the plexus, (3) basement membrane on capillary wall, (4) the intermediate interstitium and ECM, (5) the permselective subepithelial basement membrane, (6) the paracellular pathway, including tight junctions, (7) basolateral membrane specializations,

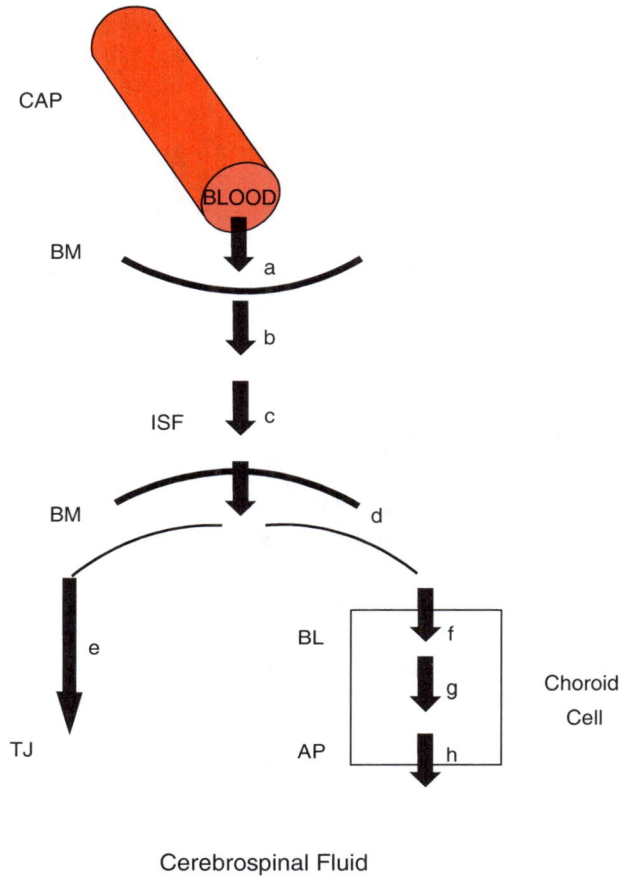

Cerebrospinal Fluid

Fig. 4. Solute migration pathways, from plasma to CSF across CP compartments: Leaky choroidal capillaries (CAP) permit diffusion of even large plasma-delivered molecules across wall (step **a**). Minimal restriction to diffusing solutes is offered by the thin basement membrane (BM) (step **b**) under the CAP wall. A healthy interstitial matrix allows relatively unhindered diffusion (step **c**) up to the subepithelial BM. Immune complexes are trapped in BM adjacent to the epithelium [64], but most substances transported into CSF are not impeded at step **d**. Following passage through BM (step **d**), a given solute is then distributed either paracellularly (step **e**) between epithelial cells up to the tight junction, TJ, where it may be blocked, or transcellularly (steps **f–h**) by membrane-lodged carriers that facilitate transport of water-soluble micro-nutrients or peptides into cell (step **f**). Solute distribution through cytoplasm (step **g**) varies, whether a substance is metabolized, e.g., glucose, or broken down by lysosome as an enzymatic barrier to some xenobiotic agents, or prepared by Golgi for further transport, e.g., of peptides across apical membrane (step **h**) into CSF.

(8) numerous organelles involved in protein synthesis and secretion, (9) apical membrane microvilli specializations, and (10) the CSF-contacting cilia. The histological features of these various tissue compartments are delineated below in the context of molecular traffic across the BCSFB.

5.1. Choroidal Vasculature

The reddish cast of CP reveals its great vascular perfusion. Approximately 25% of the choroid tissue weight is residual blood content (40). To support the considerable epithelial cell metabolism and secretion, there is a substantial choroidal blood flow of

4–5 mL/min/g (41, 42). Normalized for weight, the vascular supply to the plexus is nearly 10 times that of the average blood flow to brain. This highlights the enormous substrate provision to the BCSFB interface for driving CSF secretion. Anterior and posterior choroidal arteries ramify into numerous arterioles that supply the plexus tissues of the lateral and 4th ventricles. Sheath-like capillary networks surround the densely-packed parallel configuration of arterioles and venules (43).

5.2. Capillary Networks

Plexus angioarchitecture has been studied by microscopy of vascular corrosion casts in various mammals. In rabbits, for example, capillary diameters are typically 10–15 µm (44). Regular sinusoid-like nodular bulgings, up to 24 µm, stud the capillary walls. Glomerular-type capillaries are found in both rodent and human CP plexuses (45). Choroidal capillaries have attenuated endothelium, similar to "leaky" microvessel counterparts in the intestine and kidney. The plexus microvessels contain numerous fenestrae, i.e., wall thinnings, each of which is closed by a thin diaphragm (46). Plasma-borne horseradish peroxidase (HRP; ~44,000 Da) readily escapes the capillaries, through fenestrae; HRP also undergoes endothelial pinocytosis (46). Even ferritin (~474,000 Da) can extravasate CP capillaries through fenestrae and via temporary openings in the interendothelial junctions (46). Thus, in contradistinction to the brain microvessels, the choroidal capillaries are highly permeable to macromolecules (Fig. 4, step a). Although there are pentalaminar junctions in CP capillaries, the outer leaflets are not fused (47) and thus do not restrain diffusion of most proteins. Therefore, the BCSFB is *not* localized to CP microvessels, but rather resides in tight junctions between *epithelial* cells adjacent to CSF, i.e., considerably distant from plasma. Accordingly, the largest proteins in plasma diffuse out of choroidal capillaries into the interstitium, and then diffuse all the way to the interepithelial tight junction where restriction to diffusion finally occurs (46). This fundamental difference in CP vs. brain capillaries has pharmacological, immunological, and pathological implications due to unrestricted access of plasma-borne macromolecules into and throughout the CP interstitial space.

5.3. Capillary Basement Membrane

The basement membrane overlying the choroidal capillary walls is devoid of astrocyte process input, a factor that partially explains endothelial leakiness (48). Thus, the capillary basement membrane is highly permeable to most solutes (Fig. 4b). However, differential collagen composition at the capillary basement membrane, vs. the subepithelial basement membrane, gives rise to possible differences in filtering or permeability properties (Table 1). Cationic ferritin, compared to anionic or neutral, has a greater ability (electrostatic) to penetrate the negatively charged capillary wall and underlying basement membrane (49). Cationic dye studies with ruthenium red (50) to analyze polyanionic sites concluded that the capillary basement membrane in CP is substantially more polyanionic than the counterpart renal peritubular capillary basement membrane. Electrostatic anionic charge interaction would repel polyanionic proteins in transit via diffusion from blood to interstitium. The incompletely developed capillary wall-basement membrane in the fetus is fragile. However, choroidal capillary walls stabilized by dexamethazone undergo increased basement membrane width, intactness, and protein particle thickness (51).

5.4. Interstitial Space

The interstitium in adult mammalian CP has a volume of about 15–18% of the tissue (52), equivalent to the compartmental size of brain extracellular space but considerably larger than that of muscle (10%). Inert extracellular markers such as inulin (~5,000 Da) and HRP rapidly permeate the interstitial space from plasma, revealing that large water-soluble molecules readily penetrate the capillary wall and subendothelial basement membrane in plexus, but not brain (53). Plasma-borne radiolabeled mannitol (182 Da) and raffinose (54) (540 Da) also indicate a steady-state interstitial space of 15–18% in CP. Normally the interstitium offers little resistance to molecular diffusion (Fig. 4, step c). Anatomically situated between the vascular endothelium and parenchymal epithelium, the ECM/interstitium is strategically positioned to biochemically modulate cells on either compartmental side. Vascular endothelial growth factor (VEGF), secreted by choroid epithelial cells into the basal lamina type ECM, induces endothelial fenestrations that enhance capillary permeability (55). On the other hand, vascular permeability can be decreased by the ligand Angiopoietin-1 (Ang-1) through its tyrosine kinase-1 (Tie-1) receptor; these proteins (as well as Ang-2, that antagonizes Ang-1) are expressed in CP vessels and epithelium (56), suggesting potential for stabilizing permeability at the BCSFB.

Moreover, enzymes secreted by fibroblasts into ECM affect the epithelium. Accordingly, ecto-5'-nucleotidase, a CP stromal enzyme not present inside epithelial cells, catalyzes generation of extracellular adenosine for receptor modulation at the BCSFB (57). By stabilizing the matrix metalloproteinase MMP-3 in CP, dexamethasone, after meningitis induction, stabilizes the ECM and BCSFB. This minimizes penetration of bacteria and leucocytes from blood to CSF (58). ECM of CP is thus a potential pharmacological target for regulating migration of endogenous substances and drugs through the BCSFB (59).

5.5. Subepithelial Basement Membrane

A filtering function has been ascribed to the basement membrane subjacent to the choroidal epithelium (Fig. 4, step d), perhaps protecting the basal surface of the BCSFB. Even very early in development (mouse embryonic day12), the epithelial basement membrane is continuous (60). The type IV collagen (α3, α4, and α5) in CP membrane is similar to that of renal glomeruli, prompting the idea that both systems function as a permeability-selective barrier to filter fluid transfer (61). Consistent with this notion is the loss of kidneyfiltering capacity, and the onset of hydrocephalus, in mice having collagen XVIII knockout; the CP basement membrane, lacking the XVIII component, undergoes thickening from 61 to 86 nm (62). Thickening of the basement membrane, with aging (63) or after retention trapping of immunoglobulins (IgG and IgA) (64), is generally associated with less efficient epithelial-CSF transfer functions. Our hypothesis is that solute diffusion, between interstitium and epithelial cells, is decreased by basement membrane thickening and destabilization.

5.6. Paracellular Space

Following diffusion through the subepithelial basement membrane, a solute or water molecule migrates to the basolateral membrane for multistep transport *transcellularly* across the choroid epithelium (Fig. 4, steps f–h) or is convected *paracellularly* between epithelial cells up to the tight junction at the apical pole of the cells (Fig. 4, step e). The relative amount of transport via transcellular vs. paracellular routes, especially for water and ions, is difficult to assess experimentally (65) and is the subject of considerable debate (66). Water-soluble molecules the size of mannitol (mw 182) or larger penetrate adult choroid epithelial cells to a negligible extent (52); therefore, hydrophilic solutes such as mannitol, sucrose (342 Da), inulin (~5,000 Da), and proteins that appear in CSF after systemic administration (53, 67) likely have gained access to the ventricles by slow leakage through CP tight junctions. Tight junction permeability is a function of claudin expression. Rodent CP tight junctions express claudin 1, 2, and 11 (68). Some epithelial tight junctions, like those in CP, are more leaky than counterparts in tighter barrier systems such as the BBB. Claudin-2, a barrier weakener or permeability enhancer, is a prototypical paracellular, channel-forming, tight junction protein responsible for specific transfer of solutes across the epithelium without entering the cells (69). Even though the CP is less tight than the BBB by an order of magnitude, in terms of impedance (70), still, relatively small molecules such as vitamin C need to be actively transported into CSF by a stepwise process that starts with sodium-ascorbate cotransport at the basolateral membrane (8).

5.7. Basolateral Membrane

The basolateral membrane regulates the uptake of interstitial molecules by CP (Fig. 4, step f) and therefore access to the CSF beyond. Organic substrates such as vitamin C and glucose enter into CP epithelium by sodium cotransporters at the basolateral (8, 18), but not at the apical membrane. A key feature of the CP epithelium, as in other epithelia, is the structural and functional asymmetry of the basolateral vs. apical membranes. This polarization of the plasma- vs. CSF-facing membranes enables the strong unidirectional secretory flow of materials and water from blood to ventricles. It also permits net solute flux in the reverse direction. Thus, the asymmetric distribution of transporters enables the smaller, but important, reabsorptive clearance of certain catabolites from CSF to blood (71). Accordingly, following solute uptake from CSF by apical membrane transporters (Fig. 1, 12) and transfer of the particular solute across the cell to the basolateral face, certain efflux transporters (72) then complete basolateral extrusion of the catabolite from the epithelium to the interstitial-venous drainage. Concurrent secretory and reabsorptive fluxes across the BCSFB (Fig. 1, 11 and 12) require distinctive structural features at opposite poles of the epithelial cells.

The main feature of the basolateral membrane is the basal labyrinth, a complex intertwining of the basal membranes of adjacent epithelial cells (Fig. 5). Extensive infolding of the basolateral membrane amplifies the membrane area for molecular exchanges between interstitial fluid and cytoplasm. The basolateral membrane, by impeding diffusion, is an integral part of the BCSFB to hydrophilic solutes. Consequently, specific carrier membrane-bound transport proteins (8, 73, 74) are necessary to translocate micronutrient substrates and CSF-bound ions (Na^+ and HCO_3^-) across the plasma-face of the CP epithelium.

The basolateral membrane anchoring protein, utrophin A, plays a role in structural stability, transmembrane signaling, and ion/water homeostasis in both the CP and glomerulus (75, 76). Utrophin, along with the homologous dystrophin glycoprotein complex, physically links the ECM to the actin cytoskeleton (76). Utrophin expression is established by early postnatal life (77) and may be independently regulated from dystrophin. Utrophin A in the CP and pia mater, like utrophin B in the cerebrovascular BBB (75), stabilizes the barriers to enable homeostatic functions. Syntrophin, a structurally related anchoring protein, in both the α-1 and β-2 isoforms, colocalizes with utrophin in CP (78). Dystrophin-related protein is also enriched at the CP and other CNS transport interfaces (79). Dystrophin and utrophin differentially affect the structural dynamics of actin (80). Clearly, a precise functional understanding of basolateral transporters and aquaporins (81) is predicated on new information about the role of utrophin/dystrophin/syntrophin complexes in regulating transfer of molecules at the plasma-facing membrane of CP (Fig. 4, step f). Upon basolateral transport into the epithelium, a given molecule encounters various organelles in transit through the cell on the way to CSF.

5.8. Intracellular Organelles

Electron micrographs suggest that choroid epithelial cells engage in a high level of transport and synthetic/secretory activities (Fig. 2). Abundant mitochondria, Golgi apparatus, and ribosomes support the great work load. The mitochondrial volume

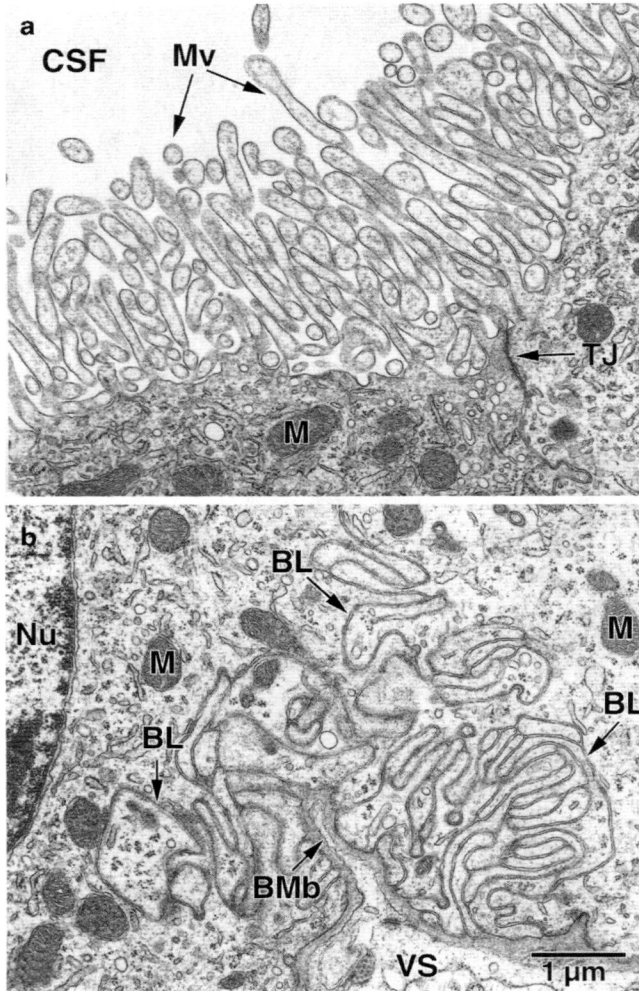

Fig. 5. Ultrastructural features of transporting membranes at the CSF- and plasma-facing surfaces of CP. (a) Adult rat lateral ventricle shows the apical portion of two CP epithelial cells. The profuse microvilli (MV) along the cell surface provide a great surface area for the transport processes delineated in Fig. 3. A tight junction (TJ) is present between two epithelial cells. Numerous mitochondria (M) and other organelles are present in the cytoplasm. (b) Basal portion of the CP epithelium highlights extensive foldings of the basal labyrinth (BL) that impart substantial surface area for basolateral transporters. The serpentine BL is comprised of multiple sites at which the lateral and basal membranes of adjacent epithelial cells intertwine. Cytoplasmic elements are mitochondria (M) and part of cell nucleus (Nu). The subepithelial basement membrane (Bmb) and vascular space (VS) are also shown. Scale bar = 1 μm. Reproduced with permission (116).

is 0.12–0.15 of total choroid cell volume (82), i.e., a fraction even greater than for BBB endothelium (83). A continuous ample supply of mitochondrial-generated ATP is needed to fuel the numerous active transport systems and thereby sustain CSF formation and neurochemical homeostasis. Much of the protein/peptide processing by the Golgi is geared for extrusion across the apical membrane into ventricular CSF. Whereas water and ions taken up across the basolateral membrane are extruded eventually by circumscribed apical transporters and membrane-bound channels, the accumulated or synthesized peptides/proteins are released into CSF by a variety of membrane-separation mechanisms. Of particular interest is the apical membrane blebbing phenomenon. Some epithelial cells contain substantial mature blebs, or large secretory protrusions, as well as small uniform ones (84, 85). Autoradiographs of CP have demonstrated radiolabeled-amino acid protein synthesis in both attached and free apical bleb-like protusions, suggesting apocrine-type secretion of peptides into CSF (86). The mRNA-containing blebs, even after separation, continue to synthesize protein products (87). The CSF distribution of blebs by bulk flow is relevant to emerging models that characterize CSF nanostructures, i.e., spheres or particles in CSF that encase and protect enzymes or other bioactive material destined for CSF-brain regions that modulate specific functions, such as, prostaglandin D synthase and sleep. CSF-borne vesicles or blebs (Fig. 2) originating from the apical membrane may prove to be a significant endocrine-like aspect of signal distribution by volume transmission (88).

5.9. Apical Microvilli

Membrane surface area is a salient factor in transport physiology. By Fick's equation for diffusion, the solute flux is directly proportional to the surface area of the transport interface. Ultrastructure of the CP microvilli (Fig. 5) shows extensive infoldings of the apical membrane that maximize the surface area for transport. The lush microvillous system that contacts the CSF points to an enormous potential for solute exchange. This is consistent with the robust activity of molecular traffic into and out of the ventricular fluid. Surface area determinations for CP microvilli in young adult rats set a figure of about 75 cm^2 (89), surprisingly close to the well-known extensive area of 155 cm^2 for the entire brain capillary network. A later analysis in a whole-cell patch clamp study of rat CP (90) revealed an even greater CP surface area. Such prominent surface area determinations indicate that the BCSFB substantially affects the brain microenvironment (Fig. 1) by way of substantial fluxes and convection distribution of CP-transported substances.

A diverse array of transporters and channels is embedded in the apical microvilli. Primary active Na$^+$ pumping outward into CSF maintains the favorable transmembrane Na$^+$ gradient, directed into the cell that drives several sodium cotransporters (Fig. 3). Apical Na$^+$-K$^+$-2Cl$^-$ cotransport (91, 92) has a versatile role in maintaining the ion homeostasis that is inextricably linked to stable formation of CSF. Anion channels (93) conduct Cl$^-$ and HCO$_3^-$ into CSF, while aquaporin water channels mediate the transfer of water in response to changing hydrostatic pressure and osmotic gradients. Collectively then, the apical membrane mechanisms regulate passage to CSF of: (1) water for the bulk flow aspect of the CSF circulation, (2) ions and metabolites homeostatically regulated to be neuron compatible, and (3) peptides and proteins for conveying trophic, endocrine, and immune signals to multiple brain regions. The ability of the CNS to adapt to insults and stressors vitally depends on robust CP function to affect CSF homeostasis and orderly CSF flow.

5.10. Cilia

Regular CSF flow is critical for balanced cerebral metabolism. In aging, when optimal CSF flow is disrupted (12), adverse effects on brain occur (11). More than one mechanism propels CSF forward. Arterial pulsations in CP, and a hydrostatic pressure gradient from CSF to venous blood, are driving forces to move CSF from the lateral ventricles down the neuraxis (11). Ciliary motion also moves fluid along. Extending from the apical surface of CP and ependymal cells are cilia (Fig. 2) that promote CSF flow by a beating action related to intraflagellar transport motion. Intimate ciliary contact with the CSF system allows extensive functional interactions among CP, CSF, ependyma, and brain.

Disrupted cilia cause brain maldevelopment. This stems from altered CSF flow patterns as well as damage to CP-CSF homeostatic ion transport mechanisms (94, 95). Impaired CSF flow, leading to expansion of the ventricles, results from ciliary stasis caused by knockdown of the G protein G αi2 (96). Beating of cilia is also required for the CSF flow that guides neuroblasts in adults from the lateral ventricle to forebrain regions (97). It is now clear that the CP-CSF system carries out a plethora of functions for brain well-being. A better understanding of the basic transport and permeability features of the BCSFB in health (11) will allow more effective strategies for maintaining CP functions in the face of challenges from aging and disease.

6. Choroid Plexus Preparations to Study Transport and Permeability

It has been particularly challenging to develop CP preparations/models that closely simulate the barrier properties in the intact organism. The in vitro counterpart should, as much as possible, mimic the natural in vivo system. How can the true physiological significance of in vitro experimental data be properly gauged? For a size range of water-soluble test agents, the relative values of the permeability or influx coefficients determined in vitro should correlate closely with in vivo assessments. Also, the steady-state concentration ratios, CSF/plasma or CP/CSF, should be comparable when relating data from an artificial environment to the organism.

Each CP preparation has advantages as well as drawbacks (1, 29, 98). In vivo experimentation provides a natural context but has complex variables, whereas the transwell approach with a cultured monolayer offers the capability of adjusting apical vs. basolateral medium composition (99). It is sound to corroborate in vitro data, i.e., cell culture with possible altered transporter expression (100) and modified tissue structure, with BCSFB information procured from more intact systems (in situ CP (101), in vivo CP (102), and isolated CP (34)). It is desirable to have a CP preparation with tight junction integrity, normal transepithelial electrical resistance, cellular energy reserve, and adequate CSF-forming ability. Reliable CP transport models can be built on data obtained from a battery of in vitro and intact preparations.

7. Permeation of Water, Nonelectrolytes, and Proteins Across BCSFB

For molecules distributing passively, the rapidity of movement across CP is inversely proportional to the molecular weight of the migrating species. Diffusion through membranes of BCSFB is promoted by lipid solubility, as indexed by the oil/water partition coefficient of a given molecule. Water-soluble molecules, as discussed here, traverse most efficiently along diffusion pathways with the least steric hindrance. Thus, the smallest molecules, such as water, penetrate the BCSFB rapidly by moving through small pores or channels. As the largest molecules, proteins encounter considerable impediment to diffusion and therefore have the highest reflection coefficients and lowest penetration rates. The discussion below treats the permeations of water (quantitatively the most preponderant molecule in formation of CSF, which is 99% water), urea (an endogenous prototype nonelectrolyte moderately sieved at the BCSFB), and proteins (which are mainly excluded from CSF, but undergo slow leakage across CP in reverse rank order to size). The BCSFB is presented here as a screen or sieve that decreases the diffusion of hydrophilic molecules in proportion to their size.

7.1. Water

The BCSFB is uniquely equipped to transfer a lot of fluid, hence it displays fast water permeation. Brisk CP blood flow, substantial Na^+,K^+-ATPase, and carbonic anhydrase activities, along with a rapid metabolic rate that reflects extensive transport and syntheses, all support voluminous CSF turnover. Fluid throughput across CP, when normalized for mass of tissue, is equivalent to glomerular filtration rate in the kidneys. Human CP secretes about 400,000 µL/day. Efficient transfer of water across the BCSFB is assured by Aquaporin-1 (AQP-1), a channel conducive to the osmotic flow of water. Rapid and extensive permeation from blood into the CP-CSF system is demonstrated by kinetic analysis of uptake curves for tritiated-water, which penetrates the entire water compartment (extracellular plus intracellular) within minutes (31). This is consistent with water moving transcellularly through epithelial cells of CP.

AQP-1, present in both the red blood cell and CP epithelium, facilitates water movement across the plasma membrane. There is nearly universal agreement that AQP-1 is on the CSF-facing apical membrane, and that water efflux (Fig. 4, step h) is passively conducted through AQP-1 in response to elevated osmotic pressure set up by solute extrusion (Fig. 4, step h) into the ventricles (Fig. 3). Less conclusive is how water is taken up at the basolateral membrane (Fig. 4, step f) from the plasma ultrafiltrate. Basolateral staining for AQP-1 in CP was reported in a study of early development (103). Also, AQP-3, the glyceroaquaporin in the CP basolateral membrane (81), deserves further study in regard to water loading of the choroid epithelial cells from the blood side.

Findings from several studies point to CP apical expression of AQP-1 as being intimately related to CSF formation rate (Table 2). Relatively low levels of AQP-1 are found in early fetal development (104), before CSF formation has reached a peak level (103), and then again in late stages of life (105) when CP secretion of CSF is dwindling (12). AQP-1 null mice have lowered CSF formation and pressure (106). Acute hyponatremia, when water enters too rapidly into the CNS, causes up-regulation of AQP-1 channels in CP (107); down-regulation of BCSFB AQP-1 in hyponatremia might alleviate ICP elevation by reducing osmotic flow into the ventricles. Basic research is creating new opportunities for controlling CSF flow (Table 2), either by regulating transcription factors (108) that modulate AQP-1 expression, applying fluid-homeostatic peptides to alter AQP-1 conductance (109), or synthesizing new drugs, e.g., bumetanide-related agents, that inhibit AQP-1 water channels (110).

Does Starling's law of filtration apply to water distribution across AQP-1 and AQP-4? Clearly, AQP-1 is osmosensitive. In response to osmolar gradients between plasma and CSF, there is substantial transfer of fluid across the BCSFB (107, 111). But what about elevations in ICP that reduce the cerebral perfusion pressure? Under these conditions of changing hydrostatic pressure gradients between CSF and blood, there may be greater clearance of fluid from the CSF across aquaporins in CP. Still, experimental evidence is needed to verify this point. The differential expression of AQP-1 and AQP-4, respectively, at the BCSFB and BBB prompt the pursuit of new agents that can target water channels at each barrier interface separately. Such pharmacologic selectivity would be a step forward in attaining fluid homeostasis by more finely controlling water movements among various CNS compartments.

7.2. Urea

Transport and partitioning data for urea, which is smaller than sucrose and mannitol, are useful for evaluating changes in barrier permeability to nonelectrolytes. Steady-state concentrations of

Table 2
Aquaporin 1 (AQP-1), choroid plexus, and fluid turnover across the BCSFB

Experiment or state	Observation	Comments	Species
Fetal choroid plexus development of AQP-1 channels	Exclusive apical staining from 14th gestational week	By 18th week, cytoplasm as well as CP apical membrane stained	Human (14–40 week of gestation) (104)
Effects of natural aging on AQP-1 expression	Reduced AQP-1 and Na$^+$,K$^+$-ATPase in CP of old (20 mo) vs. young animals	The decreased CSF formation in old rats (12) is consistent with less AQP-1 and Na$^+$ pumps	Sprague-Dawley rats (105)
Peptidergic inhibition of AQP-1 function	Atrial natriuretic peptide (ANP) reduced basal to apical fluid transport	ANP altered the conductance of CP AQP-1 channels (cGMP gated)	Rat choroid plexus primary culture (109)
Effects of acute hyponatremia on AQP-1 channels	Hyponatremia increased CP AQP-1 expression by 28%	AQP-1-mediated excessive water flux across BCSFB may elevate the ICP	Rats (injected with hypotonic dextrose and dDAVP for 2 h) (107)
Deletion of AQP-1 channels in null mice	Deleted AQP-1 in CP led to a 25% decline in CSF formation rate	Secondary to reduced CSF turnover, the ICP fell by 55%	Wild-type control and AQP-1 null mice (106)
Thyroid hormone transcription factor-1 (TTF-1) regulates AQP-1 expression	TTF-1 is coexpressed with AQP-1 in CP and enhances AQP-1 gene transcription	Antisense to TTF-1 given i.c.v. lowered AQP-1 synthesis and CSF formation rate	Rat (108)
Inhibition of AQP-1 by a derivative of the bumetanide loop diuretic	The AQP-1 inhibitor, AqB013 (bumetanide analog), reduced H$_2$O permeability in oocytes, with an IC50 of 20 µm	Bumetanide inhibits CSF production by effects on CP apical membrane (139); AqB013 might be useful in brain H$_2$O homeostasis	Oocytes (110)

urea in adult human and animal fluids reveal a CSF level of urea only 0.7–0.8 of that in plasma (112). Since there is not active transport of urea by carrier from CSF to blood (113), this indicates that the CSF/plasma ratio, R_{urea}, of <1 is due to normal blood-to-CSF molecular sieving of urea by adult CP. Urea therefore permeates BCSFB more slowly than water due to CP restriction on diffusion, even when factoring for the effect of molecular size on unrestricted diffusion. In states in which the barrier is more effectively permeable to nonelectrolytes, e.g., in pig fetal life, R_{urea} approaches unity. Such equilibrium distribution, i.e., the R_{urea} of 1.0 in the CP-CSF system of fetal pigs (114) and neonatal rats (102), points to early-life lack of BCSFB tightness/secretion (67, 115) and therefore attenuated molecular sieving. It is expected that many water-soluble drugs would be similarly sieved by the CP epithelial membranes in transit to the ventricles (116), thereby resulting in lower agent levels in CSF.

In mature mammals, the CSF sink action or drain effect (102, 112) occurs when CP gears up to actively form CSF (11). With BCSFB tightening and greater fluid turnover across adult CP, hydrophilic solutes such as urea undergo more sieving, relative to water movement, in diffusion pathways as reflected by decreasing values (102) for R_{urea}. Urea, like water, diffuses transcellularly (102) via membrane channels or pores as well as paracellularly (117) across tight junctions. When CSF in adult animal investigations is sampled from the cisterna magna, which is a mixture of subarachnoid and ventricular fluid, R_{urea} values of 0.7–0.8 are observed. However, a lower R_{urea} value of 0.6, indicating greater sieving, is obtained for ventricular fluid sampled *upstream* of the basal cisterns. This more accurately reflects CSF-inward molecular fluxes across the BCSFB, when the ventricular CSF sample is taken closer to the secreting CP tissues.

Transfer and influx coefficients, analogous to the permeability-surface area product, quantify the penetration of radiolabeled urea and other nonelectrolytes (67) from blood to CSF (31, 102, 113, 117). As in other transport investigations, the assumption is made that the radiolabeled test molecule penetrates in the same manner as the nonlabel. The influx coefficient K_{in} in mL/g/min is calculated from the initial slope of the curve for test solute volume of distribution, V_d vs. time allowed for distribution. Permeability is proportional to K_{in}. To determine BCSFB permeability in rats, the V_d data for early uptake into CSF at 30 m are strongly associated with a K_{in} value across CP. Thus, there is anatomical, developmental (67), and pharmacologic evidence (118) that K_{in} for the BCSFB can be obtained from the fast component of CSF uptake (119).

K_{in} for urea transfer across rat BCSFB decreases with postnatal age from 1 week to adulthood (67). This is likely due to CP tightening/maturation, as manifested by a substantial *increase* after 1 week in the K_{in} for Na⁺ actively transported into the ventricles (67, 102). Moreover, the K_{in} for CSF uptake of mannitol, a nonelectrolyte larger than urea, also *decreases* over this same postnatal interval (Fig. 6). Slower penetrations of water-soluble urea and mannitol with advancing age are expected as the CP tightens up for more efficient CSF secretion into the ventricles. Barrier tightening should be reflected by an increase in the BCSFB K_{in} ratio of urea/mannitol. This is the case as the K_{in} ratio triples, from 4.4 to 14.3, between 2 weeks and adulthood (Fig. 6). Another instance of an augmented K_{in} ratio for urea/mannitol occurs after acute hypertension when CP tight junctions are damaged, consequently allowing greater molecular leakage into CSF (117). Thus, by comparing permeation rates of two solutes, as with the K_{in} ratio above, and relating this to the ratio of their free diffusion coefficients, D, cm^2/s (117), insight is gained on the degree of restriction offered to molecules diffusing through barriers (Fig. 6).

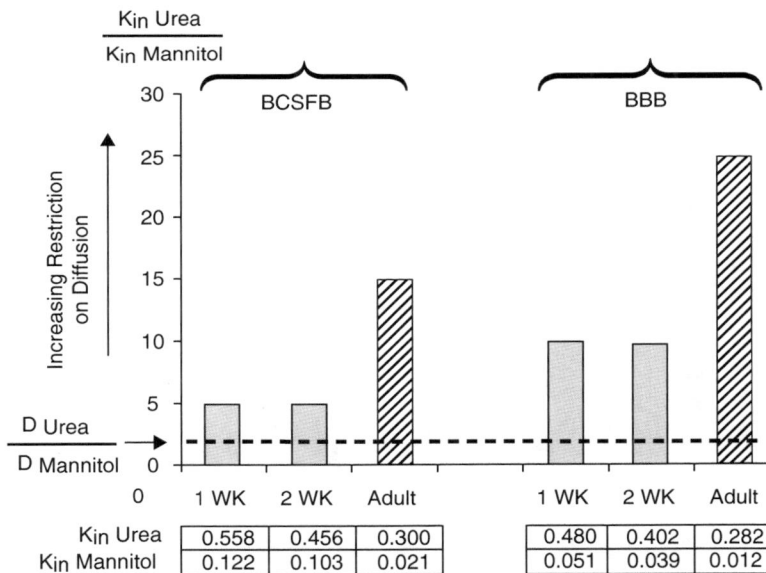

	K_{in} Urea	1 WK	2 WK	Adult	1 WK	2 WK	Adult
	K_{in} Urea	0.558	0.456	0.300	0.480	0.402	0.282
	K_{in} Mannitol	0.122	0.103	0.021	0.051	0.039	0.012

Fig. 6. Nonelectrolyte influx coefficient ratio analysis: K_{in} values for ^{14}C-urea (31, 102) and ^{3}H-mannitol (67) were obtained for fast-component CSF uptake across CP, the BCSFB, vs. cortical BBB uptake in Sprague-Dawley rats ($n = 6$ for 1 week, 2 weeks, and adult > 5 weeks). Mean values for K_{in} (mL/g/m) are presented below the graph; slopes for the early linear part of uptake curve, V_d vs. time, were statistically compared and, for each region, the adult values were significantly different ($P < 0.05$) from those at both 1 and 2 weeks. The ratio of the free diffusion coefficients (D), urea/mannitol, in agar gel is 2, as indicated by horizontal broken line. In biological systems, e.g., capillaries or epithelial membranes, the larger solute will undergo proportionally more sieving due to greater restriction to diffusion, thus increasing the K_{in} (urea/mannitol) ratio. Therefore, the extent to which K_{in} (urea/mannitol) exceeds D (urea/mannitol) presumably reflects the degree of restriction to diffusion across the barrier interfaces. With ongoing maturation, there was tightening of both the BCSFB and BBB, i.e., more molecular sieving after 2 week, as indicated by the sharply rising values in the ratio of K_{in}urea/K_{in}mannitol; after 2 week the CSF formation in rats also rapidly increases (67). Thus, in infant animals at 1–2 week with evidently less barrier hindrance to diffusion into CSF and cortex, the radiolabeled urea and mannitol were cleared from blood at rates closer to free unhindered diffusion.

Polar molecules with low-permeability properties have also been tested for penetrability across in situ and in vitro preparations of CP. A classical experiment (120) analyzed permeation rates of several nonelectrolytes of graded molecular weights, diffusing from isolated CP (bathed in CSF containing the test molecules) to venous blood draining the plexus preparation; this enabled calculation of BCSFB permeability coefficients, P, cm/s, for urea, mannitol, sucrose, and inulin (120). More recently, Strazielle et al. (98) used primary cultures of rat CP as a transwell monolayer system to quantify fluxes of extracellular markers that penetrate tight junctions vs. drugs that permeate CP epithelial cells. Their determination of P values for hydrophilic nonelectrolyte penetration across the rat monolayer compares favorably with previous P values for in vitro rabbit CP (120) (Fig. 7). Comparable P values and impedance, found for in vitro vs. in vivo preparations, augurs well for using cultured CP preparations to pharmacologically assess the BCSFB. Because CP cell culture monolayer permeability varies, however, according to differential expression and localization of tight junction proteins (121), it is imperative in transport and permeability analyses to characterize each cultured CP preparation for its particular barrier properties.

7.3. Proteins

CSF protein concentration in adult mammals is much lower than plasma, by 2–3 orders of magnitude. This substantial protein gradient, from blood to CSF, reflects the degree of the BCSFB

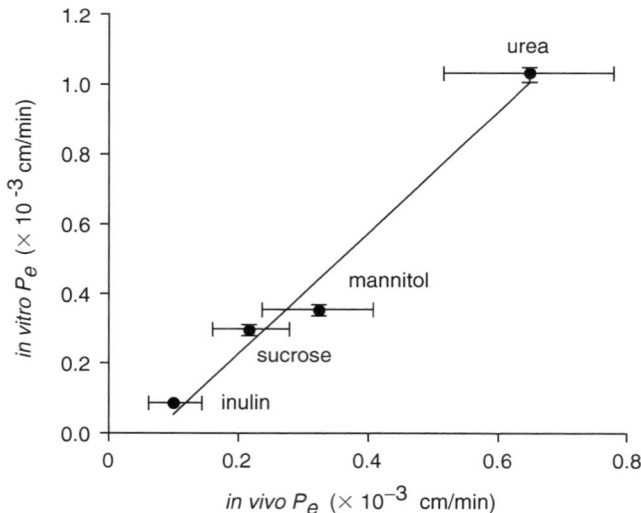

Fig. 7. A comparison of in vitro vs. in vivo CP permeability to water-soluble substances: Permeability coefficients, Pe, were determined for a series of graded molecular weights of hydrophilic nonelectrolytes. The in vitro coefficients were measured in a Transwell system, with means ± SEM for $n = 3$ or 4 [98]. Pe values were similar in the in vivo rabbit CP epithelium [120] vs. an in vitro primary culture monolayer of rat CP cells [98]. The straight line is from linear regression analysis, $r = 0.98$. Thus, the four polar molecules assessed displayed similar permeativities in the in vitro and intact systems. Reproduced with permission [98].

permeability to macromolecules. Even though the "leaky" tight junctions of the BCSFB (70), like the proximal tubule, are substantially more permeable than the BBB, still, plasma proteins penetrate the BCSFB very slowly compared to Na$^+$, urea, and water. Even small proteins like microperoxidase (~2,000 Da) are blocked by CP tight junctions (122). This molecular sieving of plasma proteins and polypeptides by CP is important for homeostasis of CSF osmotic pressure and sink action (102), as well as for regulation of CSF-mediated immune and endocrine signaling (11). Protein trafficking across the epithelial BCSFB is finely controlled by mechanisms differing from the endothelial BBB.

Flux regulation for most proteins at the BCSFB does *not* begin at the choroidal capillary wall. In marked contrast to microvascular/interstitial functional relationships in the brain, where plasma protein diffusion is constrained by the cerebral capillary wall, many plasma proteins in CP readily escape the plexus capillaries and gain entrance to the interstitium (Fig. 4). Lacking a lymphatic drainage system, the CP disposes of unneeded protein by phagocytosis via fibroblasts and macrophages in the interstitium (46). In addition, other extravasated protein molecules are either endocytosed by epithelial cells/lysosomes for breakdown or diffuse into the basolateral clefts up to the tight junctions (46). *Zonulae occludentes* at the apex of CP epithelial cells, together with the basolateral membrane, are the true anatomic substrates of the BCSFB and therefore greatly restrict diffusion passage of most proteins and polypeptides into CSF. CP tight junctions appear early in fetal development (123), indicating that protein access to CSF is carefully regulated at all life stages.

Due to very slow penetration rates of proteins from blood to CSF, it has not been feasible to determine influx coefficient, K_{in}, or permeability-surface area products for plasma-borne radiolabeled proteins, as has been the case for ions and small nonelectrolytes (67). Rather, protein steady-state ratio data, CSF/plasma, along with biochemical information about protein size, yield deductive information about macromolecular transport across BCSFB (124, 125). CSF sampling *source* is significant. The closer the CSF specimen is to CP, e.g., ventricular or nascent CSF, the more valid it is to attribute BCSFB modeling, using CSF/plasma ratios, to transport and permeability mechanisms in CP. Due to anatomical and functional complexity of CP, and the wide variety of protein species (ranging from α1-antitrypsin ~45,000 Da to β-lipoprotein ~2,239,000 Da) translocated across BCSFB, it is not surprising that multimechanistic heterogeneous transport models (125) provide better data fit than homogeneous ones.

There are several routes/mechanisms in CP of adults, for plasma proteins to reach CSF involving diffusion as well as nondiffusion processes. At least one CSF accession route involves diffusion through pore-like structures or channels (124). Evidence for diffusion stems from observations that the concentration of many plasma proteins in CSF is inversely proportional to their respective molecular weights (124). It is challenging with EM to unequivocally identify tight junction pores or discontinuities relating to protein permeation (126), but a tight junction discontinuity of even 0.08% of the total perimeter (124) would enable a pore-like route for a small leak of protein into CSF. A dual-mechanism model of protein permeation across the CP epithelium, incorporating both diffusion of smaller proteins across pores of uniform size and putative pinocytosis of larger proteins, was proposed (124). CSF protein levels have been correlated (125) with the CP tight junction "pores" of varying diameters as described previously (127). The latter findings were supported by 3-D tight junctional analyses (128).

Both in vitro (120) and in vivo (129) models of CP-CSF have furnished evidence for bulk flow or convective, nondiffusion components of transport through CP. Pinocytotic uptake of protein occurs at the basolateral membrane, but it is difficult to ultrastructurally demonstrate subsequent apical membrane exocytosis into CSF (46). Transcellular transfer of protein, especially early in development, likely occurs in CP by the tubulocisternal endoplasmic reticulum; the selectivity of this process for transport to fetal CSF has been demonstrated for different forms of albumin (123), suggesting specific endocytotic mechanisms that initiate the translocation. Transcellular protein transport at the BCSFB awaits further analyses in adult mammals, including the possible role of cytoplasmic tubular networks in conveying proteins across CP to CSF.

The steady-state CSF/serum albumin ratio, Qalbumin, has long been used as an endogenous index of BBB permeability status. Normally Qalbumin is about 0.005 (130). An elevated Qalbumin ratio reflects BBB breakdown. In a generic sense, "BBB" collectively denotes barriers in brain capillaries as well as elements of the BCSFB. Although changes in Qalbumin manifest altered brain capillary function, they may also indicate damage to CP as well. Indeed, a study of Qalbumin in aging patients revealed increased permeability of the BCSFB in older subjects (131).

For IgG (~160,000 Da), a much larger molecule than albumin (~68,000 Da), the steady-state CSF/plasma ratio (QIgG) is only 0.0027 (130). IgG penetration of CP, at least transcellularly (132), was not detectable by peroxidase or electron microscopy. Leakage of IgG from plasma to CSF is so slow that IgG is used as a reference protein in CSF to indicate central IgG synthesis, i.e., from analysis of elevated values for the ratio of QIgG/Qalbumin. CSF protein ratio analyses, then, provide valuable information about central protein metabolism as well as BCSFB permeability.

8. Future Investigations

Future investigations of transport and permeability, particularly in isolated preparations of CP (101), should shed further insight on how BCSFB integrity or breakdown, and CSF composition, impact brain metabolism. Although protein gains access to CNS from extrachoroidal sources (133), the CP is likely the preponderant source of CSF protein. With normal aging (101), there is increased BCSFB permeability, reduced CP synthesis of protein, and slower CSF turnover which alters protein concentrations. The isolated lateral ventricle CP, located in the proximal part of the CSF system, is an excellent model for studying how transport of proteins across the BCSFB is affected by the altered CSF hydrodynamics and choroid epithelial metabolism of aging. A promising area is to analyze the effects of extracellular chaperone proteins in CSF to stabilize the folding of peptides (134). Isolating the CP eliminates the potential confounding CSF variables associated with extrachoroidal protein input to the ventricles and subarachnoid space. The isolated CP model, assessed longitudinally with age (101), affords a unique opportunity to precisely assess the effects of BCSFB transport and permeability on the global metabolic economy of the brain.

References

1. Johanson CE (1988) The choroid plexus-arachnoid-cerebrospinal fluid system. In: Boulton A, Baker G, Walz W (eds) Neuromethods: Neuronal microenviron-ment-electrolytes and water spaces, 1st edn, Humana Press, Clifton.

2. Johanson C (1998) Arachnoid membrane, subarachnoid CSF and pia-glia. In: Pardridge W (ed) An introduction to the blood-brain barrier: methodology and biology, Cambridge University Press, Cambridge

3. Harbut RE, Johanson CE (1986) Third ventricle choroid plexus function and its response to acute perturbations in plasma chemistry. Brain Res 374: 137–146

4. Levine S, Saltzman A, Ginsberg SD (2008) Different inflammatory reactions to vitamin D3 among the lateral, third and fourth ventricular choroid plexuses of the rat. Exp Mol Pathol 85: 117–121

5. Netsky M, Shuangshoti S (1975) The choroid plexus in health and disease. University Press of Virginia, Charlottesville

6. Saunders NR, Habgood MD, Dziegielewska KM (1999) Barrier mechanisms in the brain, II. Immature brain. Clin Exp Pharmacol Physiol 26: 85–91

7. Redzic ZB, Preston JE, Duncan JA, Chodobski A, Szmydynger-Chodobska J (2005) The choroid plexus-cerebrospinal fluid system: from development to aging. Curr Top Dev Biol 71: 1–52

8. Spector R, Johanson C (2006) Micronutrient and urate transport in choroid plexus and kidney: implications for drug therapy. Pharm Res 23: 2515–2524

9. Spector R, Johanson CE (2007) Vitamin transport and homeostasis in mammalian brain: focus on Vitamins B and E. J Neurochem 103: 425–438

10. Costa R, Ferreira-da-Silva F, Saraiva MJ, Cardoso I (2008) Transthyretin protects against A-beta peptide toxicity by proteolytic cleavage of the peptide: a mechanism sensitive to the Kunitz protease inhibitor. PloS one 3: e2899

11. Johanson CE, Duncan JA III, Klinge PM, Brinker T, Stopa EG, Silverberg GD (2008) Multiplicity of cerebrospinal fluid functions: New challenges in health and disease. Cerebrospinal Fluid Res 5: 10

12. Preston JE (2001) Ageing choroid plexus-cerebrospinal fluid system. Microsc Res Tech 52: 31–37

13. Battle T, Preisser L, Marteau V, Meduri G, Lambert M, Nitschke R, Brown PD, Corman B (2000) Vasopressin V1a receptor signaling in a rat choroid plexus cell line. Biochem Biophys Res Commun 275: 322–327

14. Nakamura S, Milhorat TH (1978) Nerve endings in the choroid plexus of the fourth ventricle of the rat: electron microscopic study. Brain Res 153: 285–293

15. Mignini F, Bronzetti E, Felici L, Ricci A, Sabbatini M, Tayebati SK, Amenta F (2000) Dopamine receptor immunohistochemistry in the rat choroid plexus. J Auton Pharmacol 20: 325–332

16. Brightman MW, Reese TS (1969) Junctions between intimately apposed cell membranes in the vertebrate brain. J Cell Biol 40: 648–677

17. Vorbrodt AW, Dobrogowska DH (2003) Molecular anatomy of intercellular junctions in brain endothelial and epithelial barriers: electron microscopist's view. Brain Res Brain Res Rev 42: 221–242

18. Kumagai AK, Dwyer KJ, Pardridge WM (1994) Differential glycosylation of the GLUT1 glucose transporter in brain capillaries and choroid plexus. Biochim Biophys Acta 1193: 24–30

19. Johanson CE, Jones HC, Stopa EG, Ayala C, Duncan JA, McMillan PN (2002) Enhanced expression of the Na-K-2 Cl cotransporter at different regions of the blood-CSF barrier in the perinatal H-Tx rat. Eur J Pediatr Surg 12 Suppl 1: S47–S49

20. Itokazu Y, Kitada M, Dezawa M, Mizoguchi A, Matsumoto N, Shimizu A, Ide C (2006) Choroid plexus ependymal cells host neural progenitor cells in the rat. Glia 53: 32–42

21. Kuchler-Bopp S, Ittel ME, Dietrich JB, Reeber A, Zaepfel M, Delaunoy JP (1998) The presence of transthyretin in rat ependymal cells is due to endocytosis and not synthesis. Brain Res 793: 219–230

22. Dohrmann GJ (1970) Dark and light epithelial cells in the choroid plexus of mammals. J Ultrastruct Res 32: 268–273

23. Johanson CE, Szmydynger-Chodobska J, Chodobski A, Baird A, McMillan P, Stopa EG (1999) Altered formation and bulk absorption of cerebrospinal fluid in FGF-2-induced hydrocephalus. Am J Physiol 277: R263–271

24. Johanson CE, Preston JE, Chodobski A, Stopa EG, Szmydynger-Chodobska J, McMillan PN (1999) AVP V1 receptor-mediated decrease in Cl- efflux and increase in dark cell number in choroid plexus epithelium. Am J Physiol 276: C82–90

25. Johanson CE, Donahue JE, Spangenberger A, Stopa EG, Duncan JA, Sharma HS (2006) Atrial natriuretic peptide: its putative role in modulating the choroid plexus-CSF system for intracranial pressure regulation. Acta Neurochir Suppl 96: 451–456

26. Nag S (1991) Effect of atrial natriuretic factor on permeability of the blood-cerebrospinal fluid barrier. Acta Neuropathol 82: 274–279

27. Bishop GA, Berbari NF, Lewis J, Mykytyn K. (2007) Type III adenylyl cyclase localizes to primary cilia throughout the adult mouse brain. J Comp Neurol 505: 562–571

28. Singhrao SK, Neal JW, Rushmere NK, Morgan BP, Gasque P (1999) Differential expression of individual complement regulators in the brain and choroid plexus. Lab Invest 79: 1247–1259

29. Cserr HF (1971) Physiology of the choroid plexus. Physiol Rev 51: 273–311

30. Pollay M, Stevens FA, Roberts PA (1983) Alteration in choroid-plexus blood flow and cerebrospinal-fluid formation by increased ventricular pressure. In: Wood JH (ed) Neurobiology of cerebrospinal fluid, Raven Press, New York

31. Johanson CE, Woodbury DM (1978) Uptake of [14C]urea by the in vivo choroid plexus–cerebrospinal fluid–brain system: identification of sites of molecular sieving. J Physiol 275: 167–176

32. Spector R, Johanson CE (1989) The mammalian choroid plexus. Sci Am 261: 68–74.

33. Husted RF, Reed DJ (1977) Regulation of cerebrospinal fluid bicarbonate by the cat choroid plexus. J Physiol 267: 411–428

34. Reed DJ, Yen MH (1978) The role of the cat choroid plexus in regulating cerebrospinal fluid magnesium. J Physiol 281: 477–485

35. Ghersi-Egea JF, Strazielle N (2001) Brain drug delivery, drug metabolism, and multidrug resistance at the choroid plexus. Micros Res Tech 52: 83–88

36. Vercellino M, Votta B, Condello C, Piacentino C, Romagnolo A, Merola A, Capello E, Mancardi GL, Mutani R, Giordana MT, Cavalla P (2008) Involvement of the choroid plexus in multiple sclerosis autoimmune inflammation: a neuropathological study. J Neuroimmunol 199: 133–141

37. Nilsson C, Lindvall-Axelsson M, Owman C (1992) Neuroendocrine regulatory mechanisms in the choroid plexus-cerebrospinal fluid system. Brain Res Brain Res Rev 17: 109–138

38. Chodobski A, Szmydynger-Chodobska J (2001) Choroid plexus: target for polypeptides

and site of their synthesis. Microsc Res Tech 52: 65–82

39. Smith DE, Johanson CE, Keep RF (2004) Peptide and peptide analog transport systems at the blood-CSF barrier. Adv Drug Deliv Rev 56: 1765–1791

40. Johanson CE, Reed DJ, Woodbury DM (1974) Active transport of sodium and potassium by the choroid plexus of the rat. J Physiol 241: 359–372

41. Faraci FM, Heistad DD (1992) Does basal production of nitric oxide contribute to regulation of brain-fluid balance? Am J Physiol 262: H340–344

42. Szmydynger-Chodobska J, Chodobski A, Johanson CE (1994) Postnatal developmental changes in blood flow to choroid plexuses and cerebral cortex of the rat. Am J Physiol 266: R1488–1492

43. Zagorska-Swiezy K, Litwin JA, Gorczyca J, Pitynski K, Miodonski AJ (2008) The microvascular architecture of the choroid plexus in fetal human brain lateral ventricle: a scanning electron microscopy study of corrosion casts. J Anat 213: 259–265

44. Weiger T, Lametschwandtner A, Hodde KC, Adam H (1986) The angioarchitecture of the choroid plexus of the lateral ventricle of the rabbit. A scanning electron microscopic study of vascular corrosion casts. Brain Res 378: 285–296

45. Motti ED, Imhof HG, Janzer RC, Marquardt K, Yasargil GM (1986) The capillary bed in the choroid plexus of the lateral ventricles: a study of luminal casts. Scan Elec Micros (Pt 4): 1501–1513

46. Hurley JV, Anderson RM, Sexton PT (1981) The fate of plasma protein which escapes from blood vessels of the choroid plexus of the rat–an electron microscope study. J Pathol 134: 57–70

47. Brightman MW (1975) Ultrastructural characteristics of adult choroid plexus: Relation to the blood-cerebrospinal fluid barrier to proteins. In: Netsky MG, Shuangshoti S (eds) The choroid plexus in health and disease, University Press of Virginia, Charlottesville, VA

48. Bouchaud C, Le Bert M, Dupouey P (1989) Are close contacts between astrocytes and endothelial cells a prerequisite condition of a blood-brain barrier? The rat subfornical organ as an example. Biol Cell 67: 159–165

49. Peress NS, Tompkins D (1981) Effect of molecular charge on choroid-plexus permeability: Tracer studies with cationized ferritins. Cell Tiss Res 219: 425–431

50. Schmidley JW, Wissig SL (1986) Basement membrane of central nervous system capillaries lacks ruthenium red-staining sites. Microvas Res 32: 300–314

51. Liu J, Feng ZC, Yin XJ, Chen H, Lu J, Qiao X (2008) The role of antenatal corticosteroids for improving the maturation of choroid plexus capillaries in fetal mice. Eur J Ped 167: 1209–1212

52. Smith QR, Pershing LK, Johanson CE (1981) A comparative analysis of extracellular fluid volume of several tissues as determined by six different markers. Life Sci 29: 449–456

53. Johanson CE (1980) Permeability and vascularity of the developing brain: cerebellum vs cerebral cortex. Brain Res 190: 3–16

54. Murphy VA, Johanson CE (1990) Na(+)-H(+) exchange in choroid plexus and CSF in acute metabolic acidosis or alkalosis. Am J Physiol 258: F1528–1537

55. Esser S, Wolburg K, Wolburg H, Breier G, Kurzchalia T, Risau W (1998) Vascular endothelial growth factor induces endothelial fenestrations in vitro. J Cell Biol 140: 947–959

56. Nourhaghighi N, Teichert-Kuliszewska K, Davis J, Stewart DJ, Nag S (2003) Altered expression of angiopoietins during blood-brain barrier breakdown and angiogenesis. Lab Invest 83: 1211–1222

57. Braun JS, Le Hir M, Kaissling B (1994) Morphology and distribution of ecto-5'-nucleotidase-positive cells in the rat choroid plexus. J Neurocytol 23: 193–200

58. Tenenbaum T, Matalon D, Adam R, Seibt A, Wewer C, Schwerk C, Galla HJ, Schroten H (2008) Dexamethasone prevents alteration of tight junction-associated proteins and barrier function in porcine choroid plexus epithelial cells after infection with Streptococcus suis in vitro. Brain Res 1229: 1–17

59. Haselbach M, Wegener J, Decker S, Engelbertz C, Galla HJ (2001) Porcine choroid plexus epithelial cells in culture: regulation of barrier properties and transport processes. Micros Res Tech 52: 137–152

60. Oda Y, Nakanishi I (1987) Ultrastructural observations of the development of the fourth ventricular roof in the mouse brain. J Comp Neurol 263: 282–289

61. Urabe N, Naito I, Saito K, Yonezawa T, Sado Y, Yoshioka H, Kusachi S, Tsuji T, Ohtsuka A, Taguchi T, Murakami T, Ninomiya Y (2002) Basement membrane type IV collagen molecules in the choroid plexus, pia mater and capillaries in the mouse brain. Arch Histol Cytol 65: 133–143

62. Utriainen A, Sormunen R, Kettunen M, Carvalhaes LS, Sajanti E, Eklund L, Kauppinen R, Kitten GT, Pihlajaniemi T

(2004) Structurally altered basement membranes and hydrocephalus in a type XVIII collagen deficient mouse line. Hum Mol Genet 13: 2089–2099

63. Serot JM, Foliguet B, Bene MC, Faure GC (2001) Choroid plexus and ageing in rats: a morphometric and ultrastructural study. Eur J Neurosci 14: 794–798

64. Pittella JE, Bambirra EA (1991) Immune complexes in the choroid plexus in liver cirrhosis. Arch Pathol Lab Med 115: 220–222

65. Larsen EH, Willumsen NJ, Mobjerg N, and Sorensen JN (2009) The lateral intercellular space as osmotic coupling compartment in isotonic transport. Acta Physiol (Oxf) 195: 171–186

66. Hill AE (2008) Fluid transport: A guide for the perplexed. J Mem Biol 223: 1–11

67. Smith QR, Woodbury DM, Johanson CE (1982) Kinetic analysis of [36Cl]-, [22Na]- and [3H]mannitol uptake into the in vivo choroid plexus-cerebrospinal fluid brain system: ontogeny of the blood brain and blood-CSF barriers. Brain Res 255: 181–198

68. Wolburg H, Wolburg-Buchholz K, Liebner S, Engelhardt B (2001) Claudin-1, claudin-2 and claudin-11 are present in tight junctions of choroid plexus epithelium of the mouse. Neurosci Lett 307: 77–80

69. Amasheh S, Milatz S, Krug SM, Markov AG, Gunzel D, Amasheh M, Fromm M (2009) Tight junction proteins as channel formers and barrier builders. Ann NY Acad Sci 1165: 211–219

70. Rapoport SI (1976) The blood-brain barrier in physiology and medicine, Raven, New York

71. Miller DS (2004) Confocal imaging of xenobiotic transport across the choroid plexus. Adv Drug Del Rev 56: 1811–1824

72. Rao VV, Dahlheimer JL, Bardgett ME, Snyder AZ, Finch RA, Sartorelli AC, Piwnica-Worms D. (1999) Choroid plexus epithelial expression of MDR1 P glycoprotein and multidrug resistance-associated protein contribute to the blood-cerebrospinal-fluid drug-permeability barrier. Proc Natl Acad Sci U S A 96: 3900–3905

73. Praetorius J, Nejsum LN, Nielsen S (2004) A SCL4A10 gene product maps selectively to the basolateral plasma membrane of choroid plexus epithelial cells. Am J Physiol Cell Physiol 286: C601–610

74. Markovic I, Segal M, Djuricic B, Redzic Z (2008) Kinetics of nucleoside uptake by the basolateral side of the sheep choroid plexus epithelium perfused in situ. Exp Physiol 93: 325–333

75. Weir AP, Burton EA, Harrod G, Davies KE (2002) A- and B-utrophin have different expression patterns and are differentially up-regulated in mdx muscle. J Biol Chem 277: 45285–45290

76. Haenggi T, Fritschy JM (2006) Role of dystrophin and utrophin for assembly and function of the dystrophin glycoprotein complex in non-muscle tissue. Cell Mol Life Sci 63: 1614–1631

77. Knuesel I, Bornhauser BC, Zuellig RA, Heller F, Schaub MC, Fritschy JM (2000) Differential expression of utrophin and dystrophin in CNS neurons: an in situ hybridization and immunohistochemical study. J Comp Neurol 422: 594–611

78. Gorecki DC, Abdulrazzak H, Lukasiuk K, Barnard EA (1997) Differential expression of syntrophins and analysis of alternatively spliced dystrophin transcripts in the mouse brain. Eur J Neurosci 9: 965–976

79. Khurana TS, Watkins SC, Kunkel LM (1992) The subcellular distribution of chromosome 6-encoded dystrophin-related protein in the brain. J Cell Biol 119: 357–366

80. Prochniewicz E, Henderson D, Ervasti JM, Thomas DD (2009) Dystrophin and utrophin have distinct effects on the structural dynamics of actin. Proc Natl Acad Sci U S A 106: 7822–7827

81. Mobasheri A, Wray S, Marples D (2005) Distribution of AQP2 and AQP3 water channels in human tissue microarrays. J Mol Histol 36: 1–14

82. Cornford EM, Varesi JB, Hyman S, Damian RT, Raleigh MJ (1997) Mitochondrial content of choroid plexus epithelium. Exp Brain Res. Experimentelle Hirnforschung 116: 399–405

83. Oldendorf WH, Cornford ME, Brown WJ (1977) The large apparent work capability of the blood-brain barrier: a study of the mitochondrial content of capillary endothelial cells in brain and other tissues of the rat. Ann Neurol 1: 409–417

84. Mathew TC (2007) Diversity in the surface morphology of adjacent epithelial cells of the choroid plexus: an ultrastructural analysis. Mol Cell Biochem 301: 235–239

85. De Spiegelaere W, Casteleyn C, Van den Broeck W, Simoens P (2008) Electron microscopic study of the porcine choroid plexus epithelium. Anat Histol Embryol 37: 458–463

86. Agnew WF, Alvarez RB, Yuen TG, Crews AK (1980) Protein synthesis and transport by the rat choroid plexus and ependyma: an autoradiographic study. Cell Tiss Res 208: 261–281

87. Gudeman DM, Brightman MW, Merisko EM, Merril CR (1989) Release from live choroid plexus of apical fragments and electrophoretic characterization of their synthetic products. J Neurosci Res 24: 184–191

88. Johanson C, McMillan P, Palm D, Stopa E, Doberstein C, Duncan JA (2003) Volume transmission-mediated protective impact of choroid plexus-CSF growth factors on forebrain ischemic injury. In: Sharma H, Westman J (eds) Blood-spinal cord and brain barriers in health and disease, Academic Press, San Diego

89. Keep RF, Jones HC (1990) A morphometric study on the development of the lateral ventricle choroid plexus, choroid plexus capillaries and ventricular ependyma in the rat. Brain Res Dev Brain Res 56: 47–53

90. Speake T, Brown PD (2004) Ion channels in epithelial cells of the choroid plexus isolated from the lateral ventricle of rat brain. Brain Res 1005: 60–66

91. Keep RF, Xiang J, Betz AL (1994) Potassium cotransport at the rat choroid plexus. Am J Physiol 267: C1616–1622

92. Wu Q, Delpire E, Hebert SC, Strange K (1998) Functional demonstration of Na+-K+-2Cl- cotransporter activity in isolated, polarized choroid plexus cells. Am J Physiol 275: C1565–1572

93. Brown PD, Davies SL, Speake T, Millar ID (2004) Molecular mechanisms of cerebrospinal fluid production. Neurosci 129: 957–970

94. Banizs B, Komlosi P, Bevensee MO, Schwiebert EM, Bell PD, Yoder BK (2007) Altered pH(i) regulation and Na(+)/HCO3(-) transporter activity in choroid plexus of cilia-defective Tg737(orpk) mutant mouse. Am J Physiol 292: C1409–1416

95. Banizs B, Pike MM, Millican CL, Ferguson WB, Komlosi P, Sheetz J, Bell PD, Schwiebert EM, Yoder BK (2005) Dysfunctional cilia lead to altered ependyma and choroid plexus function, and result in the formation of hydrocephalus. Development 132: 5329–5339

96. Monkkonen KS, Hakumaki JM, Hirst RA, Miettinen RA, O'Callaghan C, Mannisto PT, Laitinen JT (2007) Intracerebroventricular antisense knockdown of G alpha i2 results in ciliary stasis and ventricular dilatation in the rat. BMC Neurosci 8: 26

97. Sawamoto K, Wichterle H, Gonzalez-Perez O, Cholfin JA, Yamada M, Spassky N, Murcia NS, Garcia-Verdugo JM, Marin O, Rubenstein JL, Tessier-Lavigne M, Okano H, Alvarez-Buylla A (2006) New neurons follow the flow of cerebrospinal fluid in the adult brain. Science 311: 629–632

98. Strazielle N, Belin MF, Ghersi-Egea JF (2003) Choroid plexus controls brain availability of anti-HIV nucleoside analogs via pharmacologically inhibitable organic anion transporters. AIDS 17: 1473–1485

99. Strazielle N, Ghersi-Egea JF (2005) In vitro investigation of the blood-cerebrospinal fluid barrier properties: Primary cultures and immortalized cell lines of the choroidal epithelium. In: Zheng W, Chodobski A (eds) The blood-cerebrospinal fluid barrier, 1st ed. Taylor & Francis, Boca Raton

100. Qiao H, May JM (2008) Development of ascorbate transporters in brain cortical capillary endothelial cells in culture. Brain Res 1208: 79–86

101. Chen RL, Kassem NA, Redzic ZB, Chen CP, Segal MB, Preston JE (2009) Age-related changes in choroid plexus and blood-cerebrospinal fluid barrier function in the sheep. Exp Gerontol 44: 289–296

102. Parandoosh Z, Johanson CE (1982) Ontogeny of blood-brain barrier permeability to, and cerebrospinal fluid sink action on, [14C] urea. Am J Physiol 243: R400–407

103. Johansson PA, Dziegielewska KM, Ek CJ, Habgood MD, Mollgard K, Potter A, Schuliga M, Saunders NR (2005) Aquaporin-1 in the choroid plexuses of developing mammalian brain. Cell Tissue Res 322: 353–364

104. Gomori E, Pal J, Abraham H, Vajda Z, Sulyok E, Seress L, Doczi T. (2006) Fetal development of membrane water channel proteins aquaporin-1 and aquaporin-4 in the human brain. Int J Dev Neurosci 24: 295–305

105. Masseguin C, LePanse S, Corman B, Verbavatz JM, Gabrion J (2005) Aging affects choroidal proteins involved in CSF production in Sprague-Dawley rats. Neurobiol Aging 26: 917–927

106. Oshio K, Watanabe H, Song Y, Verkman AS, Manley GT (2005) Reduced cerebrospinal fluid production and intracranial pressure in mice lacking choroid plexus water channel Aquaporin-1. FASEB J 19: 76–78

107. Moon Y, Hong SJ, Shin D, Jung Y (2006) Increased aquaporin-1 expression in choroid plexus epithelium after systemic hyponatremia. Neurosci Lett 395: 1–6

108. Kim JG, Son YJ, Yun CH, Kim YI, Nam-Goong, IS, Park JH, Park SK, Ojeda SR, D'Elia AV, Damante G, Lee BJ (2007) Thyroid transcription factor-1 facilitates cerebrospinal fluid formation by regulating aquaporin-1 synthesis in the brain. J Biol Chem 282: 14923–14931

109. Boassa D, Stamer WD, Yool AJ (2006) Ion channel function of aquaporin-1 natively

expressed in choroid plexus. J Neurosci 26: 7811–7819

110. Migliati E, Meurice N, DuBois P, Fang JS, Somasekharan S, Beckett E, Flynn G, Yool AJ (2009) Inhibition of aquaporin-1 and aquaporin-4 water permeability by a derivative of the loop diuretic bumetanide acting at an internal pore-occluding binding site. Mol Pharmacol 76: 105–112

111. Johanson CE, Foltz FM, Thompson AM (1974) The clearance of urea and sucrose from isotonic and hypertonic fluids perfused through the ventriculo-cisternal system. Exp Brain Res 20: 18–31

112. Davson H, Segal M (1996) Physiology of the CSF and blood-brain barriers, CRC, Boca Raton

113. Bradbury MW, Davson H (1964) The transport of urea, creatinine and certain monosaccharides between blood and fluid perfusing the cerebral ventricular system of rabbits. J Physiol 170: 195–211

114. Flexner LB (1938) Changes in the chemistry and nature of the cerebrospinal fluid during fetal life in the pig. Am J Physiol 124: 131–135

115. Johanson C, Woodbury D (1974) Changes in CSF flow and extracellular space in the developing rat. In: Vernadakis A, Weiner N (eds) Drugs and the developing brain, Plenum, New York

116. Johanson CE, Duncan JA, Stopa EG, Baird A (2005) Enhanced prospects for drug delivery and brain targeting by the choroid plexus-CSF route. Pharm Res 22: 1011–1037

117. Murphy VA, Johanson CE (1985) Adrenergic-induced enhancement of brain barrier system permeability to small nonelectrolytes: choroid plexus versus cerebral capillaries. J Cereb Blood Flow Metab 5: 401–412

118. Smith QR, Johanson CE (1980) Effect of carbonic anhydrase inhibitors and acidosis in choroid plexus epithelial cell sodium and potassium. J Pharmacol Exp Ther 215: 673–680

119. Johanson CE (1989) Potential for pharmacological manipulation of the blood-cerebrospinal fluid barrier. In: Neuwelt E (ed) Implications of the blood-brain barrier and its manipulation: Basic science aspects, Plenum, New York

120. Welch K, Sadler K (1966) Permeability of the choroid plexus of the rabbit to several solutes. Am J Physiol 210: 652–660

121. Szmydynger-Chodobska J, Pascale CL, Pfeffer AN, Coulter C, Chodobski A (2007) Expression of junctional proteins in choroid plexus epithelial cell lines: a comparative study. Cerebrospinal Fluid Res 4: 11

122. Reese TS, Feder N, Brightman MW (1971) Electron microscopic study of the blood-brain and blood-cerebrospinal fluid barriers with microperoxidase. J Neuropathol Exper Neurol 30: 137–138

123. Johansson PA, Dziegielewska KM, Ek CJ, Habgood MD, Liddelow SA, Potter AM, Stolp HB, Saunders NR (2006) Blood-CSF barrier function in the rat embryo. Eur J Neurosci 24: 65–76

124. Rapoport SI, Pettigrew KD (1979) A heterogenous, pore-vesicle membrane model for protein transfer from blood to cerebrospinal fluid at the choroid plexus. Microvas Res 18: 105–119

125. Kluge H, Hartmann W, Mertins B, Wieczorek V (1986) Correlation between protein data in normal lumbar CSF and morphological findings of choroid plexus epithelium: a biochemical corroboration of barrier transport via tight junction pores. J Neurol 233: 195–199

126. Mollgard K, Lauritzen B, Saunders NR (1979) Double replica technique applied to choroid plexus from early foetal sheep: completeness and complexity of tight junctions. J Neurocytol 8: 139–149

127. van Deurs B, Koehler JK (1979) Tight junctions in the choroid plexus epithelium. A freeze-fracture study including complementary replicas. J Cell Biol 80: 662–673

128. Bundgaard M (1984) The three-dimensional organization of tight junctions in a capillary endothelium revealed by serial-section electron microscopy. J Ultrastruct Res 88: 1–17

129. Liddelow SA, Dziegielewska KM, Ek CJ, Johansson PA, Potter AM, Saunders NR (2009) Cellular transfer of macromolecules across the developing choroid plexus of Monodelphis domestica. Eur J Neurosci 29: 253–266

130. Keir G, Thompson EJ (1986) Proteins as parameters in the discrimination between different blood-CSF barriers. J Neurol Sci 75: 245–253

131. Pakulski C, Drobnik L, Millo B (2000) Age and sex as factors modifying the function of the blood-cerebrospinal fluid barrier. Med Sci Monit 6: 314–318

132. Aleshire SL, Hajdu I, Bradley CA, Parl FF (1985) Choroid plexus as a barrier to immunoglobulin delivery into cerebrospinal fluid. J Neurosurg 63: 593–597

133. Broadwell RD, Sofroniew MV (1993) Serum proteins bypass the blood-brain fluid barriers for extracellular entry to the central nervous system. Exp Neurol 120: 245–263

134. Yerbury JJ, Wilson MR (2010) Extracellular chaperones modulate the effects of Alzheimer's patient cerebrospinal fluid on Abeta(1-42) toxicity and uptake. Cell Stress Chaperones 15: 115–121

135. Kato M, Soprano DR, Makover A, Kato K, Herbert J, Goodman DS (1986) Localization of immunoreactive transthyretin (prealbumin) and of transthyretin mRNA in fetal and adult rat brain. Differentiation 31: 228–235

136. Gath U, Hakvoort A, Wegener J, Decker S, Galla H (1997) Porcine choroid plexus cells in culture: expression of polarized phenotype, maintenance of barrier properties and apical secretion of CSF-components. Eur J Cell Biol 74: 68–78

137. Montonen O, Aho M, Uitto J, Aho S (2001) Tissue distribution and cell type-specific expression of p120ctn isoforms. J Histochem Cytochem 49: 1487–1496

138. Shi LZ, Li GJ, Wang S, Zheng W (2008) Use of Z310 cells as an in vitro blood-cerebrospinal fluid barrier model: tight junction proteins and transport properties. Toxicol In Vitro 22: 190–199

139. Javaheri S, Wagner KR (1993) Bumetanide decreases canine cerebrospinal fluid production. In vivo evidence for NaCl cotransport in the central nervous system. J Clin Invest 92: 2257–2261

140. Praetorius J (2007) Water and solute secretion by the choroid plexus. Pflugers Arch 454: 1–18

141. Johanson CE, Parandoosh Z, Smith QR (1985) Cl-HCO3 exchange in choroid plexus: analysis by the DMO method for cell pH. Am J Physiol 249: F478–484

142. Johanson CE, Parandoosh Z, Dyas ML (1992) Maturational differences in acetazolamide-altered pH and HCO3 of choroid plexus, cerebrospinal fluid, and brain. Am J Physiol 262: R909–914

143. Johanson CE, Reed DJ, Woodbury DM (1976) Developmental studies of the compartmentalization of water and electrolytes in the choroid plexus of the neonatal rat brain. Brain Res 116: 35–48

144. Smith QR, Johanson CE (1985) Active transport of chloride by lateral ventricle choroid plexus of the rat. Am J Physiol 249: F470–477

145. Johanson CE (1984) Differential effects of acetazolamide, benzolamide and systemic acidosis on hydrogen and bicarbonate gradients across the apical and basolateral membranes of the choroid plexus. J Pharmacol Exp Ther 231: 502–511

146. Millar ID, Brown PD (2008) NBCe2 exhibits a 3 HCO3(-):1 Na+ stoichiometry in mouse choroid plexus epithelial cells, Biochem Biophys Res Commun 373: 550–554

147. Johanson CE, Murphy VA (1990) Acetazolamide and insulin alter choroid plexus epithelial cell [Na+], pH, and volume. Am J Physiol 258: F1538–1546

148. Bondy C, Chin E, Smith BL, Preston GM, Agre P (1993) Developmental gene expression and tissue distribution of the CHIP28 water-channel protein. Proc Natl Acad Sci U S A 90: 4500–4504

149. Praetorius J, Nielsen S (2006) Distribution of sodium transporters and aquaporin-1 in the human choroid plexus. Am J Physiol Cell Physiol 291: C59–67

Chapter 5

The Blood-Retinal Barrier: Structure and Functional Significance

E. Aaron Runkle and David A. Antonetti

Abstract

Formation and maintenance of the blood-retinal barrier is required for proper vision and loss of this barrier contributes to the pathology of a wide number of retinal diseases. The retina is responsible for converting visible light into the electrochemical signal interpreted by the brain as vision. Multiple cell types are required for this function, which are organized into eight distinct cell layers. These neural and glial cells gain metabolic support from a unique vascular structure that provides the necessary nutrients while minimizing interference with light sensing. In addition to the vascular contribution, the retina also possesses an epithelial barrier, the retinal pigment epithelium, which is located at the posterior of the eye and controls exchange of nutrients with the choroidal vessels. Together the vascular and epithelial components of the blood-retinal barrier maintain the specialized environment of the neural retina. Both the vascular endothelium and pigment epithelium possess a well-developed junctional complex that includes both adherens and tight junctions conferring a high degree of control of solute and fluid permeability. Understanding induction and regulation of the blood-retinal barrier will allow the development of therapies aimed at restoring the barrier when compromised in disease or allowing the specific transport of therapies across this barrier when needed. This chapter will highlight the anatomical structure of the blood-retinal barrier and explore the molecular structure of the tight junctions that provide the unique barrier properties of the blood-retinal barrier.

Key words: Retina, Tight junctions, Blood-retinal barrier, Claudin, Occludin, Zonula occludens, Endothelial cell, Retinal pigmented epithelium

1. Introduction

1.1. Overview

The retina is a highly specialized, neural tissue responsible for conversion of visible light into the electrochemical signal interpreted by the brain as vision. A unique vascular structure supports the neural retina, providing necessary nutrients but also causing minimal interference with light sensing. This retinal vasculature, like other vasculature in the central nervous system (CNS),

Sukriti Nag (ed.), *The Blood-Brain and Other Neural Barriers: Reviews and Protocols*, Methods in Molecular Biology, vol. 686, DOI 10.1007/978-1-60761-938-3_5, © Springer Science+Business Media, LLC 2011

necessitates a well-developed blood-neural barrier. In addition to the vascular contribution, the retina also possesses an epithelial barrier, the retinal pigment epithelium (RPE), which controls the flow of fluid and nutrients from the highly vascularized choroid into the outer retina. Together the vascular and epithelial components of the blood-retinal barrier (BRB) maintain the specialized environment of the neural retina. Both the vascular endothelium and pigment epithelium possess well-developed junctional complexes that includes both adherens and tight junctions. While the adherens and tight junctions collectively form the BRB, this chapter will focus on the tight junctions as this complex greatly contributes to the unique barrier of the retinal microvasculature. For more details on the adherens junctions, see reviews in the literature (1, 2).

This chapter will describe the BRB in relation to the retinal structure and examine the molecular components of the barrier by highlighting the most well studied proteins of the junctional complex. Mechanisms of barrier induction and important differences between the epithelial and endothelial components of the barrier will be highlighted. This chapter should provide the reader with a fundamental understanding of the BRB structure and function and provide a framework for the study of this barrier in normal physiology and disease.

1.2. The Retina

An understanding of the structure of the retina provides a contextual perspective for understanding the function of the BRB. The retina is located at the posterior of the eye between the vitreous body and the choroid and is composed of vascular cells (pericytes and endothelial cells), macroglia (Müller cells and astrocytes), neurons (photoreceptors, bipolar cells, amacrine cells, horizontal cells, and ganglion cells), pigment epithelium and microglia or resident macrophages (3, 4). These cells are organized into eight distinct layers that make up the retina: the nerve fiber, ganglion cell, inner plexiform, inner nuclear, outer plexiform, outer nuclear, photoreceptors (rods and cones), and the RPE (Fig. 1a) (4). Thus, light must pass through the largely transparent retina before it reaches the light sensitive rods and cones at the outer retina. After conversion of light to the electrochemical signal by the neurons, it travels from the rods and cones through bipolar, horizontal, and amacrine cells to the ganglion cells in the inner retina, which then conducts the signal via the optic nerve, formed by the ganglion cell axons, out of the retina to eventually reach the visual cortex. The outer segments of the rods and cones are continually degraded and resynthesized, resulting in a high metabolic demand. Further, the neural retina requires metabolic support that offers minimal interference to the passage of light. In humans, this is achieved by two independent vascular systems; the retinal and the choroidal (5). The choroidal vessels include a dense, highly permeable capillary network that supports the outer retina, including the rods and

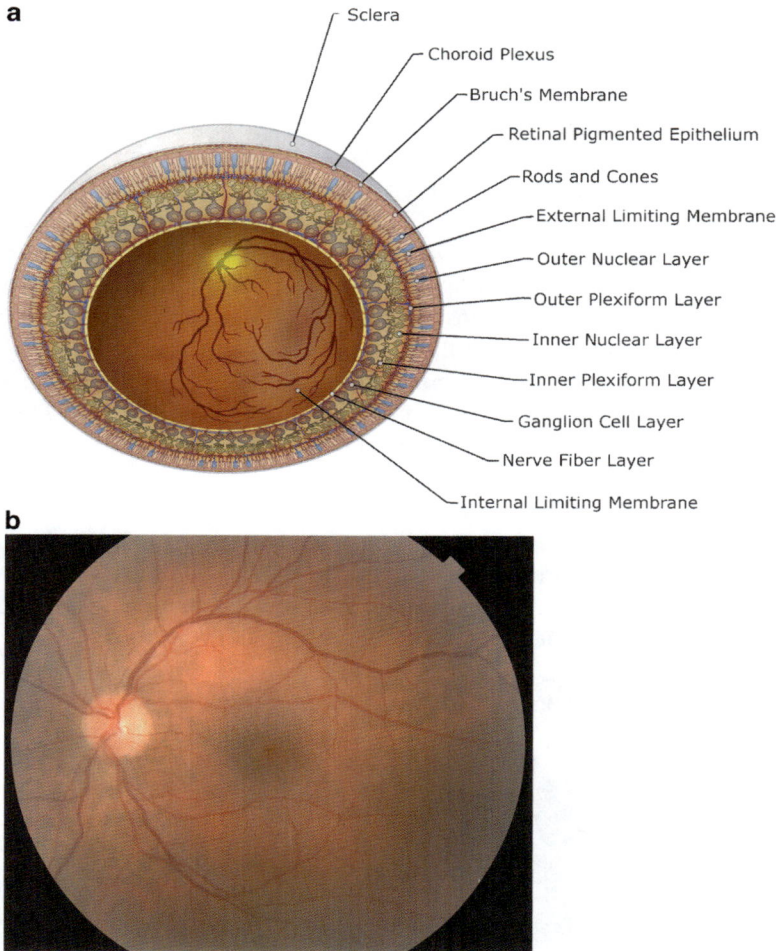

Fig. 1. (**a**) Depiction of the multiple cell types composing the retina as a whole (not drawn to scale). (**b**) Fundus photograph of a healthy retina taken during a routine ophthalmology exam. A physician can use a fundus photograph to evaluate the integrity of the blood-retinal barrier (BRB). The retinal vasculature can be seen to radiate outward from the optic disk (*left, center,* shown in *yellow*).

cones, by diffusion and transport of metabolites across the RPE. The RPE possess well-developed tight junctions and contribute to the BRB. The inner retina, including the ganglion cell layer, is supported by the retinal vascular system that emanates from the central retinal artery in the optic nerve and radiates to the four retinal quadrants (4, 5) (Fig. 1b). These four branches further extend to three capillary plexuses, composed of the inner capillary and two outer capillary beds. The inner capillary bed resides within the nerve fiber layer and ganglion cell layer while the outer capillary networks are located along the inner and outer nuclear layers (6). Primates have an avascular region known as the macula that is highly enriched with cones necessary for the high contrast central vision. Loss of integrity in the vascular barrier and subsequent macular edema is closely associated with vision loss in diabetic retinopathy (7).

1.3. The Blood-Retinal Barrier

The BRB controls the exchange of metabolites and waste products between the vascular lumen and the neural retina and is formed by the interaction of retinal glia and pericytes with the endothelium (Fig. 2). The vascular BRB controls permeability from the retinal blood vessels and consists of a well-developed junctional complex in the vascular endothelial cells as well as limited or no fenestration. Fenestrations are regions of capillary wall thinning, where solutes diffuse out of the capillaries. In fact, loss of endothelial fenestration is one of the initial steps in the development of the blood-brain barrier (BBB) (8). The BRB allows the neural retina to establish and maintain the appropriate environment necessary for neuronal function.

Pericytes support endothelial cells by secreting angiopoietin 1, which induces tight junction protein expression (9). Additionally, studies using a rat brain endothelial cell line and a primary rat pericyte co-culture system reduce permeability to sodium fluorescein, in a manner that is partially-dependent on transforming growth factor β (10). Further evidence for the role of pericytes in barrier induction comes from the platelet-derived growth factor B (PDGF B)

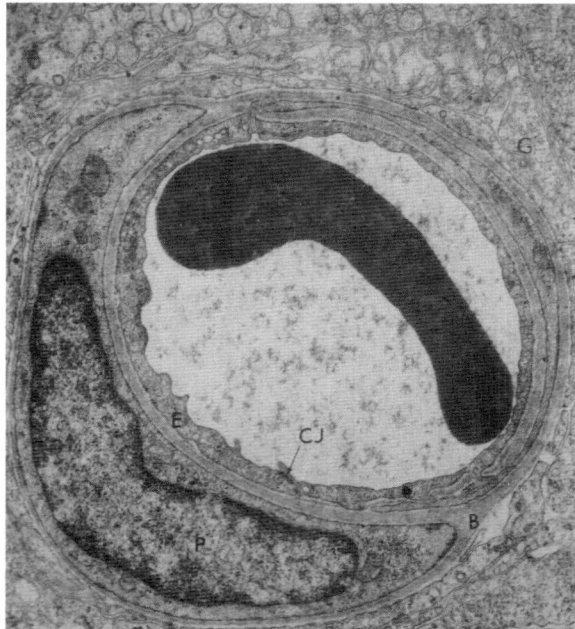

Fig. 2. Electron micrograph showing the cross-section of a rat retinal capillary consisting of the basement membrane (B), the endothelial cells (E), intramural pericyte (P), the glial cells (G), and the cell junction (CJ). The location of the essential cell types contributing to the BRB complete with a red blood cell within the capillary is depicted. Reproduced with permission (16).

and PDGF B receptor β (PDGFR β) knock-out study. In the brain, endothelial cells express PDGF B as a means to recruit pericytes, which express PDGFR β. Knock-out of either receptor or ligand is lethal as a result of hemorrhage and edema (11, 12). Phenotypes associated with this gene deletion include loss of pericytes surrounding brain capillaries, increased capillary diameter, abnormal ultrastructure, and increased vascular endothelial growth factor-A (VEGF-A) content. These studies lend support to the concept that pericytes contribute to the induction of barrier properties.

The induction of the BRB in vascular endothelium can be recapitulated in vitro using cell culture by incubating isolated primary endothelial cells with astrocyte conditioned media together with cAMP analogs, or corticosteroids to induce barrier properties (13–15). Taken together, these observations suggest that the BRB in the retinal vasculature is dependent on production of factors from various retinal cell types. While a number of studies have focused on factors that increase endothelial permeability in disease, the factors that promote barrier induction and the changes in these factors during disease remain ill defined. Interestingly, the RPE appears to form a barrier without external signals, as evidenced by multiple cell lines and primary cultures of RPE, which are able to recapitulate the in vivo barrier. This difference in barrier induction between vascular endothelium and RPE may reflect unique mechanisms by which these two tissues form the BRB during development.

1.4. Molecular Constituents of the Blood-Retinal Barrier

The BRB is composed of both tight and adherens junction complexes (Fig. 3). The retinal vasculature and RPE both have well-developed tight junctions that confer a high degree of control of solute and fluid permeability. The tight junctions are the apical-most junction complex in epithelial cells. However, in the vascular endothelium of the brain and retina, the tight junction and adherens junction complexes are indistinguishable at the ultrastructural level (16). The tight junctions restrict flux of a wide variety of substances such as lipids and protein (17, 18). The retinal capillaries are relatively impermeable, even to particles as small as sodium ions (19). The adherens junctions are essential to development of the barrier and influence the formation of the tight junction (20–23). Together the adherens junctions and tight junctions create the resistance barrier to the neural parenchyma.

In the human retina, expression of tight junction proteins begins at the 24th week of gestation (24). In rat brain, expression of tight junction proteins in the vasculature is detected as early as 1 week postnatal (25). Over time, tight junction protein expression increases in the brain, retinal vasculature, and RPE resulting in decreased permeability and establishment of barriers in brain and retina (25–31).

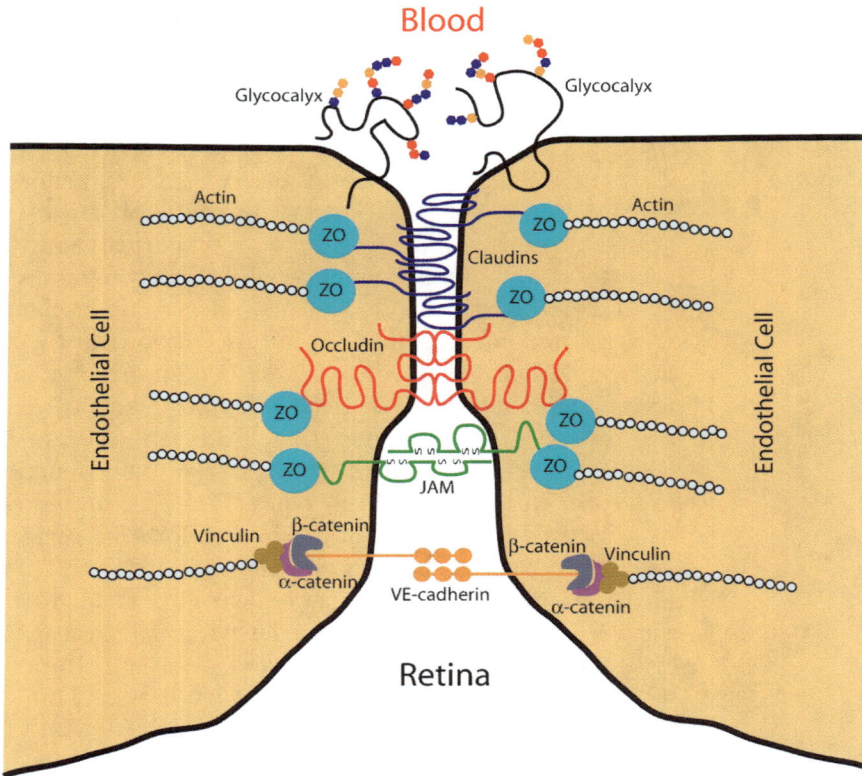

Fig. 3. Model of the BRB which is composed of tight junctions and the adherens junctions is shown. The glycocalyx is depicted within the lumen of the capillary. The tight junction, shown by the transmembrane proteins claudin, occludin, and junctional adhesion molecule (JAM) complexed to the scaffolding protein ZO connected to actin, is the apical most junction complex. Beneath it is the adherens junction, depicted by vascular endothelial (VE) cadherin, complexed with β-catenin, followed by α-catenin, which then link to vinculin, and then to actin.

1.5. Markers for Retinal Pigment Epithelium Cells

Successful BRB cell culture studies are dependent upon a complete, pure preparation of the cell type in question. To this end, it is essential to ensure homogeneity and removal of any contaminating cell types. This section will address markers that can be used to detect and verify RPE cells (Table 1) in culture. One of the more common markers of RPE are the cytokeratin (CK) intermediate filament proteins 8 and 18, although positive expression of these markers only defines an epithelial lineage (32). More specific to the RPE are the expression of the cellular retinaldehyde binding protein (CRALBP) and RPE65. These proteins are essential to the biochemical modification of retinoid in the synthesis of visual pigments for photoreceptors (32). Another RPE-specific marker is bestrophin-1 but a complete understanding of the function of this protein remains ambiguous. A recent paper suggests that bestrophin-1 functions to maintain appropriate intracellular (Ca^{2+}) and regulates pH homeostasis (33). Further, the enzymes tyrosinase or tyrosinase-related proteins,

Table 1
Retinal pigmented epithelial markers

Protein	Function	Reference
Cellular retinaldehyde binding protein (CRALBP)	Retinoid processing for the regeneration of visual pigments used by photoreceptors	(32)
Bestrophin-1	Maintenance of intracellular (Ca^{2+}) and pH homeostasis	(33)
Tyrosinase or Tyrosinase-related proteins	Oxidation of tyrosine in melanin synthesis	(34)
Cytokeratin (CK) 8 and 18	Cell is derived from an epithelium	(32)

which function in the oxidation of tyrosine in melanin synthesis, can be used as RPE markers (34). Finally, when RPE are grown in culture, verification of the presence of tight and adherens junction proteins should be completed prior to experimental analysis of the BRB.

2. Tight Junctions

2.1. Components

Tight junctions form the apical most junctional complex and restrict flux of solute into the tissue parenchyma. At the ultrastructural level, tight junctions appear as regions of fused plasma membranes between adjacent cells (35) and freeze-fracture microscopy shows tight junctions as anastomosing strands (36). In the membrane, tight junctions have two functions: the barrier function, in which they restrict solute flux, and the fence function, in which they restrict the movement of proteins and lipids from the apical to the basolateral surface (17, 18).

Tight junctions are composed of over 40 proteins encompassing transmembrane proteins and intracellular scaffolding proteins, acting in concert to influence barrier properties as well as regulatory proteins. The transmembrane proteins include occludin and tricellulin, claudin family members, and the junctional adhesion molecules (JAMs). The transmembrane proteins are linked to the cytoskeleton via an interaction with the zonula occludens (ZO) family of scaffolding proteins.

2.2. Claudins

The claudin family consists of 24 distinct proteins of size 20–27 kDa that have four membrane spanning domains, two extracellular loops, and an N- and C-terminus within the cytoplasm (37). The C-terminus of claudin is essential for both stability and membrane targeting (38, 39). Claudin family expression patterns vary by tissue, and most claudins localize at tight junctions and contribute to barrier properties (40). For more information on the claudin family see the following reviews (41, 42).

In the RPE, the expression of claudins-1, -2, and -5 have been detected in the developing chick embryo by embryonic day 14 (43). Claudin-1, claudin-5, and claudin-15 have been localized in endothelial cells and contribute to the BRB (40, 44). Specifically, gene deletion of claudin-5 in mice leads to normal development in utero and the tight junctions appear normal by electron microscopy. However, the mice die within 10 h after birth due to brain hemorrhage (45). Under conditions of hypoxia, retinal claudin-5 is reduced by 59% and permeability to 10 kDa dextran is increased (46).

2.3. Occludin

Occludin was the first transmembrane tight junction protein identified and, like the claudins, is composed of a cytoplasmic N-terminus, four membrane spanning domains, two extracellular loops, one intracellular loop, and a cytoplasmic C-terminus; however, they share no sequence conservation with the claudins. Occludin is a 522 amino acid protein with a predicted molecular weight of approximately 59 kDa, but on Western blot the protein runs with an apparent molecular weight of 62 kDa. Occludin is well conserved and has been cloned in human, rat, mouse, chicken, and dog (47).

Occludin contributes to control of barrier properties and tight junction protein trafficking in retinal endothelial cells. A number of laboratories have identified changes in occludin phosphorylation (48–51); most recently occludin phosphorylation is associated with invasion of mononuclear cells across the BBB in human immunodeficiency virus-1 encephalitis (48). Using mass spectrometric analysis, multiple occludin phosphorylation sites were identified in bovine retinal endothelial cells (BRECs) treated with VEGF-A (51). This phosphorylation leads to occludin ubiquitination and tight junction endocytosis. Mutational analysis of one of these phosphorylation sites, Ser490, completely prevented the VEGF-induced endocytosis of tight junction proteins and blocked the VEGF-induced increase in permeability (52). In the RPE, occludin reduction by small interfering RNA (siRNA) increased permeability to small molecules (467 Da tetramethylrhodamine) under hydrostatic pressure when compared to controls, while minimally effecting larger molecules, suggesting that occludin also regulates the RPE barrier in response to hydrostatic pressure changes (53). Together, these studies indicate that occludin contributes to the cells response to external signals and the subsequent regulation of barrier properties. These results are consistent with the complex phenotype of animals with occludin gene deletion, which still form tight junctions but demonstrate a host of abnormalities (54). Similarly, siRNA studies of occludin in Madin Darby canine kidney cells demonstrate an increase in permeability to mono- and divalent inorganic cations and monovalent organic cations and an altered cytoskeletal response to cholesterol depletion (55).

2.4. Junctional Adhesion Molecules

The junctional adhesion molecules, or JAMs, are glycosylated single pass transmembrane proteins, with the C-terminus located intracellularly, and an extracellular N-terminus which contains two immunoglobulin-like domains (37). The JAMs, composed of JAM-A, -B, and -C, directly interact with ZO-1 and Partition-Defective 3 (Par-3). Par-3 forms the polarity complex by interacting with atypical protein kinase C (aPKC) and binding Par-6, through a C-terminal class II postsynaptic density (PSD95), drosophila disc large tumor suppressor (DlgA), and zonula occludens-1 (ZO-1) or PDZ-binding domain motif (56–58). This polarity complex is necessary for the formation or maintenance of the junctional complex in epithelial cells.

In the RPE, JAM-A, AF-6, Par-3 and Par-6 are expressed during the early development of the tight junction (59). JAM-C is expressed in both the RPE and at adherens junctions in the outer limiting membrane of the mouse retina (60). The outer limiting membrane separates the nuclei of the rods and cones from their outer segments, which undergo rapid turnover in association with the RPE. Studies in epithelial cells suggest that JAM-A contributes to the establishment of polarity and junctional assembly (23). However, the function of the JAM proteins in the retina remains to be clarified.

2.5. ZO Proteins

The zonula occludens, or ZO, family members connect the transmembrane tight junction proteins to the cytoskeleton. By binding other ZO proteins they contribute to the creation of the junctional network. ZO-1 (210–225 kDa) was the first tight junction protein identified, and subsequent studies using co-immunoprecipitation identified the other ZO family members, ZO-2 (180 kDa) and ZO-3 (130 kDa) (61–65). In cells lacking tight junctions, ZO-1 and ZO-2 associate with the adherens junctions (66). ZO proteins are characterized by the presence of three PDZ domains, one Src homology 3 (SH3) domain, a guanylate-kinase-like (GUK) domain, and are members of the membrane-associated guanylate kinase (MAGUK) family (61, 63, 67–69). Members of the MAGUK family contribute to the organization of transmembrane signaling complexes such as PSD95 at synaptic termini (70) and InaD in rod photoreceptors (71, 72).

The ZO proteins are essential for organization of tight junctions. Silencing ZO-1 results in delayed formation of tight junctions in a calcium switch assay (73, 74). In this assay, calcium is removed to induce disassembly of the junctional complex, and is then replaced, to allow reassembly. ZO-2 also contributes to junction assembly and permeability as its depletion impeded junction reassembly and increased permeability to 70 kDa dextran (75). Use of a cell system that lacks ZO-1, ZO-2, and ZO-3 reveals that the ZO family is essential for barrier formation by directing insertion of claudins into the membrane (76). In vivo, ZO-1 (77) and ZO-2 (78) gene deletions are both lethal very early in mouse embryogenesis. However, distinct phenotypes suggest non-redundant function for these isoforms. ZO-1 gene deletion results in developmental defects in mouse embryo, yolk sac, and allantoic membrane vasculature suggesting a role for ZO-1 in angiogenesis (77).

3. Breakdown of the BRB in Response to VEGF-A

Elevated VEGF alters the integrity of the BRB in a number of pathological conditions. Much of what is known about this phenomenon relates to the disruption of the tight junction complex. In BREC culture, occludin is phosphorylated and ubiquitinated in response to VEGF-A (52, 79) in a PKC-dependent manner resulting in redistribution of the tight junction complex and increased permeability (80). In vivo, increased paracellular permeability is associated with reduction or redistribution of occludin (81) and models of diabetic complications demonstrate decreased retinal occludin content and increased vascular permeability (82).

In addition to the action of VEGF-A on tight junctions of endothelial cells, VEGF also increases permeability through increased caveolae (83). Hofman et al. injected VEGF-A into the vitreous of monkey eyes and examined vascular permeability using fluorescein angiography. They found that VEGF-A induced fluorescein accumulation increased the number of caveolae. Interestingly, electron microscopy revealed no fenestrations or vesiculo-vacuolar organelles in the VEGF-A-injected eyes. Further, Feng et al. using BREC observed VEGF-A-induced permeability by an endothelial nitric oxide synthase (eNOS) dependent mechanism (84). These studies suggest that VEGF-A induces multiple routes of permeability across retinal endothelial cells increasing both paracellular and transcellular permeability.

4. Induction of the Blood-Retinal Barrier

Astrocytes (85), Müller cells (86), and pericytes cooperate to induce the retinal vascular component of the BRB. Experimental evidence demonstrates vascularization of astrocytes or Müller cells injected into the anterior chamber of the rat eye leads to vessels which have tight barrier properties (85, 86). Perhaps what separates the microvasculature of the CNS from non-neural vasculature is its dependence on supporting cells such as astrocytes and pericytes, as opposed to vasculature in other parts of the body, which gain support from smooth muscle cells (87, 88). A-kinase anchor protein 12 (AKAP12) expression in astrocytes enhances retinal barrier formation by increasing angiopoietin 1 (Ang1) and decreasing VEGF-A (24). In this study, Choi et al. demonstrated that in human astrocytes, AKAP12 promoted a stronger association of hypoxia inducible factor 1α (HIF-1α) with the von Hippel-Lindau (vHL) protein. Under normoxic conditions vHL promotes degradation of HIF-1α through the ubiquitin-proteasome system (89, 90). HIF-1α is an important transcription factor associated with increased hypoxia-mediated VEGF-A expression (91). Ang1 is a ligand for the Tie2 receptor and the association of Ang1 and Tie2 stabilizes blood vessels and protects against VEGF-A-induced neovascularization and permeability (92–94). Choi et al. also demonstrated that conditioned media from AKAP12 astrocytes induced an increase in ZO-1, ZO-2, claudin-1, claudin-3, claudin-5, and occludin in human retinal microvascular endothelial cells and this significantly reduced permeability to 10 kDa dextran (24), while an antibody against Ang1 blocked this effect. These data suggest molecular signal from astrocytes, controlled in part by AKAP12, induce tight junction expression in vascular endothelium and is necessary for a functional barrier.

Glucocorticoids also have a strong effect on endothelial barrier properties. The glucocorticoid receptor (GR) is maintained in an inactive state in the cytoplasm by interacting with heat shock proteins (hsp) to conceal nuclear localization signals (NLS). Upon binding of glucocorticoid with GR, hsps dissociate and unmask NLSs allowing nuclear translocation of the GR to influence gene transcription (15). Activation of the GR induces barrier function by increasing the content of tight junction proteins (14) and redistribution of these proteins to the cell border (95). Glucocorticoid-induced barrier properties has been verified in vivo (96). Interestingly, glucocorticoids reverse occludin phosphorylation, a modification shown to increase permeability (14).

A recent study suggests a transcriptional response to glucocorticoids. In this report, Felinski et al. identified a novel *cis*-acting element, termed the occludin enhancer element (OEE) through mutational analysis of the occludin promoter, and showed that this sequence is both necessary and sufficient to mediate glucocorticoid-induced transcription of occludin and a homologous sequence is found in the claudin-5 promoter (97). These studies suggest a transcriptional mechanism that coordinates expression of tight junction proteins. It remains to be determined whether the same or similar *cis*-elements are utilized in vivo in response to glial signals.

Finally, a recent study showed that Wnt/β-catenin signaling controls claudin-3 expression and modulates barrier properties in vitro and in vivo (98). Canonical Wnt/wingless signaling involves the stabilization of β-catenin, allowing it to translocate to the nucleus where it binds the T-cell factor (TCF) transcription factor to regulate gene expression. In their study, Liebner et al. demonstrated that β-catenin activation results in an increase in claudin-3 expression along with an increased maturation of the BBB (98). These studies point to a transcriptional regulation of the BBB and BRB that 1 day may be amenable to therapeutic intervention.

5. Conclusion

The retina is a highly specialized neural tissue that requires a unique vascular structure and tight control of permeability to allow proper visual function. The regulation of flux of blood-borne metabolites into the retina is controlled by the RPE and specialized blood vessels, which utilize well-developed junctional complexes to regulate permeability and maintain the neural environment of the retina. The junctional complexes that create these barriers are composed of distinct proteins, which have both structural and regulatory functions contributing to barrier properties. Our current understanding of the barrier implicates the claudins in forming the barrier, occludin in regulating permeability, and ZO in assembling the barrier by interacting with the transmembrane proteins. However, a host of additional junctional related proteins have been identified and their role in the BRB remains unclear. Further, the necessity of glia and pericytes to induce the barrier has been clearly demonstrated but we are just beginning to understand the complex interplay between these multiple cell types. Future studies elucidating BRB induction and maintenance will provide a framework for rational therapies to restore this barrier when compromised in disease or allow specific transport of therapies across this barrier when needed.

References

1. Dejana E, Orsenigo F, Lampugnani MG (2008) The role of adherens junctions and VE-cadherin in the control of vascular permeability. J Cell Sci 121:2115–2122

2. Bazzoni G, Dejana E (2004) Endothelial cell-to-cell junctions: molecular organization and

role in vascular homeostasis. Physiol Rev 84:869–901

3. Gardner TW, Antonetti DA, Barber AJ et al (2002) Diabetic retinopathy: more than meets the eye. Surv Ophthalmol 47 Suppl 2:S253–262

4. Erickson KK, Sundstrom JM, Antonetti DA (2007) Vascular permeability in ocular disease and the role of tight junctions. Angiogenesis 10:103–117

5. Pournaras CJ, Rungger-Brandle E, Riva CE et al (2008) Regulation of retinal blood flow in health and disease. Prog Retin Eye Res 27:284–330

6. Gariano RF, Iruela-Arispe ML, Hendrickson AE (1994) Vascular development in primate retina: comparison of laminar plexus formation in monkey and human. Invest Ophthalmol Vis Sci 35: 3442–3455

7. Caldwell RB, Bartoli M, Behzadian MA et al (2003) Vascular endothelial growth factor and diabetic retinopathy: pathophysiological mechanisms and treatment perspectives. Diabetes Metab Res Rev 19:442–455

8. Fenstermacher J, Gross P, Sposito N et al (1988) Structural and functional variations in capillary systems within the brain. Ann N Y Acad Sci 529:21–30

9. Hori S, Ohtsuki S, Hosoya K et al (2004) A pericyte-derived angiopoietin-1 multimeric complex induces occludin gene expression in brain capillary endothelial cells through Tie-2 activation in vitro. J Neurochem 89:503–513

10. Dohgu S, Takata F, Yamauchi A et al (2005) Brain pericytes contribute to the induction and up-regulation of blood-brain barrier functions through transforming growth factor-beta production. Brain Res 1038:208–215

11. Hellstrom M, Gerhardt H, Kalen M et al (2001) Lack of pericytes leads to endothelial hyperplasia and abnormal vascular morphogenesis. J Cell Biol 153:543–553

12. Lindahl P, Johansson BR, Leveen P et al (1997) Pericyte loss and microaneurysm formation in PDGF-B-deficient mice. Science 277:242–245

13. Wolburg H, Neuhaus J, Kniesel U et al (1994) Modulation of tight junction structure in blood-brain barrier endothelial cells. Effects of tissue culture, second messengers and cocultured astrocytes. J Cell Sci 107 (Pt 5):1347–1357

14. Antonetti DA, Wolpert EB, DeMaio L et al (2002) Hydrocortisone decreases retinal endothelial cell water and solute flux coincident with increased content and decreased phosphorylation of occludin. J Neurochem 80:667–677

15. Felinski EA, Antonetti DA (2005) Glucocorticoid regulation of endothelial cell tight junction gene expression: novel treatments for diabetic retinopathy. Curr Eye Res 30:949–957

16. Cunha-Vaz JG, Shakib M, Ashton N (1966) Studies on the permeability of the blood-retinal barrier. I. On the existence, development, and site of a blood-retinal barrier. Br J Ophthalmol 50:441–453

17. Farquhar MG, Palade GE (1963) Junctional complexes in various epithelia. J Cell Biol 17:375–412

18. van Meer G, Simons K (1986) The function of tight junctions in maintaining differences in lipid composition between the apical and the basolateral cell surface domains of MDCK cells. Embo J 5:1455–1464

19. Tornquist P, Alm A, Bill A (1990) Permeability of ocular vessels and transport across the blood-retinal-barrier. Eye 4 (Pt 2):303–309

20. Miyoshi J, Takai Y (2005) Molecular perspective on tight-junction assembly and epithelial polarity. Adv Drug Deliv Rev 57:815–855

21. Suzuki A, Ishiyama C, Hashiba K et al (2002) aPKC kinase activity is required for the asymmetric differentiation of the premature junctional complex during epithelial cell polarization. J Cell Sci 115:3565–3573

22. Fukuhara A, Irie K, Nakanishi H et al (2002) Involvement of nectin in the localization of junctional adhesion molecule at tight junctions. Oncogene 21:7642–7655

23. Fukuhara A, Irie K, Yamada A et al (2002) Role of nectin in organization of tight junctions in epithelial cells. Genes Cells 7:1059–1072

24. Choi YK, Kim JH, Kim WJ et al (2007) AKAP12 regulates human blood-retinal barrier formation by downregulation of hypoxia-inducible factor-1alpha. J Neurosci 27:4472–4481

25. Hirase T, Staddon JM, Saitou M et al (1997) Occludin as a possible determinant of tight junction permeability in endothelial cells. J Cell Sci 110 (Pt 14):1603–1613

26. Dermietzel R, Krause D (1991) Molecular anatomy of the blood-brain barrier as defined by immunocytochemistry. Int Rev Cytol 127:57–109

27. Stewart PA, Hayakawa K (1994) Early ultrastructural changes in blood-brain barrier vessels of the rat embryo. Brain Res Dev Brain Res 78:25–34

28. Rizzolo LJ (1997) Polarity and the development of the outer blood-retinal barrier. Histol Histopathol 12:1057–1067

29. Williams CD, Rizzolo LJ (1997) Remodeling of junctional complexes during the development of the outer blood-retinal barrier. Anat Rec 249:380–388

30. Rubin LL, Staddon JM (1999) The cell biology of the blood-brain barrier. Annu Rev Neurosci 22:11–28

31. Wolburg H, Lippoldt A (2002) Tight junctions of the blood-brain barrier: development, composition and regulation. Vascul Pharmacol 38:323–337

32. Burke JM (2008) Epithelial phenotype and the RPE: is the answer blowing in the Wnt? Prog Retin Eye Res 27:579–595

33. Marmorstein AD, Cross HE, Peachey NS (2009) Functional roles of bestrophins in ocular epithelia. Prog Retin Eye Res 28(3):206–226

34. Dryja TP, O'Neil-Dryja M, Pawelek JM et al (1978) Demonstration of tyrosinase in the adult bovine uveal tract and retinal pigment epithelium. Invest Ophthalmol Vis Sci 17:511–514

35. Fanning AS, Little BP, Rahner C et al (2007) The unique-5 and -6 motifs of ZO-1 regulate tight junction strand localization and scaffolding properties. Mol Biol Cell 18:721–731

36. Staehelin LA (1974) Structure and function of intercellular junctions. Int Rev Cytol 39:191–283

37. Chiba H, Osanai M, Murata M et al (2008) Transmembrane proteins of tight junctions. Biochim Biophys Acta 1778:588–600

38. Ruffer C, Gerke V (2004) The C-terminal cytoplasmic tail of claudins 1 and 5 but not its PDZ-binding motif is required for apical localization at epithelial and endothelial tight junctions. Eur J Cell Biol 83:135–144

39. Arabzadeh A, Troy TC, Turksen K (2006) Role of the Cldn6 cytoplasmic tail domain in membrane targeting and epidermal differentiation in vivo. Mol Cell Biol 26:5876–5887

40. Morita K, Furuse M, Fujimoto K et al (1999) Claudin multigene family encoding four-transmembrane domain protein components of tight junction strands. Proc Natl Acad Sci U S A 96:511–516

41. Angelow S, Ahlstrom R, Yu AS (2008) Biology of claudins. Am J Physiol Renal Physiol 295:F867–876

42. Krause G, Winkler L, Mueller SL et al (2008) Structure and function of claudins. Biochim Biophys Acta 1778:631–645

43. Rahner C, Fukuhara M, Peng S et al (2004) The apical and basal environments of the retinal pigment epithelium regulate the maturation of tight junctions during development. J Cell Sci 117:3307–3318

44. Morita K, Sasaki H, Furuse M et al (1999) Endothelial claudin: claudin-5/TMVCF constitutes tight junction strands in endothelial cells. J Cell Biol 147:185–194

45. Nitta T, Hata M, Gotoh S et al (2003) Size-selective loosening of the blood-brain barrier in claudin-5-deficient mice. J Cell Biol 161:653–660

46. Koto T, Takubo K, Ishida S et al (2007) Hypoxia disrupts the barrier function of neural blood vessels through changes in the expression of claudin-5 in endothelial cells. Am J Pathol 170:1389–1397

47. Ando-Akatsuka Y, Saitou M, Hirase T et al (1996) Interspecies diversity of the occludin sequence: cDNA cloning of human, mouse, dog, and rat-kangaroo homologues. J Cell Biol 133:43–47

48. Persidsky Y, Heilman D, Haorah J et al (2006) Rho-mediated regulation of tight junctions during monocyte migration across the blood-brain barrier in HIV-1 encephalitis (HIVE). Blood 107:4770–4780

49. Elias BC, Suzuki T, Seth A et al (2009) Phosphorylation of Tyr-398 and Tyr-402 in occludin prevents its interaction with ZO-1 and destabilizes its assembly at the tight junctions. J Biol Chem 284:1559–1569

50. Suzuki T, Elias BC, Seth A et al (2009) PKC eta regulates occludin phosphorylation and epithelial tight junction integrity. Proc Natl Acad Sci U S A 106:61–66

51. Sundstrom JM, Tash BR, Murakami T et al (2009) Identification and analysis of occludin phosphosites: a combined mass spectrometry and bioinformatics approach. J Proteome Res 8(2):808–817

52. Phillips BE, Cancel L, Tarbell JM et al (2008) Occludin independently regulates permeability under hydrostatic pressure and cell division in retinal pigment epithelial cells. Invest Ophthalmol Vis Sci 49:2568–2576

53. Saitou M, Furuse M, Sasaki H et al (2000) Complex phenotype of mice lacking occludin, a component of tight junction strands. Mol Biol Cell 11:4131–4142

54. Yu AS, McCarthy KM, Francis SA et al (2005) Knockdown of occludin expression leads to diverse phenotypic alterations in epithelial cells. Am J Physiol Cell Physiol 288:C1231–1241

55. Ebnet K, Suzuki A, Horikoshi Y et al (2001) The cell polarity protein ASIP/PAR-3 directly associates with junctional adhesion molecule (JAM). Embo J 20:3738–3748

56. Itoh M, Sasaki H, Furuse M et al (2001) Junctional adhesion molecule (JAM) binds to PAR-3: a possible mechanism for the recruitment of PAR-3 to tight junctions. J Cell Biol 154:491–497

57. Ebnet K, Aurrand-Lions M, Kuhn A et al (2003) The junctional adhesion molecule

(JAM) family members JAM-2 and JAM-3 associate with the cell polarity protein PAR-3: a possible role for JAMs in endothelial cell polarity. J Cell Sci 116:3879–3891

58. Luo Y, Fukuhara M, Weitzman M et al (2006) Expression of JAM-A, AF-6, PAR-3 and PAR-6 during the assembly and remodeling of RPE tight junctions. Brain Res 1110:55–63

59. Daniele LL, Adams RH, Durante DE et al (2007) Novel distribution of junctional adhesion molecule-C in the neural retina and retinal pigment epithelium. J Comp Neurol 505:166–176

60. Haskins J, Gu L, Wittchen ES et al (1998) ZO-3, a novel member of the MAGUK protein family found at the tight junction, interacts with ZO-1 and occludin. J Cell Biol 141:199–208

61. Stevenson BR, Siliciano JD, Mooseker MS et al (1986) Identification of ZO-1: a high molecular weight polypeptide associated with the tight junction (zonula occludens) in a variety of epithelia. J Cell Biol 103:755–766

62. Jesaitis LA, Goodenough DA (1994) Molecular characterization and tissue distribution of ZO-2, a tight junction protein homologous to ZO-1 and the Drosophila discs-large tumor suppressor protein. J Cell Biol 124:949–961

63. Gumbiner B, Lowenkopf T, Apatira D (1991) Identification of a 160-kDa polypeptide that binds to the tight junction protein ZO-1. Proc Natl Acad Sci U S A 88:3460–3464

64. Balda MS, Gonzalez-Mariscal L, Matter K et al (1993) Assembly of the tight junction: the role of diacylglycerol. J Cell Biol 123:293–302

65. Itoh M, Morita K, Tsukita S (1999) Characterization of ZO-2 as a MAGUK family member associated with tight as well as adherens junctions with a binding affinity to occludin and alpha catenin. J Biol Chem 274:5981–5986

66. Willott E, Balda MS, Heintzelman M et al (1992) Localization and differential expression of two isoforms of the tight junction protein ZO-1. Am J Physiol 262:C1119–1124

67. Beatch M, Jesaitis LA, Gallin WJ et al (1996) The tight junction protein ZO-2 contains three PDZ (PSD-95/Discs-Large/ZO-1) domains and an alternatively spliced region. J Biol Chem 271:25723–25726

68. Woods DF, Bryant PJ (1993) ZO-1, DlgA and PSD-95/SAP90: homologous proteins in tight, septate and synaptic cell junctions. Mech Dev 44:85–89

69. Aartsen WM, Kantardzhieva A, Klooster J et al (2006) Mpp4 recruits Psd95 and Veli3 towards the photoreceptor synapse. Hum Mol Genet 15:1291–1302

70. Kumar R, Shieh BH (2001) The second PDZ domain of INAD is a type I domain involved in binding to eye protein kinase C. Mutational analysis and naturally occurring variants. J Biol Chem 276:24971–24977

71. Lemmers C, Medina E, Delgrossi MH et al (2002) hINADl/PATJ, a homolog of discs lost, interacts with crumbs and localizes to tight junctions in human epithelial cells. J Biol Chem 277:25408–25415

72. Umeda K, Matsui T, Nakayama M et al (2004) Establishment and characterization of cultured epithelial cells lacking expression of ZO-1. J Biol Chem 279:44785–44794

73. McNeil E, Capaldo CT, Macara IG (2006) Zonula occludens-1 function in the assembly of tight junctions in Madin-Darby canine kidney epithelial cells. Mol Biol Cell 17:1922–1932

74. Hernandez S, Chavez Munguia B, Gonzalez-Mariscal L (2007) ZO-2 silencing in epithelial cells perturbs the gate and fence function of tight junctions and leads to an atypical monolayer architecture. Exp Cell Res 313:1533–1547

75. Umeda K, Ikenouchi J, Katahira-Tayama S et al (2006) ZO-1 and ZO-2 independently determine where claudins are polymerized in tight-junction strand formation. Cell 126:741–754

76. Katsuno T, Umeda K, Matsui T et al (2008) Deficiency of zonula occludens-1 causes embryonic lethal phenotype associated with defected yolk sac angiogenesis and apoptosis of embryonic cells. Mol Biol Cell 19:2465–2475

77. Xu J, Kausalya PJ, Phua DC et al (2008) Early embryonic lethality of mice lacking ZO-2, but Not ZO-3, reveals critical and nonredundant roles for individual zonula occludens proteins in mammalian development. Mol Cell Biol 28:1669–1678

78. Antonetti DA, Barber AJ, Hollinger LA et al (1999) Vascular endothelial growth factor induces rapid phosphorylation of tight junction proteins occludin and zonula occluden 1. A potential mechanism for vascular permeability in diabetic retinopathy and tumors. J Biol Chem 274:23463–23467

79. Harhaj NS, Felinski EA, Wolpert EB et al (2006) VEGF activation of protein kinase C stimulates occludin phosphorylation and contributes to endothelial permeability. Invest Ophthalmol Vis Sci 47:5106–5115

80. Barber AJ, Antonetti DA (2003) Mapping the blood vessels with paracellular permeability in the retinas of diabetic rats. Invest Ophthalmol Vis Sci 44:5410–5416

81. Antonetti DA, Barber AJ, Khin S et al (1998) Vascular permeability in experimental diabetes is associated with reduced endothelial occludin content: vascular endothelial growth factor decreases occludin in retinal endothelial cells. Penn State Retina Research Group. Diabetes 47:1953–1959

82. Hofman P, Blaauwgeers HG, Tolentino MJ et al (2000) VEGF-A induced hyperpermeability of blood-retinal barrier endothelium in vivo is predominantly associated with pinocytotic vesicular transport and not with formation of fenestrations. Vascular endothelial growth factor-A. Curr Eye Res 21:637–645

83. Feng Y, Venema VJ, Venema RC et al (1999) VEGF-induced permeability increase is mediated by caveolae. Invest Ophthalmol Vis Sci 40:157–167

84. Janzer RC, Raff MC (1987) Astrocytes induce blood-brain barrier properties in endothelial cells. Nature 325:253–257

85. Tout S, Chan-Ling T, Hollander H et al (1993) The role of Muller cells in the formation of the blood-retinal barrier. Neuroscience 55:291–301

86. Iadecola C (2004) Neurovascular regulation in the normal brain and in Alzheimer's disease. Nat Rev Neurosci 5:347–360

87. Abbott NJ, Ronnback L, Hansson E (2006) Astrocyte-endothelial interactions at the blood-brain barrier. Nat Rev Neurosci 7:41–53

88. Salceda S, Caro J (1997) Hypoxia-inducible factor 1alpha (HIF-1alpha) protein is rapidly degraded by the ubiquitin-proteasome system under normoxic conditions. Its stabilization by hypoxia depends on redox-induced changes. J Biol Chem 272:22642–22647

89. Maxwell PH, Wiesener MS, Chang GW et al (1999) The tumour suppressor protein VHL targets hypoxia-inducible factors for oxygen-dependent proteolysis. Nature 399:271–275

90. Ikeda E, Achen MG, Breier G et al (1995) Hypoxia-induced transcriptional activation and increased mRNA stability of vascular endothelial growth factor in C6 glioma cells. J Biol Chem 270:19761–19766

91. Maisonpierre PC, Suri C, Jones PF et al (1997) Angiopoietin-2, a natural antagonist for Tie2 that disrupts in vivo angiogenesis. Science 277:55–60

92. Asahara T, Chen D, Takahashi T et al (1998) Tie2 receptor ligands, angiopoietin-1 and angiopoietin-2, modulate VEGF-induced postnatal neovascularization. Circ Res 83:233–240

93. Thurston G, Rudge JS, Ioffe E et al (2000) Angiopoietin-1 protects the adult vasculature against plasma leakage. Nat Med 6:460–463

94. Romero IA, Radewicz K, Jubin E et al (2003) Changes in cytoskeletal and tight junctional proteins correlate with decreased permeability induced by dexamethasone in cultured rat brain endothelial cells. Neurosci Lett 344:112–116

95. Edelman JL, Lutz D, Castro MR (2005) Corticosteroids inhibit VEGF-induced vascular leakage in a rabbit model of blood-retinal and blood-aqueous barrier breakdown. Exp Eye Res 80:249–258

96. Felinski EA, Cox AE, Phillips BE et al (2008) Glucocorticoids induce transactivation of tight junction genes occludin and claudin-5 in retinal endothelial cells via a novel cis-element. Exp Eye Res 86:867–878

97. Liebner S, Corada M, Bangsow T et al (2008) Wnt/beta-catenin signaling controls development of the blood-brain barrier. J Cell Biol 183:409–417

98. Moon RT (2005) Wnt/beta-catenin pathway. Sci STKE 2005:cm1

Chapter 6

The Blood-Nerve Barrier: Structure and Functional Significance

Ananda Weerasuriya and Andrew P. Mizisin

Abstract

The blood–nerve barrier (BNB) defines the physiological space within which the axons, Schwann cells, and other associated cells of a peripheral nerve function. The BNB consists of the endoneurial microvessels within the nerve fascicle and the investing perineurium. The restricted permeability of these two barriers protects the endoneurial microenvironment from drastic concentration changes in the vascular and other extracellular spaces. It is postulated that endoneurial homeostatic mechanisms regulate the *milieu intérieur* of peripheral axons and associated Schwann cells. These mechanisms are discussed in relation to nerve development, Wallerian degeneration and nerve regeneration, and lead neuropathy. Finally, the putative factors responsible for the cellular and molecular control of BNB permeability are discussed. Given the dynamic nature of the regulation of the permeability of the perineurium and endoneurial capillaries, it is suggested that the term blood–nerve interface (BNI) better reflects the functional significance of these structures in the maintenance of homeostasis within the endoneurial microenvironment.

Key words: Blood–nerve barrier, Blood–nerve interface, Capillary, Endoneurial fluid flow, Endoneurium, Homeostasis, Lead neuropathy, Nerve development, Nerve regeneration, Permeability, Perineurium, Peripheral nerve, Wallerian degeneration

1. Introduction

Vertebrate peripheral axons and their associated Schwann cells function within a specialized milieu intérieur – the endoneurial microenvironment. Exchange of material between this intrafascicular physiological space and the general extracellular space is restricted and regulated by the blood–nerve barrier (BNB), also known as the blood–nerve interface (BNI), which consists of the endoneurial vascular endothelium and the multilayered ensheathment, the perineurium. The relative impermeability of the BNB to blood-borne material protects the endoneurial

Sukriti Nag (ed.), *The Blood-Brain and Other Neural Barriers: Reviews and Protocols*, Methods in Molecular Biology, vol. 686, DOI 10.1007/978-1-60761-938-3_6, © Springer Science+Business Media, LLC 2011

microenvironment from potentially harmful plasma constituents, and rapid fluctuations of plasma solute concentrations that influence Schwann cell and axonal functions. The absence of lymphatic drainage in the endoneurial space (1) further emphasizes the protective nature of the BNB. The composition of the endoneurial fluid and its physical properties are regulated by homeostatic mechanisms located at the BNB and in the endoneurial cellular elements. The concept of the endoneurial microenvironment as a regulated physiological space was first proposed about 30 years ago (2, 3).

Early morphological studies with several tracers described the permeability of the BNB as either leaky or impermeant and thus downplayed its graded permeability to solutes of different sizes as shown by subsequent physiological studies. When interpreting results from morphological studies on BNB permeability, it is more prudent to limit the description to what is observed, i.e., a change in permeability to a particular tracer, rather than to misleadingly postulate a "breakdown" of the BNB or BNI. Numerous studies have demonstrated increases in endoneurial microvessel permeability without any accompanying ultrastructural evidence of endothelial cell damage or death. At the same time, physiological studies have measured initial increases and subsequent decreases in endoneurial microvessel and perineurial permeability. These results point to a BNI that reflects the operation of endoneurial homeostatic mechanisms, much as in the blood–tissue interfaces of other mammalian organs, rather than a passive restrictive barrier that impedes access of hydrophilic solutes to the endoneurium. Thus, for reasons outlined above and in anticipation of the increasing significance of emerging concepts such as the neurovascular unit and immunologically driven dynamic permeability changes, it is proposed that the BNI represents a more powerful and rigorous description of the regulatory dynamics governing the blood–nerve exchange of various solutes and water.

2. Structural Components of the BNB

The BNB delimits the endoneurial microenvironment of the peripheral nervous system (PNS), a space that extends from the proximal root attachment zones of cranial and spinal nerves to the distal sensory and motor end-organs. Numerous studies in a variety of species, including rats, mice, rabbits, cats, dogs, humans, and frogs have demonstrated that nerve fascicles and the endoneurial compartment therein are anatomically circumscribed by a connective tissue ensheathment within which and through which travel the anastomosing plexuses of the vasa nervorum. With a few important exceptions that may be a consequence of the absence of lymphatic capillaries (1), entry into and exit from the endoneurial microenvironment is tightly regulated.

2.1. Spinal Roots

The connective tissues ensheathing the peripheral nerves merge with the meninges of the central nervous system (CNS) at the subarachnoid angle (4) (Fig. 1). Here the dense irregularly arranged collagen fibers and scattered fibroblasts of the dura mater blend with the epineurium, the outermost covering of peripheral nerves that binds fascicles together, while the arachnoid layer merges with the outer lamellae of the perineurium, the multilayered cellular investment of nerve fascicles that circumscribe the endoneurial space (4, 5). The inner layers of the perineurium are reflected onto the nerve roots as the root sheath. The outer root sheath is reminiscent of the pia mater (6) and consists of loosely arranged cell layers with intermittent junctional contacts, allowing communication between the subarachnoid and intercellular spaces, and containing collagen and small elastic fibers (4). The inner cell layers of the root sheath are closely opposed and contain squamous cells with a polygonal outline (5) and a basal lamina that is intermittent except for that which separates the innermost cells from the endoneurium of the root (4).

From the subarachnoid angle to the attachment zones, spinal roots traverse the subarachnoid space (Fig. 1). As noted above, the cellular layers of the inner root sheath are a centrally directed continuation of a portion of the perineurium that ends as an open-ended tube in the vicinity of the root attachment zone (6). In this region, continuity between the CSF-containing subarachnoid space and the endoneurium of the roots via pial tissue space has been demonstrated ultrastructurally (6). The continuity between subarachnoid and endoneurial extracellular spaces represents a pressure-driven egress pathway for CSF, which is a source of endoneurial fluid. The root attachment zones contain the regions where rootlets are connected to the spinal cord. The structural basis for the interface between CNS and PNS parenchyma is the glia limitans, the CNS surface-limiting membrane composed of astrocytic foot processes that is subjacent to the pia mater and crosses each rootlet at its point of attachment to the spinal cord (reviewed by Fraher (7)). The glia limitans thickens as it crosses the rootlets and the interdigitating, layered astrocytic foot processes become perforated by channels that contain either individual myelinated fibers or groups of unmyelinated axons. CNS and PNS tissues interpenetrate in the rootlets or spinal cord surface forming transition zones with a length of rootlet containing both tissues (7). Central to the glia limitans in the transitional zone, myelinated fibers are ensheathed by oligodendrocytes, while peripheral to it, myelinated fibers are ensheathed by Schwann cells and unmyelinated fibers by Remak cells. The abrupt change in ensheathment at the glia limitans is in marked contrast to the

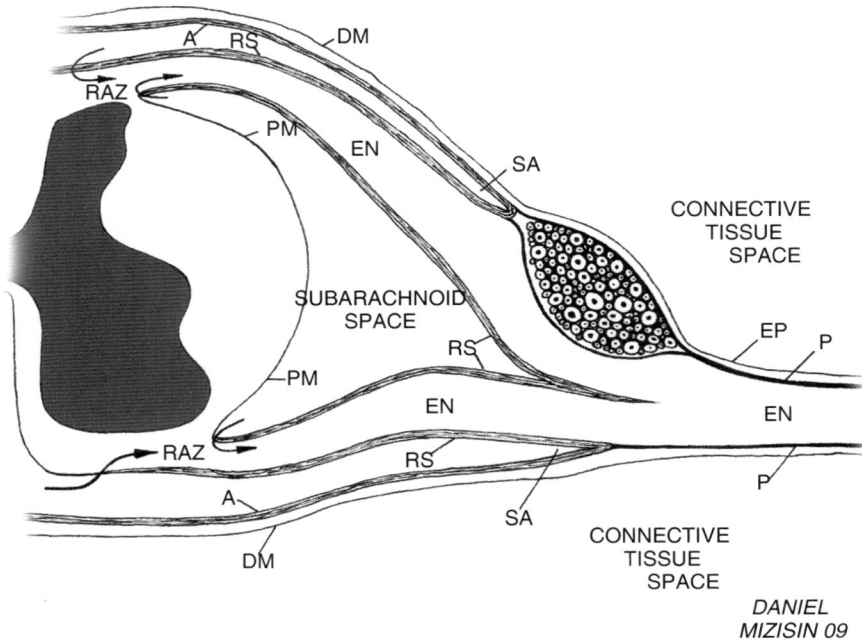

Fig. 1. The relationship of peripheral nerve and spinal root connective tissue ensheathments to the meningeal coverings of the spinal cord are shown. The outermost connective tissue of peripheral nerve, the epineurium (*EP*), is continuous with the outermost meningeal covering, the dura mater (*DM*). At the subarachnoid angle (*SA*), the perineurium (*P*), the multilayered connective tissue ensheathment that defines nerve fascicles, divides, with the innermost layers continuing on to become the inner layers of the root sheath (*RS*) and the outermost layers merging with the arachnoid layer (*A*). As the dorsal and ventral spinal roots pass through subarachnoid space, some of the arachnoid layer is reflected onto the root sheath at the subarachnoid angle, becoming the outermost layers of this connective tissue ensheathment. At the root attachment zone (*RAZ*), the pia mater (*PM*) of the spinal cord is reflected onto the spinal root and merges with the outer layers of the root sheath, while the glia limitans continues across the attachment zone to form the interface between the central and peripheral nervous systems. The innermost layers of the root sheath terminate on the spinal root side of the glia limitans. At the dorsal and ventral root attachment zone, continuity between the subarachnoid space and endoneurium of the spinal roots (*arrows*) has been demonstrated ultrastructurally (modified from (4)).

continuity of myelinated and unmyelinated axons across this surface-limiting membrane, which defines the most proximal region of the endoneurial microenvironment.

The vasculature of the spinal roots has been examined in alkaline phosphatase preparations and shown to have large, more sparsely distributed vessels comparable to spinal cord white matter as well as most of the more distal portions of the peripheral neuraxis (8). Capillary density is greater in sensory ganglia and spinal cord gray matter (8), which is in line with the greater metabolic demands of nerve-cell-body-containing nervous tissue (9). The vessels in the spinal roots are permeable to morphological tracers of various molecular weights (9). The structural basis of tracer permeability in endoneurial vessels of spinal roots appears to be fenestrations in endothelial cells as well as open interendothelial clefts (10).

2.2. Peripheral Nerves

Nerve trunks are ensheathed by two distinct types of connective tissue: epineurium and perineurium. The epineurium is the outermost connective tissue and consists of irregularly arranged collagen fibers, scattered fibroblasts and occasional clusters of adipocytes. The epineurium surrounds mono-fascicular nerves or loosely binds fascicles of multi-fascicular nerves together and contains the outermost portion of the intrinsic component of the vasa nervorum. The epineurial vasculature is an anastomosing vascular plexus fed by radicular vessels that connect to the extrinsic component of the vasa nervorum, which are segmental regional vessels that approach the nerve trunk at multiple levels (11). The epineurial arterioles and venules are large longitudinally oriented vessels that are evident superficially in the nerve trunk as well as in the deeper regions between fascicles. Extravasation of a variety of morphological tracers has been observed in the epineurial vasculature and attributed to diffusion through both open interendothelial clefts and fenestrated vessels (9).

The perineurium circumscribes the endoneurium in concentric cellular layers, each one cell thick. In larger nerve fascicles, the perineurium consists of as many as 15 cellular layers, with fewer layers in smaller fascicles and only one cell layer in the vicinity of sensory and motor end-organs. Perineurial collagen fibrils, which are smaller in diameter than those in the epineurium (12), are interposed in the extracellular space between cell layers arranged in circular, longitudinal and oblique bundles along with occasional elastic fibers (9) and likely provide the perineurium with its passive compliance properties. Perineurial cells are flattened squamous cells with a serrated polygonal border considered by some to be epithelial cells (5, 13) but later shown to have a fibroblast origin (14). As in the root sheath, perineurial cells lack a polarized architecture and are bounded on both sides by basal laminae (15, 16), which distinguishes perineurial cells from epineurial and endoneurial fibroblasts. In addition to abundant caveolae, filaments and dense bodies have been observed in the perineurial cells of mouse sciatic nerve, suggesting a contractile capacity (17) and possibly representing the structural basis for active compliance properties of the perineurium. Within a sleeve, tight junctions join perineurial cells together, while desmosomes and gap junctions are occasionally observed linking cells in adjacent layers together (18). These tight junctions, particularly those of the innermost sleeves, provide a barrier to the diffusion of various larger molecular weight tracers and restrict diffusion of low molecular weight tracers (19).

Anastomotic connections between the epineurial and perineurial vascular plexuses occur at various levels in the perineurium, with the longitudinally oriented vessels penetrating the cellular layer gradually in an oblique fashion before connecting to the endoneurial plexus (11). Single perineurial layers accompany these penetrating vessels as they enter the endoneurium and represent a restricted site of continuity between the epineurial and endoneurial spaces (18). The oblique penetration of the perineurial vasculature leaves these vessels vulnerable to compression when the endoneurial fluid pressure is increased (20). The endoneurial vascular plexus and the obliquely penetrating perineurial vessels comprise a fascicular vascular unit, a system of longitudinally oriented vessels that remains intact when an individual fascicle is isolated (11). Endoneurial vessels consist of a network of arterioles, capillaries and venules that are sparsely distributed as are vessels of the spinal cord white matter and roots. The arterioles are thin-walled with a fragmentary internal elastic membrane (8). The diameters of the continuous, nonfenestrated capillaries are larger than the diameters of capillaries found in adjacent skeletal muscle (6–10 vs. 3–6 μm) (8). The larger size and unusually complete pericyte investment of endoneurial capillaries are features shared with endoneurial postcapillary venules, although the presence of tight junctions and alkaline phosphatase activity distinguish these vessels (8). In contrast to leaky endoneurial vessels in new-borne mouse sciatic nerve, tight interendothelial junctions in adult endoneurial vessels represent the structural basis for restricted permeability to vascular tracers of various molecular weights including Evans Blue albumin and fluorescein isothiocyanate (9, 21–23), while open interendothelial gaps distinguish these vessels from the consistently tight vessels in most regions of the brain (8).

2.3. DRG and other Ganglia

The connective tissue capsule surrounding sensory and other ganglia is continuous with the ensheathment of peripheral nerves with an outer epineurium surrounding an inner perineurium. The perineurial cellular lamellae of ganglia are bounded by a basal lamina on both sides and have numerous tight junctions and desmosomes, as well as occasional gap junctions (9, 24). Interposed between the cellular layers are collagen-containing extracellular spaces. The 2–4 cellular lamellae of the perineurium of the superior cervical ganglion act as a diffusion barrier, with the innermost layers preventing transperineurial movement of larger molecular weight tracers (24). In contrast to the perineurium and the endoneurial vessels of the distal nerve trunk, the endoneurial vasculature of cranial, spinal, and autonomic ganglia is permeable to a variety of low and high molecular weight tracers (24, 25). As in endoneurial vessels of the spinal roots, blood-borne tracer is extravasated through open interendothelial clefts, as well as through vessels with fenestrated endothelia, within minutes of injection (24).

2.4. Sensory and Motor End-Organs

As nerve fascicles become smaller and approach sensory and motor end-organs, the number of surrounding perineurial layers decrease to a single layer (5, 18). The decrease in the number of perineurial layers occurs by the termination of the innermost layer or by loss of the innermost layer as it accompanies nerve fibers separating from the parent fascicle, in which case tight junctions ensure an effective seal between the branching sleeve and the remaining perineurial layers (18). Whether single perineurial layers maintain their ensheathment as individual fibers reach sensory and motor end-organs has been a subject of controversy. Some maintain that a complete perineurial investment is continuous with the capsules of sensory organs, such as Meissner's, Krause's, and Pacinian corpuscles and also covers the motor end plate (5, 13). Others have provided ultrastructural evidence that the perineurial layer ends just before reaching the motor end plate, providing an open-ended termination with continuity between epineurial and endoneurial space (18). While ultrastructural evidence is lacking, there is likely an open-ended perineurial sleeve for the naked nerve endings of the intraepidermal innervation (5). The open-ended perineurial sleeve of motor and epidermal nerves provides distal continuity of the endoneurial microenvironment with the surrounding extracellular tissue space and may ensure maintenance of proximodistal endoneurial fluid flow (EFF) by providing distal drainage sites.

3. Dynamics of Blood–Nerve Exchange

Two routes are available for blood–nerve exchange of material. One pathway, the direct one, is across the endothelium of the endoneurial microvasculature. The other is an indirect route requiring passage of material through a third compartment interposed between the vascular space and the endoneurial extracellular space. This compartment is the epineurial extracellular space that exchanges directly with the vascular compartment through the relatively leaky epineurial capillaries. Evidence from various physiological and morphological studies strongly indicates that blood–nerve exchange occurs predominantly via endoneurial capillaries and that transperineurial exchange is a minor contributor. Hence, the terms BNB permeability and endoneurial capillary permeability are used interchangeably. Strictly speaking, BNB permeability to a given solute is slightly higher than the endoneurial capillary permeability to that same solute, and the difference is due to the minor contribution from the relatively impermeable perineurium.

3.1. Relative Contributions of Perineurium and Microvessels

Theoretically, blood-borne substances can reach the endoneurial extracellular space either by traversing the endoneurial vascular endothelium or by crossing the multilayered perineurium after gaining access to the perineurial extracellular space. Multiple observations, including the following, favor the former pathway as the major route of blood–nerve exchange. The perineurium is impermeable to ionic lanthanum, but the endoneurial capillaries are not (13, 24). Intravascular perfusion with a hyperkalemic solution inactivates peripheral nerve much more rapidly than when it is bathed by the same hyperkalemic solution (26). Histamine increases the permeability of endoneurial capillaries to macromolecular tracers, but is without effect on the perineurium (27, 28). In leprosy, the endoneurial blood vessels become permeable to ferritin, whereas the perineurium remains impermeable to this tracer (29). In the frog sciatic nerve as well as rat tibial and sciatic nerve, where perineurial permeability has been measured independently, the endoneurial capillaries are more permeable (2, 3, 28, 30–32). During the second to sixth week of Wallerian degeneration, while the perineurial permeability increases about fourfold, the permeability-surface area product (PS) of the frog sciatic nerve decreases by more than 60% reflecting the greater sensitivity of PS to permeability of the capillaries than to that of the perineurium (33). PS of the adult rat sciatic nerve perineurium to (^{125}I) albumin was measured to be $1.48 \pm 0.28 \times 10^{-7}$ ml. $g^{-1}.s^{-1}$ ($n = 6$) (Weerasuriya, unpublished observations), which is about two orders of magnitude less than the corresponding value for the endoneurial vessels (34). All these studies clearly emphasize the much more restrictive diffusion barrier properties of the perineurium.

3.2. Two Routes of Transcapillary Permeation

Before adequate structural evidence was available, physiologists, on the basis of permeability measurements to hydrophilic solutes of various sizes, postulated that the capillary endothelium contained a set of large and a set of small hydrophilic pores (35, 36). The ratio of large to small pores, though variable, seems to be in the range of about 1:30,000 (37, 38). Recent evidence from electron microscopic studies suggest that the small pores are the interendothelial clefts, which have a width of about 20 nm and occupy about 0.4% of the capillary surface area (39). Because of their rarity, the identity of large pores is less certain. Potential candidates are widened intercellular junctions, transendothelial channels formed by the fusion of plasmalemmal invaginations, and fenestrae. Properties of the small and large pore pathways are exhaustively reviewed elsewhere (40–42).

3.3. Route of Transperineurial Permeation

In contrast to epithelial tissue, the perineurial cells do not display an apicobasal polarity. Thus, the major route for transperineurial passage of material is a paracellular pathway, which consists of a number of belts of intercellular tight junctions arranged in series due to the multilayered nature of the perineurium. The relative impermanence of this route is illustrated by the fact that local anesthetics administered for nerve blocks are injected via a needle that penetrates the perineurial sheath. The number of layers in the perineurium decreases in a proximodistal direction with the fine terminals of sensory and motor nerves having only one or two layers of perineurial cells (18). In comparison to connective tissue ensheathments of the CNS, the perineurium combines the mechanical strength of the dura mater and the impermanence of arachnoid. A recent report of the presence of VE-cadherin in perineurial cells (43) suggests that perineurial permeability is regulated by mechanisms similar to those operating on the vascular endothelial cells of the endoneurium. The presence of tight junction proteins, occludin, and zonula occludens-1 in perineurial and nerve endothelial cells (44) is consistent with similar mechanisms regulating permeability at these sites. Attempts to examine the consequences of perineurial removal are confounded by a compromise of the endoneurial vasculature that receives nutrient branches from transperineurial arteries (45–48). Nevertheless, this remains an important question especially with regard to perineurial regeneration associated with nerve repair.

3.4. Transporters at the BNB

Given the relative impermeability of the perineurium and endoneurial capillaries, it is not surprising that blood–nerve exchange of several solutes depends on the presence of specific transporter molecules at the interface. The endoneurial microenvironment, unlike the cerebral microenvironment, does not require a moment-to-moment regulation of nutrients and oxygen. Several transporters designed to meet the metabolic requirements of endoneurial constituents have been described. The earlier radioisotopic demonstration of facilitated transport of D-glucose (49), have been complemented by reports on the presence of GLUT-1 in endothelial cells and perineurial cells (50–52). In keeping with the earlier postulate that the perineurium is a specialized connective tissue, the above studies did not demonstrate an apicobasal polarity of GLUT-1 transporters in perineurial cells. Thus, the role of perineurial GLUT-1 transporters appears to be the nourishment of perineurial cells.

3.5. Assessment of Blood–Nerve Exchange of Solutes

Concentrations of endoneurial constituents are affected by exchange across the endoneurial capillary interface and the barrier imposed by the inner layers of the perineurium, as well as the metabolic processes of endoneurial cells. Permeability of the two exchange sites are assessed by morphological and physiological methods. Both techniques have their advantages and limitations. Historically, morphological techniques establish the initial broad parameters and then physiological methods provide quantitative measurements of blood–nerve transfer coefficients. The major advantage of morphological techniques is the localization of barriers to the penetrance of markers and the array of tracers of varying molecular weights, charges, and sizes. On the other hand, these methods are limited by the sensitivity of the histological staining techniques. For example, histological methods have not detected blood–nerve transfer of macromolecules, but physiological methods using radiotracers estimate blood–nerve transfer of albumin at about 10^{-5} cm/s. Thus, physiological methods not only allow a quantification of transfer rates, but also are more sensitive especially to macromolecules and other larger species. Furthermore, transfer rates of amino acids, sugars, and other small molecules can only be studied by physiological radioisotopic methods. However, physiological techniques do not allow localization of the sites of transfer or transport. Immunostaining techniques and in situ hybridization complement these transport studies by localizing the putative transporters and diffusion sites. Technical details of measurement of blood–nerve transfer rates are described elsewhere (34, 53) as are theoretical foundations and limitations of blood–tissue transfer studies (54).

4. Endoneurial Fluid Dynamics

The extracellular space within a nerve fascicle is about 20–25% of the total intrafascicular volume. The endoneurial hydrostatic pressure (EHP) exerted by this fluid is about 2–3 mmHg. In terms of fluid dynamics, the endoneurial contents consist of a noncompressible aqueous solution, and a somewhat compressible cellular content constrained by an elastic sheath, the perineurium. Hence, the EHP will be determined by the volume of the endoneurial contents and the compliance of the perineurial sheath. Fluid enters and leaves a given segment of the endoneurial space across the walls of the endoneurial microvessels, and by convective proximodistal EFF. Given the slight positive tissue hydrostatic pressure of the endoneurial space, fluid does not normally enter this space across the perineurium. Thus, endoneurial extracellular fluid exchanges material with blood directly across the endoneurial vascular endothelium, indirectly across the multilayered perineurium, and is turned over by convective EFF.

4.1. Continuity Between CSF and Endoneurial Fluid

The meninges of the CNS are continuous with sheaths of peripheral nerve (4). The epineurial connective tissue layers are continuous with the dura mater at the subarachnoid angle (Fig. 1), while the relationship between the other meninges (pia mater and arachnoid layer) and perineurium is more complex. At the root attachment zone (Fig. 1), there appears to be continuity of the subarachnoid and endoneurial spaces and thus continuity between CSF and endoneurial fluid, providing a conduit through which material passes from CSF to the endoneurial fluid (55). The embryological aspects of this continuity do not appear to have been investigated.

4.2. Proximodistal Gradient

In addition to the blood–nerve exchange discussed above, another putative source of input to and output from the endoneurium is convective EFF. In an elegant series of experiments, the presence of a proximodistal flow of fluid was demonstrated in rat and guinea pig sciatic nerve (56). As indices of fluid movement, endoneurial injections of dyes, crystals, and radioactive mineral salts were used, with conclusions based on comparisons of proximodistal spread of the indicators in dead and living tissue. The two major limitations of this study are (1) the use of injection volumes no smaller than 100 µL and (2) the use of small hydrophilic tracers that could have migrated down the nerve by entering the vascular compartment and later reentering the endoneurial space. Therefore, they were unable to calculate a precise rate of convective fluid flow from their data but suggested an approximate rate of 3 mm per hour, similar to results by others studying traumatized chicken sciatic nerves (57). Low (58), injecting 10 µl of tetrodotoxin into the endoneurium and monitoring the rate and spread of inactivation, concluded that convective fluid flow is about 4–8 mm per hour. Morphological studies demonstrating continuity between the spinal subarachnoid space and the endoneurial space at the root attachment zone suggest (4, 59) that CSF contributes to endoneurial fluid. Earlier, it was suggested that the diphtheric toxin enters the CNS through peripheral nerves (60), probably exploiting the continuity between subarachnoid and endoneurial space.

Experiments with radiotracers have clearly demonstrated a proximodistal convective flow of endoneurial fluid (Fig. 2). Here, the pattern of ^{22}Na distribution along the length of the nerve at the three different survival times clearly demonstrates a progressive asymmetrical proximodistal movement of the isotope.

4.3. Driving Force of Proximodistal Fluid Flow

There does not appear to have been any mechanistic studies addressing the driving force of proximodistal fluid flow. The descriptive studies on EHP (61) are consistent with the CSF pressure in the spinal cord being the pressure head for the proximodistal flow of endoneurial fluid. The hydrostatic pressure in the cord is about 10 mmHg, in the dorsal root ganglia it is about 3–5 mmHg, and in the peripheral nerves it is about 2–3 mmHg (62). The presence of this pressure gradient from the interstitial spaces of the spinal cord to the endoneurial interstitium of peripheral nerves supports the contention that the CSF pressure is the main propulsive force of the endoneurial fluid. Unfortunately, this postulate is not without theoretical limitations. For example, what would be the fate of this CSF pressure driven endoneurial fluid when it reaches the distal ends of sensory nerves that do not have open perineurial sleeves is not clear (63). One possibility is that the perineurium in such nerves is more permeable distally than proximally to allow for a transperineurial dissipation of endoneurial fluid. Alternatively, one has to postulate a slower turnover of endoneurial fluid at the distal end of a sensory nerve, which would account for the greater vulnerability of sensory nerves to pyridoxine toxicity.

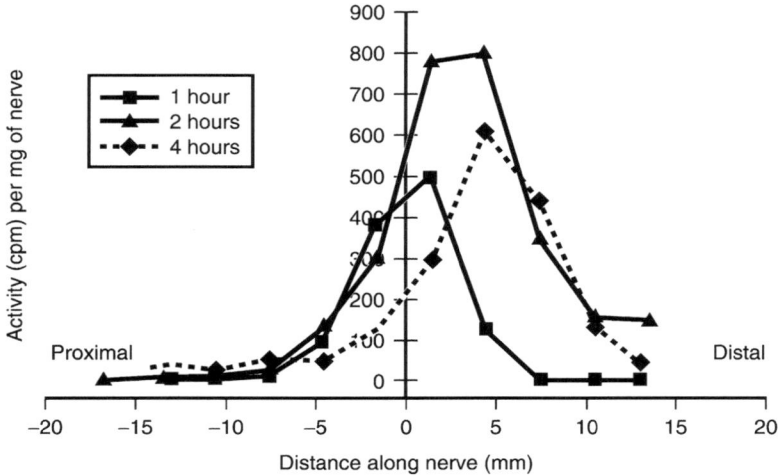

Fig. 2. The rate of endoneurial convective fluid flow of ^{22}Na as a function of time is shown. In all three experiments, 70 nL of saline with ^{22}Na were microinjected into a rat sciatic nerve. The nerves were harvested either 1, 2 or 4 h later, cut into 3 mm segments, counted for ^{22}Na activity, dried and weighed. Negative numbers indicate distances proximal to the site of injection and positive numbers represent distances distal to the site of injection. The data in this figure are consistent with a convective flow of endoneurial fluid. The shift of the curve to the *right* (proximal to distal) from the first to the second hour is clear-cut, while that from the second to the fourth hour is somewhat more subtle, but nevertheless evident upon closer inspection of the two curves. While both curves seem to peak at the 5 mm, the 2-h curve has a bigger shoulder at lengths less than 5 mm, and the 4-h curve has a shoulder at lengths greater than 5 mm (Weerasuriya, unpublished data).

4.4. Endoneurial Fluid Turnover

The albumin content of desheathed human sural nerve was reported to be 8.7 µg/mg of dry weight (64). If it is assumed that endoneurial wet/dry weight ratio is 3.0, endoneurial albumin is extracellular and free, and endoneurial extracellular space is about 25%, then the above value can be converted to an albumin concentration of 11.6 mg/mL in the endoneurial fluid. Plasma albumin concentration is 33.1 mg/mL (64). From these two concentration terms and the PS to albumin (65), the calculated rate of blood–nerve albumin transfer is about 1.2 mg.g^{-1}.day^{-1}. At this rate of transfer, and assuming relatively constant albumin concentrations in endoneurium and plasma, about 30% of the endoneurial albumin is turned over each day. By comparison, CSF and its constituents are turned over about four times each day (66).

The rate of removal of albumin from the endoneurium is also about 1.2 mg.g^{-1}.day^{-1}. In nonneural tissues, lymphatic drainage plays a role in clearing interstitial albumin, and, in the CNS, CSF, and perivascular drainage are the sinks for extracellular albumin (27). In peripheral nerve, in the absence of both lymphatics and an active CSF circulation, the route of removal for albumin and other macromolecules remains to be identified. The metabolic breakdown of albumin to supply amino acids to axons and glia is one possibility; another is the long suspected proximodistal convective flow of endoneurial fluid.

**4.5. Endoneurial
Protein Concentration**

From the equation proposed by Landis and Pappenheimer (67), endoneurial albumin can be expected to exert an interstitial oncotic pressure of about 3 mmHg. The recorded EHP of 2–3 mmHg will oppose the endoneurial oncotic pressure and thus minimize net fluid filtration from the endoneurial vasculature. However, this balance of forces can be disturbed if the capillary permeability to albumin increases slightly allowing the endoneurial albumin concentration to rise and thus draw in more fluid from the vascular compartment into the endoneurial interstitium. The resulting endoneurial edema together with the low compliance and hydraulic conductivity of the perineurium will elevate EHP. The net fluid gain by the endoneurial space will cease when a new equilibrium is established among the hydrostatic and oncotic pressures of the endoneurial and intravascular compartments. It is quite likely that this is the mechanism for edema formation observed in experimental diabetic and lead neuropathy, as well as in early Wallerian degeneration. On the other hand, the edema present in galactose-induced neuropathy is likely to be due to the presence of an excess of nonvascularly derived osmolytes in the endoneurium, corroborated by the absence of change in the PS of BNB to mannitol in galactose-intoxicated rats (68).

**4.6. Perineurial
Compliance Decreases
with Age**

The developmental increase of EHP has two plateaus (69). The first one is from about 3 to 13 weeks of age, and the second one is from 6 months onward. Elevations of EHP can be produced either by increased EFF or by reduced perineurial compliance. These observations (69) demonstrated that reduced perineurial compliance with aging contributes to the second increase in EHP.

**4.7. Implications
for Entrapment
Neuropathies**

Extrafascicular mechanical compression and the resultant ischemia certainly contribute to the symptomatology of entrapment neuropathies. But, it is quite likely that there are other factors more closely tied to the endoneurial microenvironment that affect the susceptibility and evolution of the nerve injury. For example, it is hypothesized that the initial event in the evolution of an entrapment neuropathy is the limitation and reduction of EFF due to externally applied mechanical forces. Secondly, elevation of EHP due to obstruction of EFF and continued application of these external forces leads to endoneurial ischemia and its attendant pathology. Additionally, the reduced compliance of the aged nerve (69) makes it more susceptible to externally applied mechanical pressures causing an elevation of EHP. This hypothesis is consistent with the paucity of carpal tunnel syndrome (CTS) in arcade game-playing teenagers, higher incidence of CTS in conditions with increased tissue water content or edema, such as pregnancy and diabetes, and the presence of symptoms quite proximal to the site of entrapment. However, this hypothesis does not explain the higher incidence of CTS in females or the varying patterns of recovery seen after resection of the flexor retinaculum.

5. Endoneurial Homeostasis

The various compartments of the nervous system employ homeostatic mechanisms to regulate their respective internal environments. The relevant physiological functional unit for a vertebrate peripheral nerve is a fasciculus, the cylindrical structural entity encompassing axons, glial cells, and other attendant satellite cells that is defined by the perineurium. The physical limit of the perineurium is also a physiological barrier, limiting the exchange of material between the endoneurial space and the general extracellular space surrounding a peripheral nerve.

5.1. The Controlled Variables of Endoneurial Homeostasis

The specialized microenvironment of peripheral nerve fibers is maintained with the assistance of the BNI. Regulated blood–nerve exchange across the BNI and turnover of endoneurial fluid by convective fluid flow are vital for the maintenance of endoneurial physiological parameters, including blood flow, O_2 tension, pH, oncotic pressure, hydrostatic pressure, ion concentrations, within the normal range necessary for the proper functioning of nerve fibers. There are independent transendothelial pathways for the movement of ions and macromolecules. Some of these physiological parameters have been measured: PS to (^{14}C) sucrose (70, 71), (^{14}C) glucose (49), (^{125}I) albumin (34), and ^{22}Na (30); endoneurial blood flow (72, 73); EHP (74, 75); and ion concentration in endoneurial fluid (58, 76). Other parameters have been described only qualitatively or not at all: endoneurial concentration of H^+ and Ca^{2+}; rate of convective EFF; volume of endoneurial extracellular space; perineurial permeability to macromolecules; and tortuosity of the endoneurial extracellular contents. The above list of variables is neither exhaustive nor does it relate to all the pathophysiological alterations associated with the endoneurial microenvironment. However, it is hoped that discussions of these few variables with available data during development, nerve degeneration, and regeneration and in a few clinical scenarios will emphasize the relevance of considering pathophysiological alterations as perturbations of endoneurial homeostasis.

5.2. Endoneurial Homeostasis During Development

During normal development, results from studies employing morphological tracers indicate that the BNB of juvenile animals is more permeable to macromolecules than that of adult animals (23). The decrease in the PS of the BNB during development has been quantified with radiotracer studies (77). An intriguing issue arising out of these observations is how developing nerve with its highly permeable BNB avoids the pathology observed in an adult nerve with a comparably permeable BNI. For example, in the adult elevating the PS of the BNI to an extent where there is extravasation of plasma proteins inevitably leads to endoneurial edema and increased EHP. The primary event is the increase of endoneurial albumin content, which elevates the oncotic pressure in the endoneurial interstitial space. Due to the operation of Starling forces, water is drawn into the endoneurial interstitium from the vascular compartment, leading to elevated endoneurial water content and, together with the low hydraulic conductivity and compliance of the perineurium, causes an increase in the EHP. Elevated EHP can compromise the intrafascicular microcirculation leading to ischemia and its accompanying pathology.

By contrast, a highly permeable BNB during the first weeks of development does not lead to an elevated EHP and edema because the endoneurial accumulation of fluids and osmolytes is prevented by a more permeable perineurium allowing for the clearance of this material by the epineurial lymphatics. Additionally, a higher compliance of the juvenile endoneurial tissue mass could also counter a tendency for an increased EHP. Therefore, in the developing nerve two independent factors contribute to a much more permeable endoneurial microvasculature. The first is the need for a greater blood–nerve exchange of material to support the more active metabolism, and the second is a consequence of tight junction dissolution and reformation during endothelial proliferation in vasculogenesis. Increased permeability of the endoneurial microvasculature does not lead to pathological consequences due to the combination of a relatively more permeable perineurium and epineurial lymphatics, as well as metabolic clearance of plasma-derived macromolecules.

5.3. Endoneurial Homeostasis in Nerve Degeneration and Regeneration

Wallerian degeneration is probably the most drastic reorganization of the endoneurial architecture (78, 79). The fact that perineurial permeability increases only about twofold under these conditions (80) argues against a "breakdown" of the barrier properties of this structure. An alternative and more plausible interpretation is a dynamic response of the perineurium to maintain endoneurial homeostasis. Morphologically, perineurial cells in degenerating nerve show no evidence of disruption (81, 82). On the contrary, they proliferate, hypertrophy and increase their organelle content (81).

During Wallerian degeneration, the perineurium exhibits a bimodal increase in permeability (Fig. 3). The initial peak of perineurial permeability, lasting only a few days (80), is most likely related to the acute inflammatory response triggered by the trauma of nerve section, and the proliferative response of the perineurium (81).

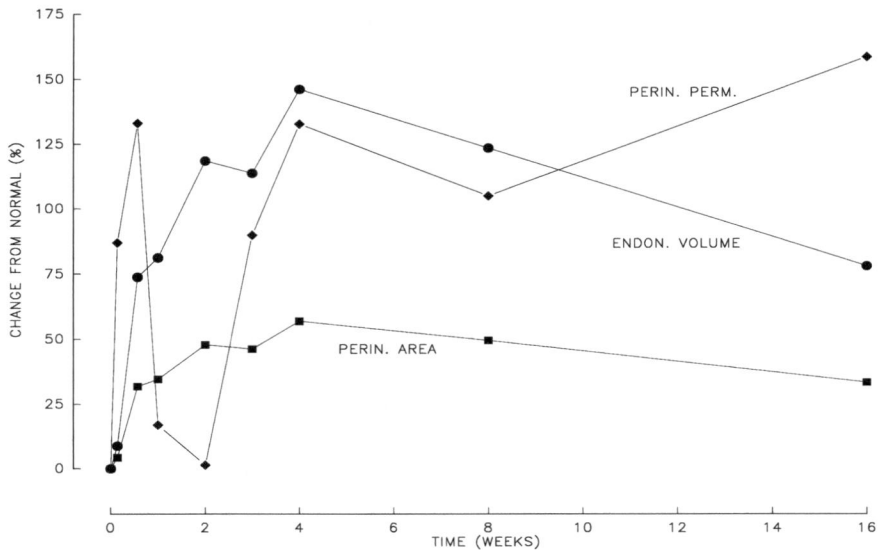

Fig. 3. Relative changes in perineurial permeability to ^{22}Na, endoneurial volume and perineurial area during Wallerian degeneration is shown. *Perin. perm.* Perineurial permeability; *Endon. volume* Endoneurial volume; and *Perin. area* Perineurial area (reproduced with permission (80)).

Sectioned degenerating nerves show an increased number of mast cells (61, 81, 82), which on degranulation release histamine, a biogenic amine with potent inflammatory properties. The later, sustained increase of perineurial permeability is probably a component of endoneurial homeostasis related to preventing elevation of EHP, and clearing myelin debris from the endoneurium (Fig. 4). The increased permeability of both components of the BNB is likely to facilitate the entry of monocytes into the endoneurium for myelin phagocytosis (83). Lipid droplets and proteinaceous material are present among perineurial layers in degenerating nerves, the cells themselves are not disrupted (81, 82).

Delineation of the properties of the BNB during nerve regeneration provides a more comprehensive picture. From day 4 up to the eighth week after the crush lesion, BNB PS to ^{22}Na was significantly greater than the normal value. The increased ability of ^{22}Na to move into the endoneurial space is compared with the measured endoneurial water content at the same time points in Figs. 5 and 6. However, by the 18th week after the crush, PS was not significantly different from the normal value.

The endoneurial water content, calculated as the wet weight to dry weight ratio increased from day 4 and remained elevated during the entire experimental period (Fig. 6). Its peak at the second week postcrush corresponds to a period of increased PS of the BNB to ^{22}Na. On the other hand, at 18 weeks post crush when the BNB PS to ^{22}Na is normal, the nerve is still edematous.

While an increased BNB PS to plasma macromolecules could elevate the osmolality of the endoneurial fluid and thus draw

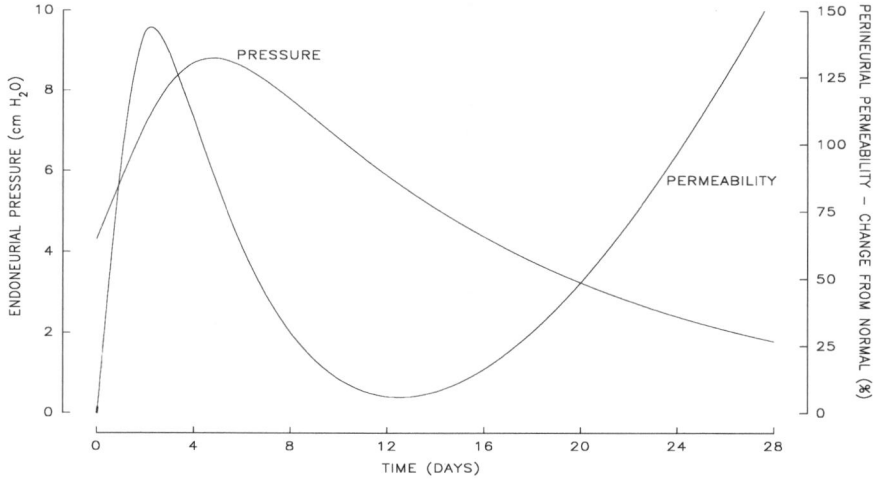

Fig. 4. Endoneurial hydrostatic pressure and perineurial permeability to ^{22}Na during the first 4 weeks of Wallerian degeneration is shown (reproduced with permission (80)).

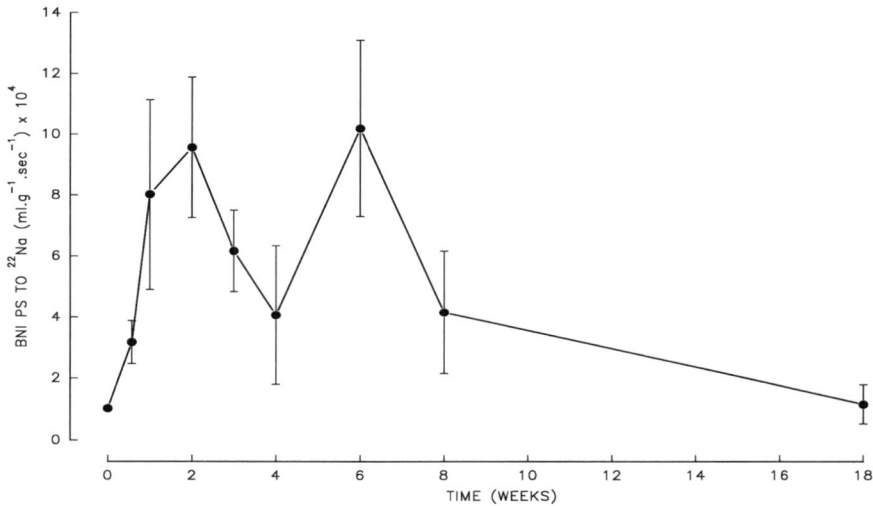

Fig. 5. Blood–nerve barrier permeability to ^{22}Na during regeneration after a crush lesion is shown. Time after the lesion is represented in the x-axis. There were four animals at the day 4 and 1 week time points, and five animals at all other time points (Weerasuriya, unpublished data).

water into this space, it is contended that a decrease in the endoneurial fluid turnover is also a factor in maintaining the increased water content of the regenerated nerve. On the other hand, the elevated water content is not associated with nerve edema; at 12 weeks after a crush injury the EHP of the regenerating nerve returns to normal (84). It is postulated that an altered compliance of the regenerated nerve helps to maintain EHP in the normal range.

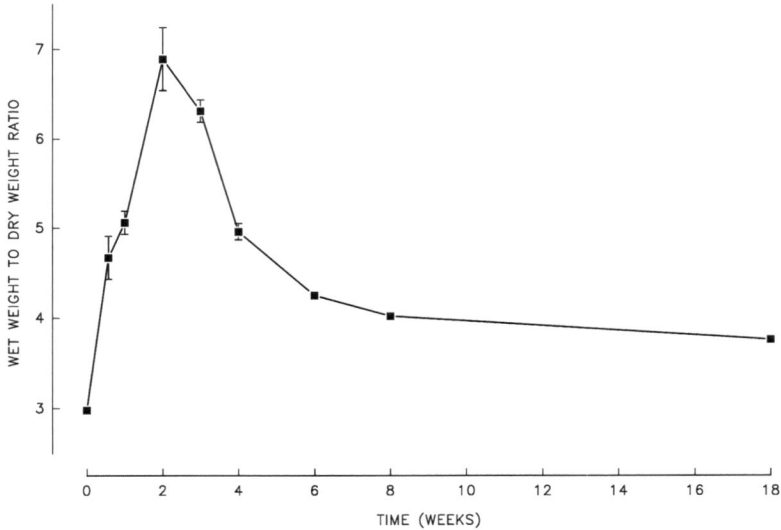

Fig. 6. Changes in endoneurial wet weight to dry weight ratio in desheathed rat sciatic nerve following a crush injury is shown. Data are presented as mean ± SEM. The SEMs of normal, and 6, 8, and 18 weeks are smaller than the size of the symbol (Weerasuriya, unpublished data).

5.4. Endoneurial Homeostasis in Lead Neuropathy

The major effects of lead on peripheral nerve are endoneurial edema, nuclear inclusion bodies, demyelination, elevation of EHP, and increased permeability of BNB (85). Of these alterations, the nuclear inclusion bodies are observed first followed by endoneurial edema. It had been suggested earlier (85, 86) that an increase in the permeability of endoneurial capillaries was the primary pathological event leading to subsequent Schwann cell damage and segmental demyelination. Later studies indicated that accumulation of lead in the endoneurium and nerve edema, precede qualitative changes in the permeability of endoneurial capillaries (87, 88). The BNB index to albumin (a measure of the rate of albumin entry and removal from the endoneurium) starts to increase only at 6 weeks in lead-intoxicated rats, and suggests that the change in BNB permeability is subsequent to the direct toxic effect of lead on Schwann cells (89). A subsequent study (90) provided clear, quantitative evidence that endoneurial pathology precedes an increase in permeability of BNB, and furthermore, that the increase in permeability is about threefold. It is further implied that the increase of BNB permeability is not a consequence of disruption of its barrier properties, but more in the nature of an adaptive response to aid in the clearing of myelin debris from the endoneurium. A scheme, which is an extension of earlier hypotheses, attempts to delineate the causal relationships among the changes in various components of the endoneurium in lead-induced neuropathy (Fig. 7). The hypothesis underlying this scheme and its implications are described in detail elsewhere (65).

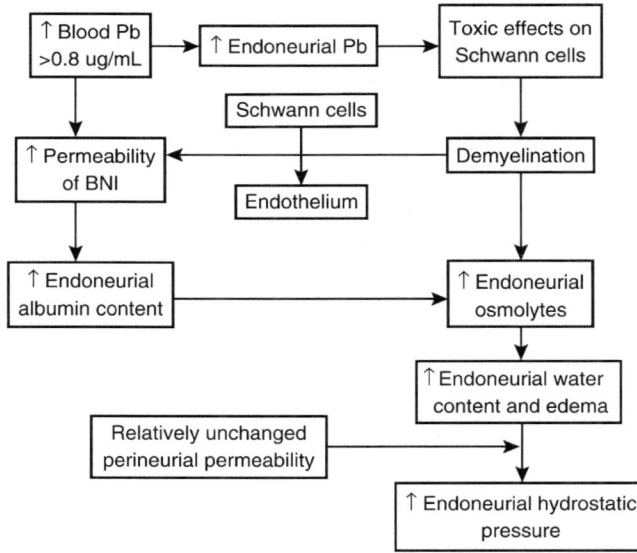

Fig. 7. Proposed sequence of events leading to demyelination and endoneurial edema in lead neuropathy are shown (reproduced with permission from (65)).

An intriguing but unresolved clinical issue is that lead toxicity predominantly leads to an encephalopathy in children, whereas in adults the major manifestation is a neuropathy. A greater vulnerability of the juvenile cerebral neurovascular unit to Pb is a likely but, as yet, unproven hypothesis.

6. Factors Regulating BNB Permeability

Microvessels in the endoneurium are the least permeable capillaries except for intracerebral capillaries. The investing perineurium is even less permeable than the endoneurial capillaries. Hence, cellular components of the endoneurium secrete soluble factors responsible for maintaining the integrity of the tight junctional complexes and associated actin cytoskeletal components of endoneurial vascular endothelial cells and of cells composing the innermost layers of the perineurium. It should be recognized that endoneurial elements not only expend metabolic energy to maintain the junctional integrity of its barrier tissues, but are at the same time obliged to devote metabolic energy to maintain the pumps and transporters for blood–nerve exchange of critical material needed by the endoneurial cellular elements. Thus, the dynamically selective permeability properties of the BNB extract a considerable metabolic cost.

6.1. Neurovascular Unit in Nerve

When considering the homeostatic mechanisms of the cerebral microenvironment, the neurovascular unit is an important concept that is emerging with increasing validity and support from experimental data (91, 92). The concept of a neurovascular unit, albeit with some modification, is equally relevant for the endoneurial microenvironment. Functionally, axons and Schwann cells can tolerate limited hypoxia and metabolic insults for seconds to perhaps a few minutes without suffering irreversible damage. Structurally, only pericytes are within a few microns of the endothelial cells and invested by a common basal lamina. Nevertheless, focal ischemic lesions in peripheral nerve are probably indicative of the presence of functional neurovascular units in the endoneurial microenvironment with two important differences from their cerebral counterparts. In peripheral nerve, given the greater separation between axons and endothelial cells and between Schwann cells and endothelial cells, signaling molecules have to diffuse through greater distances to reach their targets. Hence, the neurovascular unit in nerve, compared to that in cerebrum, is likely to operate in a larger volume and respond more slowly to perturbations.

6.2. Cellular Control of BNI Permeability

The perineurium and the endoneurial microvasculature form the BNI, which isolates the endoneurium by virtue of the "tight" intercellular junctions present in these two structures. Perineurial microvessels, which are permeable to plasma macromolecules, acquire "tight" intercellular junctions upon penetrating the innermost layers of the perineurium to enter the endoneurium (16). Thus, one or more diffusible factors present in the endoneurium are responsible for the induction of "tight" intercellular junctions in barrier tissues circumscribing the endoneurium. Such an inducer may be produced by axons, Schwann cells or both, and regulate synthesis and elaboration of intercellular junctional proteins by perineurial and endothelial cells.

In chronically transected nerves, the near normal PS of the endoneurial endothelium to small nonelectrolytes (33), and the presence of "tight" intercellular junctions in the endothelium (81) strongly argues against axons as the source of a regulatory factor for the maintenance of the tight junctions between vascular endothelial cells. However, the increased permeability of the perineurium accompanying axonal degeneration (81, 93, 94) strongly suggests that axonally derived factors might be responsible for maintaining the relatively tight barrier properties of the perineurium. A working hypothesis based on a dual source of regulatory proteins for modulating the permeability properties of the endoneurial vascular endothelium and perineurium needs to be rigorously tested by examining BNI permeability when either axons or Schwann cells are preferentially damaged by endoneurial injection of toxins.

A corollary of this hypothesis is that Schwann cells have several phenotypes, and a major function of some of these subsets is regulation of endoneurial vascular permeability. Regulation of vascular permeability would be accomplished through soluble factors released into the endoneurial extracellular space and diffusing less than 100 μm to reach the intended targets, endothelial cells. In effect, this would be a form of paracrine signaling within a confined and specialized extracellular space. On the other hand, based on reasons outlined above, perineurial junctional integrity appears to depend primarily on soluble factors secreted by axons and perhaps aided by Schwann cell-derived signaling molecules. A role for Schwann cells is postulated because of the expression of VE-cadherins by perineurial cells (43). Hence, it is hypothesized that the baseline resting permeability of the two components of the BNI are regulated by axons and Schwann cells. An integral aspect of endoneurial homeostasis is the ability of the BNI to adaptively alter its permeability properties to meet the changing needs of endoneurial elements. Such changes may be dictated by programmed and gradual changes associated with growth and maturation, as well as microenvironmental disturbances precipitated by disease and trauma.

It appears that these adaptive alterations in permeability of the BNI are initiated and effected through an immune response. Recent evidence strongly implicates the immune system as an active modulator of BNI permeability in a whole host of conditions ranging from trauma to metabolic neuropathies and to vascular disorders (95–98). Thus, it is postulated that hematogenous elements interacting directly with the endoneurial vascular elements cause increases in capillary permeability. The perineurium appears to be unusually resistant to inflammatory mediators (99). Hence, immunomodulation of the BNI is expected to be limited to endothelial cells in nerve.

6.3. Molecular Basis of BNI Permeability Regulation

Even though the molecular biology of junctional complexes of the BNI have not been investigated in detail, it is reasonable to assume that they are essentially similar to those of the cerebral microvasculature (100, 101). Thus, mediators of BNI permeability alterations have to utilize one or more of the following mechanisms: modification of the actin cytoskeleton responsible for anchoring the junctional complexes to the intracellular scaffolding; posttranslational modification of the constituent proteins (occludin, family of ZO proteins, cadherins, claudins, catenins) of the junctional complexes; and proteolytic degradation of the junctional components. While the first two modes require signaling molecules to gain intracellular access, the last mode may be initiated and completed extracellularly at the junction itself. The classical cytokines and interleukins mediate inflammation-related changes in BNB. They are also being implicated as having a role in disease-related BNB changes as well.

7. Conclusions

The term blood–nerve interface (BNI) is preferable to blood–nerve barrier (BNB). The former term encompasses the regulatory nature as well as the passive barrier properties of the perineurium and the endoneurial microvasculature. As we learn more about the structure and functions of the perineurium and endoneurial microvasculature, especially the graded nature of its permeability and its dynamic adaptive responses to alterations in the endoneurial microenvironment, it seems more appropriate to refer to them as the BNI. This conceptual thrust is likely to play a significant role in our understanding of peripheral nerve disorders and the evolution of therapeutic strategies and paradigms for the treatment of these disorders.

Acknowledgments

This work is partially supported by RO1–NS30197 and RO1-DI2078374 (NIH), IBN-9420525 (NSF), Juvenile Diabetes Research Foundation and MedCen Foundation, Macon, GA (No. 23750). The invaluable assistance of Mr. Daniel Mizisin and Mr. John Knight in the preparation of figures and discussions with Dr. Quentin Smith are gratefully acknowledged.

References

1. Sunderland S (1978) Nerves and nerve injury. Churchill Livingstone, Edinburgh

2. Weerasuriya A, Rapoport SI, Taylor RE (1979) Modification of permeability of frog perineurium to (^{14}C) sucrose by stretch and hypertonicity. Brain Res 173:503–512

3. Weerasuriya A, Rapoport SI, Taylor RE (1980) Ionic permeabilities of the frog perineurium. Brain Res 191, 405–415

4. Haller FR, Low FN (1971) The fine structure of the peripheral nerve root sheath in the subarachnoid space in the rat and other laboratory animals. Am J Anat 131:1–19

5. Shanthaveerappa TS, Bourne GH (1962) The 'perineurial epithelium', a metabolically active, continuous, protoplasmic cell barrier surrounding peripheral nerve fasciculi. J Anat 96:527–537

6. Haller FR, Haller C, Low FN (1972) The fine structure of cellular layers and connective tissue space at spinal nerve root attachment in the rat. Am J Anat 133:109–123

7. Fraher JP (1999) The transitional zone and CNS regeneration. J Anat 194:161–182

8. Bell MA, Wedell AGM (1984) A morphometric study of intrafascicular vessels in mammalian sciatic nerve. Muscle Nerve 7:524–534

9. Olsson Y (1990) Microenvironment of the peripheral nervous system under normal and pathological conditions. Critical Rev Neurobiol 5:265–311

10. Olsson Y, Kristensson K (1971) Permeability of blood vessels and connective tissue sheaths in the peripheral nervous system to exogenous proteins. Acta Neuropathol 5:61–69

11. Lundborg G (1988) Nerve injury and repair. Churchill Livingstone, Edinburgh

12. Peters A, Palay SL, DeF Webster H (1991) The fine structure of the nervous system. Oxford University Press, New York

13. Shanthaveerappa TS, Bourne GH (1966) Perineurial epithelium: a new concept of its role in the integrity of the peripheral nervous system. Science 154:1464–1467

14. Bunge MB, Wood PM, Tynan LB et al (1989) Perineurium originates from fibroblasts: demonstration in vitro with a retroviral marker. Science 243:229–231

15. Shinowara NL, Michel ME, Rapoport SI (1982) Morphological correlates of permeability in the frog perineurium: vesicles and transcellular channels. Cell Tissue Res 227: 11–22

16. Olsson Y (1984) Vascular permeability in the peripheral nervous system. In: Dyck PJ, Thomas PK, Lambert EH,Bunge R (eds) Peripheral neuropathy, 2nd edn. Saunders, Philadelphia

17. Ross MH, Reith EJ (1969) Perineurium: evidence for contractile elements. Science 165: 604–605

18. Burkel WE (1967) The histological fine structure of perineurium. Anat Rec 158:177–190

19. Olsson Y, Reese TS (1971) Permeability of vasa nervorum and perineurium in mouse sciatic nerve studied by fluorescence and electron microscopy. J Neuropathol Exp Neurol 30:105–119

20. Myers RR, Urakami H, Powell HC (1986) Reduced nerve blood flow in edematous neuropathies – a biochemical mechanism. Microvasc Res 32:145–151

21. Olsson Y (1966) Studies on vascular permeability in peripheral nerve I Acta Neuropathol 7: 1–15

22. Olsson Y (1968) Topographical differences in the vascular permeability of the peripheral nervous system. Acta Neuropathol 10:26–33

23. Kristensson K, Olsson Y (1971) The perineurium as a diffusion barrier to protein tracers; differences between mature and immature animals. Acta Neuropathol 17:127–138

24. Arvidson B (1979) A study of the perineurial diffusion barrier of a peripheral ganglion. Acta Neuropathol 46:139–44

25. Jacobs JM, Macfarlane RM, Cavanagh JB (1976) Vascular leakage in the dorsal root ganglia of the rat, studied with horseradish peroxidase. J Neurol Sci 29:95–107

26. Krnjevic K (1954) Some observations on perfused frog sciatic nerves. J Physiol 123: 338–356

27. Rapoport SI (1976) Blood-brain barrier in physiology and medicine. Raven Press, New York

28. Soderfeldt B (1974) The perineurium as a diffusion barrier to protein tracers. Influence of histamine, serotonine and bradykinine. Acta Neuropathol 27:55–60

29. Boddingius J (1984) Ultrastructural and histophysiological studies on the blood-nerve barrier and perineurial barrier in leprosy neuropathy. Acta Neuropathol 64:282–296

30. Weerasuriya A (1990) Permeabilities of endoneurial capillaries and perineurium of rat sciatic nerve to ^{22}Na during development. Physiologist 33:A-100

31. Weerasuriya A, Rapoport SI (1986) Endoneurial capillary permeability to (^{14}C) sucrose in frog sciatic nerve. Brain Res 375:150–156

32. Weerasuriya A (1987) Permeability of endoneurial capillaries to K, Na, and Cl and its relation to peripheral nerve excitability. Brain Res 419:188–196

33. Weerasuriya A (1988) Patterns of change in endoneurial capillary permeability and vascular space during Wallerian degeneration. Brain Res 445:181–187

34. Weerasuriya A, Curran GL, Poduslo JF (1989) Blood-nerve transfer of albumin and its implications for the endoneurial microenvironment. Brain Res 494:114–121

35. Grotte G (1956) Passage of dextran molecules across the blood-lymph barrier. Acta Chir Scand Suppl 211:1–84

36. Pappenheimer JR (1953) Passage of molecules through capillary walls. Physiol Rev 33:387–423

37. Karnovsky MJ (1970) Morphology of capillaries with special reference to muscle capillaries. In: Crone C, Lassen NA (eds) Capillary permeability. Munksgaard, Copenhagen

38. Renkin EM (1977) Multiple pathways of capillary permeability. Circ Res 41:735–743

39. Bundgaard M, Frokjaer-Jensen J (1982) Functional aspects of the ultrastructure of terminal blood vessels: a quantitative study on consecutive segments of the frog mesenteric microvasculature. Microvasc Res 23:1–30

40. Crone C, Levitt DG (1984) Capillary permeability to small solutes. In: Renkin EM, Michel CC (eds) Handbook of physiology. The cardiovascular system, vol 4. Microcirculation American Physiological Society, Bethesda

41. Rippe B, Haraldsson B (1987) How are macromolecules transported across the capillary wall? News Physiol Sci 2:135–138

42. Taylor AE, Granger DN (1984) Exchange of macromolecules across the microcirculation. In: Renkin EM, Michel CC (eds) Handbook of physiology, Sect 2, The cardiovascular system, vol 4. Microcirculation American Physiological Society, Bethesda

43. Smith ME, Jones TA, Hilton D (1998) Vascular endothelial cadherin is expressed by perineurial cells of peripheral nerve. Histopathology 32:411–413

44. Tserentsoodol N, Shin BC, Koyama H, et al (1999) Immunolocalization of tight junction proteins, occludin and ZO-1, and GLUT1 in

the cells of the blood-nerve barrier. Arch Histol Cytol 62:459–469

45. Nesbitt JA, Acland RD (1980) Histopathological changes following removal of the perineurium. J Neurosurg 53:233–238

46. Nukada H, Powell HC, Myers RR (1992) Perineurial window: demyelination in nonherniated endoneurium with reduced nerve blood flow. J Neuropathol Exp Neurol 51: 523–530

47. Spencer PS, Weinberg HJ, Raines CS (1975) The perineurial window – a new model of focal demyelination and remyelination. Brain Res 96:323–329

48. Terho PM, Vuorinen VS, Roytta M (2002) The endoneurial response to microsurgically removed epi- and perineurium. J Peripheral Nerv Syst 7:155–162

49. Rechthand E, Smith QR, Rapoport SI (1985) Facilitated transport of glucose from blood into peripheral nerve. J Neurochem 45:957–964

50. Froehner SC, Davies A, Baldwin SA et al (1988) The blood-nerve barrier is rich in glucose transporter. J Neurocytol 17:173–178

51. Gerhart DZ, Drewes LR (1990) Glucose transporters at the blood-nerve barrier are associated with perineurial cells and endoneurial microvessels. Brain Res 508:46–50

52. Magnani P, Cherian PV, Gould GW et al (1996) Glucose transporters in rat peripheral nerve: paranodal expression of GLUT1 and GLUT3. Metabolism 45:1466–1473

53. Rechthand E, Smith QR, Rapoport SI (1988) A compartmental analysis of solute transfer and exchange across blood-nerve barrier. Am J Physiol 255:317–325

54. Smith QR, Allen DD (2003) In situ brain perfusion technique. Methods Mol Med 89: 209–213

55. Rydevik BL, Kwan MK, Myers RR, et al (1990) Effects of acute stretching on rabbit tibial nerve: an in vitro mechanical and histological study. J Orthop Res 8:694–701

56. Weiss P, Wang H, Taylor AC et al (1945) Proximo-distal fluid convection in the endoneurial spaces of peripheral nerves, demonstrated by colored and radioactive (isotope) tracers. Am J Physiol 143:521–540

57. Mellick RS, Cavanagh JB (1967) Longitudinal movement of radio-iodinated albumin within extravascular spaces of peripheral nerves following three systems of trauma. J Neurol Neurosurg Psychiat 30:458–463

58. Low PA (1985) Endoneurial potassium is increased and enhances spontaneous activity in regenerating mammalian nerve fibers. Implications for neuropathic positive symptom. Muscle Nerve 8:27–33

59. McCabe JS, Low FN (1969) The subarachnoid angle: an area of transition in peripheral nerve. Anat Rec 164:15–33

60. Waksman BH (1961) Experimental study of diphtheritic polyneuritis in the rabbit and guinea pig. III. The blood-nerve barrier in the rabbit. J Neuropathol Exp Neurol 20: 35–77

61. Powell HC, Myers RR (1989) The blood-nerve barrier and the pathological significance of nerve edema. In: Neuwelt EA (ed) Implications of the blood-brain barrier and its manipulation, vol 1. Plenum Press, New York

62. Myers RR, Rydevik BL, Heckman HM et al (1988) Proximo-distal gradient in endoneurial fluid pressure. Exp Neurol 102:368–370

63. Low FN (1976) The perineurium and connective tissue of peripheral nerve. In: Landon DN (ed) The peripheral nerve. Chapman and Hall, London

64. Poduslo JF, Curran GL, Dyck PJ (1988) Increase in albumin, IgG and IgM blood-nerve barrier indices in human diabetic neuropathy. Proc Natl Acad Sci 88:4879–4883

65. Weerasuriya A (2005) Blood-nerve interface and endoneurial homeostasis. In: Dyck PJ, Thomas PK (eds) Peripheral neuropathy, 4th ed. Elsevier, New York

66. Kandel ER, Schwartz JH, Jessell TM (2000) Principles of neural science. 4th edn. McGraw-Hill, New York

67. Landis EM, Pappenheimer JR (1963) Exchange of substances through the capillary wall. In: Hamilton WF, Dow P (eds) Handbook of physiology, Circulation, Sect 2, vol II. American Physiological Society, Washington DC

68. Mizisin AP, Kalichman MW (1993) Permeability and surface area of the blood-nerve barrier in galactose intoxication. Brain Res 618:109–114

69. Crowley NR, Nelson SL, Weerasuriya A. et al (1996) Endoneurial hydrostatic pressure in rat sciatic nerve during development. Soc Neurosci Abstr 22:772

70. Rechthand E, Smith QR, Rapoport SI (1987) Transfer of nonelectrolytes from blood into peripheral nerve endoneurium. Am J Physiol 252:H1175–H1182

71. Schmelzer JD, Low PA (1988) The effects of hyperbaric oxygenation and hypoxia on the blood-nerve barrier. Brain Res 473: 321–326

72. Low PA, Tuck RR (1984) Effects of changes in blood pressure, respiratory acidosis and hypoxia on blood flow in the sciatic nerve of the rat. J Physiol 347:513–524

73. Rundquist I, Smith QR, Michael ME et al (1985) Sciatic nerve blood flow measured by laser Doppler flowmetry and (^{14}C) iodoantipyrine. Am J Physiol 248:H311-H317

74. Low PA, Marchand G, Knox F et al (1977) Measurement of endoneurial fluid pressure with polyethylene matrix capsule. Brain Res 122:373–377

75. Myers RR, Powell HC, Costello ML et al (1978) Endoneurial fluid pressure: direct measurement with micropipettes. Brain Res 148:510–515

76. Weerasuriya A, Coath G, Crowley N (1996) Endoneurial Na concentration and hydrostatic pressure in galactose neuropathy. FASEB J 10:A764

77. Weerasuriya A, Curran GL, Poduslo JF (1990) Developmental changes in blood-nerve transfer of albumin and endoneurial albumin content in rat sciatic nerve. Brain Res 521:40–46

78. Stoll G, Jander S, Myers RR (2002) Degeneration and regeneration in the peripheral nervous system: from August Waller's observation on neuroinflammation. J Peripher Nerv Syst 7:13–27

79. Tsao JW, George EB, Griffin JW (1999) Temperature modulation reveals three distinct stages of Wallerian degeneration. J Neurosci 19:4718–4726

80. Weerasuriya A, Hockman CH (1992) Perineurial permeability to sodium during Wallerian degeneration in rat sciatic nerve. Brain Res 587:327–333

81. Latker CH, Wadhwani KC., Balbo A et al (1991) Blood-nerve barrier in the frog during Wallerian degeneration: are axons necessary for maintenance of barrier function? J Comp Neurol 309:650–664

82. Williams PL, Hall SM (1971) Chronic Wallerian degeneration – an in vivo and ultrastructural study. J Anat 109:487–503

83. Stoll G, Trapp BD, Griffin JW (1989) Macrophage function during Wallerian degeneration of rat optic nerve: clearance of degenerating myelin and Ia expression J Neurosci 9:2327–2335

84. Weerasuriya A, Nelson SL, Crowley NR (1998) Evidence for endoneurial fluid flow from endoneurial hydrostatic pressure measurements in transected and crushed rat sciatic nerves. Soc Neurosci Abstr 24:267

85. Low PA, Dyck PJ (1977) Increased endoneurial fluid pressure in experimental lead neuropathy. Nature 269:427–428

86. Myers RR, Powell HC, Shapiro HM, et al (1980) Changes in endoneurial fluid pressure, permeability and peripheral nerve ultrastructure in experimental lead neuropathy. Ann Neurol 8:392–401

87. Powell HC, Myers RR, Lampert PW (1982) Changes in Schwann cells and vessels in lead neuropathy. Am J Pathol 109:193–205

88. Windebank AJ, McCall JT, Hunder HG et al (1980) The endoneurial content of lead related to the onset and severity of segmental demyelination. J Neuropathol Exp Neurol 39:692–699.

89. Ohi T, Poduslo JF, Curran GL et al (1985) Quantitative methods for detection of blood-nerve barrier alterations in experimental animal models of neuropathy. Exp Neurol 90:365–372

90. Weerasuriya A, Curran GL, Poduslo JF (1990) Physiological changes in the sciatic nerve endoneurium of lead intoxicated rats: a model of endoneurial homeostasis. Brain Res 517:1–6

91. Abbott NJ (2002) Astrocyte-endothelial interactions and blood-brain barrier permeability. J Anat 200:629–638

92. Lo EH, Broderick JP, Moskowitz MA (2004) tPA and proteolysis in the neurovascular unit. Stroke 35:354–356

93. Wadhwani KC, Latker CH, Balbo A et al (1989) Perineurial permeability and endoneurial edema during Wallerian degeneration of the frog peripheral nerve. Brain Res 493:231–239

94. Weerasuriya A, Rapoport SI, Taylor RE (1980) Perineurial permeability increases during Wallerian degeneration. Brain Res 192: 581–585

95. Griffin JW (2001) Vasculitic neuropathies. Rheum Dis Clin North Am 27:751–760

96. Polydefkis M, Griffin JW, McArthur J (2003) New insights into diabetic polyneuropathy. JAMA 290:1371–1376

97. Powell HC, Myers RR (2004) Impact of inflammatory disease on the nerve microenvironment. J Neurol Sci 220:131–132

98. Zhou L, Griffin JW (2003) Demyelinating polyneuropathies. Curr Opin Neurol 16:307–313

99. Abbott NJ (2000) Inflammatory mediators and modulation of blood-brain barrier permeability. Cell Mol Neurobiol 20:131–147

100. Anderson CM, Nedergaard M (2003) Astrocyte-mediated control of cerebral microcirculation. Trends Neurosci 26: 340–345

101. Gloor SM, Wachtel M, Bolliger MF et al (2001) Molecular and cellular permeability control at the blood-brain barrier. Brain Res Brain Res Rev 36:258–264

Part II

Imaging the Barriers

Chapter 7

Detection of Multiple Proteins in Intracerebral Vessels by Confocal Microscopy

Janet L. Manias, Anish Kapadia, and Sukriti Nag

Abstract

Assessment of the blood-brain barrier (BBB) may involve the localization of endothelial proteins within the context of endothelial permeability to plasma proteins. The use of antibodies conjugated to fluorescent dyes, coupled with analysis by confocal microscopy, allows for the detection of multiple proteins in components of the neurovascular unit including endothelium and astrocytes. This chapter provides a detailed protocol for detection of three proteins in fixed or frozen sections of rat brain using three fluorophores with unique excitation/emission spectra. Also included is a protocol for tyramide signal amplification, which is useful for detecting proteins of low abundance, and methods for quantitation of intracerebral vessels expressing a particular protein of interest with and without BBB breakdown to plasma proteins.

Key words: Blood-brain barrier, Brain endothelium, Immunofluorescence, Multiple labeling, Confocal microscopy, Fluorophore, Quantitation

1. Introduction

Many forms of brain injury such as infarcts, trauma, inflammation, and tumors are associated with breakdown of the blood-brain barrier (BBB) to plasma proteins leading to vasogenic edema. The availability of antibodies to plasma proteins led to the detection of areas with BBB breakdown to plasma proteins in tissue sections as far back as 1984 when the unlabeled antibody peroxidase–antiperoxidase method was used (1). In a lesion area, 60–80% of vessels consisting of arterioles and corresponding-sized veins show BBB breakdown (2, 3). Therefore, to detect proteins of interest in the endothelium of vessels with BBB breakdown, serial sections of vessels were obtained and adjacent sections were used to detect plasma protein extravasation and the

Sukriti Nag (ed.), *The Blood-Brain and Other Neural Barriers: Reviews and Protocols*, Methods in Molecular Biology, vol. 686,
DOI 10.1007/978-1-60761-938-3_7, © Springer Science+Business Media, LLC 2011

protein of interest (4), a technique which was very labor-intensive. In the case of the immunogold technique, gold particles of different sizes were used to study localization of an endothelial protein along with a marker of BBB breakdown (5).

The availability of fluorescent dyes, which can be conjugated directly or indirectly to antibodies has provided significant advantages to the field (6). Many different dyes are available, each with a unique excitation/emission spectrum, which when used in combination can allow for the localization of multiple proteins at the same time. The technique of multiple labeling by immunofluorescence (IF) to study colocalization of various proteins within the cell has been used extensively in the field of BBB research (2, 3, 7–9). One disadvantage of IF studies is interpretation in cases where there is overlap of signals from surrounding cells or tissues. This has been circumvented by the use of a confocal laser scanning microscope which allows for the visualization of fluorescence in only one focal plane, thus eliminating background noise (10, 11). Confocal microscopy also uses sensitive photomultipliers and narrow bandpath filters, allowing for the capture of fluorescence from only a small range of wavelengths from visible light. This minimizes cross-talk or "bleed-through" where spectra of fluorescent dyes used may overlap (12). These unique parameters make confocal microscopy essential for analysis of colocalization of proteins within cells or tissues.

Most commonly, multiple IF labeling makes use of two different primary antibodies along with their specific secondary fluorophores, in combination with a nuclear stain. Hoescht or DAPI, two common dyes that bind to DNA, emit blue light under UV excitation. The development of a wide variety of fluorophores allows for labeling of three unique proteins at once, while still allowing cells to be counterstained with a nuclear dye. Although the procedure of multiple labeling by IF is relatively simple, careful thought must be given when determining which primary and secondary antibodies and which fluorophores are to be used. The use of highly specific, affinity purified primary antibodies will decrease any background signal (13).

In this chapter, methods for multiple labeling by IF are described. The protocol is designed to make use of fluorophores from the green, red, and infrared portion of the light spectrum (see Table 1). A protocol for tyramide signal amplification (TSA) for immunofluorescent detection of antigens using very low levels of primary antibody is also included. It can be used alone, or in combination with multiple immunolabeling protocols. The final section of this chapter describes quantitative methods for determining vessel numbers in a lesion area showing expression of a protein of interest, in the context of BBB breakdown.

Table 1
Information on the secondary antibodies and streptavidin used and the laser required for the fluorophore detection is given

Secondary antibody	Excitation (nm)	Emission (nm)	Dilution and source	Laser required for detection
Biotinylated goat-anti mouse			1:900, Jackson Immunoresearch Labs Inc	
Streptavidin AlexaFluor488	494	519	1:200, Molecular Probes®	Argon 488
Goat Antirabbit Cy3	548	562	1:175, Jackson Immunoresearch Labs Inc	Helium–Neon 543
Donkey Antirabbit Cy5	650	670	1:500, Jackson Immunoresearch Labs Inc	Helium–Neon 643

2. Materials

2.1. Fixation and Preparation of Paraffin-Embedded Sections

1. Isoflurane (Pharmaceutical Partners of Canada Inc, Richmond Hill, ON).
2. Three percent paraformaldehyde in 0.1M phosphate buffer, pH 7.4.
3. Solvents: Ethanol, Xylene, Acetone.
4. Colorfrost slides (Fisher Scientific, Nepean, ON), or other glass slides.
5. 3-aminopropyltriethoxysilane (Sigma-Aldrich, St Louis, MO).
6. Paraplast Plus Tissue Embedding medium (Oxford Labware, St. Louis, MO).
7. Sta-On Tissue Section Adhesive (Surgipath Medical Industries Inc, Richmond, IL).

2.2. Preparation of Frozen Sections

1. Detergent solution: Sparkleen1 (Fisher Scientific).
2. Thirty percent sucrose in 0.1M phosphate buffer.
3. Tissue-Tek® O.C.T. Compound 4583 (Fisher Scientific).
4. Subbing solution: 1% Type A Gelatin and 0.1% chromium potassium sulfate (chrome alum) in distilled water.
5. Poly-L-lysine hydrobromide, MW 300,000 (Sigma-Aldrich). Make a 0.005% Poly-L-lysine hydrobromide solution in 0.1M Tris HCl, pH 8.

2.3. Antigen Retrieval

1. 0.01M Na Citrate buffer, pH 6.0.
2. 0.5% pepsin in 0.01M HCl solution.
3. Microwave, hotplate, and vortexer.

2.4. Triple Labeling for Immunofluorescence Analysis

1. A humid chamber (see Note 1).
2. Glass or plastic staining dishes.
3. Pipetors: 1–10 and 10–100 μL with tips.
4. Shaker for slide washes.
5. Phosphate buffered saline (PBS): Dissolve 8.0 g of NaCl, 0.2 g of KCl, 1.44 g of Na_2HPO_4, and 0.24 g of KH_2PO_4 in 800 mL of distilled water. Adjust pH to 7.4 and the volume to 1 L. Sterilize by autoclaving. This solution can be made as a 10× stock.
6. Rubber Cement.
7. 5 mL plastic syringe with a 22gauge needle.
8. Normal goat whole serum and normal donkey whole serum, 1:20 dilution in PBS (Jackson Immunoresearch Laboratories Inc, Westgrove, PA, USA).
9. Primary antibodies diluted in PBS: Monoclonal mouse anti-claudin-5, 1:45 dilution (Zymed, San Francisco, CA, Cat# 35-2500); Polyclonal rabbit antihuman Fibronectin, 1:150 dilution (Sigma-Aldrich, Cat# F3648); Polyclonal rabbit anti-glial fibrillary acidic protein (GFAP), 1:2000 dilution (Dako Canada Inc, Mississauga, On, Cat# Z0334), (see Note 2).
10. Fluorophore-conjugated secondary antibody or fluorophore-conjugated streptavidin and secondary biotinylated link antibody (see Table 1, Note 3).
11. Mowiol mounting medium (14): Add 2.4 g of Mowiol (Calbiochem, San Diego, CA) to 6 g of glycerol. Stir to mix. Add 6 mL of water and leave for several hours at room temperature. Add 15 mL of 0.2M Tris (pH 8.5) and heat to 50°C for 10 min with occasional stirring. After the mowiol dissolves, clarify by centrifugation at $5,000 \times g$ for 15 min. For fluorescence detection, add 2.5% 1,4-diazobicyclo-(2,2,2)-octane (DABCO) to reduce fading. Place aliquots in airtight containers and store at –20°C (see Note 4).
12. Cover slips.
13. Zeiss LSM 510 Confocal laser scanning microscope (Carl Zeiss International, Germany) with appropriate lasers and filter sets (see Table 1).

2.5. Tyramide Signal Amplification Technique

1. Methanol.
2. Thirty percent Hydrogen peroxide (H_2O_2).

3. TSA™ Detection Kit (Molecular Probes, Burlington, On): Contains Streptavidin-HRP, blocking reagent, amplification buffer, dimethyl sulfoxide (DMSO), H_2O_2 and tyramide-Alexa Green488.

4. Tyramide stock solution: Dissolve tyramide in 150 μL DMSO. Mix well and store small aliquots at 4°C protected from light.

5. Prepare 1% blocking solution in PBS.

6. Streptavidin-HRP 1:650 (Jackson Immunoresearch Laboratories Inc) or use the streptavidin-HRP which comes with the kit.

7. Amplification buffer with 0.0015% H_2O_2: Add 1 μL of 30% H_2O_2 to 200 μL amplification buffer. Add 1 μL of this solution to 100 μL amplification buffer. Prepare only as needed, making up 100 μL per slide.

2.6. Quantitation of Intracerebral Vessels

1. Anti-Factor VIII antibody 1:200 dilution (Dako Canada Inc).

2. Antiglucose transporter antibody 1:300 dilution (Calbiochem).

3. Adobe Creative Suite 3 Software (Adobe Systems Inc, San Jose, CA).

4. NIH Image J Software: http://www.rsb.nih.gov/ij, 1997–2004.

3. Methods

3.1. Fixation

1. Rats are anesthetized using 3% Isoflurane for induction and 1–2% for maintenance.

2. Rats are fixed by vascular perfusion of 350–500 mL of 3% paraformaldehyde via a cannula in the ascending aorta at a pressure of 120 mmHg (see Note 5).

3. The brain is removed and immersed in the same fixative solution at room temperature.

4. After a few hours, the brain is cut into coronal slabs having a thickness of 2 mm to allow better penetration of fixative.

5. These slabs are placed in the same fixative solution overnight at room temperature and are processed for paraffin embedding the next day.

3.2. Preparation of Paraffin-Embedded Sections

3.2.1. Tissue Processing

1. An automatic processor is recommended since processing is done under vacuum and this reduces processing time.

2. The shortest processing schedule for coronal slabs of rat brain are:

   (a) Eighty percent ethanol for 30 min

   (b) Two changes of 95% ethanol for 30 min each

(c) Three changes of ethanol for 40 min each

(d) Two changes of xylene for 50 min each

(e) Two changes of wax (Paraplast®) at 60°C for 30 min each.

(f) The brain should be embedded in Paraplast® within 30 min of the completion of the processing cycle to prevent the tissue from becoming brittle and difficult to cut.

3.2.2. Slide Coating for Paraffin-Embedded Sections

1. Fisher Colorfrost slides are slightly more expensive than the regular grade slides, but their advantage is that they do not require coating.

2. All other slides should be coated with 2%, 3-aminopropyltri-ethoxysilane in acetone as follows:

(a) Open a fresh box of slides and place in a metal rack.

(b) Dip slides in 2% aminoalkylsilane for 5 min. The same solution can be reused to coat up to 500 slides.

(c) Rinse by dipping 3 times in three changes of distilled water.

(d) Dry slides for 2 h or overnight at 37°C.

(e) Store slides in a dust-free container at 4°C. These slides can be used up to 2–3 weeks when stored at either room temperature or 4°C.

3.2.3. Paraffin-Embedded Sectioning and Deparaffinization

1. Section thickness is usually 10 μm for confocal microscopy (see Note 6).

2. Sta-0n Tissue section adhesive (1%) is added to the water bath used for floating the sections during cutting to ensure that sections adhere to both Color Frost and silanized slides.

3. Allow slides to dry at room temperature for 2 h and then place in an oven at 55°C overnight.

4. Sections are deparaffinized in the specified solutions for the stated time:

(a) Xylene 2 changes, 5 min each

(b) Absolute ethanol two changes, 3 min each

(c) Ninety percent ethanol one change for 3 min

(d) Seventy percent ethanol one change for 2 min

(e) Fifty percent ethanol one change for 2 min

5. Place slides in water and leave container in a 37°C water bath and proceed to Subheading 3.4.

3.3. Preparation of Frozen Sections

1. If frozen sections are required, follow steps 1–4 described in Subheading 3.1.

3.3.1. Fixation

2. Brains slabs are placed in 30% sucrose in 0.1M phosphate buffer at 4°C up to 3 days or until the brain sinks to the bottom of the container.

3. Brains slabs are then placed in a puddle of Tissue-Tek® and frozen in a cryostat at –25°C.

4. The brain slabs are wrapped in aluminum foil and placed in a labeled jar and stored at –80°C.

3.3.2. Coating Slides for Frozen Sections

To prevent sections of fixed frozen brain from washing off during immunostaining, it is recommended to first sub slides followed by a poly-L-lysine coating (15). These steps are essential if *in situ* hybridization precedes the immunostaining.

1. Slides are placed in racks and soaked in detergent solution for 30 min or overnight.

2. Rinse for 30 min in hot running tap water and then in three changes of distilled water over a period of 30 min.

3. Drain excess water from slides by placing the slide rack on a paper towel.

4. Dip in 100% ethanol and then air dry.

5. Immerse rack for 10 min in the subbing solution at 37°C and drain slides on a paper towel.

6. Dry slides overnight at 37°–50°C and store in boxes for up to 2 months.

7. Before use, subbed slides are placed in a rack and coated with the 0.005% poly-L-lysine hydrobromide solution for 30 min.

8. Drain excess solution and dry at least 30 min in a fume hood and store in dust-free boxes.

9. Treated slides are placed in a dessicator overnight prior to cutting frozen sections.

10. Frozen section thickness can range from 15 to 20 μm.

11. Sections can be stored at –20°C or –70°C for months without loss of antigenicity.

12. Hydrate sections by placing in PBS and proceed to Subheading 3.4.

3.4. Antigen Retrieval

The mechanism of antigen retrieval, either by enzyme action or heat, is based on cleaving cross-linkages and extending polypeptides to more closely resemble the 3D structure of native proteins for better recognition by its antibody (16, 17). Enzymes such as pepsin, trypsin, pronase, and Proteinase K may be used. Heat-induced epitope retrieval may be followed by enzyme-induced epitope retrieval or either method may be used alone.

3.4.1. Enzyme-Induced Epitope Retrieval

1. In our laboratory 0.5% pepsin alone for 30 min at 37°C is used for 10 μm paraffin-embedded or frozen sections (see Note 7).

2. Heat 0.01M HCl in a microwave at full power for 10 s in a coplin jar and then add the required amount of pepsin and stir.

3. When the temperature of the pepsin is 39°C, add the slides to the container and place in a 37°C water bath for 30 min.

4. Wash slides with three changes of distilled water for 3 min followed by two changes of PBS for 3 min each.

3.4.2. Heat-Induced Epitope Retrieval (see Note 8)

Tight complexing of calcium ions or other divalent metal cations with proteins during formaldehyde fixation is suggested to be responsible for the masking of certain antigens (18). During heat-induced epitope retrieval, the citrate buffer in which the sections are immersed removes calcium from the tissues.

1. A plastic rack containing the slides is immersed in a beaker containing 500 mL of the NaCitrate buffer. The beaker is covered with loosely fitting aluminum foil.

2. Place beaker on a hot plate and boil for 10 min, then place the beaker on the counter and cool for 20 min.

3. Place sections in cool 0.1M PBS.

3.5. Triple Labeling for Analysis by Immunofluorescence

Best results are obtained when primary antibodies raised in different species are used for multiple labeling (19) (Fig. 1). The methods to detect claudin-5 in brain endothelium, fibronectin as a marker of BBB breakdown, and GFAP in astrocytes in the same brain section from a lesion site are described and illustrated (Fig. 2a–d). Two primary antibodies raised in the same species can be used if monovalent Fab fragments are used as secondary antibodies for detection of the first antigen. These fragments contain only one binding site, thereby reducing the possibility of cross-reactivity (10, 20, 21) in most but not all cases (see Fig. 2e, f).

3.5.1. Detection of the First Antigen

1. Dispense rubber cement in a syringe through a 22gauge needle to circle the tissue section. Place slides in a humid chamber.

2. Shake off excess buffer and apply normal goat serum to sections for 15 min at room temperature. Use an appropriate volume of solution to ensure full coverage of the section including at least 2 mm around the section to avoid "edge effect."

3. Shake off normal serum and apply monoclonal mouse anti-claudin-5 antibody to the section. Incubate slides for 4°C overnight (approximately 16 h). The dilution of the primary antibody and the incubation time should be determined empirically (see Notes 9 and 10).

4. Set up appropriate controls (see Notes 11 and 12).

5. Wash slides 3 times with PBS for 5 min each (see Notes 13–14).

6. Shake off excess buffer and apply biotinylated goat-anti mouse to sections for 30 min at room temperature. Again, the

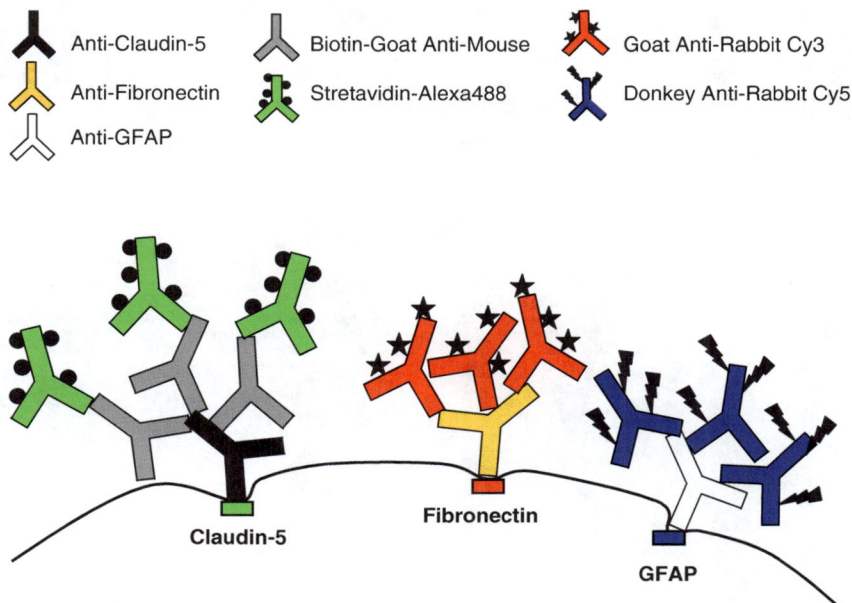

Fig. 1. Schematic showing the primary and secondary antibodies used for triple labeling immunofluorescence.

dilution of the secondary antibody should be determined empirically.

7. Repeat step 5, shake off excess buffer and apply Streptavidin-Alexa488 and incubate at room temperature for 20 min in the dark (see Note 15).

8. Repeat step 5.

3.5.2. Detection of the Second Antigen

1. Shake off excess buffer and apply normal goat serum for 15 min at room temperature.

2. Shake off normal serum and apply rabbit antihuman fibronectin antibody for 1.5 h at room temperature.

3. Wash slides 3 times with PBS for 5 min each.

4. Shake off excess buffer and apply goat antirabbit Cy3 antibody to sections for 20 min at room temperature.

5. Repeat step 3.

3.5.3. Detection of the Third Antigen

1. Shake off excess buffer and apply normal donkey serum for 15 min at room temperature.

2. Shake off normal serum and apply polyclonal rabbit anti-GFAP antibody for 2 h at room temperature.

3. Wash slides 3 times with PBS for 5 min each.

4. Shake off excess buffer and apply goat antirabbit Cy5 antibody to sections for 15 min at room temperature.

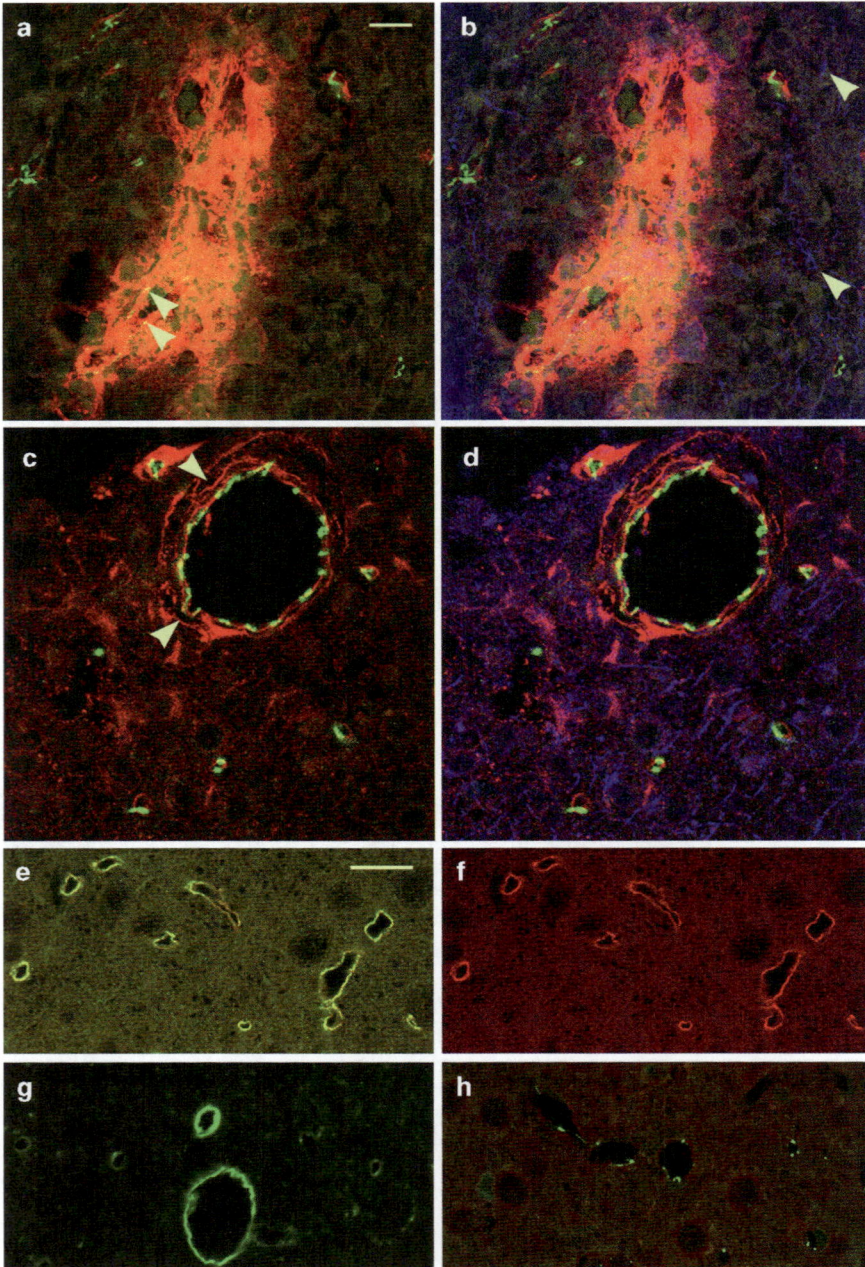

Fig. 2. Confocal images of the cerebral cortex of cold-injured (a–d) and control rats (e–h). (a) *Red* and *green* channels of a day 2 cold-lesion shows vessels with BBB breakdown to fibronectin (*red*) have reduced claudin-5 signal (*arrowheads*). The nonpermeable microvessels show claudin-5 (*green*). (b) The merged confocal image shows astrocytic processes in the surrounding neuropil which is pseudocolored in blue (*arrowheads*). (c) *Red* and *green* channels of a day 4 cold-lesion show a vein with BBB breakdown to fibronectin (*arrowheads*). There is restoration of claudin-5 at endothelial tight junctions (*green*). (d) The merged image shows astrocytic processes in the surrounding neuropil (*blue*). (e, f) This example shows the pitfalls of dual labeling using two primary antibodies from the same species despite the use of a monovalent Fab fragment of a secondary antibody conjugated with Cy3. In this case glucose transporter-1 (*green*) and fibronectin (*red*) colocalize in endothelium (e). This should not occur since normal microvessels should not show endothelial fibronectin immunostaining (f). (g) Results of tyramide enhancement detection have to be evaluated with caution. The *green* channel shows diffuse circumferential immunoreactivity for JAM-A with tyramide enhancement which is nonspecific. Specific localization of JAM-A at tight junctions is shown in (h). (a–h) Scale bars = 20 μm.

5. Remove rubber cement from around tissue and repeat step 3.

6. Mount sections using mowiol mounting medium.

7. Save slides in folders with covers at 4°C.

3.6. Tyramide Signal Amplification Technique

TSA can be used to improve the detection sensitivity of the immunofluorescent signal in many tissues including the brain (16, 22). The tyramide added in the procedure reacts with horseradish peroxidase (HRP) to create reactive tyramide radicals, which can then bind to tyrosine residues in the tissues in close proximity to where HRP is bound (22). The tyramide is conjugated to a fluorophore, which can then be visualized by fluorescence microscopy. TSA can also be used in multiple labeling using primary antibodies raised in the same species. With TSA, tenfold less primary antibody is used and this low concentration of antibody is not recognized by a fluorophore-conjugated secondary antibody used in the traditional IF method (22). We have obtained optimal results using tyramide enhancement to detect phosphorylated caveolin-1 and fibronectin as a marker of BBB breakdown at the cold-injury site (2).

1. Prepare tissue samples as described in Subheadings 3.1–3.4.

2. To quench endogenous peroxidases, place slides in 0.3% methanolic peroxide for 20–30 min depending on the number of red blood cells present in the section.

3. Wash slides in distilled water for 3 min and then in two changes of PBS for 3 min each.

4. Follow steps 1–5 in Subheading 3.5.1 to detect the antigen required.

5. Shake off buffer and apply streptavidin-HRP for 30 min.

6. Wash slides 3 times in PBS for 5 min each.

7. Prepare a working solution of tyramide-Alexa488, by diluting the tyramide stock solution 1:100 in amplification buffer/0.0015% H_2O_2. Add the diluted tyramide-Alexa488 to slides and incubate for 4 min at room temperature.

8. Repeat step 6 and continue with detection of other antigens (see Subheading 3.5.2) or proceed to mounting slides.

9. When evaluating results ensure that the signal obtained is specific (see Note 16, Fig. 1g, h).

3.7. Quantitation of Intracerebral Vessels

Captured images from an area of interest can be used to quantitate the total number of intracerebral arterioles, which typically have diameters greater than 10 μm, and microvessels such as venules or capillaries, which have a diameter of ~5 μm. These data can be used to calculate the percentage of vessels showing BBB breakdown and the percentage showing an endothelial protein of interest.

3.7.1. Quantitation of Intracerebral Arterioles with BBB Breakdown

1. Horizontal sections (10 μm) of the cerebral cortex containing the lesion area are immunostained for the protein of interest as described in Subheading 3.5.1 and for fibronectin as described in Subheading 3.5.2.

2. Images of at least five adjacent fields of the area of interest per section are captured at an objective magnification of 20× using a fluorescence or confocal microscope.

3. Obtain images from at least three to five rats per group.

4. The area of the images used for quantitation is measured using Adobe Photoshop CS3 or equivalent soft ware. This value multiplied by the number of sections gives the total lesion area assessed for each experimental group in mm^2.

5. Determine the following parameters in a blinded manner with no knowledge of the experimental groups:

    (a) The total number of vessels per section (>10 μm in diameter).

    (b) The number of these vessels showing extravasation of fibronectin indicating BBB breakdown and how many of these contain the protein of interest.

6. Express values as percentage of vessels showing BBB breakdown and the percentage positive for the protein of interest.

7. In the cold-injury model it is mainly arterioles and corresponding-sized veins that show BBB breakdown to fibronectin. In this model 66% of vessels in a 0.7 mm^2 lesion area show BBB breakdown and 100% of them show endothelial phosphorylated caveolin-1 (2).

3.7.2. Quantitation of Microvessels

1. Stain sections containing the region of interest with a marker of brain endothelium such as Factor VIII or Glucose transporter 1 as described in Subheading 3.5.1. Other options are CD31 (see Chap. 14) or fluorescein–lycopersicin esculentum lectin (see Chap. 22).

2. Sections are then immunostained for fibronectin as given in Subheading 3.5.2.

3. Follow steps 2–4 given in Subheading 3.7.1.

4. Using Adobe Photoshop CS3 or equivalent soft ware determines the following parameters in a blinded manner with no knowledge of the experimental groups:

    (a) Number of microvessels per section (<10 μm in diameter).

    (b) The number of vessels showing extravasation of fibronectin indicating BBB breakdown and how many of these contain the protein of interest.

5. Express values as percentage of vessels showing BBB breakdown and the percentage positive for the protein of interest.

4. Notes

1. A humid chamber can be purchased or made by gluing glass rods to the bottom of a plastic box. Use a leveler to ensure that the rods are level to prevent antibodies from pooling in one area of the section. Strips of paper towel moistened with an excess of water are placed at the bottom of the chamber.

2. Antibodies are stable indefinitely when stored concentrated in aliquots at −80°C. Once they are thawed, they should not be refrozen. Prior to use, make sure that the antibody is fully reconstituted in PBS and vortex before use.

3. In our experience, Jackson Immunoresearch Laboratories Inc and Molecular Probes® provide the best quality of fluorescent-tagged secondary antibodies which can be used at low dilutions and give a high signal to noise ratio.

4. An alternate approach is to purchase mounting medium from suppliers. Prolong® Antifade Kit (Molecular Probes®) reduces photobleaching with little or no quenching of the photo signal, while a range of Vectashield® mounting media is available including mounting media with DAPI or Propidium iodide for nuclear staining (Vector Laboratories Canada Inc, Burlington, On).

5. Perfusion-fixed brains give the best signal and least background as compared to immersion-fixed tissue. If the antibody fails to give a signal on fixed tissue, frozen sections can be tried.

6. Thicker sections can be used, but problems may arise in the interpretation of results. The endothelium and astrocytes are in close proximity in microvessels and in thick sections (> than 10 µm), overlapping of the signal in these cell types may give a false impression of colocalization of the proteins of interest. In such cases examination of multiple optical slices may help or the use of software such as ImarisColoc (Bitplane Inc, Saint Paul, MN, USA) or Colocalizer Pro (Colocalization Research Software, Tokyo, Japan) can be used to perform a quantitative colocalization analysis of fluorescence images. A free program available from NIH, called Image J (http://rsbweb.nih.gov/ij/), has a variety of plug-ins available for quantification of colocalization. Most of these programs make use of algorithms to calculate correlation coefficients of multiple fluorescent signals to determine the degree of colocalization.

7. The concentration of pepsin can be varied from 0.05 to 0.5% and the period of exposure to pepsin can be varied from 15 to 30 min. Both parameters have to be established for demonstration of a specific protein. In our laboratory if 0.5%

pepsin alone for 30 min at 37°C does not give an optimal result, other options are explored.

8. Alternate methods for heat-induced epitope retrieval include the use of a microwave (23–25), pressure cooker (26, 27), or autoclaving (28, 29). These methods are particularly useful for antigen retrieval in human tissue fixed with 10% formalin.

9. Purchased antibodies have data sheets stating the recommended working concentrations, which are usually in the range of 10–20 µg/mL. It is usual to set up serial dilutions of the antibody to determine the optimum dilution for the test tissue which gives a high signal with the least amount of nonspecific background staining. Typically, antibody concentrations are much higher for IF studies as compared to immunohistochemistry.

10. The time of incubation in primary antibody varies from few hours at room temperature to overnight incubations at 4°C. Overnight incubations allow the use of higher dilutions of antibody, thus decreasing nonspecific background staining. Select the shortest time which gives maximal signal with the least amount of background staining.

11. Negative control sections are set up in which apply (a) PBS in place of the primary antibody or (b) antiserum which has been preincubated with the appropriate blocking peptide for an hour.

12. It is helpful to set up positive controls as well by selecting tissue in which the test antigen is known to be present.

13. Thorough rinsing is essential for reducing background. Shake off the antibody and use a wash bottle to direct a jet of PBS directly above the section, so it flows down over the section washing it for 30 s. Then place the sections in a staining dish containing PBS and place on a shaker for 5 min. Remove slides and tap corners onto filter paper to remove excess solution. Decant the PBS and replace with fresh PBS for two more washes.

14. If a high amount of background fluorescence becomes a problem and cannot be resolved by reducing antibody concentration, add a detergent to PBS washes. Typically, 0.05% Triton X-100 is added to PBS and used for washing after application of the secondary antibody. This detergent formula can be used for the first PBS wash only, or all three washes, depending on the level of background staining present.

15. Once antibodies conjugated with fluorophores are used, antibody incubations and washes on shakers are done in the dark by covering the humid chamber or container with a cardboard box. A table lamp may be used, but overhead fluorescent lights should remain off.

16. We used tyramide-Alexa Green488 to amplify the junctional adhesion molecule-A (JAM-A) signal using a polyclonal anti-JAM-A antibody (Santa Cruz Biotechnology, CA, USA). The results showed diffuse circumferential labeling of endothelial cells (Fig. 2g), which was nonspecific since JAM-A should localize only to the tight junctions. Localization of JAM-A to tight junctions only was achieved using another primary antibody to JAM-A (R&D Systems, Minneapolis, MN) without tyramide enhancement (3) (Fig. 2h).

Acknowledgements

This work was supported by Grants from the Heart & Stroke Foundation of Ontario #5347, #6003.

References

1. Nag S (1984) Cerebral changes in chronic hypertension: combined permeability and immunohistochemical studies. Acta Neuropathol (Berl) 62:178–184

2. Nag S, Manias JL, Stewart DJ (2009) Expression of endothelial phosphorylated caveolin-1 is increased in brain injury. Neuropathol Appl Neurobiol 35:417–426

3. Yeung D, Manias JL, Stewart DJ, Nag S (2008) Decreased junctional adhesion molecule-A expression during blood-brain barrier breakdown. Acta Neuropathol 115:635–642

4. Nag S, Takahashi JL, Kilty DW (1997) Role of vascular endothelial growth factor in blood-brain barrier breakdown and angiogenesis in brain trauma. J Neuropathol Exp Neurol 56:912–921

5. Cornford EM, Hyman S, Cornford ME (2003) Immunogold detection of microvascular proteins in the compromised blood-brain barrier. Methods Mol Med 89:161–175

6. Mullins JM (1994) Overview of fluorophores. Methods Mol Biol 34:107–116

7. Adamec E, Yang F, Cole GM, Nixon RA (2001) Multiple-label immunocytochemistry for the evaluation of nature of cell death in experimental models of neurodegeneration. Brain Res Brain Res Protoc 7:193–202

8. Bausch SB (1998) A method for triple fluorescence labeling with Vicia villosa agglutinin, an anti-parvalbumin antibody and an anti-G-protein-coupled receptor antibody. Brain Res Brain Res Protoc 2:286–298

9. Nag S, Venugopalan R, Stewart DJ (2007) Increased caveolin-1 expression precedes decreased expression of occludin and claudin-5 during blood-brain barrier breakdown. Acta Neuropathol (Berl) 114:459–469

10. Miyashita T (2004) Confocal microscopy for intracellular co-localization of proteins. Methods Mol Biol 261:399–410

11. Wouterlood FG, Vinkenoog M, van den OM (2002) Tracing tools to resolve neural circuits. Network 13:327–342

12. Stelzer EH, Wacker I, De M, Jr. (1991) Confocal fluorescence microscopy in modern cell biology. Semin Cell Biol 2:145–152

13. Harvath L (1994) Overview of fluorescence analysis with the confocal microscope. Methods Mol Biol 34:337–347

14. Harlow E, Lane D (1998) Using antibodies. A laboratory manual. Cold Spring Harbor Laboratory Press, Cold Spring Harbor, New York

15. Simmons DM, Arriza JL, Swanson LW (1989) A complete protocol for in situ hybridization of messenger RNAs in brain and other tissues with radiolaveled single-stranded RNA probes. J Histotechnol 12:169–180

16. Werner M, von WR, Komminoth P (1996) Antigen retrieval, signal amplification and intensification in immunohistochemistry. Histochem Cell Biol 105:253–260

17. Yamashita S (2007) Heat-induced antigen retrieval: mechanisms and application to histochemistry. Prog Histochem Cytochem 41:141–200

18. Morgan JM, Navabi H, Schmid KW, Jasani B (1994) Possible role of tissue-bound calcium ions in citrate-mediated high-temperature antigen retrieval. J Pathol 174:301–307

19. Paddock SW (2000) Principles and practices of laser scanning confocal microscopy. Mol Biotechnol 16:127–149

20. Brelje TC, Wessendorf MW, Sorenson RL (1993) Multicolor laser scanning confocal immunofluorescence microscopy: practical application and limitations. Methods Cell Biol 38:97–181

21. Wessel GM, McClay DR (1986) Two embryonic, tissue-specific molecules identified by a double-label immunofluorescence technique for monoclonal antibodies. J Histochem Cytochem 34:703–706

22. Wang G, Achim CL, Hamilton RL, Wiley CA, Soontornniyomkij V (1999) Tyramide signal amplification method in multiple-label immunofluorescence confocal microscopy. Methods 18:459–464

23. Cattoretti G, Pileri S, Parravicini C, Becker MH, Poggi S, Bifulco C, Key G, D'Amato L, Sabattini E, Feudale E (1993) Antigen unmasking on formalin-fixed, paraffin-embedded tissue sections. J Pathol 171:83–98

24. Pileri SA, Roncador G, Ceccarelli C, Piccioli M, Briskomatis A, Sabattini E, Ascani S, Santini D, Piccaluga PP, Leone O, Damiani S, Ercolessi C, Sandri F, Pieri F, Leoncini L, Falini B (1997) Antigen retrieval techniques in immunohistochemistry: comparison of different methods. J Pathol 183:116–123

25. Shi SR, Chaiwun B, Young L, Cote RJ, Taylor CR (1993) Antigen retrieval technique utilizing citrate buffer or urea solution for immunohistochemical demonstration of androgen receptor in formalin-fixed paraffin sections. J Histochem Cytochem 41:1599–1604

26. Norton AJ, Jordan S, Yeomans P (1994) Brief, high-temperature heat denaturation (pressure cooking): a simple and effective method of antigen retrieval for routinely processed tissues. J Pathol 173:371–379

27. Miller RTEstran C (1995) Heat-induced epitope retrieval with a pressure cooker. Suggestions for optimal use. Appl Immunohistochem 3:190–193

28. Bankfalvi A, Navabi H, Bier B, Bocker W, Jasani B, Schmid KW (1994) Wet autoclave pretreatment for antigen retrieval in diagnostic immunohistochemistry. J Pathol 174:223–228

29. Shin RW, Iwaki T, Kitamoto T, Tateishi J (1991) Hydrated autoclave pretreatment enhances tau immunoreactivity in formalin-fixed normal and Alzheimer's disease brain tissues. Lab Invest 64:693–702

Chapter 8

Multiparametric Magnetic Resonance Imaging and Repeated Measurements of Blood-Brain Barrier Permeability to Contrast Agents

Tavarekere N. Nagaraja, Robert A. Knight, James R. Ewing, Kishor Karki, Vijaya Nagesh, and Joseph D. Fenstermacher

Abstract

Breakdown of the blood-brain barrier (BBB) is present in several neurological disorders such as stroke, brain tumors, and multiple sclerosis. Noninvasive evaluation of BBB breakdown is important for monitoring disease progression and evaluating therapeutic efficacy in such disorders. One of the few techniques available for noninvasively and repeatedly localizing and quantifying BBB damage is magnetic resonance imaging (MRI). This usually involves the intravenous administration of a gadolinium-containing MR contrast agent (MRCA) such as Gadolinium-diethylenetriaminepentaacetic acid (Gd-DTPA), followed by dynamic contrast-enhanced MR imaging (DCE-MRI) of brain and blood, and analysis of the resultant data to derive indices of blood-to-brain transfer. There are two advantages to this approach. First, measurements can be made repeatedly in the same animal; for instance, they can be made before drug treatment and then again after treatment to assess efficacy. Secondly, MRI studies can be multiparametric. That is, MRI can be used to assess not only a blood-to-brain transfer or influx rate constant (K_i or K_1) by DCE-MRI but also complementary parameters such as: (1) cerebral blood flow (CBF), done in our hands by arterial spin-tagging (AST) methods; (2) magnetization transfer (MT) parameters, most notably T_{1sat}, which appear to reflect brain water-protein interactions plus BBB and tissue dysfunction; (3) the apparent diffusion coefficient of water (ADC_w) and/or diffusion tensor, which is a function of the size and tortuosity of the extracellular space; and (4) the transverse relaxation time by T_2-weighted imaging, which demarcates areas of tissue abnormality in many cases. The accuracy and reliability of two of these multiparametric MRI measures, CBF by AST and DCE-MRI determined influx of Gd-DTPA, have been established by nearly congruent quantitative autoradiographic (QAR) studies with appropriate radiotracers. In addition, some of their linkages to local pathology have been shown via corresponding light microscopy and fluorescence imaging. This chapter describes: (1) multiparametric MRI techniques with emphasis on DCE-MRI and AST-MRI; (2) the measurement of the blood-to-brain influx rate constant and CBF; and (3) the role of each in determining BBB permeability.

Key words: Apparent diffusion coefficient, Arterial spin tagging, Blood-brain barrier, Cerebral blood flow, Cerebral ischemia, Gd-DTPA, Hemorrhagic transformation, Influx constant, Look-locker, Magnetic resonance contrast agents, Magnetization transfer, Patlak plot, Quantitative autoradiography, Rat, T_1, T_{1sat}, T_1WI, T_2, TOMROP

Sukriti Nag (ed.), *The Blood-Brain and Other Neural Barriers: Reviews and Protocols*, Methods in Molecular Biology, vol. 686,
DOI 10.1007/978-1-60761-938-3_8, © Springer Science+Business Media, LLC 2011

1. Introduction

The blood-brain barrier (BBB) is a structural and functional feature of the cerebral microvascular system that tightly regulates the flux of substrates into and out of the brain (1). The "barrier" in this system is mainly formed by the continuous series of tight junctions that join the endothelial cells of the brain microvessels and close the paracellular pathway. The lack of fenestrae and reduced numbers of caveolae within the endothelial cells also contribute to barrier function. Other constituents of the BBB complex are the basal lamina (BL), the pericytes that are embedded in the BL, and a surrounding cuff of astrocytic end-feet (1).

Spontaneous, mechanical or drug-induced thrombolysis after cerebral ischemia can lead to either partial or complete restitution of cerebral blood flow (CBF) resulting in reperfusion of the ischemic regions. However, if not begun within a very short time window after stroke onset, reperfusion can induce opening of the BBB and a different, more damaging lesion than that caused by permanent occlusion (2). This acute BBB damage can worsen over time, leading to hemorrhagic transformation (HT) which can be fatal (3). Furthermore, the risk of HT can increase as much as tenfold with tPA treatment (4). Presently, apart from stroke onset time, no other reliable indicator(s) of the risk of HT are available. The heightened threat of HT associated with thrombolysis has, thus, restricted the widespread application of this vital therapy.

Some noninvasive techniques are available for characterizing acute stroke in humans. Traditionally, computed tomography (CT) has been the imaging modality of choice. More recently, multiparametric MRI, which combines two or more MRI techniques such as diffusion-weighted imaging (DWI) and perfusion-weighted imaging (PWI), has been used to identify the penumbra of the ischemic lesion. This region is characterized by a so-called perfusion-diffusion mismatch and is thought to be made up of tissue that is injured, but not irreversibly, and can be saved by prompt reperfusion (5). This approach has been employed to select patients for tPA therapy within and beyond the "3-hour-after-stroke-onset" time window (6, 7). Adding to the repertoire of multiparametric MRI, MRCA-enhanced imaging has also been shown to delineate ischemic regions with acute BBB alteration (8, 9). As alluded to above, detection of acute BBB opening in stroke is important since these regions have been shown to develop HT later either with or without tPA in both experimental models (3, 10) and humans (11).

Despite the capability of MRI to detect areas of BBB opening, precise localization and quantification of such damage in stroke by dynamic contrast agent-enhanced MRI (DCE-MRI) was still lacking until several years ago. Originally, quantification of DCE-MRI data was achieved by deconvolution to determine a transfer constant, K^{trans}, which was highly model-dependent (12). We have developed a novel method for localizing *and* quantifying DCE-MRI data using Patlak plots to obtain the blood-to-brain transfer or influx rate constant (K_i) of Gadolinium- diethylenetri- aminepentaacetic acid (Gd-DTPA) and tested the spatial and quantitative accuracy of these estimates with radiotracers and quantitative autoradiography (QAR) (13). This 25-year-old Patlak plot technique, recently applied by us to DCE-MRI, has now been replicated successfully by other labs using various magnet and console combinations in both experimental models and human disease (see Note 1).

Patlak and coworkers published an analysis of tracer uptake over a period of circulation by taking into account tissue and blood concentrations of the tracer at a number of points (minimally more than four) after injection (14, 15). This multiple-time graphic method yields a curve of tissue uptake as a function of blood tracer level and time. The slope of the linear part of this curve represents K_i, which is the most useful and accurate of the several blood-brain transfer constants that can be linked to capillary permeability. These data and the Patlak plot approach for analyzing MRI findings were confirmed with the images and K_i results obtained with ^{14}C-sucrose-QAR some minutes later in the same animals (13). This work was further extended to establish the spatial resolving power of DCE-MRI after a bolus MRCA injection by comparing these results to the QAR images of ^{14}C-α-aminoisobutyric acid (AIB) distribution gained concurrently in the same rats (16).

Despite employing a very precise MRCA infusion technique, it is possible to miss contrast-enhancement in spite of BBB damage in stroke since adequate blood flow is required for the delivery of the MRCA to the tissue of interest. In acute stroke, CBF is the first function to be affected and low flow may result in little or no delivery of MRCA to the tissue. For this reason, an MRI technique for estimating the leakiness of the BBB that does not involve an MRCA was sought. In one series of experiments, the ability of magnetization transfer (MT)-MRI methods was tested against classical Gd-DTPA enhanced MRI to look for alternate quantitative MRI signatures that indicated BBB opening (17). The MT-based indices tested were: T_1 and T_{1sat}. The regions with increased T_1 and T_{1sat} changes accurately reflected BBB changes as confirmed by DCE-MRI, QAR and dual-contrast enhanced MRI (18). The MT-MRI could also be used to segment and measure tissue damage (17).

Together these techniques represent a powerful array of minimally invasive and noninvasive quantitative MRI methods for

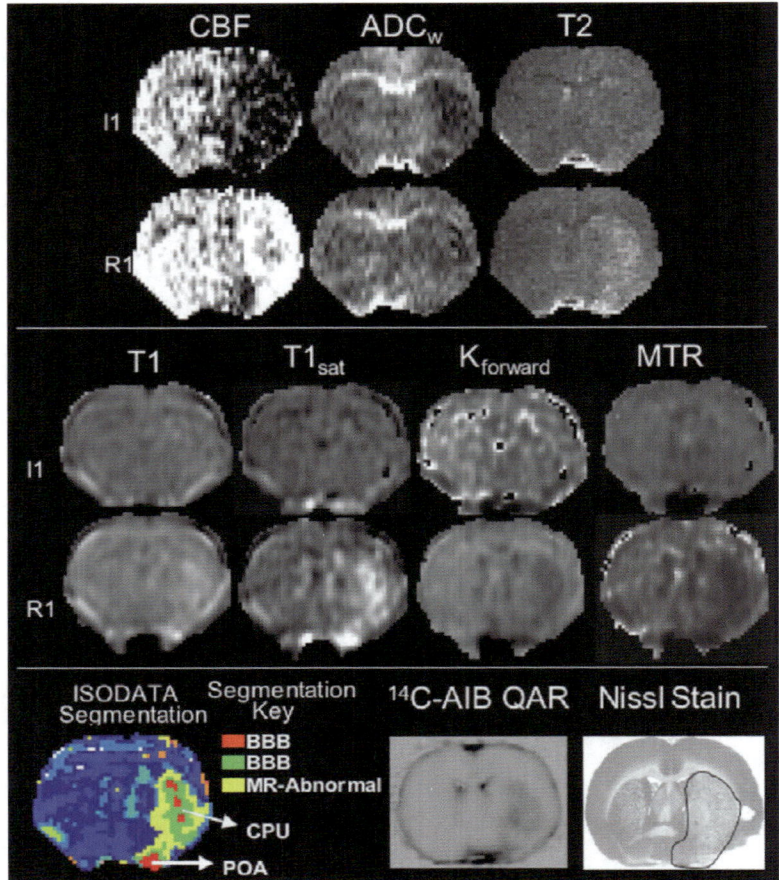

Fig. 1. A representative data set showing multiparametric MRI-derived maps obtained before (I1t) and after reperfusion (R1t) from an animal subjected to transient focal ischemia for 3 h. *Top* (*left* to *right*): cerebral blood flow (CBF), apparent diffusion coefficient (ADCw), and transverse relaxation time (T_2) maps are shown. *Middle* (*left* to *right*): spin–lattice relaxation times in the absence and presence of off-resonance saturation (T_1 and T_{1sat}, respectively), the forward rate of magnetization transfer (K_{for}), and MTR maps. *Bottom* row: a corresponding ISODATA segmentation theme map to demonstrate the regions of interests (ROIs) that were selected to represent ischemic tissue with (*red* and *green*) and without (*yellow*) BBB disruption (*left*). To the right, are a corresponding ^{14}C-AIB autoradiographic image and a cresyl violet stained histologic section that were used to confirm acute BBB damage and the region of ischemic damage, respectively (reproduced with permission from (17)).

accurate localization and quantification of tissue and BBB damage in stroke (Fig. 1) and brain tumors. The multiparametric MRI approach can also be applied to study other pathologies such as multiple sclerosis, Parkinson's disease, and neurotrauma. The greatest advantage is that it can be repeatedly applied in the same subject within several hours on the same day or employed as follow-up across different days. These measures are expected to aid in cerebrovascular disease characterization and its response to treatments.

2. Materials

2.1. Middle Cerebral Artery Occlusion (MCAO) Model

1. Young adult Wistar rats, weighing 275–300 g (Charles River Breeding Laboratories, Wilmington, MA).

2. Sterile surgical instruments, including microvascular clips (Codman, Raynham, MA), skin clamps.

3. Sterile 4–0 nylon suture, about 2 cm in length with its tip rounded using a heat source.

4. Sterile silk 4–0 sutures with needles.

5. Sterile 1 mL syringes (Becton Dickinson, Rutherford, NJ, USA).

6. Sterile 23 gauge needles (Becton Dickinson, Franklin Lakes, NJ, USA).

7. PE–50 catheters with beveled, rounded tips, about 100 cm long for venous and arterial lines. The other end of the catheter is connected to a blunted 23gauge needle attached to a 1-mL syringe (Becton Dickinson, Sparks, MD, USA).

8. Betadine solution and alcohol swabs.

9. A small, animal hair clipper.

10. Sterile saline (500 mL) bags, each containing 50 IU heparin.

11. Anesthesia apparatus and small-animal nose cones for isoflurane anesthesia (Baxter Healthcare Corpn., Deerfield, IL) and lidocaine (Abbott Labs, N. Chicago, IL) for irrigation.

12. A water-recirculating heating pad (Kent Scientific Corp., Torrington, CT, USA).

13. Sterile drapes, masks, gloves, and gauze.

14. Operating microscope (Carl Zeiss, Germany).

15. A manual or programmable syringe pump (Model 944, Harvard Apparatus, South Natick, MA).

2.2. 9L Cerebral Tumor Implantation Model

1. Male Fischer-344 rats, weighing between 200 and 240 g (Charles River Breeding Laboratories, Wilmington, MA).

2. Stereotaxic equipment with two positioning holders for a dental drill and a Hamilton syringe each (David Kopf Instruments, Tujunga, CA).

3. Sterile -surgical instruments including scalpels, silk sutures, bone wax.

4. A sterilized 10 µL Hamilton syringe with a beveled tip.

5. A foot-operated dental drill (Foredom Series F, Bethel, CT).

6. Injectable anesthetics: Cocktail containing Ketamine 80 mg/kg (Abbott Laboratories, N. Chicago, IL) and Xylazine 10 mg/kg (Phoenix Scientific Inc., St. Joseph, MO).

7. A tube of ophthalmic ointment (AKWA Tears, Akorn Inc., Buffalo Grove, IL).

8. Sterile drapes, masks, gloves, gauze.

9. A source of immortalized glioblastoma cells (Dept. of Radiation Oncology, Henry Ford Hospital, Detroit, MI).

10. Eagle's minimum essential medium (Baxter Healthcare, Deerfield, IL), supplemented with 10% fetal calf serum, 1% multivitamins and 1% nonessential amino acids.

11. Trypsin 0.05% stock solution diluted 1:1 with PBS (GIBCO, Grand Island, NY, USA).

2.3. Magnetic Resonance Imaging

1. A dedicated small animal magnet (Magnex Scientific, Inc.; Abingdon, UK) (see Note 2).

2. Magnet-compatible imaging sequences and consoles (Bruker Biospin MRI, Billerica, MA, USA).

3. Offline, postprocessing hardware (SUN Microsystems, Inc., Santa Clara, CA) and software (Eigentool, Henry Ford Health System, Detroit, MI; C-program based Unix Shell scripts).

4. A commercially available or lab-made Gadolinium-based MRCA such as Magnevist 0.1 mmol/Kg (Berlex, Montville, NJ, USA).

---

3. Methods

Animal studies were performed in accordance with the National Institutes of Health guidelines for animal research and were approved by the Institutional Animal Care and Use Committee.

3.1. Experimental Models

3.1.1. Middle Cerebral Artery Occlusion (MCAO) Model

In order to develop and test BBB permeability measurement techniques, it was necessary to utilize an animal model having a focal region of brain tissue with BBB opening. For this purpose, we used a model of transient focal cerebral ischemia in rats via intraluminal occlusion of the middle cerebral artery (MCA) with a nylon filament (3, 10).

1. Rats are anesthetized and placed supinely on the water-recirculating heating pad and the core temperature is kept at $37 \pm 1°C$ throughout all surgical and MRI procedures using a feed-back regulated rectal probe.

2. The neck area is cleaned with alcohol swabs and betadine solution and shaved to expose skin.

3. A 2-cm midline longitudinal incision is made in the ventral aspect of the neck and the right common carotid artery

(CCA), external carotid artery (ECA), and internal carotid artery (ICA) are exposed under the operating microscope. Care should be taken to avoid injuring the vagus nerve.

4. The ICA is further dissected to identify the pterygopalatine branch and intracranial ICA and a 5–0 silk suture is loosely tied at the origin of the ECA and the distal end of the ECA is ligated.

5. The CCA and ICA are temporarily clamped with microvasculature clips.

6. The 2 cm long, 4–0 surgical nylon filament, with a rounded tip round is introduced into the ECA lumen through a small puncture and the suture at the origin of the ECA tightened around the intraluminal nylon filament to prevent bleeding and the microvascular clips are removed.

7. The nylon filament is gently advanced from the ECA into the lumen of the ICA, a distance of 18.5–19.5 mm according to animal weight, until the tip of the filament blocks the origin of the (MCA) and the microvascular clip is released.

8. The skin incision is temporarily closed with either loose silk sutures or skin clips. After MR quantification of occlusion effects, the rat is taken out of the magnet and reperfusion is performed at the desired time as indicated in step 9.

9. The skin sutures/clamps are opened and the occluding filament is withdrawn until its tip is visible in the lumen of the ECA and is no longer occluding the MCA or restricting flow in the ICA and the neck incision is closed.

3.1.2. 9L Gliosarcoma
Brain Tumor Model

In brain tumors, BBB opening can be due to both BBB damage and increased angiogenesis which is typically seen at the periphery of solid experimental tumors. Owing to the tremendous leakiness of the vessels, it is also necessary to correct the Patlak model for MRCA backflux. The optimal experimental conditions that accurately reflect native leakage and indices of response to putative treatments were developed (19, 20) using the 9L gliosarcoma model (21).

1. The tumor cells are maintained as exponential cultures in Eagle's minimum essential medium (see Subheading 2.2, #10) (21).

2. Immediately prior to each implantation, cells are trypsinized, resuspended and a final dilution of 2×10^6 cells/mL is made in Eagle's medium without serum.

3. The rat is anesthetized with an intraperitoneal injection of a cocktail of ketamine and xylazine and positioned in the stereotaxic head frame.

4. The scalp is cleaned with alcohol swabs and betadine and hair is shaved from the frontal surface of the scalp with electric clippers and ophthalmic ointment is applied to the rat's eyes to keep the eyes moist.

5. A midline incision is made on the dorsal surface of the head to expose the frontal and temporal bones.

6. The position of the injection site is marked using one positioning arm of the dental drill: 2.5 mm anterior to the bregma and 2.0 mm to the right of the midline.

7. A 1.0-mm burr-hole is drilled at this point through the skull without breaking the dura for implanting tumor cells.

8. The Hamilton syringe fitted with a 26-gauge needle is filled with 5 µL of the cell suspension and fixed to the other stereotaxic positioning arm. Its tip is positioned over the burr-hole. The arm is then lowered to a depth of 3 mm and raised by 0.5 mm. The 5 µL volume is slowly injected to implant the tumor cells.

9. The syringe is left in place for 5 min and then gently retracted. The burr-hole is sealed with sterile bone wax and the skin sutured with sterile silk sutures.

10. Tumors are allowed to grow for up to 14–15 days postimplantation before the MRI studies.

3.1.3. Femoral Arterial and Venous Cannulation

1. Anesthetize the rat, place it in a supine position on the heated rubber mat. Extend its legs and fix them to the operating base with adhesive tape.

2. Clean the skin of either the left or right groin with alcohol swabs and shave hair in this region and apply betadine to the skin surface.

3. Make approximately a 1-cm long, oblique (along the direction of extended leg) skin incision in the groin.

4. Expose the femoral artery and vein, separate them and clean the fascia.

5. For each vessel, a small length (~10 cm) 4–0 silk suture is put loosely around the proximal part and the ends of the suture are weighed down by a hemostat to pull the vessel taut. The distal end (approximately 1 cm apart) is ligated using a 4–0 silk suture with enough length for two more knots.

6. An opening (approximately one-half the vessel diameter) is made in the middle of this 1 cm segment of the vessel.

7. The beveled end of a PE-50 catheter is introduced into this opening, threaded about 1.5 cm proximally without kinking the vessel and the proximal suture is used to ligate the vessel over the catheter. Care must be exercised not to ligate beyond the catheter tip to avoid occluding the vessel.

8. The remaining part of the distal suture is used to tie the catheter to the vessel to prevent it from slipping out.

9. The catheters are tested for smooth blood flow. Blood samples can be now collected or the vessels can be used for physiological monitoring, drug injections etc.

3.2. Basic MRI Experimental Procedures

1. The rat is placed in a plastic holder following the surgical procedures. It can be either in a supine or prone position depending on the design of the MR coil. This holder should have provisions for supporting the animal's body and for administering anesthetic gases if they are being used.

2. A feedback-regulated water blanket and nonmagnetic stereotaxic ear bars minimize head movement during imaging.

3. The animal holder is placed inside a 5 cm diameter quadrature driven transmit/receiver birdcage coil. The entire assembly is placed inside the MRI unit. The coil is then tuned to the resonant proton frequency of ~300 MHz (see Note 2).

4. Scout images are obtained to adjust the position of the animal's head until the central image slice is located at the bregma.

5. Shimming, and setting up central frequency, the radio frequency (RF) gain and receiver gain are done either automatically or manually.

3.2.1. Measurement of Cerebral Blood Flow (CBF)

Blood flow, an important function of the cerebral vascular system, can be measured at the local level by MRI. It is important that CBF be measured in studies of BBB permeability because the actual parameter determined in the latter experiments is an influx rate constant, K_i, depending not only on capillary permeability (P) and surface area (S) but also CBF. To get from K_i to PS product, the physiological expression of capillary permeability and the rate of blood flow are required. Hence, CBF is estimated in a single slice in all of all of our MRI experiments by PWI using the arterial spin tagging (AST) technique as follows:

1. Cerebral perfusion can be estimated using AST with a variable tip-angle gradient-refocused echo (VTA-GRE) imaging technique (22).

2. In this procedure, the protons within the blood passing through the arteries of the neck are magnetically inverted thereby serving as an autologous label. The decrease in the net magnetization of the perfused tissue, a function of the rate of blood flow, is then regionally detected in the brain.

3. AST uses gradient and frequency offsets to create a plane of inverted spins in the neck of the animal approximately 18 mm below the center of the imaging plane. AST is performed by

labeling inflowing arterial protons via a continuous wave (CW) RF adiabatic inversion pulse applied in the presence of a magnetic field gradient.

4. Labeling of inflowing arterial water protons is achieved via an axial gradient of ±0.3 kHz/mm using a CW RF pulse at a power of 0.3 kHz at a frequency offset of ±6 kHz followed by a VTA-GRE sequence with repetition time (TR)/echo time (TE) = 11/5 ms (22).

5. To allow for equilibration of inverted protons in the imaging plane, the inversion pulse is turned on for a period of 4.5 s preceding the production of each VTA-GRE image.

6. A corresponding set of control scans is acquired using gradient and frequency offsets that put the inversion plane 18 mm above the imaging plane (i.e., outside the head).

7. The images are acquired in sets of four with the frequency offset and gradient polarities permuted through all four combinations to remove any gradient asymmetries in the axial direction and to balance off-resonance MT effects. The images are acquired over a 32-mm field of view (FOV) and reconstructed using a 64×64 image matrix.

3.2.2. Magnetization Transfer (MT) Parameters

This involves the adiabatic transfer of magnetization between the free or mobile pool of protons, which in biological specimens is mostly water, and the bound or immobile pool of protons, mainly those associated with hydration layers of water surrounding large macromolecules and proteins. Of importance, changes in the MT parameters often correlate well with BBB opening, probably as the result of increased tissue water (i.e., vasogenic edema). The MT related parameters are M_0, M_{0sat}, T_1, T_{1sat}, K_{for}, and the magnetization transfer ratio (MTR). Calculation of the apparent forward magnetization transfer rate, K_{for}, requires estimates of T_1 in both the absence and presence of off-resonance irradiation of the "bound" proton fraction (T_1 and T_{1sat}, respectively).

1. Estimates of T_1 and T_{1sat} are obtained from a single slice using the Phase Incremented Progressive Saturation (PIPS) method (TR/TE = 40/7 ms, tip angle ~18°) (23).

2. The sequence is run in sets of two to produce estimates of M_{0sat}, T_{1sat}, M_0, and T_1. To estimate M_{0sat} and T_{1sat}, a CW RF saturation pulse of approximately 0.3 kHz amplitude, off-resonance by 6 kHz, is turned on for 4.5 s before the beginning of each image of the PIPS sequence, and for 30 ms (of the 40 ms TR) between each gradient-echo line in k-space to partially saturate the macromolecular proton pool.

3. The value of K_{for} is calculated from the equation:

$$K_{for} = \frac{1}{T_{1sat}}\left(1 - \frac{M_{0sat}}{M_0}\right),$$ (1)

where K_{for} represents the product of $1/T_{1sat}$ and the magnetization transfer ratio ($MTR = 1 - M_{0sat}/M_0$), which is often used alone to examine MT effects (If the transfer of magnetization decreases, then T_{1sat} may have increased or MTR decreased; K_{for} from Eq. 1, thus, provides a useful summary of MT effects). The total imaging time for both variations of the PIPS sequence is approximately 24 min.

3.2.3. Measurement of T_2

The biophysical meaning of T_2, the transverse relaxation time, is unclear, but changes in T_2 within areas of the brain often indicate underlying tissue injury or dysfunction with elevated T_2 values being highly correlated with increased tissue water content.

1. Sets of T_2-weighted image data are obtained using a multislice Carr-Purcell-Meiboom-Gill (CPMG) sequence (TR = 1,200 ms, TE = 30, 60, 90 and 120 ms, with interleaved slice acquisition, 13 slices, slice thickness = 1 mm). Total imaging time for the T_2WI series is approximately 13 min.

2. T_2 maps (quantitative images obtained by pixel-by-pixel processing of raw imaging data) are produced from a straight line least squares estimate of the slope from a plot of the natural logarithm of the normalized image intensity $\ln(S/S_0)$ vs. TE.

3.2.4. Measurement of the Apparent Diffusion Coefficient of Water (ADC_w)

The ADC_w in brain is a function of the size and tortuosity of the extracellular space, the rate of water exchange between intracellular and extracellular fluid (ECF), the volume of intracellular water, and the rate of bulk flow of extracellular and edema fluid, if present.

1. A pulsed gradient spin-echo imaging sequence with progressively incremented diffusion weighting (b-value) is used to measure ADC_w (TR/TE = 1,500/40 ms, b-value = 0, 400, 800 s/mm², 128 × 64 image matrix and interleaved slice acquisition). The total scan time for this DWI is approximately 10 min for the complete data set.

2. Maps of ADC_w are produced from a straight line least squares estimate of the slope from a plot of $\ln(S/S_0)$ vs. gradient b-value.

3.2.5. Look-Locker MRI Quantification of T_1 after Gd-DTPA Injection

Gadolinium-DPTA changes the longitudinal relaxation rate (symbolized by ΔR_1), of the nearby water protons. The MRCA concentration is proportional to ΔR_1 (R_1 is the inverse of the longitudinal relaxation time, T_1) when this parameter is quantitatively determined.

1. Rapid quantitative estimates of brain tissue T_1 relaxation times are necessary to accomplish an estimate of MRCA tissue concentration as it varies with time, and can be obtained using an imaging adaptation of the Look-Locker method (24).

2. The technique employs the T_1 by Multiple Read-Out Pulses (TOMROP) sequence (25) to produce efficient and unbiased pixel-by-pixel estimates of T_1 (26).

3. Baseline high-resolution T_1-weighted spin-echo images are collected prior to Gd-DTPA injection. Then, after obtaining one or two baseline L-L T_1 estimates, Gd-DTPA is injected as a bolus into the femoral vein in <5 s and L-L data sets are acquired sequentially at approximately 2.5 min intervals for up to 30 min (13).

4. Data are obtained for five interleaved 2 mm thick slices. At the conclusion of the L-L series, a final T_1-weighted multislice SE image set is obtained.

3.3. Estimating Transfer Constants by Patlak Plots

1. The L-L T_1 maps over a series of time points are acquired as described above to estimate arterial plasma and tissue concentrations of the MRCA.

2. The plasma and tissue MRCA concentrations are determined by measuring MRCA induced T_1 changes in the superior sagittal sinus (SSS) and brain-tissue, respectively according to the relationships: $C_{pa}(t_n) \propto \dfrac{\Delta R_{1a}(t_n)}{(1 - Hct)}$ and $C_{tis}(t_n) \propto \Delta R_{1tis}(t_n)$, where $\Delta R_{1tis}(t_n)$ and $\Delta R_{1a}(t_n)$ are differences in R_1 ($=1/T_1$) measured in the tissue and sagittal sinus, respectively, at time point $t = t_n$ and $t = 0$ (baseline or pre-CA time point) and Hct is the hematocrit (see Note 3).

3. Patlak plot technique is employed to estimate K_i and $V_p + V_o$ by graphing $\dfrac{(1 - Hct)\Delta R_{1tis}(t)}{\Delta R_{1a}(t)}$ vs. $\dfrac{\int_0^t \Delta R_{1a}(t)\, dt}{\Delta R_{1a}(t)}$ as described below. The solution of the equations for the Patlak model of blood-brain exchange is given by the expression (15):

$$\frac{C_t(t)}{C_{pa}(t)} = K_i \frac{\int_0^t C_{pa}(t')\, dt'}{C_{pa}(t)} + (V_p + V_o)$$
.

4. A graph of the ratio of the tissue CA concentration at the times of sampling (C_t) to the arterial plasma concentration at the respective times (C_{pa}) vs. a concentration weighted time parameter referred to as "plot time" (integral of C_{pa} over time up to the time point of measurement divided by C_{pa} at that time) can be drawn to estimate K_i and $V_p + V_o$ (Fig. 2). In the case of bolus injections, the "plot time" is greater than the actual scan time

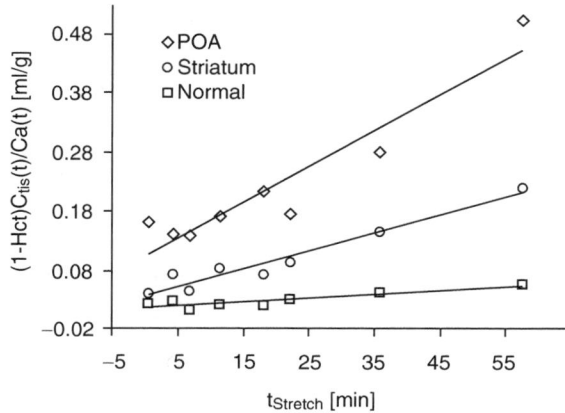

Fig. 2. Representative Patlak plots of the ratio of brain tissue (C_{tis}) concentrations of Gd-DTPA to plasma ($C_{pa} = C_a(t)/(1-Hct)$) vs. concentration ratio-stretched time ($t_{stretch}$) for three brain ROI from one rat. The Gd-DTPA K_i values are obtained from the slope of the lines. Among the regions with leaky capillaries, the influx appears to be greater in the preoptic area (PoA) than the striatum. The slopes in these two regions are significantly different from zero and indicate an appreciable, but small, Gd-DTPA influx. The slope of the line is flat (and statistically not different from zero) for the contralateral ROI and indicates a normal K_i. The units on the abscissa are plasma concentration ratio-stretched time, not real time (which was about 24 min in this case) (reproduced with permission from (16)).

and hence is called "stretched time". In MRI, linear least squares estimates of K_i and $V_p + V_o$ are performed pixel-by-pixel to construct the corresponding maps (see Note 4).

3.4. MRI Data Analysis and the Selection of Regions of Interest

1. The MRI data are processed offline using a system such as a SUN workstation.

2. A semi-automated segmentation algorithm is employed to minimize user bias in identifying tissue areas of MR abnormality indicative of processes such as BBB opening or hyperemia.

3. The MRI data are then reconstructed and baseline corrected using in-house software before applying the segmentation procedure. Most of the data from our studies were obtained from what is referred to as the central slice, 2.0 mm thick and extending from bregma +0.15 to –1.85 mm. Comparable image analyses techniques can be applied to adjacent slices if necessary and the data merged. One of the several available image segmentation protocols can be used to identify the regions of interest (ROI) (see Note 5).

3.5. MRCA Administration by Step-Down Infusion (SDI)

To compensate for the exponential decay of blood MRCA level after a bolus injection, an experimental step-down infusion (SDI) technique that maintains the blood tracer level constant for the duration of DCE-MRI can be used (27). The SDI protocol improves the signal-to-noise ratio, enhances the spatial resolution of the images, and yields a more accurate blood concentration time course (the arterial input function or AIF) of the MRCA (28). For the usual bolus injection method, 60 μL of the stock solution is diluted to 0.1 mL with normal saline and injected via the femoral vein in 4–5 s. The SDI, however, is done with a syringe pump and a prescribed infusion schedule (Table 1).

1. For the infusate, 240 μL of the stock solution is diluted in 4.0 mL of saline (the osmolality of the Gd-DTPA preparation is very high (>1,000 mOsm/kg), and this dilution brings it into the physiological range for infusion).

2. Approximately 3.5–3.6 mL of infusate is used per step-down experiment following the protocol given in Table 1. The rest of L-L T_1WI methods are identical to those explained above.

3.6. Estimating Acute Treatment Efficacy in a Brain Tumor Model

1. At the desired time-point after tumor implantation, the rat is anesthetized and a venous arterial and cannulations are performed as described in Subheading 3.1.3.

2. The rat is then positioned in the MRI holder and placed inside the magnet.

3. The scout images and T_2 images are obtained to locate and measure the tumor.

Table 1
The step-down infusion protocol developed and employed in the authors' laboratories (27)

Rate(mL/min)	Pump speed (%)	Time	Duration (min)	Volume(mL)
0.68	100	0–30 s	0.5	0.34
0.53	77.7	30–60 s	0.5	0.26
0.39	57.4	1–2 min	1.0	0.39
0.28	41.6	2–3 min	1.0	0.28
0.22	33.3	3–4 min	1.0	0.22
0.19	28.7	4–5 min	1.0	0.19
0.17	25.0	5–7 min	2.0	0.34
0.14	20.4	7–10 min	3.0	0.42
0.12	17.6	10–15 min	5.0	0.60
0.10	14.8	15–20 min	5.0	0.50
Total volume infused = 3.54 mL				

4. The chosen contrast media (normally, a large MRCA) is then injected and SE- and L-L T_1WI-images are obtained as described in Subheading 3.2.5, steps 1–4.

5. The drug being tested is injected at the appropriate dose after ensuring that the MRCA signal returns to the base-line.

6. After waiting for the drug action, one more MRCA injection is given and another set of SE- and L-L T_1WI- images are obtained.

7. The contrast enhancement before and after drug injections are compared quantitatively to assess drug effects (Fig. 3; see Note 6).

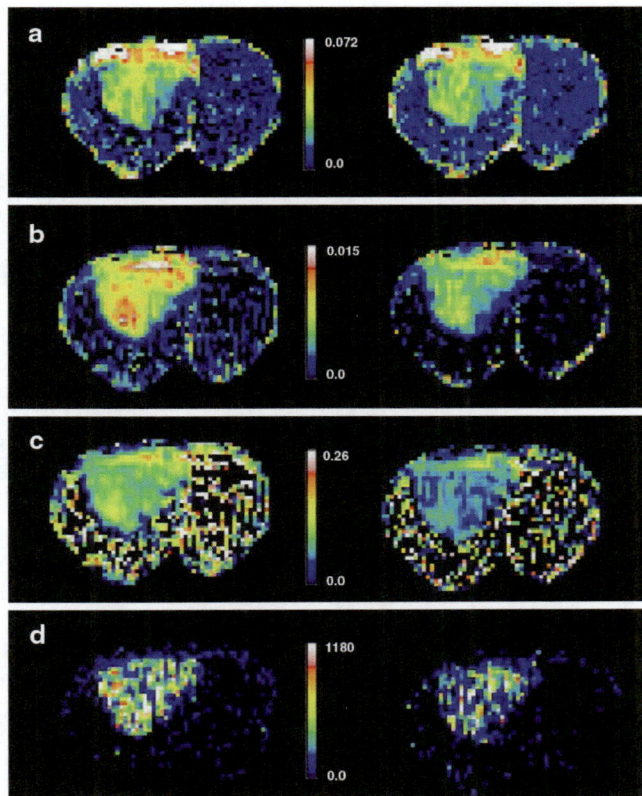

Fig. 3. Vascular parameters pre- (*left*, test) and post-dexamethasone (*right*, retest) administration showing (**a**) Vascular volume v_p, (**b**) Transfer constant K_1 (min^{-1}), (**c**) Efflux constant k_b (min^{-1}), and (**d**) F-test for model 3 vs. model 2, with a high value resulting in rejection of model 2. (**c**) Only those regions with high F-test values have valid results for the estimate of k_b. A widespread decrease in K_1 and k_b are easily visualized. Less visible is a moderate decrease in v_p. Bright spots in the maps of v_p correspond to vascular pools. Note the decrease in the F-test from pre- to postdexamethasone studies (reproduced with permission from (20)).

3.7. Sizing BBB Opening with Two Different MRCAs

1. The MCA occulusion model with acute reperfusion is employed for this purpose. Therefore, perform the MCA occlusion and reperfusion following the protocols given in Subheading 3.1.1.

2. Measure, following the MRI protocols, the ischemic status of the tissue and the extent of reperfusion.

3. Quantify the spatial extent of the BBB opening by quantitative Patlak-K_i maps in the ischemic hemisphere using Gd-DTPA-enhanced L-L T_1WI. Wait for the Gd-DTPA signal to return to baseline.

4. Then quantify BBB opening and generate an identical data set using a larger MRCA such as Gd-BSA following the same protocols.

5. Compare the spatial enhancement patterns and K_i maps for the two contrast agents to determine the differences, and,

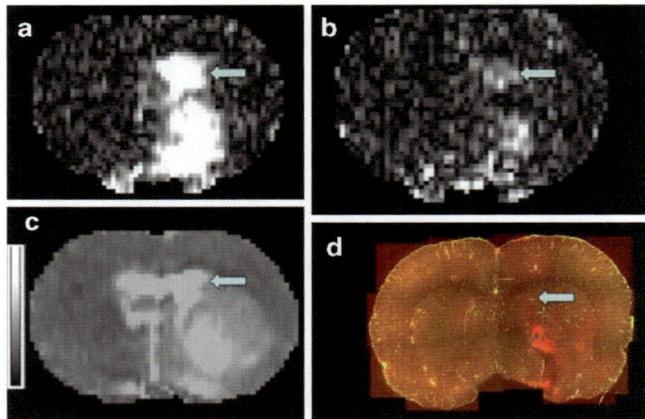

Fig. 4. A representative collage of Gd-DTPA and Gd-BSA-Evans Blue (EB) enhancements during MRI and corresponding T_{1sat} maps and fluorescence images from one experiment are shown. (a) Gd-DTPA enhancement in a rat 24 h after MCA occlusion of 3 h duration is shown. A large area of brightness is seen in the preoptic area and striatum. The small, bilateral, bright regions below this enhancement are parts of medium eminence and hence, naturally leaky. Contrast enhancement in such regions, ventricles (*large arrows*), and pial vasculature surrounding the brain are routinely observed. (b) Subsequent Gd-BSA-EB enhancement in this rat is in a much smaller area. Other normally leaky regions are also visible. (c) The T_{1sat} map for this slice is shown in gray scale. On the calibration bar on the left, hyperintense/bright areas indicate increased values. The ventricles and circumventricular organs appear bright on this map due to the inherent sensitivity of T_{1sat} to water/proton shifts. The low-magnification, reconstructed fluorescence image shown (d) has extravascular red fluorescence in the same ROI as the Gd-BSA-EB–enhancing area in (b). The greenish-yellow hue in the rest of the vasculature is due to the combination of red (EB) and green (fluorescein isothiocyanate–dextran) fluorescence. This image was constructed by collecting the coronal brain section as a series of low magnification (×2.5) images and tiling them together (reproduced with permission from (18)).

thus, relative sizes of BBB opening in the ischemic tissue (Fig. 4; see Note 7).

4. Notes

1. A major caveat to the methods presented in this chapter is that conditions described were optimal to the magnet and console used and the particular set of MR sequences that were developed for this system. It should be noted, however, that they are easily adaptable to other MRI systems as well.

2. The radiofrequency (RF) coil and animal holder assembly for MRI are designed so that the RF coil section remains fixed within the bore of the magnet and the animal holder assembly can be removed separately if necessary. This setup allows the animal to be removed from the magnet for reperfusion and then returned to exactly the same position within the magnet.

3. The term (1-Hct) is used to adjust the Gd-DTPA concentration measured in whole blood to the concentration in plasma, since Gd-DTPA distributes only in plasma. In each experiment, the arterial hematocrit (Hct) should be measured prior to injecting Gd-DTPA, and the resulting value used for the graphical estimation. It has been assumed that the constant of proportionality (i.e., the relaxivity) between concentration and ΔR_1 does not change across tissues, although there are exceptions to this assumption (29).

4. Experimental data demonstrating the applicability of the model and the Patlak plot were produced by Blasberg et al. (30) with radiotracers such as ^{14}C-AIB and ^{51}Cr-DTPA. It should be noted that the abscissa of the Patlak plot has the units of time but is an arterial concentration weighted time.

5. For our earlier work, we identified and segmented the ischemic brain regions with BBB injury based on direct subtraction of pre- and post-MRCA spin-echo T_1-weighted image (T_1WI) data or application of the Iterative Self-Organizing Data Analysis Technique Algorithm (ISODATA) to the temporal Gd-DTPA L-L T_1WI data set (Fig. 1) (31). Another method we have recently employed for segmenting ischemic brain regions with BBB opening is based on Patlak plot-derived F-test maps (Fig. 3) (19).

6. The well-characterized, solid, 9L brain tumor model lends itself nicely to the identification of drug effects on the leaky BBB and the speed of onset of such effects. Of the several parameters examined, K_1 turned out to be the most sensitive to the effects of dexamethasone (Fig. 3) (20). Acute effects of dexamethasone on brain tumor permeability shown by earlier

studies (32) provided us with a known response and an acceptable drug-tumor system for examining the sensitivity of our test-retest protocol.

(a) For a drug efficacy study, the MRCA and the protocol should have several physiological and pharmacological features and constraints. First, it must be cleared from the blood fairly rapidly after bolus administration so that the second test is not strongly influenced by CA remaining from the first test. Secondly, it should have a characteristic transfer rate to the tissue that is much slower (ideally a factor of 10 or more) than the tissue sampling rate. This latter constraint is necessary because the rate of transfer has to be inferred from the concentration-time trace, and this is impossible if the CA equilibrates in the tissue on a time scale that is rapid compared to the sampling rate.

(b) A small MRCA such as Gd-DTPA (~560 Da) diffuses quickly through the blood-tumor barrier of the 9L model, which is fairly leaky at this stage. This makes it a less than ideal contrast agent for this tumor model with respect to the second constraint above. In addition, the linear phase of uptake can be very short or even lacking on a simple Patlak plot of Gd-DTPA data. Therefore, larger MRCAs that permeate the brain-tumor barrier more slowly than Gd-DTPA, do not accumulate greatly, but move appreciably in the field of observation are more useful.

(c) In a typical experiment, the influx constant for an MRCA is determined first and then the drug is injected. Depending on the expected time-course of drug action, the MRCA is injected again at an appropriate time and L-L T_1 measurements are made thereafter as done in the test period. The predrug results are then compared to postdrug results.

(d) Though simple in design, the constraints listed above need to be considered in choosing the contrast agent, but it is quite easy to make successive studies on the same animal when appropriate care has been taken in selecting the proper combination of MRCA and drug response time. The time of retesting after drug administration, can be varied among groups of animals and the period of efficacy is determined over one to many hours after treatment.

7. BBB opening is generally not an all-or-none phenomenon but is graded with a small tracer leaking at a particular time after ictus but not a larger one. A protocol similar to the test-retest one just presented can be used to measure blood-to-brain influx of two MRCAs of different sizes within a short duration of each other and obtain nearly congruent results to evaluate the "size" of the opening.

(a) The same constraints and consideration indicated above in the test-retest protocol apply to a BBB sizing study. Perhaps the most limiting one in this instance is the plasma half-life of the MRCA. This means that the first MRCA injected is the one with the shorter half-life and the longer half-life one is given sometime later (18).

Acknowledgments

This work was supported by National Institutes of Health grants 1RO1NS38540 and 1RO1HL70023; American Heart Association grants 0270176N and 0635403N; and research funds from the Henry Ford Health System. The authors thank Polly A. Whitton, Jun Xu, Kelly A. Keenan, Richard L. Croxen and Swayamprava Panda for their technical contributions.

References

1. Fenstermacher JD, Nagaraja T, Davies KR (2001) Overview of the structure and function of the blood-brain barrier in vivo. In Kobiler D, Lustig S, Shapira S (ed) Blood-Brain Barrier, Kluwer Academic/Plenum Publishers, New York, pp 1–7

2. Aronowski J, Strong R, Grotta JC (1997) Reperfusion injury: demonstration of brain damage produced by reperfusion after transient focal ischemia in rats. J Cereb Blood Flow Metab 17:1048–1056

3. Knight RA, Barker PB, Fagan SC, LiY, Jacobs MA, Welch KM (1998) Prediction of impending hemorrhagic transformation in ischemic stroke using magnetic resonance imaging in rats. Stroke 29:144–151

4. The NINDS. (The NINDS t-PA Stroke Study Group) (1997) Intracerebral hemorrhage after intravenous t-PA therapy for ischemic stroke. Stroke 28:2109–2118

5. Albers GW, Thijs VN, Wechsler L, Kemp S, Schlaug G, Skalabrin E, Bammer R, Kakuda W, Lansberg MG, Shuaib A, Coplin W, Hamilton S, Moseley M, Marks MP (2006) Magnetic resonance imaging profiles predict clinical response to early reperfusion: the Diffusion and Perfusion Imaging Evaluation for Understanding Stroke Evolution (DEFUSE) study. Ann Neurol 60:508–517

6. Köhrmann M, Jüttler E, Fiebach JB, Huttner HB, Siebert S, Schwark C, Ringleb PA, Schellinger PD, Hacke W (2005) MRI versus CT-based thrombolysis treatment within and beyond the 3 h time window after stroke onset: a cohort study. Lancet Neurol 5:661–667

7. Schellinger PD, Thomalla G, Fiehler J, Köhrmann M, Molina CA, Neumann-Haefelin T, Ribo M, Singer OC, Zaro-Weber O, Sobesky J (2007) MRI-based and CT-based thrombolytic therapy in acute stroke within and beyond established time windows: an analysis of 1210 patients. Stroke 38:2640–2645

8. Latour LL, Kang DW, Ezzeddine MA, Chalela JA, Warach S (2004) Early blood-brain barrier disruption in human focal brain ischemia. Ann Neurol 56:468–477

9. Warach S, Wardlaw J (2006) Advances in imaging 2005. Stroke 37:297–298

10. Fagan SC, Nagaraja TN, Fenstermacher JD, Zheng J, Johnson M, Knight RA (2003) Hemorrhagic transformation is related to the duration of occlusion and treatment with tissue plasminogen activator in a non-embolic stroke model. Neurol Res 25: 377–382

11. Hjort N, Wu O, Ashkanian M, Solling C, Mouridsen K, Christensen S, Gyldensted C, Andersen G, Ostergaard L (2008) MRI detection of early blood-brain barrier disruption: parenchymal enhancement predicts focal hemorrhagic transformation after thrombolysis. Stroke 39:1025–1028

12. Tofts PS, Kermode AG (1991) Measurement of the blood-brain barrier permeability and leakage space using dynamic MR imaging. 1. Fundamental concepts. Magn Reson Med 17:367–367

13. Ewing JR, Knight RA, Nagaraja TN, Yee JS, Nagesh V, Whitton PA, Li L, Fenstermacher JD (2003) Patlak plots of Gd-DTPA MRI

data yield blood-brain transfer constants concordant with those of ^{14}C-sucrose in areas of blood-brain opening. Magn Reson Med 50:283–292

14. Patlak CS, Blasberg RG (1985) Graphical evaluation of blood-to-brain transfer constants from multiple-time uptake data. Generalizations. J Cereb Blood Flow Metab 5:584–590

15. Patlak CS, Blasberg RG, Fenstermacher JD (1983) Graphical evaluation of blood-to-brain transfer constants from multiple-time uptake data. J Cereb Blood Flow Metab 3:1–7

16. Knight RA, Nagaraja TN, Ewing JR, Nagesh V, Whitton PA, Bershad E, Fagan SC, Fenstermacher JD (2005) Quantitation and localization of blood-to-brain influx by magnetic resonance imaging and quantitative autoradiography in a model of transient focal ischemia. Magn Reson Med 54:813–821

17. Knight RA, Nagesh V, Nagaraja TN, Ewing JR, Whitton PA, Bershad E, Fagan SC, Fenstermacher JD (2005) Acute blood-brain barrier opening in experimentally induced focal cerebral ischemia is preferentially identified by quantitative magnetization transfer imaging. Magn Reson Med 54:822–832

18. Nagaraja TN, Karki K, Ewing JR, Croxen RL, Knight RA (2008) Identification of variations in blood-brain barrier opening after cerebral ischemia by dual contrast-enhanced magnetic resonance imaging and T_{1sat} measurements. Stroke 39:427–432

19. Ewing JR, Brown SL, Lu M, Panda S, Ding G, Knight RA, Cao Y, Jiang Q, Nagaraja TN, Churchman JL, Fenstermacher JD (2006) Model selection in magnetic resonance imaging measurements of vascular permeability: Gadomer in a 9L model of rat cerebral tumor. J Cereb Blood Flow Metab 26:310–320

20. Ewing JR, Brown SL, Nagaraja TN, Bagher-Ebadian H, Paudyal R, Panda S, Knight RA, Ding G, Jiang Q, Lu M, Fenstermacher JD (2008) MRI measurement of change in vascular parameters in the 9L rat cerebral tumor after dexamethasone administration. J Magn Reson Imaging 27:1430–1438

21. Kim JH, Khil MS, Kolozsvary A, Gutierrez JA, Brown SL (1999) Fractionated radiosurgery for 9L gliosarcoma in the rat brain. Int J Rad Oncol Biol Phys 45:1035–1040

22. Ewing JR, Wei L, Knight RA, Pawa S, Nagaraja TN, Brusca T, Divine GW, Fenstermacher JD (2003) Direct comparison of local cerebral blood flow rates measured by MRI arterial spin-tagging and quantitative autoradiography in a rat model of experimental cerebral ischemia. J Cereb Blood Flow Metab 23:198–209

23. Ewing JR, Jiang Q, Boska M, Zhang L, Zhang ZG, Brown SL, Li GH, Divine GW, Chopp M (1999) T_1 and magnetization transfer at 7 Tesla in acute ischemic infarct in the rat. Magn Reson Med 41:696–705

24. Look DC, Locker DR (1970) Time saving in measurement of NMR and EPR relaxation times. Rev Sci Instruments 41:250–251

25. Brix G, Schad LR, Deimling M, Lorenz WJ (1990) Fast and precise T_1 imaging using a TOMROP sequence. Magn Reson Imaging 8:351–356

26. Crawley AP, Henkelman MR (1988) A comparison of one-shot and recovery methods in T1 imaging. Magn Reson Med 7:23–34

27. Nagaraja TN, Nagesh V, Ewing JR, Whitton PA, Fenstermacher JD, Knight RA (2007) Step-down infusions of Gd-DTPA yield greater contrast-enhanced magnetic resonance images of BBB damage in acute stroke than bolus injections. Magn Reson Imaging 25:311–318

28. Knight RA, Karki K, Ewing JR, Divine GW, Fenstermacher JD, Patlak CS, Nagaraja TN (2009) Estimating blood and brain concentrations and blood-to-brain influx by magnetic resonance imaging with step-down infusion of Gd-DTPA in focal transient cerebral ischemia and confirmation by quantitative autoradiography with Gd-[^{14}C]DTPA. J Cereb Blood Flow Metab 29:1048–1058

29. Yankeelov TE, Rooney WD, Li X, Springer CS Jr. (2003) Variation of the relaxographic "shutter-speed" for transcytolemmal water exchange affects the CR bolus-tracking curve shape. Magn Reson Med 50:1151–1169

30. Blasberg RG, Fenstermacher JD, Patlak CS (1983) Transport of α-aminoisobutyric acid across brain capillary and cellular membranes. J Cereb Blood Flow Metab 3:8–32

31. Ball G, Hall D (1965) ISODATA, A novel method of data analysis and pattern classification. Stanford Research Institute, Menlo Park

32. Nakagawa H, Groothuis DR, Owens ES, Fenstermacher JD, Patlak CS, Blasberg RG (1987) Dexamethasone effects on [^{125}I]albumin distribution in experimental RG-2 gliomas and adjacent brain. J Cereb Blood Flow Metab 7:687–701

Chapter 9

Detection of Brain Pathology by Magnetic Resonance Imaging of Iron Oxide Micro-particles

Daniel C. Anthony, Nicola R. Sibson, Martina A. McAteer, Ben Davis, and Robin P. Choudhury

Abstract

Contrast agents are widely used with magnetic resonance imaging (MRI) to increase the contrast between regions of interest and the background signal, thus providing better quality information. Such agents can work in one of two ways, either to specifically enhance the signal that is produced or to localize in a specific cell type of tissue. Commonly used image contrast agents are typically based on gadolinium complexes or super-paramagnetic iron oxide, the latter of which is used for imaging lymph nodes. When blood-brain barrier (BBB) breakdown is a feature of central nervous system (CNS) pathology, intravenously administered contrast agent enters into the CNS and alters contrast on MR scans. However, BBB breakdown reflects downstream or end-stage pathology. The initial recruitment of leukocytes to sites of disease such as multiple sclerosis (MS), ischemic lesions, or tumours takes place across an intact, but activated, brain endothelium. Molecular imaging affords the ability to obtain a "non-invasive biopsy" to reveal the presence of brain pathology in the absence of significant structural changes. We have developed smart contrast agents that target and reversibly adhere to sites of disease and have been used to reveal activated brain endothelium when images obtained by conventional MRI look normal. Indeed, our selectively targeted micro-particles of iron oxide have revealed the early presence of cerebral malaria pathology and ongoing MS-like plaques in clinically relevant models of disease.

Key words: Contrast agent, MPIO, MRI, Inflammation, Brain, Molecular imaging, VCAM-1, Selectin, Endothelium

1. Introduction

The aim of molecular imaging is to visualize pathological processes at the cellular level, often long before disease symptoms become clinically apparent. This information will help clinicians to identify specific pathological processes, enabling the most appropriate therapeutic approach to be applied and to monitor

Sukriti Nag (ed.), *The Blood-Brain and Other Neural Barriers: Reviews and Protocols*, Methods in Molecular Biology, vol. 686, DOI 10.1007/978-1-60761-938-3_9, © Springer Science+Business Media, LLC 2011

the outcome of therapy. In particular, molecular imaging techniques that can accurately identify markers of early inflammation in the brain are needed to accelerate accurate diagnosis and guide-specific therapy.

We have detected adhesion molecules for immune cells on blood vessels in the brain such as vascular cell adhesion molecule-1 (VCAM-1) or CD62 in a mouse model of acute inflammatory disease using magnetic resonance imaging (MRI), at a time when pathology is undetectable by conventional MRI (1, 2). In our studies we have conjugated targeting ligands such as antibodies or carbohydrates to micro-particles of iron oxide (MPIO), which yields exquisitely sensitive molecular probes with high MR-detectable contrast that delineate the architecture of activated cerebral blood vessels. This method detects the presence of pathology in vivo in animal models of multiple sclerosis (MS) (3), brain metastases, stroke (1), and cerebral malaria (4) at a time when no structural abnormalities can be detected with conventional MRI. This molecular imaging approach manifests highly potent contrast effects, with the considerable advantage of the exceptional spatial resolution afforded by MRI compared with other tomographic techniques. The reason why this technology offers significant advantages over current diagnostic practice is that conventional MRI is limited in two important respects: (a) only relatively advanced or gross pathology is detected at which point therapeutic intervention may be of limited benefit and (b) it cannot assess disease activity at a molecular level.

The key to molecular imaging with MRI is the use of conjugated 1 μm micro-particles containing iron oxide (MPIO), which provide potent and quantifiable contrast effects that delineate the architecture of activated brain blood vessels. MPIO are superparamagnetic particles consisting of a magnetite (Fe_2O_3) and/or maghemite (Fe_3O_4) core surrounded by a polymer coat. Due to the high iron content, MPIO create potent negative contrast effects on T2*-weighted images that extend to a distance roughly 50 times the physical diameter of the MPIO. The potency of MPIO contrast has recently enabled in vivo cell tracking (5) and in vivo detection of single MPIO-labelled cells by MRI (6). The rapid clearance of the micro-particles from the blood results in minimal background signal for endovascular imaging. Indeed, it is possible to repeatedly image the brain as the bound contrast agent is removed from brain blood vessels after a few hours. Conventional non-targeted, gadolinium-based contrast agents appear bright in most MRI images. However, due to the small quantities that can be delivered to the surface of a blood vessel in an area of disease, the contrast effects are relatively modest. Indeed, our study was the first to show that a targeted gadolinium-based contrast agent (Gd-DTPA-B(sLeX)) can be used to identify activated brain blood vessels; however, the increase in detectable signal was mild (7). By comparison, magnetic dextran-coated iron oxide nanoparticles give greater signal and are licenced for use in the clinic. However, particles in the nano range still require that many have to be delivered to a given target to be detectable. Another potential drawback is that signal from nanoparticles can be difficult to distinguish from surrounding tissue. Furthermore, since these nanoparticles can be taken up by cells lining blood vessels and by immune cells, there is potential compromise to the specificity. MPIO have a high iron content, orders of magnitude greater than that contained in ultrasmall particles of iron oxide (USPIO) commonly used for MRI contrast, and owing to their size, MPIO are less susceptible than USPIO to non-specific uptake by endothelial cells and therefore, they retain specificity for endovascular molecular targets (8). Finally, for preclinical work, MPIO are commercially available with a range of reactive surface groups, providing the opportunity for covalent conjugation of protein, antibodies, or small peptides. The micron-sized particles can also be dextran coated and can be conjugated to targeting molecules. Such particles provide a versatile and safe platform for the incorporation of multiple copies of targeting molecules. Indeed, we have recently developed biodegradable MPIO, which will facilitate their use in the clinic.

This chapter describes the methodology for the generation of antibody and sugar-conjugated targeted-MPIO probes, for detection of acute brain inflammation in vivo, at a time when pathology is undetectable by conventional MRI. Protocols are included for the conjugation of MPIO and the application of targeted-MPIO for the in vivo MRI-based detection of acute brain inflammation in rodents. MPIO (1 μm diameter), with reactive tosyl or amine groups, are used for covalent conjugation of mouse monoclonal VCAM-1 antibodies or sugars (glycans), respectively. We have calculated that the MPIO have a capacity to bind covalently 1.8×10^9 IgG molecules per MPIO. The attachment of glycans to NH_2-MPIO required an amine-reactive linker group. However, the manipulation of a linker precursor to generate an active linker after synthesis of a complex glycan often results in low efficiencies and reduced modification yields. We discovered that the S-cyanomethyl (SCM) ($S-CH_2-CN$) functional group could be used at the anomeric centre of sugars, not only as a protecting group, to aid control in the synthesis, but also a masked linker. Its dual chemical character allowed the SCM group to be introduced early in a given synthetic scheme and then selectively "unmasked" or activated by conversion to the corresponding reactive chemical group 2-imido-2-methoxy-ethyl (IME) ($S-CH_2-C(NH)OCH_3$). This was accomplished cleanly, at will, before any amine-modification reaction simply by pre-treatment with sodium methoxide. Importantly, this method was made possible by the discovery that substoichiometric levels of methoxide would cleanly allow manipulation of glycan protecting groups (acetyl groups) while leaving the masked linker cyanomethyl group intact.

Activated mouse endothelial cells, stimulated with tumour necrosis factor-α (TNF-α), are used to test the capacity of the MPIO constructs for specific and quantitative binding in vitro. For in vivo studies, acute brain inflammation is induced by stereotactic injection of interleukin-1 (IL-1) into the left corpus striatum of mice. To model MS, we use a focal EAE model, and to model stroke, we employ a focal stereotaxic injection of endothelin-1. MPIO are administered intravenously and MRI of the brain is performed at 7 T using a T2*-weighted 3D gradient-echo sequence, with a final isotropic resolution of 88 μm³. The targeted-MPIO generate highly specific, potent hypointense contrast effects that delineate the architecture of activated cerebral blood vessels, with minimal background contrast (Fig. 1).

The commercial availability of functionalised MPIO and antibodies supports a straightforward protocol for producing targeted-MPIO probes for early detection of brain inflammation in vivo. An additional advantage of this protocol is that MPIO are readily adaptable for diagnostic imaging of other endothelial-specific targets, simply by modifying the protein ligand. We have

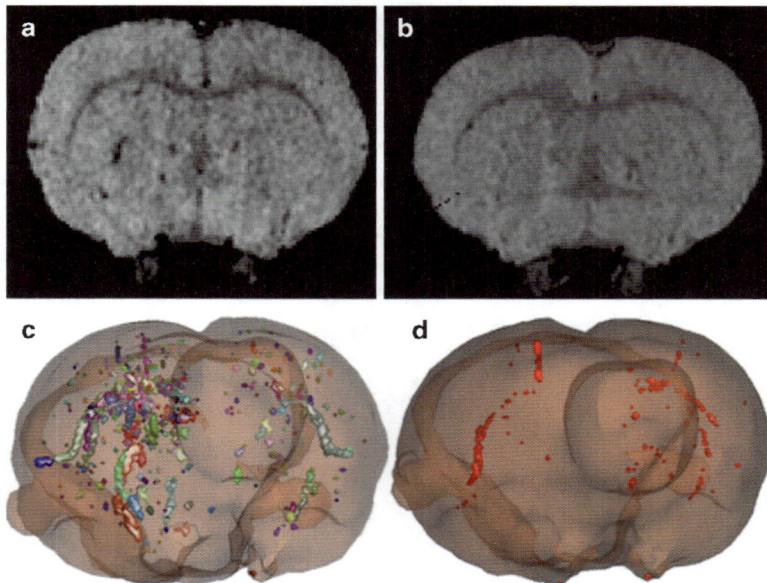

Fig. 1. sLe^x-MPIO binding pattern in the brain of animals with EAE. T$_2$*-weighted MR image of rat brain from a 7 T magnet 1–2 h post injection of sLe^x-MPIO (**a**) or control MPIO (**b**). Low signal contrast effects were segmented in contiguous MR slices using an automated signal intensity histogram tool using ImagePro Plus. Masks of low signal areas were merged and 3D reconstructed to create a 3D volumetric map of low signal voxels sLe^x-MPIO (**c**) or control MPIO (**d**). The brain has been surface rendered and, in (**c**), each voxel volume has been automatically coloured to aid identification of each volume measurement.

recently used similar MPIO-based constructs to image adhesion molecules in atherosclerosis (9) and activated platelets in mouse models of cerebral malaria (4) and atherothrombosis (10, 11).

2. Materials

2.1. Conjugation of Antibody to MyOne™ Tosylactivated MPIO

1. MyOne™ Tosylactivated super-paramagnetic polystyrene Dynabeads® (1.08 μm diameter) (Invitrogen Inc, Paisley, UK). Store vial at 4°C (see Note 1). Other sources of MPIO are Bangs Laboratories (Fishers, IN, USA) and Miltenyi Biotec Ltd (Surrey, UK).

2. Dynal MPC®-S magnetic particle concentrator (magnet) (Invitrogen Inc).

3. Pre-washing and Coating buffer: Sodium borate buffer 0.1 M, pH 9.5. Store at 4°C (see Note 2).

4. Ammonium sulphate stock solution 3 M.

5. Antibody Ligand (see Notes 3 and 4): Purified monoclonal rat anti-mouse CD106/VCAM-1 antibody (clone M/K2) (Cambridge Bioscience, Cambridge, UK). Store at 4°C (see Note 5). Purified isotype negative control IgG-1 antibody (clone Lo-DNP-1) (AbD Serotec, Oxford, UK). Store at 4°C.

6. Blocking buffer: PBS (pH 7.4) with 0.5% bovine serum albumin (BSA) and 0.05% Tween 20. Store at 4°C.

7. Washing and Storage buffer: PBS (pH 7.4), with 0.1% BSA, and 0.05% Tween 20. Store at 4°C.

2.2. In Vitro MPIO Binding to TNF-α Stimulated sEND-1 Cells

1. Dulbecco's modified Eagle's medium (DMEM) (Sigma-Aldrich, Dorset, UK) supplemented with 10% foetal bovine serum (Sigma-Aldrich), 2 mM l-glutamine (Sigma-Aldrich), and 100 U penicillin and 0.1 mg/mL streptomycin (Sigma-Aldrich). Store at 4°C.

2. Mouse endothelial cell line, sEND-1 (Obtained from Dr Choudhury, Department of Cardiovascular Medicine, University of Oxford).

3. Trypsin/EDTA solution (Sigma-Aldrich). Store at 4°C.

4. Round microscope cover slips (19 mm) (VWR, Lutterworth, UK).

5. Murine recombinant TNF-α (R&D systems, Abingdon, UK). Reconstitute with sterile PBS and store in aliquots at –20°C. Reconstituted TNF-α is stable at –20°C for 3 months.

6. Recombinant mouse VCAM-1 Fc chimera (FcVCAM-1) and recombinant mouse ICAM-1 Fc chimera (FcICAM-1)

(R&D systems). Reconstitute with sterile water (50 μg/ mL) and store in aliquots of 100 μL at –20°C. Once reconstituted, FcVCAM and FcICAM are stable at –20°C for 4 weeks.

7. Paraformaldehyde (Sigma). Prepare a 1% (w/v) solution fresh for each experiment. The solution should be heated carefully using a stirring hot-plate in a fume hood. Cool to room temperature for use.

8. Purified monoclonal rat anti-mouse CD106/VCAM-1 antibody (clone M/K2) (Cambridge Bioscience). Store at 4°C.

9. Alexa Fluor 488 conjugated rabbit secondary antibody to rat IgG, 1:100 dilution in PBS (Vector Laboratories, Peterborough, UK). Store at 4°C.

10. Vectashield mounting media containing DAPI nuclear stain (Vector Laboratories). Store at 4°C.

11. Nikon E800 (Nikon, Kingston upon Thames, UK).

12. Nikon inverted TE2000 microscope.

2.3. In Vivo Mouse Protocol

1. Mice (NMRI, ~35 g, Charles River, Margate, UK).

2. Isofluorane anaesthesia (2.0–2.5% in 70% N_2O: 30% O_2).

3. Wild M650 operating microscope (Leica Microsystems, Milton Keynes, UK).

4. Stereotaxic frame (David Kopf, Tujunga, USA).

5. Glass 1 μL-graduated haematocrit tubes pulled with Narishige forge (Narishige International, London, UK) to a tip diameter <50 μm.

6. Quadrature bird cage coil (In-house design).

7. Mouse recombinant IL-1β (1 mg/mL) (R&D systems). Reconstitute with sterile PBS and store in aliquots at –20°C. Once reconstituted, IL-1β is stable at –20°C for 3 months.

8. Low endotoxin saline containing BSA, 0.1% (Sigma-Aldrich).

9. Fluorescence Microscope (Nikon).

10. Zeiss LSM150Laser scanning confocal microscope.

11. 7-Tesla horizontal bore magnet with a Varian Inova spectrometer (Varian, Palo Alto, USA).

2.4. MR Image Analysis

1. ImagePro Plus Image analysis software (Media Cybernetics, Marlow, UK).

2. 3D Constructor plug-in for ImagePro Plus (Media Cybernetics).

2.5. Targeted Experimental Allergic Encephalomyelitis (EAE) Lesions for Evaluating Unilateral Binding of the Contrast Agents

1. 3-week old Lewis rats, 60–100 g (Charles River).

2. Myelin oligodendrocyte glycoprotein (MOG) (35–55) peptide (Activotec, Comberton UK): 25 μg diluted in 50 μL saline and emulsified with 50 μL incomplete Freund's adjuvant (IFA; Sigma-Aldrich).

2.6. ET-1-Induced Focal Ischemic Stroke Lesions

1. Adult male Wister Rats 200–250 g (Harlan-Olac, Bicester, UK).

2. Endothelin-1 (ET-1) (Sigma-Aldrich).

3. Gadolinium-DTPA-BMA (Gd[DTPA/BMA]), Omniscan, (GE Healthcare, Amersham, UK).

3. Methods

MyOne™ Tosylactivated MPIO (1 μm diameter) are used for direct covalent conjugation of mouse monoclonal antibodies. Tosylactivated MPIO do not require surface activation, unlike, for example, iron oxide particles with reactive carboxylic acid surface groups, which require activation prior to conjugation, using either carbodiimide, most commonly 1-ethyl-3-(3-dimethylaminopropyl) (EDC), or a combination of EDC and N-hydroxyl succinimide ester (NHS). Furthermore, the hydrophobic properties of tosylactivated MPIO facilitate optimal antibody orientation since Fc regions of the antibody, which are generally more hydrophobic than the Fab portion, will adsorb to the hydrophobic surface of MPIO followed by rapid covalent bond formation. This exposes the Fab-regions of the antibody, thus maximizing the binding potential of antibody-conjugated MPIO to the target protein. The antibody-conjugated MPIO are stable at 4°C for several months without loss of antigen binding. Amine-activated MPIO (1 μm diameter) were used to attach selectin-binding sugars, which are also stable for several months. We have subjected the sugar-conjugated particles to a number of freeze thaw cycles and retained in vivo binding capability.

It is important to test and validate the capacity of adhesion-molecule-binding MPIO constructs for specific and quantitative binding in vitro prior to in vivo studies. This can be accomplished by incubating the targeted-MPIO or negative control MPIO with activated mouse endothelial cells, stimulated with graded doses of TNF-α. After extensive washing to remove unbound MPIO, specific MPIO binding to cells can be visualised and quantified using differential interference confocal microscopy. To further validate binding specificity, MPIO can be pre-blocked with soluble chimeric protein such as E-selectin-Fc, ICAM-1-Fc, or VCAM-Fc.

Fc-blocked MPIO can be incubated with activated endothelial cells as above. Cells can be assessed by confocal microscopy for co-localisation of MPIO binding and adhesion molecule immunofluorescence on the cell surface.

All MPIOs are injected intravenously into a tail vein and allowed to circulate for 1.5–2 h prior to MRI. This allows time for specific MPIO binding in the brain and clearance of unbound MPIO from the blood. Control groups of mice may undergo identical treatments with substitution of IgG-MPIO.

3.1. Conjugation of Antibody to MyOne™ Tosylactivated MPIO

1. MyOne™ Tosylactivated MPIO (50 µL; 5 mg of 5×10^9 MPIO) are transferred into a 1.5 mL microcentrifuge tube (see Note 6). The tube is placed in a Dynal MPC-S magnet (see Note 7) until MPIO have formed a pellet at the side of the tube and the liquid is clear. The supernatant is discarded.

2. The tube is removed from the magnet and MPIO resuspended in 1 mL of pre-washing and coating buffer. The tube is placed in the magnet to pellet MPIO. The supernatant is removed and this step is repeated once more.

3. The MPIO pellet (5 mg) is thoroughly resuspended in antibody solution (200 µg) (see Note 8). For the VCAM-1 antibody (stock concentration: 0.5 mg/mL), 400 µL is used, and for IgG-1 (stock concentration: 1 mg/mL), 200 µL is used (see Note 9).

4. Ammonium sulphate (3 M) is immediately added to give a concentration of 1 M in the final coating solution. For VCAM-MPIO, 200 µL ammonium sulphate is added (3 M); for IgG-1 MPIO, 200 µL ammonium sulphate (3 M) and 200 µL pre-washing and coating buffer are added to give the same final coating solution volume (600 µL).

5. The tube is placed in a rotating wheel and incubated, with constant head-over-head rotation, at 37°C for 20 h.

6. MPIO are pelleted and the supernatant discarded to remove unbound antibody.

7. Blocking buffer is added at the same volume used for coating the MPIO, i.e., in this example 600 µL.

8. The tube is placed in a rotating wheel and incubated, with constant head-over-head rotation, at 37°C overnight to block any remaining unbound active tosyl sites.

9. MPIO are pelleted using the magnet and the supernatant discarded.

10. The tube is removed from the magnet and MPIO resuspended in 1 mL of washing and storage buffer. The tube is placed in a rotating wheel and incubated, with constant head-over-head rotation, at 4°C for 5 min. MPIO are pelleted

using the magnet and the supernatant discarded. This step is repeated three times.

11. Antibody-conjugated MPIO are stored in Washing and Storage buffer at concentration of 2.5×10^{10} MPIO/mL. The solution is stable at 4°C for several months without loss of antigen binding (see Note 10).

3.2. Conjugation of Sugars to Amine-Activated MPIO

1. 1 µm amine-terminated MPIO (see Note 11) (50 mg/mL, 1 mL) are washed with 1.5 mL of water three times and lyophilised overnight to yield 40 mg of solid particle.

2. Thio-S-cyanomethyl-(5-acetimido-3,5-dideoxy-d-glycero-α-d-galacto-2-nonulo pyranosylonic acid-(2→3)-β-d-galactopyranosyl-(1→4)-(1→3)-(α-l-fucosyl)-2-acetimido-2-deoxy-β-d-glucopyranoside) [sLex-SCM]:

   (a) 4 mg (5 µM) is dissolved in 1 mL of anhydrous methanol.

   (b) 1 mg of sodium methoxide (18 µM) is added and the mixture is stirred on an orbital shaker for 2 days.

   (c) The particles are washed with methanol three times, water three times, and reconstituted in 2 mL PBS (20 mg/mL).

3.3. In Vitro MPIO Binding to TNF-α Stimulated sEND-1 Cells

1. Cells of a mouse endothelial cell line, sEND-1, are passaged when approaching confluency with trypsin/EDTA. Cells are plated at a density of 8×10^5 per 35 mm well in a 6 well plate, each well containing a sterile 19 mm round cover slip.

2. Cells are stimulated for 20 h at 37°C with graded doses of mouse recombinant TNF-α (0–10 ng/mL DMEM).

3. The TNF-α media is then removed by aspiration.

4. Stimulated cells are incubated with targeted-MPIO or IgG-MPIO (2.5×10^7 MPIO in 2 mL DMEM) in duplicate. The cell plate is immediately placed onto a sample rocker to avoid sedimentation of MPIO and incubated for 30 min at room temperature with continual mixing.

5. The media is removed by aspiration and unbound MPIO is removed by extensive washing with PBS.

6. Paraformaldehyde solution (1%) is added for 30 min at room temperature to fix cells.

7. The cell coverslip is mounted onto a glass slide by slowly inverting the coverslip onto mounting medium (13 µL) on a microscope slide. Nail varnish is used to seal the sample (see Note 12). The cells can be viewed as soon as the varnish is dry.

8. MPIO binding to cells are viewed using differential interference contrast microscopy. Four fields of view are acquired per sample. The number of bound MPIO per field is quantified using ImagePro plus.

9. For blocking experiments, targeted-MPIO are pre-blocked with 5 X mouse recombinant Fc–VCAM-1(5 μg) or Fc–ICAM-1(5 μg) per μg MPIO for 1 h at room temperature (12).

10. Fc-blocked MPIO (2.5×10^7 MPIO in 2 mL DMEM) are incubated with cells stimulated with 50 ng/mL TNF-α or unstimulated cells, extensively washed with PBS and fixed as described above (see steps 4–6). Experiments are performed in triplicate.

11. Fc-blocked MPIO binding to cells are viewed using an inverted microscope (×20 objective). Four fields of view are acquired per sample. The number of bound MPIO per field is quantified using ImagePro plus.

12. For immunofluorescent staining of cells, MPIO-bound cells are incubated with rat anti-mouse antibody (5 μg/mL PBS) for 1 h at room temperature.

13. The primary antibody is removed and the cells washed three times for 5 min each with PBS.

14. The secondary Alexa Fluor 488 conjugated rabbit antibody to rat IgG is prepared at 1:100 dilution in PBS and incubated with the cells for 1 h at room temperature (see Note 13).

15. The secondary antibody is removed and the cells washed three times for 5 min each with PBS.

16. The cell coverslip is carefully mounted onto a glass slide as described above (see step 7) using mounting media containing DAPI nuclear stain. Slides are viewed on the same day of preparation. Cells are kept in the dark at 4°C until imaged using confocal microscopy.

17. The cells are assessed for MPIO binding (see Note 14) and co-localisation using confocal microscopy (see Note 15).

3.4. In Vivo Mouse Protocol

1. Mice are deeply anesthetised using isofluorane.

2. Mice are positioned in a stereotaxic frame under the operating microscope.

3. Using a glass pipette with a tip <50 μm, mouse recombinant IL-1β (1 ng in 1 μL of low endotoxin saline containing 0.1% BSA) is stereotactically injected into the left striatum, 0.5 mm anterior and 2 mm lateral to the bregma, at a depth of 2.5 mm, over a 10 min period.

4. After 3 h, a cannula is inserted into the tail vein for administration of MPIO (4×10^8 MPIO in 200 μL low endotoxin saline containing 0.1% BSA) (see Note 16). Control groups of mice either injected intracerebrally with low endotoxin saline or not injected intracerebrally are also administered VCAM-MPIO (4×10^8 MPIO).

5. To block VCAM binding sites in vivo, a further group of mice may be injected with VCAM-1 antibody (0.2 mg/kg), 3 h after IL-1β injection and VCAM-MPIO is administered 15 min later.

6. Following MPIO injection, mice are positioned in a quadrature birdcage coil with an in-built stereotaxic frame. All mice are closely monitored for any signs of ill-health or toxicity (see Note 17). During MRI, anaesthesia is maintained with 1.0–1.5% isofluorane in 70% N_2O:30% O_2, ECG is monitored via subcutaneous electrodes, and body temperature maintained at 37°C by a circulating warm-water system.

7. MR images of the brain are acquired using a T2*-weighted 3D gradient-echo sequence with the following parameters; flip angle 35°, repetition time (TR) = 50 ms, echo time (TE) = 5 ms, field of view (FOV) $22.5 \times 22.5 \times 31.6$ mm, matrix size $192 \times 192 \times 360$, two averages. The total acquisition time is approximately 1 h (see Note 18). The data are zero-filled to $256 \times 256 \times 360$ and reconstructed off-line, to give a final isotropic resolution of 88 μm^3.

3.5. MR Image Analysis

1. Extra-cerebral structures in each MR image are masked semi-automatically using Image ProPlus.

2. Low signal areas are segmented in ten evenly spaced slices per brain using the automated signal intensity histogram-based tool in ImagePro Plus to obtain the median low signal intensity value (see Note 19).

3. Low signal areas are segmented in 41 contiguous slices of the brain, spanning a depth of 3.6 mm from the dorsal hippocampus ventrally. To ensure true laterality, the left and right hemispheres are segmented simultaneously, 1 mm from the midline outwards.

4. The median signal intensity value is applied to the 41 slice sequence to correct for minor variations in absolute signal intensity between individual scans.

5. Masks of the segmented low signal areas in 41 contiguous slices are merged and reconstructed using the 3D Constructor plug-in for ImagePro Plus to visualize MPIO binding patterns in the inflamed (left) and non-inflamed (right) cerebral hemispheres.

3.6. Targeted EAE Lesions for Evaluating Unilateral Binding of the Contrast Agents

1. Lewis rats are anaesthetised with 1.5–3% isoflurane in a mixture of nitrous oxide/oxygen (70/30%) and injected s.c. at the base of the tail with a total volume of 100 μL of MOG.

2. For control experiments, rats were injected with the same volume of saline emulsified in IFA.

3. Animals were anaesthetised as above and 2 μL of a cytokine mixture containing 1.45 μg of recombinant rat TNF-α and

1 μg of recombinant rat IFNγ dissolved in sterile saline was injected stereotaxically at a depth of 3 mm from the cortical surface over a 10-min period.

4. After injection, the wound was closed and the animals were allowed to recover from anaesthesia with no overt clinical signs.

5. A lesion with all of the hallmarks of MS develops at the injection site and is associated with BBB breakdown and an increased T2 at day 14.

6. At day 28, the lesion no longer displays BBB breakdown or T2 hyperintensity. However, histological analysis reveals ongoing leukocyte recruitment at this time point.

7. The MPIO are injected on day 28 and 3D datasets are collected as described in Subheading 3.4, step 7. The unilateral focal lesions allow an objective assessment to be made of the contrast agents in an MS-like lesion.

3.7. ET-1-Induced Focal Ischemic Stroke Lesions

1. Adult male Wistar rats (Harlan-Olac, Bicester, UK), weighing 200–250 g, were anaesthetised with 2.0% isofluorane in 70%N_2O/30%O_2.

2. Using a <50 μm-tipped glass pipette, animals were stereotaxically injected in the left striatum with 1 μL of saline containing 10 pmoles ET-1 to induce a focal subcortical infarct (see Note 20) or vehicle (coordinates: 1 mm anterior to Bregma, 2.6 mm lateral, and 4.5 mm deep from cortical surface). The injection was performed over a 2 min period.

3. After 2 h animals were injected with the MPIO (4 mg of Fe).

4. A T2*-weighted 3D gradient-echo dataset encompassing the entire brain is acquired.

5. At the end of the acquisition, cerebral blood volume (CBV) maps were obtained from a time series of gradient-echo images (TR=20 ms, TE=10 ms, flip angle=20°, 128×64 matrix, 5×4 cm FOV) during a 30 μL bolus injection of the intravascular contrast agent Gd[DTPA/BMA].

6. Spin-echo T1-weighted images (TR=500 ms, TE=20 ms, 128×64 matrix, 5×5 cm FOV) are acquired both before and 10 min after the Gd[DTPA/BMA] injection to look for image enhancement due to increased BBB permeability.

4. Notes

1. MPIO should be kept in liquid suspension during storage at 4°C as drying will reduce the performance of MPIO. We found it useful to store the MPIO vial on a sample roller at 4°C in order to prevent sedimentation of MPIO. MPIO

should never be frozen as this will cause irreversible aggregation.

2. The pre-washing and coating sodium borate buffer must not contain any protein or amino groups (e.g., glycine, Tris) as these will bind to the surface of the MPIO particles, inhibiting specific antibody binding. A higher pH favours optimal antibody binding.

3. MyOne™ Tosylactivated MPIO can be conjugated to any ligand containing amino or sulphdryl groups (i.e., antibody, protein, peptide, or glycoprotein). However, the antibody or protein must be purified since all proteins or amino groups will bind to the MPIO surface.

4. Preservatives such as sodium azide may disturb antibody conjugation to MPIO. In addition, sodium azide is cytotoxic and therefore not suitable for in vivo application. Therefore, antibodies free from stabilizers should be used or else remove stabilizers from the antibody/protein solution prior to conjugation.

5. The VCAM-1 antibody is commercially supplied in sodium borate buffer, pH 8.0, free from stabilizers. We found this antibody to be excellent for immunofluorescence imaging.

6. MPIO should be thoroughly resuspended prior to use by vortexing.

7. The Dynal MPC-S magnet can be used to prepare up to six microcentrifuge tubes of antibody-conjugated MPIO simultaneously.

8. For antibody conjugation, a concentration of 40 µg antibody/mg of MPIO is optimal. Conjugating less than recommended amounts of antibody may cause aggregation of MPIO. We found it best to use a stock purified antibody concentration of 0.5–1 mg/mL in order to achieve a sufficient antibody coating concentration. More dilute antibodies may require methods to concentrate the antibody prior to conjugation with MPIO.

9. MPIO should be suspended in the antibody solution with very efficient mixing using a vortex mixer and transferred immediately to the sample rotating wheel for incubation. MPIO should not be allowed to come out of suspension at any stage.

10. We store our antibody-conjugated MPIO, continually mixing on a sample roller at 4°C to avoid sedimentation and drying of MPIO, which would reduce their binding capacity.

11. 2.8 µm beads are available commercially, but we manufactured our own to achieve a comparable 1 µm-sized particle. We expect that the 2.8 µm beads would probably work successfully.

12. Air bubbles are undesirable in the mounting medium. Therefore, the coverslip should be inverted slowly onto the mounting media using fine forceps.

13. We found Alexa Fluor 488 conjugated rabbit secondary antibody to be excellent for localizing VCAM-1 by confocal microscopy as it was not prone to bleaching. This is particularly important when constructing z-stacks.

14. MPIO autofluoresce under confocal microscopy due to their high iron content and can be viewed using either red or green emission.

15. Due to the diameter of the MPIO, it may be difficult to find a focal plane that is suitable for simultaneously visualizing bound MPIO and immunofluorescence staining using confocal microscopy. We found it useful to create a z-stack of images throughout the depth of the MPIO and use the merged image to view co-localisation of immunofluorescence and MPIO.

16. For in vivo administration, we found it best to inject mice, via tail vein, with MPIO away from the MR magnet. We previously tried positioning mice inside the magnet first and then injecting MPIO. However, MPIO rapidly came out of solution as soon as they were within the magnetic field and sedimented prior to administration.

17. We have found antibody-conjugated MPIO to be well tolerated in all mice, with no animals showing any signs of ill effect during close observation for up to 5 h postinjection.

18. We have serially imaged the same mouse and found maximal contrast at 1–2 h with diminution by 4 h.

19. Image analysis should be performed by an operator blinded to the origin of data.

20. Rats should be used for this lesion as mice are resistant to the ET-1-induced ischemia.

Acknowledgments

This work is funded by the Wellcome Trust (R.P. Choudhury) and the Medical Research Council (N.R. Sibson and D.C. Anthony).

References

1. van Kasteren SI, Campbell SJ, Serres S, Anthony DC, Sibson NR, Davis BG (2009) Glyconanoparticles allow pre-symptomatic in vivo imaging of brain disease. Proc Natl Acad Sci U S A 106:18–23

2. McAteer MA, Sibson NR, von Zur Muhlen C, Schneider JE, Lowe AS, Warrick N, Channon KM, Anthony DC, Choudhury R P (2007) In vivo magnetic resonance imaging of acute brain inflammation using microparticles of iron oxide. Nat Med 13:1253–1258

3. Serres S, Anthony DC, Jiang Y, Broom KA, Campbell SJ, Tyler D J, van Kasteren SI, Davis BG, Sibson NR (2009) Systemic inflammatory response reactivates immune-mediated lesions in rat brain. J Neurosci 29:4820–4828

4. von Zur Muhlen C, Sibson NR, Peter K, Campbell SJ, Wilainam P, Grau GE, Bode C, Choudhury R P, Anthony DC (2008) A contrast agent recognizing activated platelets reveals murine cerebral malaria pathology undetectable by conventional MRI. J Clin Invest 118:1198–1207

5. Shapiro EM, Skrtic S, Koretsky AP (2005) Sizing it up: cellular MRI using micron-sized iron oxide particles. Magn Reson Med 53:329–338

6. Shapiro EM, Sharer K, Skrtic S, Koretsky AP (2006) In vivo detection of single cells by MRI. Magn Reson Med 55:242–249

7. Sibson NR, Blamire AM, Bernades-Silva M, Laurent S, Boutry S, Muller RN, Styles P, Anthony DC (2004) MRI detection of early endothelial activation in brain inflammation. Magn Reson Med 51:248–252

8. Briley-Saebo KC, Johansson LO, Hustvedt SO, Haldorsen AG, Bjornerud A, Fayad ZA, Ahlstrom HK (2006) Clearance of iron oxide particles in rat liver: effect of hydrated particle size and coating material on liver metabolism. Invest Radiol 41:560–571

9. McAteer MA, Schneider JE, Ali ZA, Warrick N, Bursill CA, von zur Muhlen C, Greaves DR, Neubauer S, Channon KM, Choudhury RP (2008) Magnetic resonance imaging of endothelial adhesion molecules in mouse atherosclerosis using dual-targeted microparticles of iron oxide. Arterioscler Thromb Vasc Biol 28:77–83

10. von Zur Muhlen C, Peter K, Ali ZA, Schneider JE, McAteer MA, Neubauer S, Channon K M, Bode C, Choudhury RP (2008) Visualization of activated platelets by targeted magnetic resonance imaging utilizing conformation-specific antibodies against glycoprotein IIb/IIIa. J Vasc Res 46:6–14

11. von Zur Muhlen C, von Elverfeldt D, Moeller JA, Choudhury RP, Paul D, Hagemeyer C E, Olschewski M, Becker A, Neudorfer I, Bassler N, Schwarz M, Bode C, Peter K (2008) Magnetic resonance imaging contrast agent targeted toward activated platelets allows in vivo detection of thrombosis and monitoring of thrombolysis. Circulation 118:258–267

12. Kelly KA, Allport JR, Tsourkas A, Shinde-Patil VR, Josephson L, Weissleder R (2005) Detection of vascular adhesion molecule-1 expression using a novel multimodal nanoparticle. Circ Res 96:327–336

Chapter 10

Measuring the Integrity of the Human Blood–Brain Barrier Using Magnetic Resonance Imaging

Andrea Kassner and Rebecca Thornhill

Abstract

The evaluation of blood–brain barrier (BBB) integrity with contrast-enhanced magnetic resonance imaging (MRI) may prove valuable in the setting of certain brain pathologies, such as brain tumors and acute ischemic stroke. Various MRI protocols have been developed to explore the integrity of the BBB by monitoring the leakage of intravenously administered contrast medium into the brain parenchyma. In its simplest form, BBB integrity is assessed qualitatively, by determining the presence or absence of contrast-enhancement on a structural MR image. When a dynamic contrast-enhanced (DCE) MRI protocol is combined with a suitable pharmacokinetic model, DCE-MRI can map the spatial distribution of BBB integrity throughout the brain and assist with evaluating the effects of therapy. Several model-free surrogate measures of BBB permeability have been recently proposed, all of which can be readily computed from standard dynamic susceptibility contrast MRI perfusion scans. Contrast-enhanced MRI offers multiple strategies for evaluating BBB integrity.

Key words: MRI, Brain, Dynamic contrast-enhanced imaging, Pharmacokinetic modeling, Dynamic susceptibility contrast imaging, Contrast agents, Permeability, Stroke, Hemorrhagic transformation, Cancer

1. Introduction

Medical imaging techniques are rapidly replacing highly invasive in vivo procedures like microdialysis (1) or plasma protein determination in cerebrospinal fluid (2) for the assessment of blood–brain barrier (BBB) integrity. In particular, the clinical role of contrast-enhanced magnetic resonance imaging (MRI) for characterizing pathological tissue continues to expand, providing new opportunities to monitor microvascular dynamics and offering the potential to assist in both clinical decision-making and the evaluation of

Sukriti Nag (ed.), *The Blood-Brain and Other Neural Barriers: Reviews and Protocols*, Methods in Molecular Biology, vol. 686, DOI 10.1007/978-1-60761-938-3_10, © Springer Science+Business Media, LLC 2011

treatment outcomes. MRI is a highly versatile imaging modality, capable of producing images with strong soft-tissue contrast. In MRI, image acquisition parameters can be manipulated to emphasize or "weight" the image gray-levels according to spin-lattice (T_1) or spin-spin (T_2) MR relaxation times. Occasionally, the discrepancy between the relaxation times of neighboring tissues, such as normal brain parenchyma vs. lesion, is too weak to enable a confident diagnostic assessment. If there is a clinical suspicion that the pathological process affecting the brain is one that is known to disrupt BBB integrity, then a paramagnetic contrast agent can be administered to exaggerate the inherent T_1 difference between lesion and healthy tissue.

The introduction of paramagnetic contrast agents such as the gadolinium chelate gadopentetate dimeglumine (Gd-DTPA) in the 1980s was a crucial advancement in neuroradiology and opened the doors to the possibility of investigating BBB abnormalities with MRI. Gd-DTPA is a freely diffusible, extracellular tracer with a molecular size of 550 Da. The chelator moiety, DTPA, governs the kinetics of the entire compound, and clearance occurs primarily via glomerular filtration. The contrast-enhancement is produced by the paramagnetic Gd^{3+} core of the agent, which is known to reduce T_1 in a concentration-dependent manner (3). Clinical gadolinium contrast agents are administered intravenously, usually as a bolus injection manually or via a power-injector. As the contrast agent passes through the microvasculature of the brain, it is almost entirely confined to the intravascular space. In regions of BBB breakdown, however, the contrast agent can escape the vasculature and begin to accumulate in the interstitial space ("extravasation"). Once in the extravascular space, tissue pixels with higher concentrations of contrast agent than their neighbors will appear bright on T_1-weighted MR images. In the MRI literature, the term "permeability imaging" generally refers to any MR technique that exploits the T_1 enhancement caused by contrast extravasation, via a disrupted BBB. As such, permeability imaging can include either qualitative techniques, such as postcontrast T_1-weighted imaging or quantitative methods, such as dynamic contrast-enhanced MRI (DCE-MRI) or dynamic susceptibility contrast MRI (DSC-MRI).

Whereas qualitative permeability imaging involves the acquisition of high-resolution T_1-weighted images at least 5 min after the intravenous bolus injection of contrast agent, DCE-MRI typically consists of a T_1-weighted sequence repeated dozens of times over the course of several minutes following the bolus injection. The advantage of DCE-MRI over qualitative postcontrast T_1-weighted imaging is that we can measure the extent of contrast accumulation as a function of time, apply an appropriate pharmacokinetic model, and ultimately produce estimates of BBB permeability in mL/100 g/min. Furthermore, qualitative or visual evidence of gadolinium enhancement is generally considered a specific, but not sensitive indicator of BBB disruption. As such, visual assessment of gadolinium enhancement is generally unsuitable for the longitudinal study of permeability required for clinical trials. For example, there is a recognized need to identify incremental changes in microvascular permeability to evaluate the effectiveness of new antiangiogenic agents for brain tumor treatment (4). However, DCE-MRI requires familiarity with pharmacokinetic modeling or at least the interpretation of the parameter estimates provided by the analysis software, so the technique of choice also depends on institutional expertise.

Recent reports have suggested that there are model-free surrogate measures of BBB permeability that can be readily extracted from standard MRI perfusion scans. Perfusion imaging with dynamic susceptibility contrast (DSC) MRI is already a mainstay of most tumor and acute ischemic stroke evaluation protocols and can be acquired in less than 2 min. Like DCE, DSC imaging involves tracking the passage of an intravenous administered bolus of gadolinium contrast agent with a series of MR images. Following injection, the high magnetic moment of gadolinium will alter the local magnetic field of blood or change the susceptibility or T_2^* relative to the surrounding brain parenchyma. In principle, as long as the BBB remains intact, this change in susceptibility will be visualized as signal loss on T_2^*-weighted images. In the context of BBB disruption, however, the initial signal-drop associated with gadolinium passing through the blood vessels within the voxel is followed by an increase in signal intensity (SI) as contrast begins to extravasate (i.e., there is a competing T_1 effect). Evidence of this secondary SI increase forms the basis for exploring DSC-measures as potential surrogate markers of BBB disruption.

This chapter gives our recommendations for protocol design and highlights the promise and pitfalls of various MRI methods for assessing BBB permeability. While there are many variants of each of the three MRI methods discussed below, we have enumerated the materials and methods for the protocols implemented at our own institution.

2. Materials

1. Intravenous catheter, preferably 18 g: BD Angiocath (Becton Dickinson Canada, Oakville, ON, Canada).
2. Two 60–65 mL syringes.
3. MR-compatible power-injector (Medrad Spectris, Pittsburgh, PA).

4. Approximately 15 mL of Gadopentetate dimeglumine (Gd-DTPA), "Magnevist," (Bayer Healthcare, Wayne, NJ), or 0.1 mM/kg for a contrast concentration of 0.5 mM/mL.

5. Saline.

6. Sterile tubing: Spectris Solaris injection system (Medrad Spectris).

7. Clinical 1.5T MRI scanner: GE Signa EXCITE equipped with Echo-Speed gradients and an eight-channel neurovascular head coil (GE Healthcare, Milwaukee, WI).

8. Image J Software: http://rsbweb.nih.gov/ij/.

3. Methods

3.1. Postcontrast T_1-Weighted MRI

3.1.1. Data Acquisition Protocol

1. Introduce an intravenous catheter into the antecubital vein.

2. Insert both syringes into the power-injector head and follow the manufacturer's instructions for proper loading of contrast and saline.

3. Connect sterile tubing to the power injector and ensure that the tubing is free of dead air. Connect the opposite end of the tubing to the intravenous catheter.

4. Follow the power-injector manufacturer's instructions for programming the flow rate and volume. Preset the rate and volume of contrast injection for 5 mL/s and 15 mL (i.e., approximately 0.1 mM/kg), respectively. The rate and volume of saline-flush injection should be preset for 4 mL/s and 20 mL, respectively.

5. Assist the patient to lie supine on the scanner table. Position the patient's head in the center of the head-coil, and provide gentle padding such as bundled hospital linens to assist the patient in maintaining a stable head position for the duration of the scan.

6. Convey the patient table into the magnet bore, using the scanner's infrared guidance system (if available) to position the patient in the isocenter of the magnet.

7. Enter the patient's name/code, weight, and other details into the MRI console (computer).

8. Acquire three-plane gradient-recalled echo (GRE) localizer images to determine the position of the brain and the lesion within the head coil.

9. Inject contrast and saline-flush.

10. Allow at least 5 min to elapse following contrast injection. Acquire 2D postcontrast T_1-weighted images of the lesion

with the following parameters: TR/TE 750/20 ms, FOV 240 mm, matrix 256×192, slice thickness 5 mm, 22 slices, flip angle 90°.

3.1.2. Qualitative Data
Analysis and Interpretation

1. Retrieve postcontrast T_1-weighted images at an offline workstation for qualitative analysis.

2. *Brain tumors:* In most cases, evidence of regional hyperpermeability within a tumor can be considered a strong indicator of ongoing angiogenesis and neovascularization (5, 6). Necrotic constituents appear hypo-intense, surrounded by a hyper-intense rim corresponding to dense vascularization (7). Solid tumors also appear hyper-intense in most cases, although solid tumors do not always show evidence of contrast extravasation. The heterogeneous contrast-enhancement patterns typically observed in glioblastoma multiforme are shown in Fig. 1 (see Note 1).

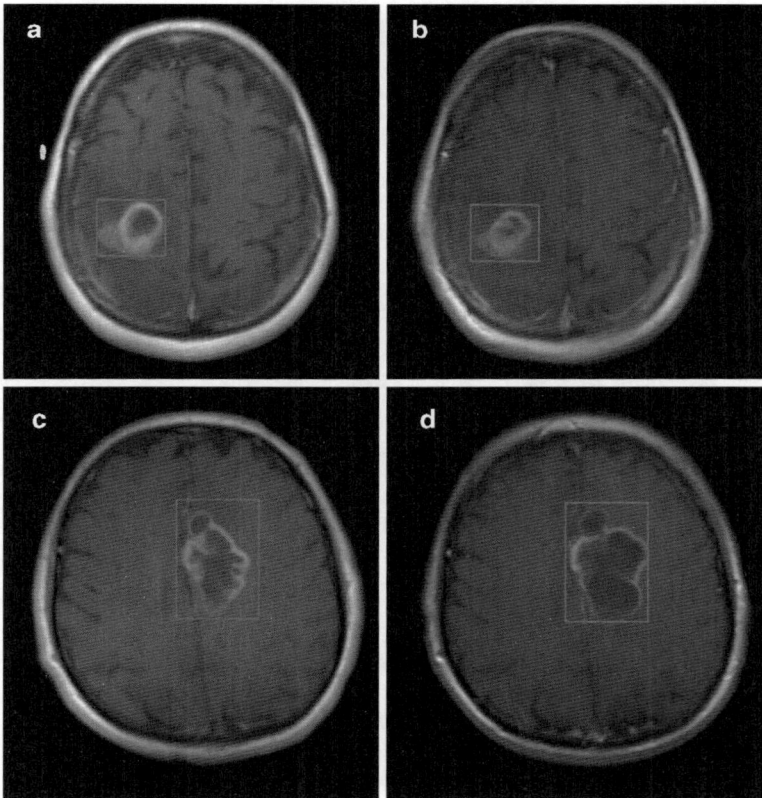

Fig. 1. Postcontrast T_1-weighted images obtained from two patients with glioblastoma multiforme (WHO grade IV gliomas). Images were acquired following surgery, but prior to radiotherapy (**a–c**) and after 6 weeks of 3D conformal radiotherapy with concurrent temozolomide chemotherapy (**b–d**). Note that the enhancing tumor appears to regress in one patient ((**b**), stable disease), but continues to grow in the second patient ((**d**), progressive disease).

3. *Acute Ischemic Stroke:* Evidence of postcontrast T_1-enhancement in hyperacute infarcts is significantly associated with subsequent hemorrhagic transformation (HT) (8–10). Parenchymal Gd-enhancement is infrequent during the crucial hours following stroke onset (9, 11–13) as it provides only a "snapshot" of a dynamic process (i.e., contrast extravasation via breaches in BBB integrity), and is therefore problematic during the hyperacute or acute phase of injury in acute ischemic stroke (see Note 2). Postcontrast T_1-weighted images of two HT cases imaged <6 h post-stroke at our institution are provided in Fig. 2 to illustrate this problem.

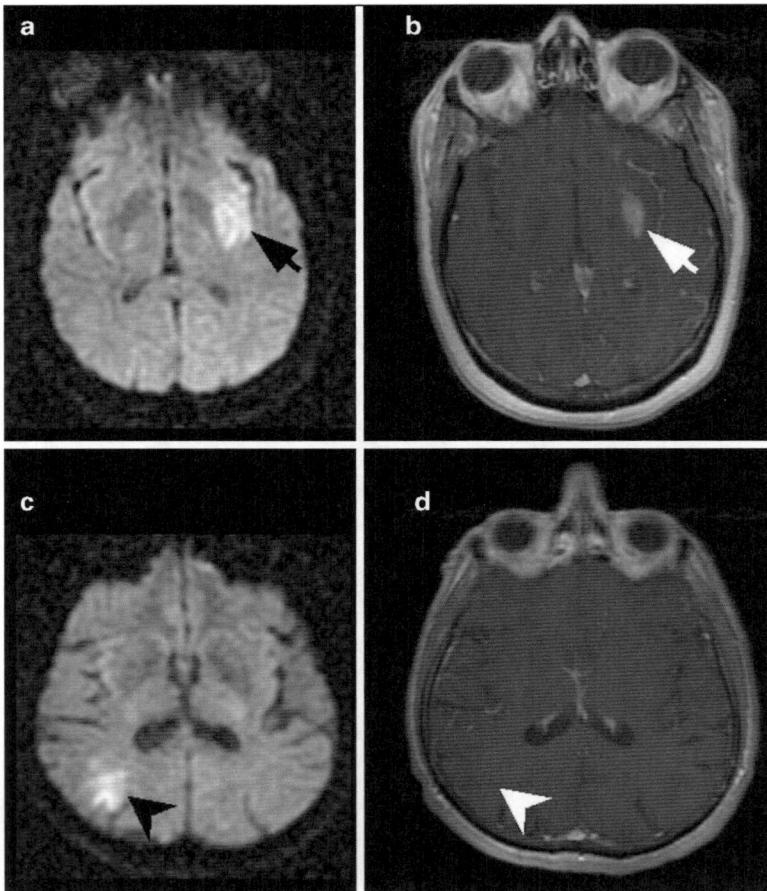

Fig. 2. Diffusion-weighted and postcontrast T_1-weighted images from two acute ischemic stroke patients who proceeded to hemorrhage are shown. (a, b) Images acquired at 3 h post-symptom onset from an 81-year-old-female patient show hyperintensity (*arrow*) in the diffusion-weighted image (a), which corresponds to an area of visible enhancement in the equivalent postcontrast T_1-weighted image (b). (c, d) A 69-year-old-male patient with acute ischemic stroke, visible as hyperintensity (*arrowhead*) on the diffusion-weighted image acquired at 5 h 40 min post-stroke (c). Unlike the previous case, the area of ischemia depicted by the diffusion-weighted image did not appear to correspond to any visible gadolinium-enhancement in the equivalent T_1-weighted image for this patient ((d), *arrowhead*).

3.2. Dynamic Contrast-Enhanced MRI

Many of the initial steps recommended for postcontrast T_1-weighted MRI are similar to the following DCE-MRI procedure.

3.2.1. Data Acquisition Protocol

1. Follow steps 1–7 as outlined in Subheading 3.1.1.

2. Set-up the 3D GRE sequence with the following parameters: TR/TE 5.9/1.5 ms, FOV 240 mm, matrix 128×128, slice thickness 5 mm, 32 slices, flip angle 20°, 31 dynamics, temporal resolution 9 s, total scan-time: 3–5 min (see Notes 3 and 4).

3. Begin the 3D GRE sequence as set-up in step 2. Allow the 3D GRE sequence to run for 30 s before injecting the contrast and saline flush.

3.2.2. Data Analysis Protocol

1. Retrieve 3D GRE images for offline post-processing and analysis.

2. Correct each set of DCE images for intra-scan motion (see Note 5).

3. Load image data-sets into the image analysis software (see Note 6). Use the interactive region of interest (ROI)-selection tool provided by the software to select a large cerebral artery or sagittal sinus. Save the coordinates of this ROI for use in all slices of the same 3D data set. Select tissue ROI(s) for each slice in which the lesion is present.

4. Ascertain the relationship between MR SI and contrast-agent concentration. Convert time-varying changes in SI values to contrast-agent concentration values for each ROI selected in step 3 (see Note 7).

5. Select a suitable pharmacokinetic model and fit the plasma (C_p) and tissue (C_t) contrast-concentration vs. time data to estimate the transfer constant (K^{trans}, see Note 8) or permeability coefficient (KPS, see Note 9) for each tissue ROI.

3.2.3. Interpretation of Data

1. *General*: The physiological interpretation of K^{trans} will be specific to the tissue or pathology of interest. Furthermore, K^{trans} will depend on whether the inter-compartmental movement of tracer is restricted primarily by capillary permeability or by regional blood flow (F) (see Note 9).

2. *Brain tumors*: DCE-MRI followed by graphical Patlak fitting has revealed that the transendothelial permeability coefficient, KPS, is significantly correlated with tumor grade (14) (see Note 10). A KPS map is provided in Fig. 3 to illustrate an example of the heterogeneous permeability distribution characteristic of high-grade brain tumors.

3. *Acute Ischemic Stroke*: Elevations in KPI are possible, even in the absence of visible postcontrast T_1-weighted enhancement in patients who have subsequently suffered a hemorrhage

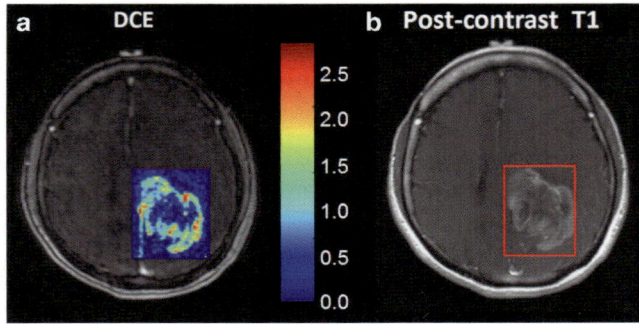

Fig. 3. KPS map superimposed on the equivalent DCE image obtained from a 53-year-old-patient with glioblastoma multiforme prior to surgery (**a**). The *KPS* map was constructed using DCE-MRI data acquired following intravenous injection of low molecular weight gadolinium contrast agent (Gd-DTPA). Note the heterogeneous distribution of *KPS* values within the tumor, with *KPS* values in excess of 2 mL/100 g/min in tumor-rim voxels, and as low as 0 mL/100 g/min in tumor-core voxels. The equivalent postcontrast *T1*-weighted image is provided in (**b**).

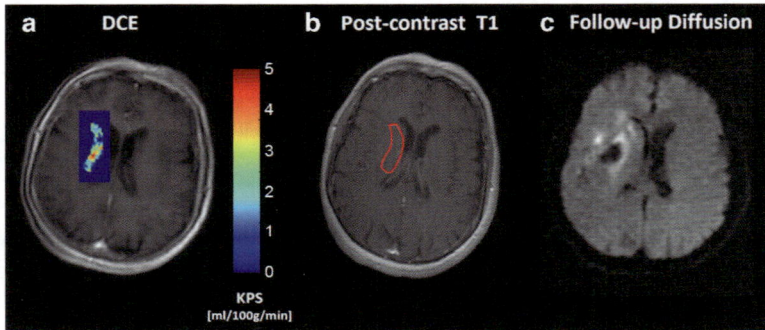

Fig. 4. *KPS* map superimposed on the equivalent DCE image obtained from a 73-year-old patient with acute ischemic stroke who subsequently hemorrhaged (**a**). There was no visible evidence of gadolinium enhancement on the equivalent static postcontrast T_1-weighted image (**b**). Both (**a, b**) were acquired approximately 3 h following stroke onset. (**c**) Diffusion-weighted MR imaging performed 48 h later revealed a region of very low signal intensity within a region of hyperintensity, characteristic of HT.

(see Fig. 4). The KPS in HT infarcts is elevated relative to the contralateral hemisphere and greater than in non-HT infarcts (15, 16) (see Note 11). This is consistently observed in HT, regardless of whether or not the subject has been treated with recombinant tissue plasminogen activator (rtPA) (17) (see Note 12).

3.3. Dynamic Susceptibility Contrast MRI

Many of the initial steps recommended for postcontrast T_1-weighted MRI are similar to the following DSC-MRI procedure (refer to the protocol in Subheading 3.1.1).

3.3.1. Data Acquisition Protocol

1. Follow steps 1–7 as outlined in Subheading 3.1.1.

2. Setup the 2D T_2^*-weighted single-shot EPI sequence with the following parameters: TR/TE 1,725/31.5 ms, FOV

240 mm, matrix 96×64, slice thickness 5 mm, 17 slices (or at least as many as needed to cover the entire lesion), flip angle 90°, 50 dynamics, temporal resolution 1.7 s, total scan-time: 1.5 min.

3. Inject the contrast and saline and immediately initiate the sequence as set-up in step 2.

3.3.2. Data Analysis Protocol

1. Retrieve 2D T_2^*-weighted EPI images for offline analysis and load image data sets into the analysis software (see Note 6). Use the interactive ROI-selection tool provided by the software to select tissue ROI(s) (repeat for each slice in which the lesion is present).

2. Convert the time-varying changes in *SI* values to ΔR_2^* for each tissue ROI $(\Delta R_2^* = \Delta(1/T_2^*) = -ln(SIt/SI_0)/TE)$ (18).

3. Fit the ΔR_2^* data to a gamma-variate function and estimate the relative recirculation (*rR*) of contrast for each tissue ROI (19) (see Fig. 5).

4. Determine the "percent recovery" of the ΔR_2^* data (*% Recovery*), to find the difference between the peak height of the ΔR_2^* vs. time curve and the average post-bolus tissue ΔR_2^*, expressed as a percentage of the peak height (20) (see Fig. 5).

3.3.3. Interpretation of Data

1. *General*: It is not presently known whether *rR* is strictly reflective of BBB permeability or rather just microvascular "abnormalities" (see Note 13).

2. *Brain tumors*: In brain tumors, elevations in *rR* correspond to regions of hyperpermeability and increased vessel tortuosity, such as those associated with high-grade gliomas (19).

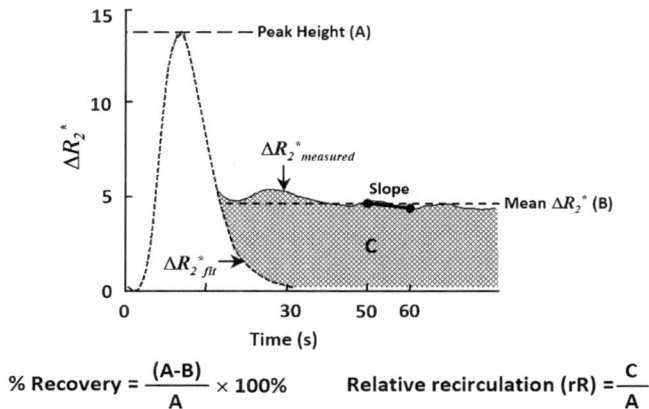

$$\% \ Recovery = \frac{(A-B)}{A} \times 100\% \qquad Relative \ recirculation \ (rR) = \frac{C}{A}$$

Fig. 5. A ΔR_2^* vs. time curve ($\Delta R_{2 \ measured}^*$), as well as its gamma-variate fit ($\Delta R_{2 \ fit}^*$), depicting four DSC surrogate measures of BBB permeability: *rR* (C/A), *peak height* (A), *% Recovery* = [100 × (A–B)/A], and *slope* = slope of $\Delta R_{2 \ measured}^*$ (t) between 50 and 60 s post-injection.

Conversely, the *% Recovery* is lower in grade IV than in grade III tumors (20) (see Note 14).

3. *Acute Ischemic Stroke*: Preliminary work suggests that *rR* is also related to the extent of BBB leakage in acute ischemic stroke, with increased *rR* in patients who subsequently hemorrhaged (21) (see Note 15).

4. Notes

1. The loss of BBB integrity is generally related to vascular proliferation and the degree of malignancy (22). However, postcontrast T_1-weighted images may lack the sensitivity required to delineate malignant from benign brain tumors. For example, in a contrast-enhanced MRI study of glioblastomas, four of the 21 tumors evaluated either failed to enhance or enhanced only minimally (23).

2. The BBB disruption thought to precede HT (13) should theoretically enable MR contrast accumulation within the brain parenchyma. While visual evidence of contrast extravasation or enhancement on postcontrast T_1-weighted MRI is a highly specific (~85%) predictor of HT, it is very weakly sensitive (~35%), which could make the test unsuitable for urgent treatment decision-making (8–10). In part, the low sensitivity of postcontrast T_1-weighted MRI for secondary hemorrhage may be a consequence of persistent ischemia or microvascular obstruction in the acute phase (24) such that the T_1-enhancement associated with that region will be effectively indistinguishable from infarcts that are not prone to HT.

3. The most common MRI sequence used in brain DCE-MRI studies is a 3D GRE sequence with a short TR, a short TE, and a flip angle of approximately 20° at 1.5 T (25). Three-dimensional volume acquisitions are favored over two-dimensional equivalents in DCE-MRI of the brain, as they permit the simultaneous sampling of the *SI* in tissue and in blood and are more likely to encompass a large vessel from which one can generate the required $C_p(t)$ curves (a considerable challenge in brain DCE (26)). In addition to coverage, a 3D acquisition offers the advantage of maximizing the saturation of inflowing blood while minimizing precontrast inflow enhancement (25). Finally, the 3D GRE sequence is less sensitive to geometric distortions or magnetic susceptibility than equivalent echo-planar sequences.

4. Our pharmacokinetic models are "expecting" the input data to be in the form of tracer concentration, i.e., $C_t(t)$ and $C_p(t)$, so we will have to confirm that the relationship between ΔSI

and $C(t)$ is linear for the sequence and contrast doses utilized in the protocol. addition to performing a precontrast measurement R_1 (R_{10}) (25, 26), it is crucial to plan the DCE protocol so that at least some of the initial dynamics are acquired prior to the arrival of contrast agent. In addition to flip angle and TR, the R_{10} values should provide us with all of the information required to calculate ΔR_1 from ΔSI data for most of the GRE sequences commonly used in DCE imaging (27, 28). The choice of flip angle is a nontrivial consideration in DCE-MRI, as the relationship between ΔR_1 from ΔSI departs further from linearity as the flip angle becomes too shallow to capture the maximum ΔR_1 anticipated for the concentration of contrast, particularly in the blood (for flip angles $\ll 90°$). When the flip angle approaches $90°$, however, then any gain realized in terms of R_1-sensitivity will be at least partially offset by a penalty in SNR. The temporal resolution of the DCE sequence must be sufficient to capture the blood–brain flux of contrast anticipated for the pathology of interest. The temporal resolutions reported for brain tumor DCE studies have ranged widely ~5–55 s (14, 29–31). By definition, the temporal resolution of the acquisition will improve with shorter TRs (i.e., more images can be collected per unit time); TE should also be short, but this choice is more directly related to minimizing susceptibility-related signal loss caused by the transit of the paramagnetic contrast agent through the vasculature (32). The TR prescribed for most brain DCE-MRI studies is typically <10 ms.

5. There are many software programs available online. We have incorporated one such tool provided by the Functional MRI of the Brain (FMRIB) Laboratory at the University of Oxford (33).

6. At minimum, the image analysis software should allow the user to (a) select regions of arbitrary shape and size, (b) save the regions for later use, and (c) sample and save the SI of the selected regions for use in subsequent modeling or computation. The *ImageJ* program will meet these minimum requirements and is freely available online. More specialized programs are required for pharmacokinetic modeling (DCE) and for the extraction of DSC parameters, but this software can be readily obtained or developed in consultation with a medical imaging physicist.

7. It is broadly assumed that ΔR_1 is related to the contrast concentration, $C(t)$ by a linear scaling factor, r_1 (i.e., T_1 relaxivity, in s^{-1} mM^{-1}). Given that, local tissue environments can modulate tracer relaxivity, this simplification may introduce errors in the final parameter estimates (34). Many sequences have been developed for acquiring dynamic T_1 "maps," whereby

the ΔSI measured in each voxel is directly related to ΔR_I. While these sequences are helpful for mitigating other sources of error (35–40), they can not address the issue of nonuniform T_I relaxivity.

8. In many cases, the pharmacokinetic models that are applied to DCE-MRI data were originally developed for nuclear medicine tracers (41–45). Most of these are compartmental models, which define the tissue space as a volume with both intravascular (or "plasma") and extravascular (or "tissue") compartments, with contrast concentration $C_p(t)$ and $C_t(t)$, respectively (see Fig. 6a). The generalized two-compartment analysis proposed by Tofts et al. (46) defines two tracer parameters of physiological interest: the transfer constant, K^{trans} [s^{-1}] and the distribution volume, v_e (i.e., the fraction of the extravascular-extracellular space occupied by tracer in [mL/g]):

$$\frac{dC_t(t)}{dt} = K^{trans}\left[C_p(t) - \left(\frac{C_t(t)}{v_e}\right)\right] \tag{1}$$

When the concentration of contrast in the tissue prior to injection is 0, $C_t(0) = 0$, we can find a solution to (1) and therefore determine the C_t for each time point post-injection (t):

$$C_t(t) = K^{trans}\int_0^t C_p(\tau)e^{-\frac{K^{trans}}{v_e}.(t-\tau)}, \tag{2}$$

$$d\tau = K^{trans}C_p(t)\otimes e^{-\frac{K^{trans}}{v_e}.t}, \tag{3}$$

where the $\otimes$ symbol indicates the convolution operator.

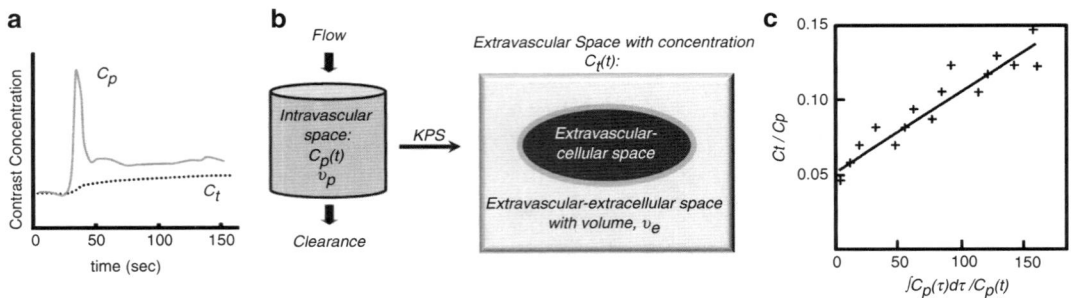

Fig. 6. The contrast concentration vs. time curves for both blood plasma (C_p, *solid line*) and tissue (C_t, *dotted line*) are depicted in (**a**). Permeability can be estimated using a unidirectional two-compartment tracer-kinetic model (**b**), such as the Patlak model (44). Plotting the ratio $C_t(t)/C_p(t)$ vs. $\int C_p(\tau)d\tau/C_p(t)$ yields a linear relationship, where permeability (*KPS*) is the slope of best fit (**c**).

9. $K^{trans} = EF\rho$ $(1-rHct)$, where E represents the fraction of contrast agent that extravasates during the first circuit through the vasculature ("extraction fraction"), ρ indicates the density of the tissue in (g/mL) and Hct is the hematocrit (the fraction of blood volume occupied by cells). In addition to flow, E is also related to the permeability surface-area product (PS), such that $E = 1-exp(-PS/F)$. Taken together, we can see that K^{trans} will predominantly reflect regional blood flow when PS/F is high ($K^{trans} \sim F\rho$ $(1-Hct)$). Conversely, if $PS/F < 1$, as is typically the case for clinical MRI contrast media in normal brain parenchyma (and intact BBB), then K^{trans} will be equivalent to $PS\rho$ (25). K^{trans} is sometimes referred to as "KPS" under these conditions. If we further assume that there is no efflux of contrast – i.e., the tracer becomes effectively "trapped" in the extravascular compartment, at least for the timescale involved in the tracer-kinetic study – then we can simplify the pharmacokinetic model in the manner proposed by Patlak et al. and depicted in Fig. 6b (44). The Patlak model is a graphical approach for estimating KPS that models the relationship between C_t and C_p using linear regression:

$$\frac{C_t(t)}{C_p(t)} = KPS \int_0^t \frac{C_p(\tau)d\tau}{C_p(t)} + V, \qquad (4)$$

where V represents the fractional blood volume in each voxel. Plotting the ratio $C_t(t)/C_p(t)$ vs. $\int C_p(\tau)d\tau/C_p$ yields a linear relationship, where KPS is the slope of best fit and V is the y-intercept (see Fig. 6c). This approach obviates the need for deconvolution and its sensitivity to noise. The Patlak model has been adapted and successfully applied to DCE-MRI data obtained in both brain tumors (14) and acute ischemic stroke (16, 47) for the estimation of BBB permeability. In order to respect the assumptions inherent in the Patlak model, the tracer concentration vs. time data must be acquired after blood–tissue equilibrium has been attained, when the transfer of contrast between compartments is no longer dominated by the rapid changes associated with contrast wash-in (44). Deviation from this assumption results in nonlinearity of the Patlak plot.

10. Tumor composition is heterogeneous, particularly with respect to vascularity (48, 49), so some degree of intra-voxel heterogeneity is probably inevitable, even with exquisitely high-spatial resolution. This heterogeneity will result in inaccurate K^{trans} (or KPS) estimates (50), but this can be minimized by acquiring the finest spatial resolution achievable for the desired TR, SNR, and total scan-time perhaps aided by parallel imaging strategies (51).

11. In a feasibility study of ten acute ischemic stroke patients assessed within 24 h from symptom onset and untreated with

rtPA, the results showed that *KPS* could be measured in the acute ischemic stroke setting (16). Follow-up imaging indicated that three of these patients converted to hemorrhage within 48 h and a retrospective analysis found significantly elevated *KPS* in infarcts that progressed to hemorrhage, compared to those in the non-HT group (16).

12. Comparing acute ischemic stroke patients who received rtPA with untreated patients who presented to the emergency department within a mean of 4 h of symptom onset (52), 13 of 36 patients showed progressive enhancement associated with increased permeability in the acute phase (six received rtPA, seven were untreated), all of whom proceeded to HT at 24–72 h post-stroke. These findings suggested the potential benefit of assessing permeability in acute ischemic stroke patients.

13. A more prudent interpretation of these results is that *rR* reflects a superposition of microvascular features, including blood volume, vessel tortuosity and, perhaps, permeability (19, 53). As such, *rR* may be less suitable than *KPS* for the investigation of new therapies and their effects on BBB integrity where absolute quantification is desirable. Whether or not *rR* will ultimately prove valuable in either tumor grading or treatment, decision-making in acute ischemic stroke remains to be demonstrated.

14. The *rR* measure, for example, involves separating the intravascular (T_2^*) and extravascular/recirculation phases (T_1) by fitting a theoretical first-pass curve to the ΔR_2^* vs. time curve and measuring the difference in the areas encompassed by the measured and the fitted curves (Fig. 5). The *rR* metric was first introduced as a means of characterizing tumor microvasculature (19). Specifically, the authors demonstrated that significant elevations in *rR* were unique to the four high-grade glioma tumors studied, and localized to the leakiest aspects of the tumor periphery. Like *rR*, the *% Recovery* metric (Fig. 5) has been shown to be capable of distinguishing between grade III and IV tumors (20).

15. We estimated *rR* in patients with acute ischemic stroke and compared this DSC metric with *KPS* estimates produced by DCE-MRI and Patlak modeling (21). Preliminary results indicated a strong and significant correlation between *rR* and *KPS*, as well as significant increases in both measures in stroke lesions that proceeded to hemorrhage. Taken together, these findings suggest that *rR* is related to the extent of BBB leakage and that *rR* may provide a reasonable surrogate for *KPS*. Bang and coworkers discovered that the change in ΔR_2^* associated with the last 10 s of a 60-s bolus tracking study ("negative slope") could predict HT in recanalized acute ischemic stroke patients with a sensitivity of 83% (54).

References

1. de Lange EC, Danhof M, de Boer AG, Breimer DD (1997) Methodological considerations of intracerebral microdialysis in pharmacokinetic studies on drug transport across the blood-brain barrier. Brain Res 25:27–49

2. Blezer E (2005) Techniques for measuring the blood–brain barrier integrity, in The blood-brain barrier and its microenvironment: basic physiology to neurological disease (de Vries, E., Prat, A Ed.), pp 441–456, Taylor and Francis, New York

3. Weinmann HJ, Laniado M, Mutzel W (1984) Pharmacokinetics of GdDTPA/dimeglumine after intravenous injection into healthy volunteers, Physiological chemistry and physics and medical. NMR 16:167–172

4. Leach MO, Brindle KM, Evelhoch JL, Griffiths JR, Horsman MR, Jackson A, Jayson G, Judson IR, Knopp MV, Maxwell RJ, McIntyre D, Padhani AR, Price P, Rathbone R, Rustin G, Tofts PS, Tozer GM, Vennart W, Waterton JC, Williams SR, Workman P (2003) Assessment of antiangiogenic and antivascular therapeutics using MRI: recommendations for appropriate methodology for clinical trials. Brit J Radiol 76 Spec No 1:S87–S91

5. Brasch R, Pham C, Shames D, Roberts T, van Dijke K, van Bruggen N, Mann J, Ostrowitzki S, Melnyk O (1997) Assessing tumor angiogenesis using macromolecular MR imaging contrast media. J Magn Reson Imaging 7:68–74

6. Brasch RC, Li KC, Husband JE, Keogan MT, Neeman M, Padhani AR, Shames D, Turetschek K (2000) In vivo monitoring of tumor angiogenesis with MR imaging. Academic Radiol 7:812–823

7. Earnest Ft, Kelly PJ, Scheithauer BW, Kall BA, Cascino TL, Ehman RL, Forbes GS, Axley PL (1988) Cerebral astrocytomas: histopathologic correlation of MR and CT contrast enhancement with stereotactic biopsy. Radiology 166:823–827

8. Vo KD, Santiago F, Lin W, Hsu CY, Lee Y, Lee JM (2003) MR imaging enhancement patterns as predictors of hemorrhagic transformation in acute ischemic stroke. AJNR Am J Neuroradiol 24:674–679

9. Kim EY, Na DG, Kim SS, Lee KH, Ryoo JW, Kim HK (2005) Prediction of hemorrhagic transformation in acute ischemic stroke: role of diffusion-weighted imaging and early parenchymal enhancement. AJNR Am J Neuroradiol 26:1050–1055

10. Kastrup A, Groschel K, Ringer TM, Redecker C, Cordesmeyer R, Witte OW, Terborg C (2008) Early disruption of the blood-brain barrier after thrombolytic therapy predicts hemorrhage in patients with acute stroke. Stroke 39:2385–2387

11. Merten CL, Knitelius HO, Assheuer J, Bergmann-Kurz B, Hedde JP, Bewermeyer H (1999) MRI of acute cerebral infarcts, increased contrast enhancement with continuous infusion of gadolinium. Neuroradiology 41:242–248

12. Virapongse C, Mancuso A, Quisling R (1986) Human brain infarcts: Gd-DTPA-enhanced MR imaging. Radiology 161:785–794

13. Latour LL, Kang DW, Ezzeddine MA, Chalela JA, Warach S (2004) Early blood-brain barrier disruption in human focal brain ischemia. Ann Neurol 56:468–477

14. Roberts HC, Roberts TP, Brasch RC, Dillon WP (2000) Quantitative measurement of microvascular permeability in human brain tumors achieved using dynamic contrast-enhanced MR imaging: correlation with histologic grade. AJNR Am J Neuroradiol 21:891–899

15. Knight RA, Barker PB, Fagan SC, Li Y, Jacobs MA, Welch KM (1998) Prediction of impending hemorrhagic transformation in ischemic stroke using magnetic resonance imaging in rats. Stroke 29:144–151

16. Kassner A, Roberts T, Taylor K, Silver F, Mikulis D (2005) Prediction of hemorrhage in acute ischemic stroke using permeability MR imaging. AJNR Am J Neuroradiol 26:2213–2217

17. Kassner A, Roberts TPL, Moran B, Silver F, Mikulis DJ (2009) rtPA increases blood-brain barrier disruption in acute ischemic stroke: an MRI permeability study. Am J Neuroradiol 30(10):1864–1869

18. Rosen BR, Belliveau JW, Vevea JM, Brady TJ (1990) Perfusion imaging with NMR contrast agents. Magn Reson Med 14:249–265

19. Kassner A, Annesley DJ, Zhu XP, Li KL, Kamaly-Asl ID, Watson Y, Jackson A (2000) Abnormalities of the contrast re-circulation phase in cerebral tumors demonstrated using dynamic susceptibility contrast-enhanced imaging: a possible marker of vascular tortuosity. J Magn Reson Imaging 11:103–113

20. Lupo JM, Cha S, Chang SM, Nelson SJ (2005) Dynamic susceptibility-weighted perfusion imaging of high-grade gliomas: characterization of spatial heterogeneity. AJNR Am J Neuroradiol 26:1446–1454

21. Wu S-P, Thornhill RE, Chen S, Rammo W, Mikulis DJ, Kassner A (2009) Relative recirculation: a fast, model-free surrogate for the

measurement of blood-brain barrier permeability and the prediction of hemorrhagic transformation in acute ischemic stroke. Invest Radiol 44(10):662–668

22. Mikulis DJ, Roberts TP (2007) Neuro MR: protocols. J Magn Reson Imaging 26:838–847

23. Knopp EA, Cha S, Johnson G, Mazumdar A, Golfinos JG, Zagzag D, Miller DC, Kelly PJ, Kricheff, II (1999) Glial neoplasms: dynamic contrast-enhanced T2*-weighted MR imaging. Radiology 211:791–798

24. Wang X, Tsuji K, Lee SR, Ning M, Furie KL, Buchan AM, Lo EH (2004) Mechanisms of hemorrhagic transformation after tissue plasminogen activator reperfusion therapy for ischemic stroke. Stroke 35:2726–2730

25. Kassner A, Roberts TP (2004) Beyond perfusion: cerebral vascular reactivity and assessment of microvascular permeability. Top Magn Reson Imaging 15:58–65

26. Cheng HL (2008) Investigation and optimization of parameter accuracy in dynamic contrast-enhanced MRI. J Magn Reson Imaging 28:736–743

27. Haacke EM, Tkach JA (1990) Fast MR imaging: techniques and clinical applications. AJNR Am J Neuroradiol 155:951–964

28. Haase A, Frahm J, Matthaei D, Hannicke W, Merboldt KD (1986) FLASH imaging. Rapid NMR imaging using low flip-angle pulses. J Magn Reson 67:9

29. Armitage PA, Schwindack C, Bastin ME, Whittle IR (2007) Quantitative assessment of intracranial tumor response to dexamethasone using diffusion, perfusion and permeability magnetic resonance imaging. Mag Reson Imaging 25:303–310

30. Haris M, Gupta RK, Husain M, Srivastava C, Singh A, Singh Rathore RK, Saksena S, Behari S, Husain N, Mohan Pandey C, Nath Prasad K (2008) Assessment of therapeutic response in brain tuberculomas using serial dynamic contrast-enhanced MRI. Clin Radiol 63:562–574

31. Li KL, Zhu XP, Checkley DR, Tessier JJ, Hillier VF, Waterton JC, Jackson A (2003) Simultaneous mapping of blood volume and endothelial permeability surface area product in gliomas using iterative analysis of first-pass dynamic contrast enhanced MRI data. Br J of Radiol 76:39–50

32. Roberts TP (1997) Physiologic measurements by contrast-enhanced MR imaging: expectations and limitations. J Magn Reson Imaging 7:82–90

33. Jenkinson M, Bannister P, Brady M, Smith S (2002) Improved optimization for the robust and accurate linear registration and motion correction of brain images. Neuroimage 17:825–841

34. Donahue KM, Weisskoff RM, Burstein D (1997) Water diffusion and exchange as they influence contrast enhancement. J Magn Reson Imaging 7:102–110

35. Kay I, Henkelman RM (1991) Practical implementation and optimization of one-shot T1 imaging. Magn Reson Med 22:414–424

36. Gowland PA, Leach MO (1992) Fast and accurate measurements of T1 using a multi-readout single inversion-recovery sequence. Magn Reson Med 26:79–88

37. Tong CY, Prato FS (1994) A novel fast T1-mapping method. J Magn Reson Imaging 4:701–708

38. Brix G, Schad LR, Deimling M, Lorenz WJ (1990) Fast and precise T1 imaging using a TOMROP sequence. Magn Reson Imaging 8:351–356

39. Scheffler K, Hennig J (2001) T(1) quantification with inversion recovery TrueFISP. Magn Reson Med 45:720–723

40. Deoni SC, Rutt BK, Peters TM (2003) Rapid combined T1 and T2 mapping using gradient recalled acquisition in the steady state. Magn Reson Med 49:515–526

41. Larsson HB, Stubgaard M, Frederiksen JL, Jensen M, Henriksen O, Paulson OB (1990) Quantitation of blood-brain barrier defect by magnetic resonance imaging and gadolinium-DTPA in patients with multiple sclerosis and brain tumors. Magn Reson Med 16:117–131

42. Brix G, Semmler W, Port R, Schad LR, Layer G, Lorenz WJ (1991) Pharmacokinetic parameters in CNS Gd-DTPA enhanced MR imaging. J Comput Assist Tomogr 15:621–628

43. Tofts PS, Kermode AG (1991) Measurement of the blood-brain barrier permeability and leakage space using dynamic MR imaging. 1. Fundamental concepts. Magn Reson Med 17: 357–367

44. Patlak CS, Blasberg RG, Fenstermacher JD (1983) Graphical evaluation of blood-to-brain transfer constants from multiple-time uptake data. J Cereb Blood Flow Metab 3:1–7

45. Patlak CS, Blasberg RG (1985) Graphical evaluation of blood-to-brain transfer constants from multiple-time uptake data. Generalizations. J Cereb Blood Flow Metab 5:584–590

46. Tofts PS, Brix G, Buckley DL, Evelhoch JL, Henderson E, Knopp MV, Larsson HB, Lee TY, Mayr NA, Parker GJ, Port RE, Taylor J, Weisskoff RM (1999) Estimating kinetic parameters from dynamic contrast-enhanced T(1)-weighted MRI of a diffusable tracer:

standardized quantities and symbols. J Magn Reson Imaging 10:223–232

47. Ewing JR, Knight RA, Nagaraja TN, Yee JS, Nagesh V, Whitton PA, Li L, Fenstermacher JD (2003) Patlak plots of Gd-DTPA MRI data yield blood-brain transfer constants concordant with those of 14C-sucrose in areas of blood-brain opening. Magn Reson Med 50: 283–292

48. Endrich B, Reinhold HS, Gross JF, Intaglietta M (1979) Tissue perfusion inhomogeneity during early tumor growth in rats. J Natl Cancer Inst 62:387–395

49. Tozer GM, Lewis S, Michalowski A, Aber V (1990) The relationship between regional variations in blood flow and histology in a transplanted rat fibrosarcoma. Br J Cancer 61:250–257

50. Simpson NE, He Z, Evelhoch JL (1999) Deuterium NMR tissue perfusion measurements using the tracer uptake approach: I. Optimization of methods. Magn Reson Med 42:42–52

51. Pruessmann KP, Weiger M, Scheidegger MB, Boesiger P (1999) SENSE: sensitivity encoding for fast MRI. Magn Reson Med 42:952–962

52. Kassner A, Roberts TPL, Moran B, Silver FL, Mikulis DJ. rtPA increases blood-brain barrier disruption in acute ischemic stroke: an MRI permeability study. AJNR Am J Neuroradiol 2009; 30:1864–1869.

53. Jackson A, Kassner A, Annesley-Williams D, Reid H, Zhu XP, Li KL (2002) Abnormalities in the recirculation phase of contrast agent bolus passage in cerebral gliomas: comparison with relative blood volume and tumor grade. AJNR Am J Neuroradiol 23:7–14

54. Bang OY, Buck BH, Saver JL, Alger JR, Yoon SR, Starkman S, Ovbiagele B, Kim D, Ali LK, Sanossian N, Jahan R, Duckwiler GR, Vinuela F, Salamon N, Villablanca JP, Liebeskind DS (2007) Prediction of hemorrhagic transformation after recanalization therapy using T2*-permeability magnetic resonance imaging. Ann Neurol 62:170–176

Chapter 11

Assessing Blood–Cerebrospinal Fluid Barrier Permeability in the Rat Embryo

Norman R. Saunders, C. Joakim Ek, Mark D. Habgood, Pia Johansson, Shane Liddelow, and Katarzyna M. Dziegielewska

Abstract

The rat is a useful model for studies of embryonic blood–CSF function in that the embryos are large enough to collect sufficient fluid samples for analysis and exteriorized embryos can be kept viable for several hours in order to conduct longer term experiments. Both quantitative and qualitative methods that are similar to those used in adult studies can be used to assess blood–CSF function in the rat embryo; however, there are technical aspects of these studies that are more challenging. The choice of the methods to be used depends largely on the question being asked. This chapter describes in detail the precise steps that need to be taken to keep rat embryos in a good physiological state while conducting the experiments, how to administer markers into the embryonic circulation, and how to sample blood and/or CSF from embryos. How to evaluate the results obtained is outlined at the end of each method, together with notes on some limitations that are inherent in developmental studies.

Key words: Choroid plexus, Blood–CSF barrier, Development, Cerebrospinal fluid, Rat embryo

1. Introduction

The choroid plexuses are the main site of direct exchange between the blood and cerebrospinal fluid (CSF) and measurements of the concentration of molecules of interest in CSF compared to their concentration in plasma or whole blood are often used as a convenient index of blood–CSF barrier permeability in both the adult and developing brain. However, in the adult brain molecules can also enter the CSF by diffusion from the brain compartment and it is thus difficult in the adult brain to precisely determine what proportion of a marker in CSF has entered directly across the choroid plexuses (blood–CSF barrier) compared to indirectly via

Sukriti Nag (ed.), *The Blood-Brain and Other Neural Barriers: Reviews and Protocols*, Methods in Molecular Biology, vol. 686, DOI 10.1007/978-1-60761-938-3_11, © Springer Science+Business Media, LLC 2011

the cerebral vasculature (blood-brain barrier). In the embryonic brain, the relative contribution from these two routes is less of a problem because the immature brain is poorly vascularized at a time when the choroid plexuses are relatively well developed (1), and there is an additional diffusion restraint ("strap junctions") at the level of the ventricular zone CSF–brain interface (CSF–brain barrier) that is not present in the adult brain (2, 3).

The main experimental approaches that have been used are (a) steady state, (b) short duration experiments using markers whose concentration can be measured, and (c) short duration experiments using markers that can be visualized directly.

Steady-state experiments involve maintaining a stable concentration of a marker of interest in the blood for an extended period of time (several hours) and then comparing its concentration in CSF once CSF levels have approached a steady equilibrium with its concentration in the blood (CSF/blood or CSF/plasma ratios). As used in the adult, this method generally requires a continuous infusion of the marker in order to compensate for losses from the blood compartment (e.g., renal clearance), coupled with repeated sampling of blood to confirm the stability of the blood levels over time. For markers that are not rapidly cleared from the blood, a single administration can be given followed by repeated blood sampling and the concentration in CSF at steady state expressed as a ratio of the mean blood (or plasma) concentration over time (area under the curve) (4). In the very small rat embryo, continuous infusion and/or serial blood sampling are impractical, but equivalent experimental conditions can be obtained by using the litter-based model (5), in which each embryo is injected with identical amounts of tracer and blood and CSF terminally sampled at different times following injection. However, the concentration of the marker of interest in the CSF at steady state is a function of the rate of entry via the blood–CSF and blood-brain barriers, the volume of the CSF compartment into which it distributes, and the rate of removal from the CSF compartment (e.g., via the CSF-sink effect, (6)). Thus, differences in steady-state CSF/blood concentrations between different markers only reflect differences in permeability if they are measured under the same conditions and in animals at the same stage of brain development. Data on steady-state CSF/plasma or CSF/brain ratios measured at different stages of development need to be interpreted with care, since rates of entry into CSF, the volume of the CSF compartment, and rates of removal from CSF will all be different (1). Steady-state experiments are suitable for markers that enter the CSF compartment slowly.

Short duration experiments involve injecting a marker of interest into the blood or peritoneal cavity and then sampling blood and CSF at much shorter times after administration (seconds to minutes). These types of short duration experiments provide information on the initial rates of entry of compounds into the CSF and are less influenced by clearance from the CSF and blood compartments, entry into CSF via brain tissue, and differences in the volume of the CSF compartment. However, these short duration experiments are only suitable for markers that enter CSF relatively quickly as there needs to be sufficient marker in the CSF at the time of sampling in order for it to be measurable. Short duration experiments using markers that can be visualized allow direct assessment of the integrity of the blood–CSF barrier interface and can also be used to investigate the route of entry of markers into CSF, particularly those that enter CSF via transcellular pathways (e.g., plasma proteins, (7)).

Conducting blood–CSF barrier experiments in embryonic animals is essentially the same as in adult animals, but there are substantial technical problems to be addressed. In utero embryonic animals are not easily accessible, they are generally very small (depending on the species), notably fragile, and more difficult to maintain under normal physiological conditions. They cannot be monitored with standard techniques of blood pressure recording and measurement of blood gases, because these involve an undue level of interference. The problems of physiological monitoring and maintaining a constant plasma level of marker for steady-state experiments can be overcome by using species with much larger embryos (e.g., sheep). They have the advantage that relatively larger volumes of blood and CSF can be obtained and it is also possible to obtain serial blood samples to monitor blood levels of injected markers over time. Results from sheep embryos and fetuses that have been carried out under well-controlled physiological conditions (8–10) provide an important underpinning for studies in rats. However, much larger amounts of marker need to be administered in these animals, the surgery is much more difficult to perform, and there are usually only 1 or 2 embryos available in each ewe. Accordingly, blood–CSF barrier experiments in species with large embryos are much more expensive and time-consuming to perform and facilities to handle large animals are required.

Developing marsupials such as *Monodelphis domestica* are much easier to maintain in a normal physiological state because they are born at a very early stage of brain development. A newborn *Monodelphis* pup is approximately equivalent to an E14 rat embryo and lateral ventricular choroid plexuses formation and differentiation occur after birth (7). Rodents have the advantage of multiple embryos (>10) in a single animal, are readily available in most research centers, and much smaller amounts of markers can be used. The small size of rat and mouse embryos means that steady-state experiments are not possible in a single animal because repeated blood sampling is not feasible. However, this limitation can be overcome by using a litter-based approach in which each embryo is given a standardized injection of the permeability marker and individual animals are sampled at different time points. This technique, which has been described in detail for newborn rats (5), not only provides information on the profile of the marker in blood, but also in CSF over the duration of the experiment. Rat embryos have the advantage that much larger volumes of CSF and blood can be obtained for analysis compared to embryonic mice.

This chapter will concentrate on experimental approaches used in rat embryos to study blood–CSF barrier function that have proved fruitful; essentially similar methods can be used for studies of blood-brain barrier function, but correction for blood contaminating marker of brain samples is required.

2. Materials

2.1. Animal Preparation and Monitoring Equipment

1. Pregnant rats, Sprague-Dawley >10 weeks.
2. Isoflurane (Veterinary Companies of Australia, Artarmon, NSW, Australia).
3. Urethane (ethyl carbamate), 25% w/v, 1 mL/100 g body weight (Sigma-Aldrich Cat # U-2500).
4. Tracheal cannula (Microtube extrusions, North Rocks, Australia).
5. Temperature-controlled heating pad to maintain body temperature: TCAT-2LV (Physitempt Instruments Inc, Clifton, USA).
6. Blood pressure monitor: MacLab (AD Instruments, Ballavista, Australia).
7. Analyzer to monitor blood gases or end tidal CO_2 levels: Capnometer (Hewlett Packard 78354A, Germany).
8. Surgical Instruments: Scalpel handle, scalpel blades, spring-scissors, fine-tipped forceps, insect mounting pins.

2.2. Markers and Their Sampling in Blood and CSF

1. Markers for assessing blood–CSF permeability are shown in Table 1.
2. Glass micropipettes: Borosilicate Glass Capillaries 1.5 mm outer diameter, 0.86 mm inner diameter (Harvard Apparatus LTC, Edenbridge, Kent, UK).
3. Microelectrode puller, P-1000 Sutter Pipette Puller (Sutter Instrument, Novato, CA).
4. PVC Tubing 1.9 × 1.4 mm (Microtube Extrusions).
5. Heparin Sodium Injection BP, 1,000 IU/mL (Hameln Pharmaceuticals GmbH, Hameln, Germany).

Table 1
Markers for assessing blood–CSF permeability

Marker type	Markers	Molecular weight (Da)	Suppliers
Radioactive	^{14}C-sucrose ^{3}H-inulin	342 ~5,000	GE Healthcare, Piscataway, NJ
Fluorescent	Rhodamine-dextran Fluorescein-dextran	4,400–70,000 4,000–20,000	Sigma-Aldrich, St Louis, MO
Proteins	Bovine albumin Human albumin Bovine fetuin	~66,000 ~66,000 ~48,700	Sigma-Aldrich Sigma-Aldrich, CalBiochem, St Louis, MO
Biotin tracers	Biotin-ethylenediamine Biotin-dextran	287 3,000–70,000	Sigma-Aldrich Invitrogen Inc, Eugene, OR

6. Parafilm M (Pechiney Plastic Packaging, Menash, USA).

7. Household plastic film.

8. 1.5 mL Eppendorf tubes to store samples (Eppendorf, Germany) or capped glass capillaries (Harvard Apparatus LTC).

2.3. Quantitative Studies of Blood–CSF Barrier Permeability

2.3.1. Radioactive Markers (See Notes 1 and 2)

1. Liquid scintillation counter.

2. Scintillant: Ultima Gold (PerkinElmer, Boston, MA, USA).

3. Plastic vials, 5–20 mL.

4. Appropriate shielding such as Perspex shield for β-emissions.

2.3.2. Fluorescent Markers

1. Fluorescence microscope with camera: BX50 Olympus with DP70 digital camera.

2.3.3. Estimating Total Protein Concentrations in Plasma and CSF

1. Spectrophotometer.

2. Bradford reagent: Coomassie brilliant blue G250 0.01%, orthophosphoric acid 8.5%, ethanol, 4.7%, deionized water.

3. Phosphate buffered saline: PBS, NaCl 0.15 M, NaH_2PO_4 2 mM, Na_2HPO_4 8 mM, deionized water.

4. Protein standard: 80 mg/mL human Serum Albumin and gamma-globulins (Sigma-Aldrich).

2.3.4. Radial Immunodiffusion

1. Standard human or bovine proteins see Table 1.

2. Monospecific precipitating antibodies:
   (a) Rabbit antihuman albumin (Dako Australia Pty Ltd, Campbellfield, VIC).
   (b) Rabbit antibovine albumin (Dako).
   (c) Rabbit antibovine fetuin (Novus Biologicals, Littleton, CO).

3. Tris/barbitone buffer 0.02 M, pH 8.6 (44.3 g Tris, 22.4 g barbitone, 0.53 g calcium lactate, 1 g sodium azide). Dissolve in water. Dilute 1:4 before use.

4. 1% Agarose (Scientifix, Cheltenham, Victoria, Australia) solution in barbitone buffer.

5. Coomassie stainer: 5 g Coomassie Brilliant Blue R dissolved in 450 mL ethanol 96%, 100 mL glacial acetic acid, and 450 mL double distilled water.

6. 0.9% sodium chloride solution.

7. Destainer: 250 mL 96% ethanol, 100 mL glacial acetic acid, 450 mL distilled water.

8. Equipment:
   (a) Water bath up to 60°C.
   (b) Horizontal plate.
   (c) Gelbond plates 85 × 100 mm.
   (d) Gel puncher and template.
   (e) Moist incubation chamber.
   (f) Magnifying graticule.

2.4. Qualitative Studies of Blood–CSF Barrier Permeability (See Note 3)

1. 5 mL glass vials for tissue processing.

2. Fixative solution: Glutaraldehyde (2.5%) in 0.1 M phosphate buffer adjusted to pH 7.3. Made fresh and cooled to 4°C before use (see Note 4).

3. 0.1 M phosphate buffer adjusted to pH 7.3.

4. Steptavidin/Horseradish Peroxidase complex kit (Dako).

5. 3,3′-Diaminobenzidine (DAB)/Nickel ammonium sulfate solution with and without H_2O_2 (3 mL of each):

   (a) Dissolve 6, DAB tablets (Cat # D4168, Sigma-Aldrich) in 6 mL distilled water.
   (b) Add 240 μL nickel ammonium sulfate solution (1% solution in water).
   (c) Transfer 3 mL of this solution into another beaker and dissolve three urea hydroxide tablets (Cat # D4168, Sigma-Aldrich).
   (d) Filter both solutions through a 0.45 μm syringe filter before use.

6. 0.1 M cacodylate buffer, pH 7.3 (Proscitech, Brisbane, Australia), (see Note 5).

7. Osmium tetroxide, 2% solution (Proscitech), (see Note 6).

8. High-grade 100% acetone (see Note 7) and ethanol.

9. Procure 812 Epoxy resin (Proscitech), (see Note 8).

10. Diamond knife (Micro Star Technologies, Huntsville, TX, USA).

11. Uranyl acetate, 3% solution (Proscitech).

12. Lead Citrate (Proscitech).

13. Ultracut E ultramicrotome (Reichard-Jung, Vienna, Austria).

14. Transmission Electron Microscope (Phillips CM10, Munich, Germany)

3. Methods

3.1. Animal Preparation (See Note 9)

3.1.1. Anesthesia

1. For recovery or short duration experiments, anesthetize the pregnant female rat with 2–3% isoflurane in oxygen delivered via a face mask (see Note 10).

2. For nonrecovery experiments, anesthetize the pregnant female rat with a single intraperitoneal (i.p.) injection of urethane.

3. Assess the depth of anesthesia regularly (see Note 11).

3.1.2. Monitoring the State of the Mother

1. Keep anesthetized pregnant females on a heated pad and monitor body temperature via a rectal probe throughout the experiments.

2. For longer duration experiments (>45 min), insert a tracheal cannula into the mother to maintain a patent airway and to prevent fluid seeping into the lungs.

3. Regularly monitor end tidal CO_2 level by connecting the tracheal cannula to a Capnometer in intubated animals or sampling expired air in nonintubated animals.

4. Cannulate a femoral artery for blood pressure monitoring and sampling of blood for blood gas analysis.

3.1.3. Surgery and Embryonic Injections

1. Shave the lower abdomen of the pregnant female and clean the area with 70% ethanol.

2. Make a midline longitudinal incision through the skin to expose abdominal muscles.

3. Access the abdominal cavity by a longitudinal incision through the abdominal wall and exteriorize one of the uterine horns.

4. Using fine spring-scissors, make a small incision through the uterine wall next to each embryo avoiding major blood vessels and taking care not to damage the placenta or amniotic sac of the embryo.

5. Draw the glass capillaries out to a fine tip using a microelectrode puller and break the tip at a suitable outer diameter for

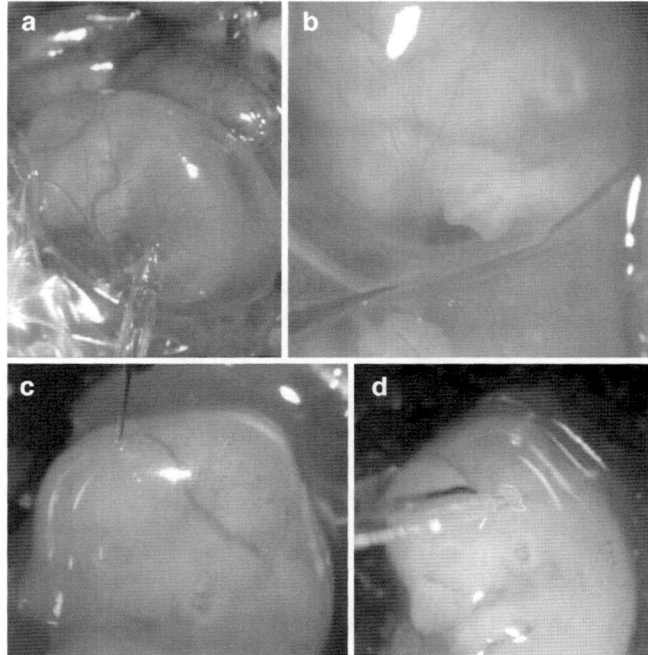

Fig. 1. Injection of markers and collection of blood and CSF in the rat embryo using glass microcapillaries with a fine tip is shown. Injection of markers is made into the intraperitoneal cavity of the embryos (**a**). In young embryos (<E18) blood is collected against the blood flow from one of the blood vessels below the surface of the amniotic sac (**b**). CSF is collected in young embryos (<E18) from either the lateral ventricles (**c**) or the fourth ventricle (**d**).

the injection/sampling site. Fit PVC tubing to the wide end of the glass micropipette.

6. Place an aliquot of the injection solution (see Note 12) onto a piece of parafilm and draw up into the end of a glass micropipette by gentle mouth suction applied to the attached PVC tubing (see Note 13).

7. Insert tip of micropipette through amniotic sac and into peritoneal cavity of embryo (Fig. 1a).

8. Inject contents by applying gentle mouth pressure to attached PVC tubing.

9. Stop injection by releasing pressure from the PVC tubing before the micropipette is completely empty in order to avoid injecting any air into the embryo.

3.2. Maintenance of Embryos

1. The embryos, placentas, and exposed uterine horn should be covered by household plastic wrap immediately after the injections to prevent dehydration and help keep embryonic sacs from deflating.

2. Position injected embryos "on top" of their placentas (propped up by parafilm and swab gauzes) to minimize strain on embryonic blood vessels.

3. A layer of cotton should be placed over the embryos to minimize heat loss.

4. Intraperitoneal injections can also be made directly through the uterine horn without exteriorizing the embryos by using transmitted light from a fiber optic to locate the position of each embryo inside the horn. However, a dye (such as India ink) should be included in the injection solution in order to visualize and confirm the injection site.

5. Intravenous administration of markers can also be made by injecting directly into side branches of blood vessels that are near the surface of the amniotic sac. Arteries in the amniotic sac carry deoxygenated blood towards the placenta and appear darker than veins which carry oxygenated blood back to the embryo. The injections should be made into veins in the direction towards the embryo.

3.3. Sampling Blood and CSF

Sampling blood from very small embryonic rats requires special sampling pipettes (see Subheading 2.2). In older embryos (E18 to birth), blood can be collected by direct cardiac puncture. In smaller embryos, blood can be collected from blood vessels close to the surface of the amniotic sac.

1. Identify a suitable artery, which is typically darker in color than the veins.

2. Insert tip of heparinized glass micropipette pointing along the vessel against the flow of blood and withdraw a small volume by very gentle mouth suction (Fig. 1b, see Note 14). This sampling method can be used for embryos as small as E13 rats (body weight 0.03 g).

3. Transfer samples to microcentrifuge tubes, seal, and centrifuge for 3 min, at $2,000 \times g$.

4. Check hematocrits of all embryonic blood samples and discard if they are abnormally low due to inclusion of amniotic or other fluid in the sample.

5. Separate the plasma into new preweighed sample tubes.

6. Open the amniotic sac using spring-scissors or fine-tipped forceps. Cut the umbilical cord and carefully transfer embryos to a dissection dish containing a layer of paraffin wax.

7. Secure the embryo to the surface using insect mounting pins. Note that the anesthetic urethane crosses the placental barrier and also anesthetizes the embryos.

8. Collect samples of CSF. In E12-E18 embryos collect samples from the fourth and/or lateral ventricles using a glass

microcapillary attached to PVC tubing and gentle suction (Fig. 1c, d). Remove the skull and dura, push the microcapillary gently through thin layer of brain tissue and remove CSF, 0.5–1.2 μL from the lateral and 1.2–2.5 μL from the fourth ventricle.

9. Examine the CSF samples microscopically for traces of blood contamination and cellular material over white background. This method can detect down to 0.1% blood (5). Discard samples that show any sign of blood or tissue contamination.

10. In older embryos and postnatal animals, collect CSF from the cisterna magna at base of skull. Expose the dorsal dura mater between the skull and first vertebra; carefully insert a fine tip glass CSF sampler at an oblique angle to avoid blood vessels on the surface of the brain stem. Draw up a small CSF sample by gentle suction to avoid damaging blood vessels.

3.4. Quantitative Studies of Blood–CSF Barrier Permeability

Quantitative methods can be used to assess the blood–CSF barrier permeability characteristics of labeled compounds. It is generally desirable to conduct qualitative studies in conjunction with a quantitative study.

3.4.1. Radioactive Markers (^3H, ^{14}C)

1. Dissolve the marker(s) in sterile saline (see Notes 12 and 15).

2. Inject the marker as outlined in Subheading 3.1.3.

3. Collect blood and CSF as in Subheading 3.3. Plasma and CSF can be stored at −20°C.

4. Transfer aliquots of the samples into preweighed scintillation tubes and reweigh.

5. Aliquot whole blood into preweighed tubes and bleach with a few drops of H_2O_2.

6. Prepare some blank tubes to measure background activity. Add nonradioactive buffer in similar volumes to the other samples being measured.

7. Add the scintillant cocktail to each tube and mix well.

8. Count each sample for at least 5 min in a liquid scintillation counter with window settings that are appropriate for the isotope(s) being used.

9. Prepare radioactive samples behind a suitable shield that blocks their emissions.

10. For accurate counting, high activity samples (e.g., plasma) should be diluted to a level similar to any lower activity samples (e.g., CSF). For dual label counting, the ^3H activity in the samples should be at least 10 times that of the ^{14}C activity in order to minimize the effect of spillover of ^{14}C activity into the ^3H counting window.

11. Calculate the concentration (dpm/g) for each sample and express the relation between the two samples as concentration ratios (CSF/plasma concentration).

3.4.2. Fluorescent-Labeled Markers in Plasma and CSF

1. Inject a fluorescent marker of choice as outlined in Subheading 3.1.3 (see Note 12).

2. Collect small samples of CSF and plasma into fine-tipped glass samplers as described in Subheading 3.3.

3. Transfer samples to clear glass 10 µL microcapillary tubes.

4. View at 10× magnification using a fluorescence microscope fitted with a digital camera and set to automatic exposure.

5. Record the exposure time for the center of the capillary tube and compare to standard curves of exposure time^{-1} vs. fluorophore concentration. A standard curve is constructed from serial dilutions of a known concentration of the fluorophore covering the expected range of the unknown samples.

6. Ensure that the exposure times of the samples are within the linear range of the standard curve (see Fig. 2). Dilute samples with buffer if necessary.

3.4.3. Proteins in Blood and CSF

Choice of the protein marker depends on the question. Protein probes should be antigenically distinguishable from the rat's endogenous proteins and be inert in its transfer properties, therefore a growth factor, enzyme, or any other protein that is actively transported is not suitable. Most commonly used protein markers are purified plasma proteins from unrelated species (5).

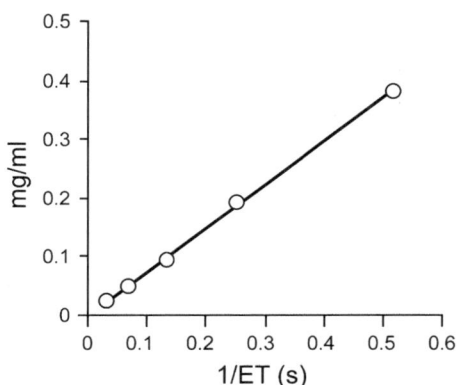

Fig. 2. Standard curve for a rhodamine-dextran using exposure time readings from a camera attached to a fluorescent microscope. Rhodamine-dextran, diluted in saline at different concentrations, is transferred to a glass capillary before readings as outlined in Subheading 3.4.2. Note that there is linear relation between the fluorescent marker concentration and the exposure time^{-1}. It is important that all plasma and CSF samples are within the linear range of the standard curve. Samples can be diluted if necessary.

Total protein levels in CSF and plasma are easily measured using the Bradford method (11), whereas individual proteins can be measured by radial immunodiffusion, crossed-immunoelectrophoresis, or semiquantitative Western blotting.

3.4.3.1. Total Protein Concentration in CSF and Plasma (Bradford Method)

1. Collect plasma and CSF samples as described in Subheading 3.3.
2. Dilute samples to appropriate protein concentration (1–10 µg protein in 100 µL PBS).
3. Add 100 µL of PBS as a reagent blank tube.
4. Add 1 mL of Bradford reagent to each tube. Leave 10 min.
5. Read absorbance in a spectrophotometer at 595 nm. Run all samples in duplicates.
6. Protein concentration is determined from standard curves constructed using protein standards between 1 and 12 µg/100 µL.

3.4.3.2. Radial Immunodiffusion

Radial immunodiffusion is a simple, rapid, quantitative, very reproducible method used to estimate concentrations of individual proteins (12). It requires availability of purified antigen for standard curves and corresponding precipitating monospecific antibody. The main advantage is that it is nonradioactive and nontoxic, thus avoiding problems with waste disposal and exposure. The drawback is that relatively large volumes of antibodies are required. Antibodies should be noncross-reacting with rat proteins and potential cross-reactivity should be removed by preabsorption with rat plasma. As a guide, an excess of ten times antigen in relation to antibody is required. Mix both together and leave for 24 h at 4°C before spinning the sample for 10 min at $5,000 \times g$. The supernatant is removed and checked for residual cross-reactivity by western blotting.

1. Place dissolved agarose in a water bath at 60°C and allow the temperature to equilibrate.
2. Mix 13 mL of agarose with the required amount of monospecific antibodies (amount determined experimentally for each antibody. Generally antibodies are incorporated into agarose at 1–3 µL/cm² of plate area). This gives 1.5 mm agarose gel thickness.
3. Place the Gelbond plate on a horizontal table with a template underneath. The template is designed to locate the position of wells. Wells are created with a puncher (2.5 mm diameter).
4. 5 µL of antigen solution (concentration of antigen is determined experimentally, but is usually in the range of 10–100 µg/mL) placed in each well with a set of standards is

used to construct a standard curve based on a known concentration of standard antigen (usually 3–5 different dilutions).

5. Place the plate in a moist, airtight incubation chamber and incubate horizontally at room temperature until the diffusion is finished, usually 24–48 h.

6. Remove the Gelbond plate from the incubation chamber and quickly soak in 0.9% saline followed by covering the gel with a piece of wet filter paper soaked in saline to avoid air bubbles. A 2–3 cm layer of cellulose tissue is placed on top and a pressure of about 10 g/cm^2 is maintained for 15–20 min. By this method, the gel is squeezed very effectively and, usually, no further washing is necessary. If the background is a problem, a 10 min extra wash in saline followed by a second squeezing of the gel can be performed.

7. Peel off the filter paper, dry the gel in warm air.

8. Stain the plate by immersion for 2–3 min in the Coomassie solution.

9. Wash in destainer until the background is light and precipitation rings are visible. Dry.

10. Diameters of the precipitation rings are measured with an accuracy of 0.1 mm using a magnifying graticule. Determine concentration from a standard curve. Run duplicate samples.

3.4.4. Interpretation of Quantitative Methods for Assessing the Blood–CSF Barrier

The rate of uptake of a marker from blood into CSF in the embryo can be calculated from blood/CSF ratios taken at multiple times after administration in a similar way to studies in the adult. However, the interpretation of quantitative experiments is more complicated in the embryo because other factors such as the surface area for exchange (e.g., choroid plexus size), CSF production rate, and the volume of the ventricles (i.e., the volume the marker distributes in) are different at different stages of development. For instance, the volume of CSF in the ventricular system of the embryonic rats more than doubles between E12 and E16 (13).

3.5. Qualitative Studies of Blood–CSF Barrier Permeability

Qualitative methods can be used to test the integrity of the blood–CSF barrier and can provide information on sites of disruption. Qualitative methods can also be used to investigate the route of entry from blood into the CSF compartment and to identify which cells are involved in selective blood–CSF transfer.

3.5.1. Biotinylated Probes

There are many probes available for barrier studies in animals. Although frequently used, Evans blue and similar dyes are not suitable. Their binding to albumin at concentrations usually used (1–4% w/v) is not as tight as generally supposed (14) and some tissues have a higher affinity for dyes than proteins in plasma (15). Thus, it is uncertain in many experiments whether the dyes are acting as small or large molecular weight probes or a combination of both. Advantages of biotinylated dextrans in vivo are: (i) A range of molecular sizes (3–500 kDa) is available, which makes it easy to compare the influence of molecular size on barrier permeability. (ii) They are well tolerated in embryos as well as in adults, with little binding to plasma proteins, which makes the interpretation of the results easier. (iii) They are also available with fluorophore conjugates (Table 1). A list of some of the more commonly used dextrans for permeability studies along with biotin-ethylenediamine (286 Da), which is a derivative of biotin and a low molecular weight marker of similar molecular size to sucrose, is shown in Table 1. They can all be fixed with standard fixation solutions and detected in tissue in similar ways as described below:

1. The biotin-labeled marker (~0.5–1 mg/g animal) of interest is injected i.p. as described in Subheading 3.1.3 (see Note 12).

2. The time the marker is left in the embryo depends on the purpose of the experiment (minutes to hours). To test for integrity of the blood–CSF barrier, 30–60 min after an i.p. injection is appropriate.

3. Dissect out brains in cold 2.5% glutaraldehyde (see Subheading 2.4). Open ventricles immediately to allow fixative into the ventricles. Carefully remove each choroid plexus from ventricles and keep in separate vials. Leave a small piece of brain tissue attached to the choroid plexus to make it easier to see and handle the tissue during processing. Leave in fixative for at least 2–3 h.

4. Wash tissue 3×, for 10 min each in 0.1 M phosphate buffer.

5. Apply the streptavidin/HRP complex solution according to the manufacturer's specifications. Incubate overnight at 4°C.

6. Wash tissue 3×, for 15 min each in phosphate buffer.

7. To detect biotin, preincubate tissue in filtered DAB/Nickel ammonium sulfate solution (see Subheading 2.4) for 10 min, followed by filtered DAB/Nickel solution containing hydrogen peroxide for 3–7 min as necessary (see Note 16).

8. Wash 3×, 5 min each in 0.1 M phosphate buffer. Transfer tissue to processing vials and use a rotating mixer for the duration of processing.

9. Wash tissue 2×, for 10 min each in 0.1 M cacodylate buffer (see Note 17).

10. Treat tissue with 2% osmium tetroxide for 1 h. Be very careful with the tissue after osmium treatment as the tissue becomes fragile.

11. Wash the tissue 3×, for 10 min each in distilled water.

12. Dehydrate in graded acetone: 70% 2×, for 5 min each, 95% 2×, for 5 min each, and finally in 100% 3×, for 10 min each.

13. Transfer to a 50/50 mixture of resin/acetone for 1 h, 75/25 mixture of resin/acetone for 1 h, and leave overnight in 100% resin.

14. Transfer to fresh resin and leave for 4–5 h and embed tissue in fresh resin.

15. Ethanol can be used for dehydration in steps 12–14, but dehydration times may have to be increased.

16. Control sections are obtained from uninjected animals and tissue processed as above.

3.5.1.1. Evaluation of the Blood–CSF Barrier

1. Semithin sections (0.5 µm) of choroid plexuses are cut and viewed under the light microscope. Brown staining of marker should be visible within blood vessels. Ultrathin sections are cut of selected areas with promising staining. Ultrathin sections are contrasted with uranyl acetate and lead citrate, but it is useful to leave some sections unstained as it is easier to identify the reaction product in these sections.

2. View ultrathin sections under an electron microscope, to determine whether tracer has reached the intercellular cleft between epithelial cells. Most tracers leak out of blood vessels into the perivascular space as vessels in the choroid plexuses have little barrier function (Fig. 3a). At the apical end of the cleft, view tight junctions under high magnification (20,000–60,000×) to assess if tracer has moved through the junction. The tight

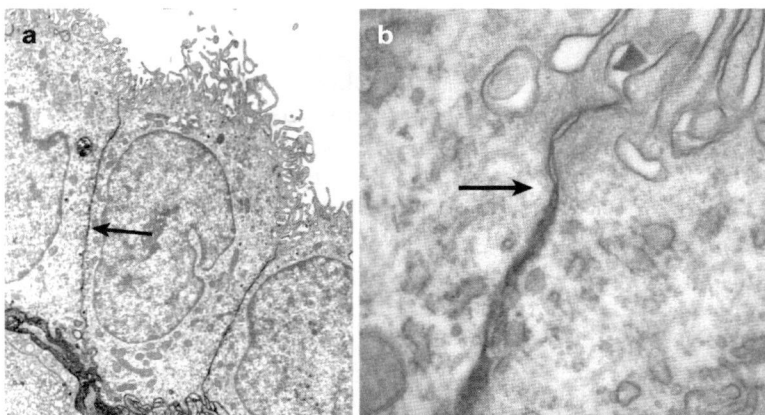

Fig. 3. Electron micrographs showing localization of the 3,000 Da biotin-dextran tracer in the developing choroid plexus. (a) The tracer is visible (dark reaction product) in the extracellular space outside blood vessels in the stroma and between epithelial cells (*arrow*) showing that there is little barrier function in the stromal blood vessels. The movement of the tracer through the tight junction (at the apical side of the epithelial cells) is assessed at high power (b). In this case the movement of the tracer appears to be restricted by the tight junction (*arrow*).

junction is visible as darker areas of the intercellular cleft where cell membranes of neighboring cells are fused at several points (Fig. 3b). Use a goniometric tilting device to get a clearer image of the tight junction. Make special note if tracer is visible in the intercellular cleft between fusion points.

3. Examine choroidal epithelial cells for tracer uptake into vesicles, lysosomes, or other structures. Collect choroid plexus tissue at different times after the marker injection to evaluate the movement of the tracer within these structures.

4. Notes

1. Data sheets supplied by manufacturers give the purity of a labeled compound. In our experience these are occasionally inaccurate as noted in the case of ^{14}C-sucrose that contained 10% ^{14}C-glucose, which gave improbably high distribution spaces because of cellular transport of glucose. Check the purity using paper chromatography or a Sephadex column.

2. For blood-brain barrier experiments using radiolabeled markers, the brain samples should be dissolved in a tissue solubilizer such as Soluene.

3. Use only high-grade chemicals and MilliQ water.

4. Glutaraldehyde generates highly toxic vapors and should be used in a fume hood.

5. Cacodylate buffer contains arsenic and is toxic. Use and store in fume hood. Waste should be disposed of by the Biohazard Department of the Institution.

6. Osmium is a highly toxic compound and should only be used in a fume hood. It should be stored in a glass container placed in another container in the fridge. Place some skim milk powder in the outer container to bind any leaking vapors which will become black in the process. After usage, dispose of in vegetable oil (to bind the osmium) or treat with potassium permanganate (to neutralize the osmium). Waste should be disposed of by the Biohazard Department of the Institution.

7. Alternatively, ethanol can be used for dehydration. Acetone should be of highest quality. 100% acetone or ethanol from bottles opened for a long time should not be used since there may be contamination by water taken up from the atmosphere.

8. Resin vapors are toxic and should only be used in a fume hood. Resin can be kept in a freezer (for at least 6 weeks), but try to avoid thawing and freezing many times as this will accelerate the polymerization (a syringe is an excellent storage

container). Waste can be polymerized at 60°C for 48 h and is harmless after that.

9. Accuracy of embryonic ages is essential for developmental studies. Most animal facilities use mating time and check for vaginal plugs as routine practice. The day on which a plug is identified is taken as E0. It is important to check the size of embryos at the end of an experiment against tables for weight and crown-rump length (16). The latter is generally considered to give a more reliable estimate of embryonic age.

10. Anesthesia can initially be induced in a closed container with isoflurane.

11. Depth of anesthesia is assessed regularly by monitoring the rate and depth of respiration and insensitivity to tail or toe pinches (noxious stimulus). A strong corneal blink reflex should be maintained; loss of this reflex indicates that the animal is too deeply anesthetized. Anesthesia via the mother is generally satisfactory for embryos.

12. High marker concentrations and large injection volumes are likely to affect the physiological state of embryos. The injection site is not as important as the volume of injection, although intracardiac injections in very small embryos have a great risk of completely disrupting the circulation. In an E16 mouse embryo, which weighs about 0.5 g, blood volume would be between 50 μL (assuming 10% of this bodyweight is blood; but there would be additional blood volume in the placental circulation). To inject 20–40 μL into the heart would greatly increase blood volume and pressure and could easily rupture the vessels causing extravasation of the tracer (17). In one set of experiments in chick (18), a very concentrated injection solution was made which, when injected into the blood, would have caused a doubling of the plasma protein concentration. Such an injection would lead to increased colloid osmotic pressure, which in turn would cause increased circulating blood volume and disruption of the tight junctions. Thus, small injections between 1 and 5 μL are recommended, but should not exceed 10% of the estimated circulating blood volume. The concentration of injected protein should be similar or lower than that of the embryo. In general, the total amount of protein injected should not exceed 10% of the total protein in plasma. Intraperitoneal injections are preferable as they are less likely to cause disruption of blood vessels due to volumetric or osmotic overload.

13. The use of the mouth tubing affords much greater control over the injections.

14. Some brands of heparin and some glass capillaries interfere with the protein concentration through breakdown or binding.

In adult samples this would not be noticeable due to higher concentration in plasma and larger volumes (plasma and CSF). It is therefore recommended to test if there is such an effect of glass capillaries and heparin on embryonic samples prior to sample collection in the experiment.

15. Passive permeability markers distribute in both vascular and extravascular fluids and a proportion will be excreted through the kidneys. Therefore, amount of marker(s) injected into the embryo should be sufficient to allow measurement in the final fluid samples. CSF levels can be as low as 1% of the plasma levels for lipid insoluble compounds.

16. Avoid long reaction times as the reaction product can diffuse which will be detected under the electron microscope. Examine the tissue under a stereomicroscope to assess the progress of the staining reaction (brown reaction product) in order to stop at a suitable time.

17. Wash tissue in cacodylate buffer to avoid precipitation of osmium, 3–4 times and for longer if needed.

References

1. Johansson PA, Dziegielewska KM, Liddelow SA, Saunders NR (2008) The blood-CSF barrier explained: when development is not immaturity. Bioessays **30**:237–48

2. Fossan G, Cavanagh ME, Evans CAN, Malinowska DH, Møllgård K, Reynolds ML, Saunders NR (1985) CSF-brain permeability in the immature sheep fetus: a CSF-brain barrier. Dev Brain Res **18**:113–24

3. Møllgård K, Balslev Y, Lauritzen B, Saunders NR (1987) Cell junctions and membrane specializations in the ventricular zone (germinal matrix) of the developing sheep brain: a CSF-brain barrier. J Neurocytol **16**:433–44

4. Ek CJ, Habgood MD, Dziegielewska KM, Potter A, Saunders NR (2001) Permeability and route of entry for lipid-insoluble molecules across brain barriers in developing Monodelphis domestica. J Physiol **536**:841–53

5. Habgood MD, Sedgwick JE, Dziegielewska KM, Saunders NR (1992) A developmentally regulated blood-cerebrospinal fluid transfer mechanism for albumin in immature rats. J Physiol **456**:181–92

6. Davson H, Welch K, Segal MB (1987). Physiology and pathophysiology of the cerebrospinal fluid. Churchill Livingstone, London

7. Liddelow S, Dziegielewska KM, Ek CJ, Johansson PA, Potter A, Saunders NR (2009) Cellular transfer of macromolecules across the developing choroid plexus of Monodelphis domestica. Eur J Neurosci **29**:253–66

8. Dziegielewska KM, Evans CAN, Malinowska DH, Møllgård K, Reynolds JM, Reynolds ML, Saunders NR (1979) Studies of the development of brain barrier systems to lipid insoluble molecules in fetal sheep. J Physiol **292**:207–31

9. Dziegielewska KM, Habgood MD, Møllgård K, Stagaard M, Saunders NR (1991) Species-specific transfer of plasma albumin from blood into different cerebrospinal fluid compartments in the fetal sheep. J Physiol **439**:215–37

10. Stonestreet BS, Patlak CS, Pettigrew KD, Reilly CB, Cserr HF (1996) Ontogeny of blood-brain barrier function in ovine fetuses, lambs, and adults. Am J Physiol **271**:R1594–601

11. Bradford MM (1976) A rapid and sensitive method for the quantitation of microgram quantities of protein utilizing the principle of protein-dye binding. Anal Biochem **72**:248–254

12. Mancini G, Carbonara AO, Heremans JF (1965) Immunochemical quantification of antigens by single radial immunodiffusion. Int J Immunochem **2**:235–54

13. Johansson PA, Dziegielewska KM, Ek CJ, Habgood MD, Liddelow SA, Potter AM, Stolp H.B, Saunders NR (2006) Blood-CSF barrier function in the rat embryo. Eur J Neurosci **24**:65–76

14. Rawson RA (1943) The binding of T-1824 and structurally related diazo dyes by the plasma proteins. Am J Physiol 138:708–17

15. Dallal MM, Chang S-W (1994) Evans blue dye in the assessment of permeability-surface area product in perfused rat lungs. J Appl Physiol **77**:1030–5

16. Butler H, Juurlink BHJ (1987) An atlas for staging mammalian and chick embryos. CRC, Boca Raton

17. Saunders NR (1992) Ontogenic development of brain barrier mechanism. In: Bradbury MWB (ed) Handbook of experimental pharmacology: physiology and pharmacology of the blood-brain barrier. Springer, Berlin, pp 327–369

18. Wakai S, Hirokawa N (1981) Development of blood-cerebrospinal fluid barrier to horseradish peroxidase in the avian choroidal epithelium. Cell Tissue Res **214**:271–8

Chapter 12

Detection of Blood–Nerve Barrier Permeability by Magnetic Resonance Imaging

Carsten Wessig

Abstract

The blood–nerve barrier (BNB) separates the endoneurium from the endovascular space and the epineurial connective tissue. An intact BNB is very important for integrity and functions of the nerve fibers within the endoneurial space. Disruption of the BNB which leads to functional and structural impairment of the peripheral nerve plays an important role in many disorders of the peripheral nerve like Wallerian degeneration, inflammatory nerve disorders, and demyelination. So far, this increased BNB permeability can only be assessed ex vivo. Assessing BNB disruption in vivo would be of great value for studying disorders of the peripheral nervous system. Gadofluorine M (Gf), a new amphiphilic contrast agent for MRI, accumulates in rat nerves with increased permeability of the BNB. After application of Gf, T1-weighted MR images show contrast enhancement of nerves with a disrupted BNB. This new tool of assessing BNB permeability in vivo is described.

Key words: Blood–nerve barrier, Contrast agent, Gadofluorine M, MRI, Permeability, PNS

1. Introduction

The blood–nerve barrier (BNB) separates the endoneurial space from the endovascular space and the epineurial connective tissue to create an immunologically and biochemically privileged territory. The BNB consists of endoneurial capillaries and the perineurial sheath, a layer of specialized cells. Both endothelial cells of the endoneurial capillaries and perineurial cells are joined by intercellular tight junctions restricting fluid movements and leukocyte transmigration into the endoneurium, the parenchyma of the peripheral nerve where the axons are located (1). Special enzymes, transporters, and receptors in the cells of the BNB enable exchange of substrates necessary for axonal integrity and metabolism. Alterations of the BNB are found in several disorders of the peripheral nerve.

Sukriti Nag (ed.), *The Blood-Brain and Other Neural Barriers: Reviews and Protocols*, Methods in Molecular Biology, vol. 686, DOI 10.1007/978-1-60761-938-3_12, © Springer Science+Business Media, LLC 2011

Increase of BNB permeability plays an important role in the pathogenesis of Guillain-Barré syndrome, chronic inflammatory demyelinating polyneuropathy, animal models of inflammatory neuropathies like experimental allergic neuritis, Wallerian degeneration, and traumatic lesions (2–5). In experimental settings, alterations of the BNB could be demonstrated in excised nerves after systemic application of tracers like Evans Blue (6). However, more research of BNB alterations is necessary for the understanding of peripheral nerve diseases. Detecting alterations of the BNB in vivo by new techniques would be of great value both for clinical application and experimental studies of nerve disorders.

On magnetic resonance imaging (MRI), normal nerves are isointense to the surrounding muscle on T1- and T2-weighted images. Lesions of the peripheral nervous system (PNS) can lead to an increase of the T2 signal and may be detectable on T2-weighted images. T2 signal changes within the nerve were attributed to breakdown of myelin and axons and endoneurial edema (7–9). However, in peripheral nerve disorders usually associated with BNB alterations, contrast enhancement is not detected consistently with the routinely used contrast agent, gadolinium-DTPA (10). Therefore, reliable assessment of BNB integrity cannot be made by MRI following administration of gadolinium-DTPA. The novel amphiphilic contrast agent Gadofluorine M (Gf), which was originally developed for lymphography, accumulates in nerves undergoing Wallerian degeneration, in inflammatory nerve disorders, and in areas of demyelination of the peripheral nerve. Gf is bound to albumin in the blood and enters the nerve via a leaky BNB (11–13). The affinity of Gf for molecules of the extracellular matrix like tenascin is similar to its affinity for albumin (14), therefore Gf accumulates in nerve segments with a disturbed BNB. If Gf is applied 24 h prior to the MRI, nerves with a disrupted BNB demonstrate contrast enhancement on T1-weighted MRI (Fig. 1a). Combining T1-weighted images with fat-saturation enables visualization of slight contrast enhancement. Using fluorescent-labeled Gf, Gf accumulation in the damaged peripheral nerve can be demonstrated by fluorescence microscopy (Fig. 1b, c). Restoration of the BNB is associated with lack of Gf enhancement.

Thus, Gf-enhanced MRI can detect alterations of BNB permeability. The method to detect BNB breakdown by Gf-enhanced MRI in vivo is described.

2. Materials

1. 1.5 Tesla MRI scanner: Magnetom Vision (Siemens, Erlangen, Germany).

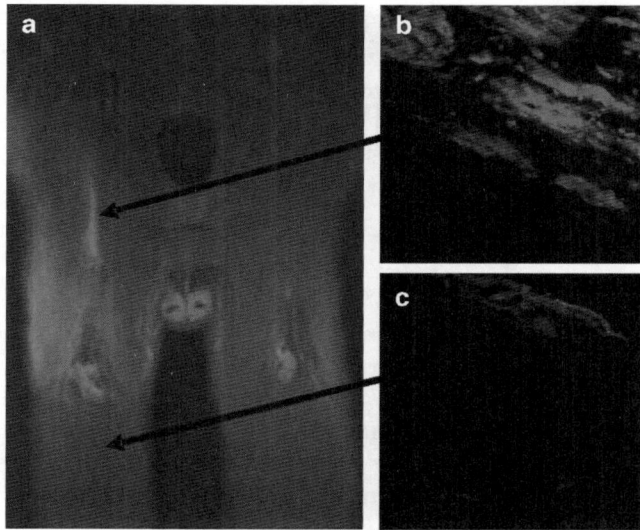

Fig. 1. Magnetic resonance imaging (MRI) shows focal Gf enhancement in a sciatic nerve segment with disruption of the BNB in a focal area of demyelination induced by injection of lysolecithin. (a) Coronal MRI through the thighs shows segmental Gf contrast enhancement within the sciatic nerve (*upper arrow*), while the distal nerve segments (*lower arrow*) and the contralateral normal sciatic nerve are spared. Gf contrast enhancement correlates with the presence of prelabeled Gf detected by fluorescence in frozen sections of nerve taken from this area (b). In keeping with the focal character of Gf enhancement on MRI, no fluorescence is seen more distally in the nonenhancing tibial nerve except for weak nonspecific fluorescence in the perineurium (c).

2. Small loop flex coil (Siemens, Erlangen, Germany). Clinically, these are used for imaging of finger joints.

3. Gadofluorine M (Gf): Amphiphilic Gadolinium complex with a molecular weight of about 1,530 g/mol (patent application no. DE 10040381) and a concentration of 250 mmol Gd/L (Bayer-Schering Pharma AG, Berlin, Germany). For histological validation of MR results, carbocyanine-prelabeled Gf (Bayer-Schering) can be used.

4. Anesthetics: Ketamine (100 mg/kg body weight; Ketanest®, Pfizer Pharma, Karlsruhe, Germany), xylazine (10 mg/kg body weight; Rompun®, Bayer HealthCare AG, Leverkusen, Germany). Inhalation anesthetics can also be used.

5. Rats (male Sprague-Dawley rats, 160–200 g; Charles River, Sulzfeld, Germany).

3. Methods

1. Rats are injected with Gf, 0.1 mmol/kg body weight, either through the tail vein or the retro-orbital plexus, 24 h prior to the MRI studies.

2. Prior to the MRI studies, rats are anesthetized with an intraperitoneal injection of ketamine and xylazine.

3. MR measurements are performed on the 1.5 Tesla MRI unit. For MRI of the lumbar spine and both legs, use the round surface coil (small loop flex coil, diameter 4 cm) (see Notes 1 and 2).

4. For imaging of the sciatic nerve, place the animals in a prone position with both legs positioned in the coil opening. For imaging the lumbar spine, place rats in a supine position with their lumbar spine centered on the coil opening.

5. The MR protocol includes a scout sequence followed by a fat-suppressed, T1-weighted sequence (TR 460 ms, TE 14 ms) in the axial plane with a slice thickness of 3 mm. Additionally, a coronal T1-weighted sequence (TR 460 ms, TE 14 ms, slice thickness 2 mm) can be applied for imaging of the legs. For best assessment, the slices should be angulated perpendicular to the axis of the nerves (see Notes 3 and 4).

6. For quantitative assessment, signal intensity can be measured in the region of interest in the nerve and compared to control nerves.

7. For histological validation of MR data, fluorescence microscopy can be performed. Animals are sacrificed after MR acquisition. After excision of sciatic nerves, nerve segments are snap-frozen in isopentane. Ten microns-thick sections are cut on a cryostat and analyzed for the presence of Gf (red fluorescence) using a fluorescence microscope.

4. Notes

1. The animal should be placed in the isocenter of the scanner with the region of interest as near as possible to the coil.

2. Symmetrical positioning of the extremities is necessary for optimal comparison between both sides.

3. T1-weighted images of animals before application of Gf are useful to study the anatomy of the region. These sequences allow a good differentiation of fat, bone, and muscle tissue. The normal nerve is isointense to muscle tissue on these T1-weighted images. Large nerves like the sciatic nerve may be differentiated from the soft tissue by a surrounding of fat tissue which is bright on T1-weighted sequences.

4. Combination with fat-saturation enables visualization of slight contrast enhancement of the nerve.

Acknowledgments

Gf was kindly provided by B. Misselwitz, Bayer-Schering Pharma AG, Berlin, Germany. Studies were supported by the Interdisziplinäre Zentrum für Klinische Forschung, University of Würzburg and the Gemeinnützige Hertie-Stiftung, Frankfurt am Main, Germany. Major contributors to all studies are Guido Stoll, Würzburg, and Martin Bendszus, Heidelberg. Excellent technical assistance was provided by Melanie Glaser, Gabi Köllner, and Virgil Michels.

References

1. Choi YK, K-W Kim (2008) Blood-neural barrier: its diversity and coordinated cell-to-cell communication. BMB Rep 41:345–352

2. Seitz RJ, Reiners K, Himmelmann F, Heininger K, Hartung H-P, Toyka KV (1989) The blood-nerve barrier in Wallerian degeneration: a sequential long-term study. Muscle Nerve 12:627–635

3. Harvey GK, Gold R, Hartung H-P, Toyka KV (1995) Non-neural specific T lymphocytes can orchestrate inflammatory peripheral neuropathy. Brain 118:1263–1272

4. Hahn AF (1996) Experimental allergic neuritis as a model for the immune-mediated demyelinating neuropathies. Rev Neurol (Paris) 152:328–332

5. Stoll G, Jander S, Myers RR (2002) Degeneration and regeneration in the peripheral nervous system: from Augustus Waller's observations to neuroinflammation. J Peripher Nerv Syst 7:13–27

6. Kristensson K, Olsson Y (1971) The perineurium as a diffusion barrier to protein tracers. Acta Neuropathol 17:127–138

7. Bendszus M, Wessig C, Solymosi L, Reiners K, Koltzenburg M (2004) MRI of peripheral nerve degeneration and regeneration: correlation with electrophysiology and histology. Exp Neurol 188:171–177

8. Peled S, Cory DG, Raymond SA, Kirschner DA, Jolesz FA (1999) Water diffusion, T(2), and compartmentation in frog sciatic nerve. Magn Reson Med 42:911–918

9. Cudlip SA, Howe FA, Griffiths JR, Bell BA (2002) Magnetic resonance neurography of peripheral nerve following experimental crush injury, and correlation with functional deficit. J Neurosurg 96:755–775

10. Lacour-Petit MC, Lozeron P, Ducreux D (2003) MRI of peripheral nerve lesions of the lower limbs. Neuroradiology 45:166–170

11. Bendszus M, Wessig C, Schütz A, Horn T, Kleinschnitz C, Sommer C, Misselwitz B, Stoll G (2005) Assessment of nerve degeneration by gadofluorine M-enhanced magnetic resonance imaging. Ann Neurol 57: 388–395

12. Wessig C, Bendszus M, Stoll G (2007) In vivo visualization of focal demyelination in peripheral nerves by gadofluorine M-enhanced magentic resonance imaging. Exp Neurol 204:14–19

13. Wessig C, Jestaedt L, Sereda MW, Bendszus M, Stoll G (2008) Gadofluorine M-enhanced magnetic resonance nerve imaging: comparison between acute inflammatory and chronic degenerative demyelination in rats. Exp Neurol 210:137–143

14. Meding J, Ulrich M, Licha K, Reinhardt M, Misselwitz B, Fayad ZA, Weinmann HJ (2007) Magnetic resonance imaging of atherosclerosis by targeting extracellular matrix deposition with Gadofluorine M. Contrast Media Mol Imaging 2:120–129

Part III

Molecular Techniques to Study the Blood-Brain Barrier

Chapter 13

Isolation of Human Brain Endothelial Cells and Characterization of Lipid Raft-Associated Proteins by Mass Spectroscopy

Romain Cayrol, Arsalan S. Haqqani, Igal Ifergan, Aurore Dodelet-Devillers, and Alexandre Prat

Abstract

The blood-brain barrier (BBB) limits the movements of molecules, nutrients, and cells from the systemic blood circulation into the central nervous system (CNS), and vice versa, thus allowing an optimal microenvironment for CNS development and function. The brain endothelial cells (BECs) form the primary barrier between the blood and the CNS. In addition, pericytes, neurons, and astrocytes that make up the neurovascular unit support the BEC functions and are essential to maintain this restrictive permeability phenotype. To better understand the molecular mechanisms underlying BBB properties, we propose a method to study the proteome of detergent resistant microdomain, namely lipid rafts, from human primary cultures of BECs. This chapter describes a robust human BECs isolation protocol, standard tissue culture protocols, ECs purity assessment protocols, lipid raft microdomain isolation method, and a mass spectrometry analysis technique to characterize the protein content of membrane microdomains.

Key words: Lipid raft, Membrane microdomains, Human brain endothelial cell, Blood-brain barrier, Proteomic, Liquid chromatography, Mass spectrometry, Detergent resistant membrane, Human primary culture

1. Introduction

The central nervous system (CNS) microenvironment is regulated by the blood-brain barrier (BBB) and the other components of the neurovascular unit (NVU) which include cell types such as, pericytes, microglia, astrocytes, and neurons (1, 2). The BBB is a property mainly of nonfenestrated brain endothelial cells (BECs) that protects the brain and regulates the movement of

Sukriti Nag (ed.), *The Blood-Brain and Other Neural Barriers: Reviews and Protocols*, Methods in Molecular Biology, vol. 686, DOI 10.1007/978-1-60761-938-3_13, © Springer Science+Business Media, LLC 2011

ions, nutrients, molecules, and cells from the blood into the CNS and vice versa. BECs form the primary barrier between the CNS and the systemic blood circulation and many other cell types such as, pericytes, microglia, astrocytes, and neurons interact with the BECs to create a tightly regulated homeostatic environment for the CNS (3). Complex networks of soluble factors and physical contacts mediate the interactions between the different cell types to regulate the entry of blood-borne molecules and immune cells into the CNS. BECs express multiprotein tight and adherens intercellular junctions to limit paracellular permeability, exhibit low levels of pinocytosis and express large amounts of metabolic enzymes and transporters to regulate CNS homeostasis (4, 5). The restrictive nature of the CNS vasculature has been known for the past 100 years and significant research is still under way to identify BBB attributes, their origin, and regulation. During pathological processes, the BBB can respond in a number of ways to limit CNS blood leakage while allowing an optimal microenvironment for repair and healing (6). The inherent complexity of the NVU has limited our understanding of this system. In the 1970s, the advent of primary in vitro tissue culture techniques for CNS cells allowed researchers to isolate different components of the mammalian NVU and study the individual cell types of the NVU under controlled conditions. Presently many laboratories use in vitro models of the BBB (7).

Due to the restrictive nature of the BBB, the movement of nutrients, molecules, and cells between the blood and the CNS requires interactions with plasma membrane proteins to signal and initiate transport across the BBB (1, 8). Detergent resistant microdomains (DRMs) enriched in cholesterol, namely lipid rafts, are small (10–100 nm) dynamic membrane microdomains that are thought to be crucial for numerous plasma membrane functions. Because of their specific lipid composition these microdomains form microplatforms, or rafts, of "ordered" plasma membrane in the loosely packed fluidic phase of the plasma membranes. Rafts float in the plasma membrane, act as important functional components of the BBB and can participate in many biological processes such as: plasma membrane signal transduction, membrane protein compartmentalization, cell polarization, cell adhesion and migration and cytoskeleton attachment (9–11). Even though the relevance of detergent resistant membranes in biological systems has been questioned, the isolation and the study of membrane microdomains have greatly increased our understanding of membrane biology and remain an important tool to study the plasma membrane (12). Current data generated on lipid rafts strongly suggest that these microdomains are biologically important and form a functional unit in the plasma membranes of cells.

Historically, lipid rafts and DRM have been defined as membrane fractions enriched in cholesterol and phospholipids isolated from detergent treated cells after sucrose gradient ultracentrifugation (10, 13). Differential solubility of the microdomains in detergent is a function of the specific biophysical properties of DRM compared to the fluidic membrane. DRM isolation has been done in many cell types and using different detergents and DRM isolation protocols have even been refined to isolate lipid rafts without the use of detergents (14, 15).

Global profiling techniques such as genomics and proteomics are powerful tools to study comprehensive gene and protein expression analysis in defined samples (16–18). Instead of looking at one, or a few genes or proteins these techniques allow the study of hundreds to thousands of molecules simultaneously. Global profiling protein expression techniques using mass spectrometry allow for the analytical identification of the chemical composition of a sample based on mass to charge ratios of charged particles. Mass spectrometry is a highly sensitive method and can be used for qualitative and quantitative purposes (protein identification, isotopic composition identification, amount quantification, etc.). The association of high sensitivity large-scale proteomic techniques (mass spectrometry) with BEC cultured in vitro allows for the global profiling of BBB protein expression (proteome) under various controlled conditions (10). An important improvement of this proteomic technique is the use of chromatography to separate the sample peptides before they are introduced in the mass spectrometer allowing for a better mass-charge resolution. The identification and the characterization of BBB proteins previously undefined will not only enhance the understanding of the molecular mechanism underlying BBB functions but might also identify potential therapeutic targets to modulate BBB phenotype. In recent years, quantitative global profiling techniques have been optimized and used to quantify protein expression differences between samples (quantitative genomics and proteomic techniques, isotope-coated affinity tag spectrometry-ICAT) (17–19). In order to identify "novel" protein effectors of the BBB, our laboratory has used tissue culture, DRM isolation and tandem liquid chromatography, and mass spectrometry proteomic techniques to probe the proteome of membrane microdomain functional units and identify novel effectors of the BBB (Fig. 1). Using this method, we have been able to identify lipid raft associated proteins previously undefined at the BBB (see Table 1) and we have described novel molecular mechanisms underlying BBB functions (16).

This chapter describes the techniques required to isolate and propagate human brain ECs. We propose robust protocols to study BECs properties and functions (immunofluorescence and permeability to small and large tracers). Furthermore this chapter contains a protocol for DRM isolation from human BECs using a discontinuous sucrose gradient and protocols to prepare and analyze DRM samples by mass spectrometry.

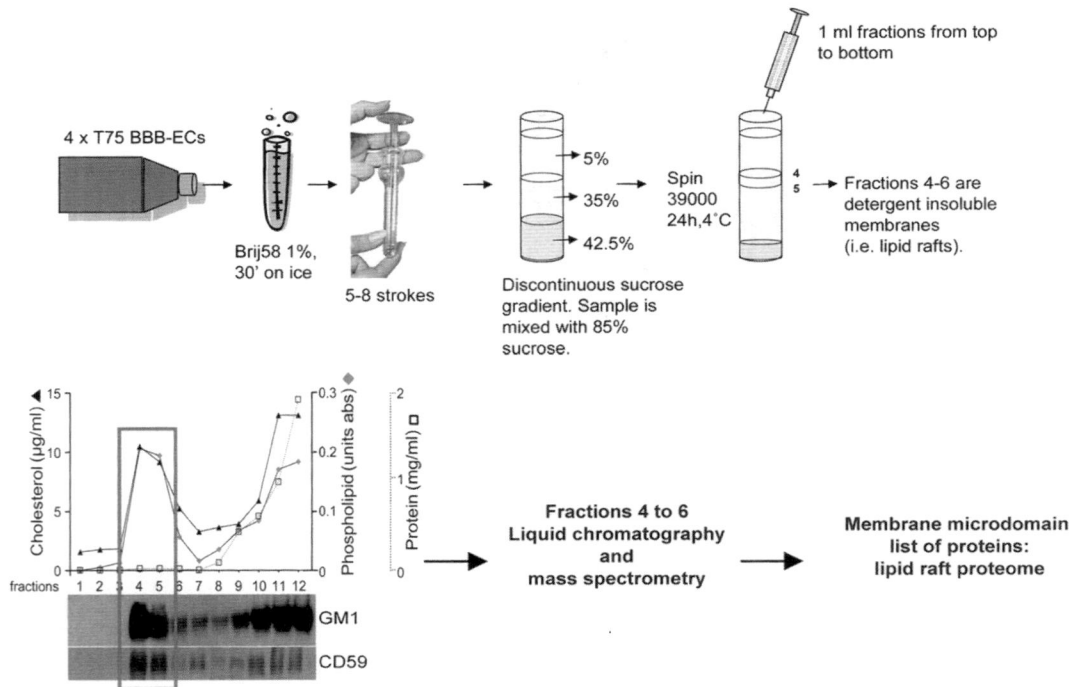

Fig. 1. Detergent resistant membrane isolation and mass spectrometry flow chart. Lipid raft or detergent resistant micro-domain isolation steps are illustrated in the *upper panel*. The *lower panel* describes the major steps that allow for proteome analysis of the lipid rafts.

2. Materials

2.1. Isolation of Primary Human BECs

1. Sterile phosphate-buffered saline (PBS): 2.7 mM KCl, 1.5 mM KH_2PO_4, 8.1 mM Na_2HPO_4, pH 7.4 in Milli-Q® water (see Note 1).

2. Sterilized Nalgene Filtration Apparatus (VWR, Mississauga, ON, cat # 28199-440).

3. Sterile 350 and 112 μm nylon or polypropene Nitex filters (BSH Thompson, Montreal, QC).

4. Sterile forceps.

5. Sterile scalpel (Paragon medical, Coral Springs, FL).

6. Six-well tissue culture plate, T25, T75, and T150 tissue culture flasks (BD Falcon, San Jose, CA).

7. Dounce tissue homogenizer, 30 mL with a tight fitting piston (Wheaton Science Products, Millville, NJ).

8. Medium 199 (Invitrogen Corp., Burlington, ON) filter sterilized with 0.22 μm filters.

9. 0.5% gelatin (BD Difco, San Jose, CA) diluted in water, and sterilized with 0.22 μm filters.

Table 1
Representative groups of lipid raft-associated proteins derived from BECs are shown

Category	Name
Signal transduction and cytoskeleton (42%)	Kv channel interacting protein 4
	Calcium channel, voltage-dependent, alpha 2/delta subunit 1
	Calcium/calmodulin-dependent serine protein kinase (MAGUK family)
	Guanine nucleotide binding protein (G protein), beta polypeptide 4
	Guanine nucleotide binding protein (G protein), alpha 13
	Guanine nucleotide binding protein (G protein), alpha inhibiting activity polypeptide 2
	Bradykinin receptor B2
	RAS-like, family 11, member A
	Guanine nucleotide binding protein (G protein), gamma 12
	RAS p21 protein activator 4
	FERM, RhoGEF (ARHGEF) and pleckstrin domain protein 1
	Membrane-spanning 4-domains, subfamily A, member 8B
	Ectonucleotide pyrophosphatase/phosphodiesterase 5
	AHNAK nucleoprotein (desmoyokin)
	Mitogen-activated protein kinase kinase 1 interacting protein 1
	Related RAS viral (r-ras)
	Protein kinase, DNA-activated
	Alpha 1 actin precursor
	Vimentin
Cell–cell adhesion (22%)	Selectin E
	Vascular cell adhesion molecule-1
	Melanoma cell adhesion molecule
	Activated leukocyte cell adhesion molecule
	Contactin 4
	Integrin, beta 1
	Integrin, beta 4
	Integrin, beta 1
	Integrin, alpha 1
	CD44
	Gap junction protein beta 6, connexin 30
	Cadherin 1, type 1and type 2
	Catenin, beta 1

(continued)

**Table 1
(continued)**

Category	Name
Protein, lipid, and carbohydrate metabolism (18%)	Gamma-glutamyltransferase-like activity 1
	UDP glucuronosyltransferase 2 family, polypeptide B11
	CD59 antigen p18-20
	Solute carrier family 3
	Mannosyl (alpha-1,6-)-glycoprotein beta-1,2-*N*-acetylglucosaminyltransferase
	Amyloid beta precursor protein binding protein 1
	5′-Nucleotidase, ecto
Intracellular protein trafficking (7%)	Caveolin 1
	Synaptophysin-like 1
	Vesicle transport through interaction with t-SNAREs homolog 1B
	Clathrin, light polypeptide
Unknowns (11%)	Hypothetical protein LOC150159
	Hypothetical protein FLJ25082
	THAP domain containing 9
	Echinoderm microtubule associated protein like 2
	Zinc finger 261
	Leucine-rich repeat-containing G protein-coupled receptor 6

10. Sterile heat inactivated human serum (Sigma-Aldrich, Oakville, ON).

11. Sterile heat inactivated fetal calf serum (BD Gibco).

12. Tissue culture grade Collagenase type IV (Sigma-Aldrich, 1 mg/mL in sterile M199).

13. Endothelial cell growth supplements (ECGS, 1.5 mg/mL in water, 0.22 μm filter sterilized, BD Bioscience).

14. Insulin transferring selenium (ITS, 1,000× concentrate, Sigma-Aldrich, 0.22 μm filter sterilized, Collaborative Medical Products, Bedford, MA).

15. Mouse melanoma conditioned medium (medium collected on day 8 from a confluent monolayer of clone M3 cells grown in DMEM supplemented with 10% fetal calf serum). The medium is 0.22 μm filter sterilized and kept at –20°C.

16. Endothelial cell medium (ECM) #1: 60% M199, 10% human serum, 10% fetal calf serum, 20% M3 conditioned medium, ECGS (6 µg/mL), ITS (1× concentrate).

17. ECM # 2: 65% M199, 5% human serum, 10% fetal calf serum, 20% M3 conditioned medium, ECGS (3 µg/mL), ITS (1× concentrate) (see Note 1).

18. ECM # 3: 65% M199, 5% human serum, 10% fetal calf serum, 20% M3 conditioned medium, ITS (1× concentrate).

2.2. Maintenance of Human BECs

1. Tissue culture grade trypsin (BD Gibco, 0.5% diluted in PBS).

2. Gelatin 0.5%.

3. Astrocyte conditioned medium (ACM). Human fetal astrocytes are isolated and grown in culture as previously described with 90% DMEM, 10% fetal calf serum (13, 20). The cultures are derived from human embryos obtained from pregnancy terminations done at 17–23 weeks for medical reasons. The latter is in accordance with ethical guidelines of the Canadian Institute of Health Research and the Royal College of Physicians and Surgeons of Canada. After 7 days in culture the supernatant from the fetal astrocyte monolayer is collected, filtered through a 0.22 µm filter and kept frozen at –20°C, until use.

2.3. Immunofluorescence Analysis of Purity of BEC Cultures

1. 4% paraformaldehyde (Sigma-Aldrich) diluted in PBS, pH 7.2.

2. Tissue culture plastic coverslips or 8-well chamber slides (Nalgene, Rochester, NY).

3. Anti-Von Willebrand factor antibody (1/100 dilution, Dako Cytomation, Glostrup, Denmark).

4. Anti-glucose transporter-1 (Glut-1) antibody (1/200 dilution, Sigma-Aldrich).

5. Ulex europaeus-1 lectin conjugated to fluorescein (1/200, Sigma-Aldrich).

6. Anti-endothelial HT-7 antibody (1/400 dilution, Sigma-Aldrich).

7. Tumor necrosis factor-α, 100 U/mL, 50 ng/mL (Biosource-Invitrogen Incorp,).

8. Antiintercellular adhesion molecule (ICAM)-1 antibody (1/100 dilution, R&D Systems, Minneapolis, MN).

9. Anti-vascular cell adhesion molecule (VCAM)-1 antibody (1/200 dilution, R&D Systems).

10. Anti-zonula occludens (ZO-1) antibody (1/150 dilution, Zymed, San Francisco, CA).

11. Anti-occludin antibody (1/75 dilution, Zymed).

12. Anti-beta-tubulin antibody (1/400 dilution, Sigma-Aldrich).

13. Anti-glial fibrillary acidic protein (GFAP) antibody (1/500 dilution, Dako).

14. Anti-alpha myosin antibody (1/500 dilution, Sigma-Aldrich).

15. Anti-CD68 antibody (1/100 dilution, Dako).

16. Nuclear stain Hoescht or Topro-3 (1/350 dilution, Invitrogen).

17. HHG blocking solution: 1 mM HEPES, 88% Hanks buffered saline solution (Sigma-Aldrich), 2% horse serum (Sigma-Aldrich), 10% goat serum (Sigma-Aldrich), and 0.01% sodium azide. PBS supplemented with 10% of the serum from the same species as the secondary antibody can also be used.

18. Triton X-100.

19. Mounting medium, gelvatol: 20% glycerol, 10% polyvinyl alcohol, 0.1 M Tris–HCl pH 8.0, incubate while stirring overnight, centrifuge $2,000 \times g$ for 20 min (some precipitate may appear, use the supernatant of the gelvatol solution).

2.4. Permeability of Human BECs to Small and Large Tracers

1. Boyden chambers and 24-well tissue culture plates (BD Falcon).

2. Bovine serum albumin coupled to fluorescein (Molecular Probes, 10 μg/mL diluted following the manufacturer's instructions).

3. ^{14}C labeled sucrose (MP Biomedicals, Aurora, OH).

2.5. Lipid Raft Isolation and Purification

1. Brij 58 detergent (1% in separating buffer, Sigma-Aldrich).

2. Cell scrapers (Sarstedt, Saint-Leonard, QC).

3. Protease inhibitors cocktail (BD Bioscience).

4. Separation buffer: 150 mM NaCl, 25 mM Tris–HCl, pH 7.4.

5. Sucrose (Sigma-Aldrich) solutions: 85, 35, and 5% diluted in separation buffer.

6. Dounce homogenizer (7 mL with a loose fitting piston, Kontes Glass company, Vineland, NJ)

7. 12 mL ultracentrifuge tubes (Beckman, Mississauga, ON).

8. Swinging bucket rotor SW41 (Beckman).

9. Ultracentrifuge (Beckman optima ultracentrifuge).

10. Cholesterol assay kit, Amplex red cholesterol assay kit, Molecular Probes.

11. Phospholipid colorimetric method kit (Wako, Richmond, VA).

12. BCA protein assay kit, Pierce.

13. Anti-CD59 antibody (R&D Systems).

14. Anti-GM1 Cholera toxin B (Molecular Probes).

2.6. Liquid Chromatography and Mass Spectrometry

1. Denaturing SDS buffer: 50 mM Tris–HCl, pH 8.5, 0.1% SDS.

2. Dithiothrietol (DTT). Freshly prepare ~250 mM in Milli-Q® water.

3. Iodoacetamide (IAA). Freshly prepare ~500 mM in Milli-Q® water and protect from light.

4. Trypsin Gold, Mass Spectrometry Grade (Promega, Nepean, ON, cat. # V5280).

5. Cartridge holder (Applied Biosystems, Foster City, CA, cat. # 4326688).

6. Cation Exchange (CE) Cartridge: POROS® 50 HS, 50-μm particle size (4.0×15 mm) (Applied Biosystems, cat. # 4326695).

7. CE Load buffer: 10 mM KH2PO4, pH 3.0, 25% acetonitrile.

8. CE elute buffer: CE load buffer + 350 mM KCl.

9. CE clean buffer: CE load buffer + 1 M KCl.

10. CE storage buffer: CE load buffer + 0.1% NaN_3.

11. MS buffer: 5% acetonitrile, 1% acetic acid.

12. SpeedVac® Concentrator SPD111V (Thermo Scientific, Waltham, MA).

13. A mass spectrometer (MS) equipped with an electrospray ionization source (ESI) and capable of tandem MS (MS/MS) analysis. For example, a hybrid quadrupole time-of-flight Q-TOF™ Ultima (Waters, Millford, MA).

14. An online nanoflow liquid chromatography (nanoLC) system. For example, CapLC HPLC pump (Waters).

15. A reverse phase nanoLC column. For example, 75 μm × 150 mm PepMap C18 nanocolumn column (Dionex/LC-Packings, San Francisco, CA).

16. Software for identifying peptide sequences from the nanoLC-MS/MS data using database searching. For example, Mascot® version 2.1.0 (Matrix Science Ltd., London), PEAKS (Bioinformatics Solutions Inc., Waterloo, ON), or PeptideProphet™ (part of the Trans-proteomic pipeline http://en.wikipedia.org/wiki/Trans-Proteomic_Pipeline).

17. Acetonitrile and formic acid (Sigma-Aldrich).

3. Methods

3.1. Isolation of Primary Human BECs (See Note 1)

1. Coat 6-well tissue culture plates with 0.5% gelatin, 2 mL per well and let the gelatin harden on plates for a minimum of 1 h.

2. Wash the human CNS specimens thoroughly using 1–2 L of PBS to remove blood clots. Human temporal lobe resections are obtained from patients undergoing surgery for the treatment of intractable temporal lobe epilepsy at CHUM-Notre-Dame Hospital. Informed consent and ethical approval was given before surgery (ethics protocol HD04.046). BECs were isolated from the nonepileptogenic resection material.

3. Remove meninges and the large pial blood vessels with sterile forceps.

4. Cut tissue into small cubes of 0.1–0.5 mm³ with sterile scalpels.

5. Aspirate pieces with a sterile pipette and transfer pieces to a 50 mL sterile conical tube.

6. Fill tube with PBS, shake, and allow pieces to sediment. Wait 30 min for the pieces to settle and discard the supernatant. At least 45 mL of PBS should be discarded after each wash.

7. Repeat washes until the specimen is free of contaminating blood (3–5 washes are necessary, see Note 2).

8. Put tissue in a dounce tissue homogeniser and homogenize using only 4–5 strokes (see Note 3).

9. Filter homogenate in Nalgene filtration apparatus previously equipped with the 350 μm filter.

10. Remove filter and place it in a 50 mL sterile tube, with the cells and debris facing toward the exterior. This will be the 350 μm fraction.

11. Pass the filtrate through a filtration unit with a 112 μm filter.

12. Remove the filter and place it in a 50 mL sterile tube, with the cells and debris facing outwards.

13. Repeat steps 12 and 13, with a new 112 μm filter. The two 112 μm filters will be the 112 μm fraction.

14. Recover the filtrate in a 50 mL tube. This will be the <112 μm fraction.

15. Fill tubes consisting the filters with PBS and shake tubes vigorously until all visible cells and debris are removed from the filter membrane. Remove filters from the tubes with sterile forceps.

16. Centrifuge all tubes (<112 μm, the two 112 μm and the 350 μm fractions) at 1,000×g for 15 min at room temperature (RT).

17. Remove supernatant and dissociate pellets vigorously. Add 5 mL of collagenase IV (1 mg/mL), use 2.5 mL for both 112 μm fractions and combine them in one tube. A total of three fractions should remain: 350, 112, and <112 μm fractions.

18. Incubate for 15 min at 37°C.

19. Inactivate collagenase with ECM # 1, and fill tubes up to 50 mL.

20. Centrifuge 10 min at $500 \times g$.

21. Remove supernatant and resuspend pellets in 12 mL of ECM # 1 each for 350 and 112 μm fractions. Resuspend <112 μm pellet in 24 mL of ECM # 1.

22. Remove gelatin from the 6-well plates, by aspiration. Do not rinse the plates with PBS.

23. Plate 350 and 112 fractions in one 6-well plate each, 2 mL per well. Plate <112 μm fraction in two 6-well plates, 2 mL per well.

24. Incubate for 48 h at 37°C, with 0.5% CO_2.

25. Remove supernatant cell culture medium from each well.

26. Wash debris off by gently adding 2–3 mL of PBS per well.

27. Remove PBS and add 2 mL of fresh ECM # 1 medium.

28. Incubate cells at 37°C, with 0.5% CO_2.

29. Every 7 days, remove 1 mL of medium from each well and add 1 mL of fresh medium. Human brain derived ECs produce autocrine growth factors important for their growth and survival.

30. Every 4–5 days endothelial cell colonies become visible. BECs are adherent and form a monolayer in culture, Fig. 2.

31. As the colonies of BECs expand, passage clones that cover over 60% of the surface of the well as a homogenous monolayer.

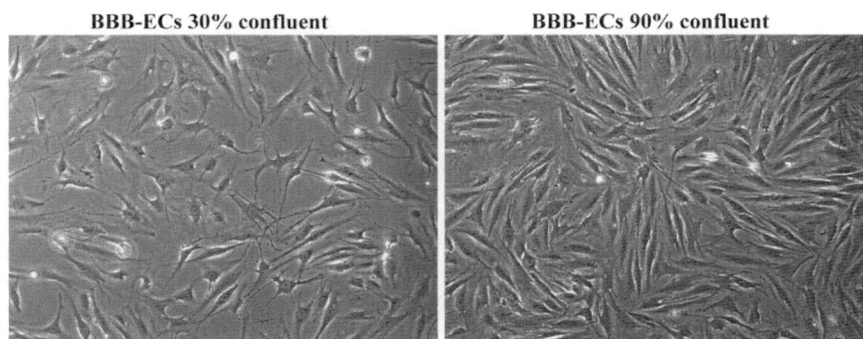

Fig. 2. Phase contrast micrographs of BEC monolayers. The *left panel* shows a monolayer after 1 day in culture and the *right panel* shows the same monolayer after 48 h in culture at around 90% confluency.

ECM # 1 is used to grow the clones in the 6-well plate. For all other BECs expansion in tissue culture, we use ECM #2 and for all experimental protocols we grow the cells in ECM # 3 (see Note 4).

32. Passage BECs in a T25 sterile flask and grow to confluence with ECM # 2 (passage 2).

33. Passage BECs in a T75 sterile flask and grow to confluence with ECM # 2 (passage 3).

3.2. Maintenance of Human BECs

1. Cells are passaged every 7 days or until the BECs form a confluent monolayer (Fig. 2).

2. Coat tissue culture flask with 0.5% gelatin for at least 1 h. Heat gelatin to 37°C for 15 min and use 5 mL for a T25 flask and 10 mL for a T75 flask.

3. Remove culture medium from the BECs and gently add PBS to the cells.

4. Remove PBS from the cells and add 0.5% trypsin previously warmed to 37°C. Add 5 mL for T25 flasks and 10 mL for T75 flasks.

5. Incubate cells at 37° C for 5 min or until cells detach. Gently tap the flask to allow cells to detach.

6. Add an equivalent volume of ECM #2 to inactivate the trypsin.

7. Collect cells and centrifuge for 10 min at $300 \times g$.

8. Discard supernatant and resuspend pellet in warmed ECM # 2.

9. Remove gelatin from the flask.

10. Plate cells in a flask, grow in ECM # 2 and incubate at 37°C with 0.5% CO_2. Cells are usually diluted 1:3 during passage but they can be diluted up to 1:5-dilution (see Note 5).

3.3. Immuno-fluorescence Analysis of Purity of BEC Cultures

1. Grow cells on plastic coverslips (Nunc) or on plastic chamber slides (Nalgene), see Note 6.

2. Gently remove medium and add PBS.

3. Gently repeat PBS wash.

4. Remove PBS and fix cells with 4% paraformaldehyde, cold methanol or cold ethanol for 10–20 min at RT.

5. Rinse cells with PBS three times.

6. Incubate cells with HHG or another suitable blocking solution supplemented with 10% serum of the same species used for preparation of the secondary antibody. For intracellular antigens, permeabilize cells with 0.3% Triton X-100 in HHG, incubate 30 min at RT.

7. Remove medium, add primary antibodies at appropriate dilutions (diluted in HHG or other appropriate solution with 3% serum of the same species as the secondary used plus 0.1% Triton X-100) and incubate for 1 h at RT.

8. Remove antibody and wash three times with PBS.

9. Add secondary antibodies at appropriate dilutions (diluted in HHG) and incubate for 1 h at RT.

10. Rinse three times with PBS.

11. Add nuclear stain if needed.

12. Rinse with PBS.

13. Mount slide with coverslip using gelvatol mounting medium or commercial aqueous mounting medium.

14. Analyze using a fluorescence microscope (see Fig. 3) (see Note 7).

3.4. Permeability of Human BECs to Small and Large Tracers

1. Coat the bottom of the Boyden chamber transwell with 0.5% gelatin and allow it to sit for at least 1 h.

2. Gently rinse BECs monolayer with 37°C warmed PBS.

3. Remove PBS and add warmed trypsin for 5 min to detach BECs.

4. Inactivate trypsin with ECM # 2 and collect cells.

5. Centrifuge at $300 \times g$ for 10 min.

6. Resuspend pellet in ECM # 2.

7. Count cells with a hemacytometer.

8. Remove gelatin from Boyden chambers.

9. Seed a defined number of BECs in each Boyden chamber and grow in ECM # 3 with or without 40% ACM. We plate 25,000 cells per Boyden chamber and allow them to grow for at least 3 days until the monolayer is confluent.

10. After 3–5 days in culture, the Boyden chambers can be used.

11. Remove treatment medium and replace with new ECM # 3. Add fluorescein conjugated BSA (50 µg/mL) or ^{14}C-sucrose (250 nCi/mL) to the upper chamber.

12. Sample the upper and lower chamber at regular time intervals and place the medium collected in a 96-well plate (1, 3, 6, 12, 24, 48, 72, 96 h for BSA-FITC and every 30 min for 6 h and then every hour for 24 h for sucrose). Collect 10–25 µL per time point.

13. Read fluorescence in a fluorescent plate reader and radioactivity in a gamma counter.

14. Calculate permeability using the following formula: (lower chamber) $\times 100/$ (upper chamber). The diffusion rate is expressed as the diffusion (%) X tracer/time (see Note 8).

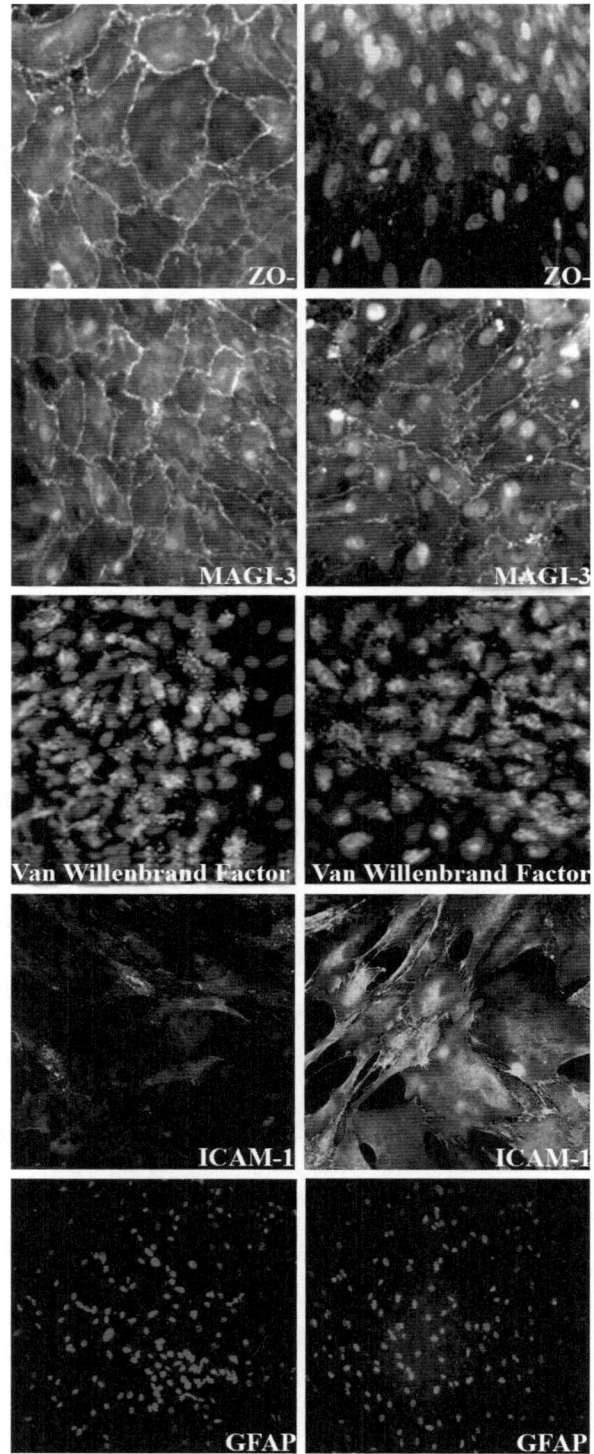

Fig. 3. Primary cultures of human BECs showing ZO-1, MAGI-3, Van Willebrand factor, ICAM-1, and GFAP by immunofluorescence. The BECs have been stained with a nuclear stain.

3.5. Lipid Raft Isolation and Purification

1. Let BECs grow as described in Subheading 3.2, until they form a confluent monolayer.

2. Treat cells as required by removing the medium and adding fresh ECM # 3 with the experimental treatment of interest. In our laboratory, we routinely compare BECs grown under normal culture conditions and BECs treated with the inflammatory cytokines: Tumor Necrosis Factor-α and Interferon-γ at 100 U/mL.

3. Incubate for 16 h overnight at 37°C with 0.5% CO_2. Use a minimum of 4, T75 flasks per conditions and up to 4, T150 flasks per condition.

4. All solutions and materials must be ice-cold and/or stored at 4°C before use.

5. Cool flasks on ice for 5 min.

6. Remove medium and wash monolayer by gently adding ice cold PBS.

7. Using a cell scraper, remove adherent cells from the flask bottom and collect cells with ice cold PBS (rinse the flasks several times and make sure that the detached cells are not forming clumps). Resuspend by moving the pipette up and down a few times to remove aggregates.

8. Centrifuge at 1,200 $\times g$ for 15 min.

9. Combine pellets of the same treatment, wash with ice-cold PBS and centrifuge for 15 min at 1,200 $\times g$.

10. Remove supernatant and carefully resuspend pellet in 1 mL of 1% Brij58 solution supplemented with protease inhibitors. Avoid the formation of air bubbles. Other detergents can be used as well. In our laboratory, Brij58 was found to be the optimal detergent to fractionate BECs membrane microdomains (14).

11. Gently rotate at 4°C for 30 min.

12. Transfer to a dounce homogenizer and homogenize sample with 10 strokes. Total volume is 1 mL.

13. Mix homogenized sample with 1 mL of 85% sucrose solution. Mix well but avoid bubble formation. Total volume is now 2 mL in a 42.5% sucrose solution.

Fig. 3. (contiuned) The micrographs on the *left* show BECs under normal culture conditions and the *right panels* show BECs treated with the inflammatory cytokines TNF and IFN-gamma for 16 h at 100 U/mL. ZO-1 (**a**) and MAGI-3 (**b**) staining patterns show an intercellular staining under normal condition and this intercellular staining is decreased with inflammatory cytokine treatment. (**c**) Shows that Van Willebrand factor expression in BECs is not influenced by inflammatory cytokine treatment, while the (**d**) shows that ICAM-1 expression is increased with cytokine treatment. In (**e**) GFAP positive cells are present in the human BECs cultures which show only nuclear staining.

14. Place the 2 mL of 42.5% sucrose cell membrane mix at the bottom of a 12 mL ultracentrifuge tube. This is the 42.5% sucrose fraction.

15. Carefully overlay 42.5% sucrose fraction with 5 mL of 35% sucrose.

16. Carefully overlay 35% sucrose fraction with 5 mL of 5% sucrose solution (the discontinuous sucrose gradient has a total volume of 12 mL).

17. Centrifuge in an ultracentrifuge having a Beckman SW41 rotor at $250,000 \times g$ for at least 16–24 h with no rotor deceleration. Avoid using the brakes to stop the rotor after the spin since it will destabilize the sucrose gradient.

18. Carefully remove tubes from rotor being careful to avoid any abrupt movements that can destabilize the sucrose gradient.

19. Harvest twelve 1.0 mL fractions from the meniscus of the liquid at the top of the tube until the bottom is reached making sure to keep the pipette tip or needle tip at the meniscus. Cholesterol enriched fractions are at the interface of the 5 and 35% sucrose solutions, usually fractions 4–6.

20. Fractions can be stored frozen at –20°C.

21. We routinely assess cholesterol, phospholipids and protein content of each fraction using commercially available kits such as the Cholesterol assay kit, Molecular Probes (see Note 9).

3.6. Liquid Chromatography and Mass Spectrometry

3.6.1. Preparation of Samples for Liquid Chromatography

1. Dialyze frozen DRM fractions (usually fractions 4–6) for 16 h against denaturing SDS buffer (see Note 10).

2. Precipitate proteins by adding 10-volumes of cold acetone and precipitating at –20°C for 1 h. Precipitation can be continued overnight if the amount of protein is expected to be low. Pellet the protein by centrifugation at $5,000 \times g$ for 5 min and redissolve the pellet in the denaturing SDS buffer.

3. Add freshly prepared DTT to a final concentration of 4 mM to each sample.

4. Vortex to mix, and spin at $10,000 \times g$ for 10 s to bring solution down.

5. Incubate samples at 95°C for 10 min to reduce the disulfide bonds of cysteine residues in proteins.

6. Vortex to mix, and spin at $10,000 \times g$ for 10 s to bring solution down.

7. Cool tubes at RT for 2 min.

8. Add freshly prepared IAA to a final concentration of 10 mM to each sample.

9. Incubate sample at RT for 20 min in the dark to alkylate cysteine residues in proteins. IAA is light sensitive.

10. Vortex to mix, and spin at $10,000 \times g$ for 10 s to bring solution down.

11. Add 50 µL of 100 ng/µL trypsin (in Milli-Q® water) to each sample.

12. Tap each sample tube to mix, and spin at $10,000 \times g$ for 10 s to bring solution down.

13. Incubate at 37°C for 12–16 h.

14. Vortex to mix, and spin at $10,000 \times g$ for 10 s to bring solution down.

15. Samples may be stored at –20°C for up to 4 weeks.

3.6.2. Cleaning Up Digested Samples Using a CE Cartridge

1. Transfer the trypsin-digested sample to a new tube labeled "PreCE" and add 2 mL CE load buffer.

2. Check the pH. If it is not <3.3, adjust by adding more CE load buffer.

3. Clean the CE cartridge by injecting 1 mL of CE clean buffer. Divert to waste.

4. Condition the CE cartridge by injecting 2 mL of CE load buffer. Divert to waste.

5. Slowly inject (~1 drop/s) all the contents of the "PreCE" tube. The flow through may be collected although not required in the subsequent steps.

6. Wash by injecting 1 mL of CE load buffer. Divert to waste.

7. Elute peptides by slowly injecting (~0.5 drop/s) 0.5 mL of CE elute buffer and collecting the eluted sample in a new tube.

8. If there are additional trypsin-digested samples, repeat steps 1–7.

9. Wash the cartridge by injecting 1 mL of CE clean buffer and 2 mL of CE storage buffer. Store the cartridge at 2–8°C.

10. Evaporate each sample to dryness in a SpeedVac®.

11. Dissolve each sample in 300 µL of MS buffer.

12. Samples may be stored at –20°C for up to 4 weeks.

3.6.3. Liquid Chromatography and Mass Spectrometry

1. Inject 5–10 µL of each sample in MS buffer onto the nanoLC online to the Q-TOF™ Ultima MS.

2. Separate peptides by gradient elution (5–75% acetonitrile, 0.2% formic acid for 90 min, 350 nL/min) and analyze in the automated MS/MS mode.

3. Acquire MS survey on ions between mass/charge (m/z) 400 and 1,600 using 30 counts/s threshold requirements to switch from MS survey to MS/MS acquisition (see Note 11).

4. Acquire MS/MS on 2+, 3+, and 4+ ions between m/z 50 and 2,000 and return back to MS survey once the total ion current reaches 2,500 counts/s or after 6.6 s regardless.

5. In-between samples, inject at least one 40-min blank run (MS buffer only) to avoid cross-contamination between samples.

6. Generate peaklist as a single text file (usually a.pkl file from Waters instruments) for each nanoLC-MS/MS run using ProteinLynx™. Each file contains m/z and charge values of all the precursor ions and their corresponding m/z and intensity values of fragment ions.

7. Submit each file to Mascot® search engine or any other probability-based engine using the following parameters: (i) specify trypsin enzymatic cleavage with one-to-two possible missed cleavages; (ii) allow variable modification for oxidation (+15.99 Da) at the methionine residues; (iii) allow variable modification for carbamidomethyl (iodoacetamide derivative; +57.02 Da) at the cysteine residues if proteins were alkylated; (iv) set parent ion tolerance 0.5 Da; (v) set fragment ion tolerance 0.2 Da.

8. The mass spectrometer spectrum of each identified peptide is then manually examined and confirmed.

9. Protein identification is considered specific if a set of peptides can only be assigned to a single protein. Protein identification is considered nonspecific if the peptides cannot be assigned to more than one possible protein.

10. Proteins are then be classified into molecular functions and biological processes using the Panther Classification System and online databases (www.pantherdb.org). Proteins can also be classified by subcellular localization (Swiss-Prot and GO databases).

11. Several validation approaches should be used to confirm the validity of the MS data. We validate candidate proteins of interest by literature mining, by using alternative detection methods in the same in vitro system (RT-PCR, Western blot or immunofluorescence) and by direct in vivo validation.

4. Notes

1. All components used for BECs isolation and tissue culture should be sterile before use. All procedures should be done aseptically in a sterile tissue culture hood. This protocol is not optimized for mouse CNS specimens.

2. The volume of tissue can be grossly estimated by looking at the volume of the CNS pieces that settle at the bottom of the tube.

At least 3 mL of CNS fragments is required to obtain endothelial cells. A maximum of up to 15 mL of minced CNS tissue (18–20 g) has been used in our laboratory.

3. A mechanical tissue grinder Dyna-mix set at medium speed, (Fisher Scientific, Ottawa, ON) is used in our laboratory.

4. Other techniques have been used to isolate and enrich primary cultures of BECs. These include the use of puromycin as a selection agent for P-glycoprotein expressing BECs (21) and the use of dextran to isolate brain microvessels from the CNS material (22). These methods have been used on murine BECs and in our hands these additional steps are not required for primary cultures of adult human BECs.

5. Occasional CD68 positive cells (microglia) are seen (less than 5% at passage 2) and their numbers decreased with passage, due to their strong adherence to plastic and their nonproliferative nature.

6. In our experience, BECs do not grow well on glass. They can grow on polylysine coated glass surfaces but their morphology is abnormal.

7. In the laboratory, BECs are routinely analyzed by flow cytometry to confirm ICAM-1 and VCAM-1 upregulation with inflammatory cytokine treatment. Due to space constraints the detailed cytometry protocol is not described here (23).

8. Human immune cells are routinely isolated from the peripheral blood of healthy donors. The immune cells are isolated using antibody coated magnetic beads (Miltenyi Biotech, Toronto, ON). A known amount of purified immune cells are added to the upper chamber of the Boyden transwell and allowed to migrate for 16 h. The immune cells of the bottom chamber are then sampled and counted to assess BECs permeability to immune cells (13, 16, 23).

9. Kits that are currently used in the laboratory are the Phospholipid colorimetric method kit (Wako), the Cholesterol assay kit (Molecular Probes) and the BCA protein assay kit (Pierce). We also assess the presence of membrane microdomain markers such as GM-1 and CD59 in each fraction by SDS-PAGE.

10. It is essential that the protein sample be free of potential contaminants for LC-MS such as high levels of sucrose, salt, acid, detergents, and/or denaturants since these may interfere with the subsequent steps. High amounts of detergents or denaturants such as SDS can deactivate trypsin activity. High salt, acid or detergents can prevent peptides binding to CE cartridge. These contaminants may be removed by dialysis against the DS buffer and precipitation with acetone.

11. For both the MS survey and MS/MS, we use 1.0 s for scan duration and 0.1 s for interscan delay. In addition, we usually do not discard MS survey scans that do not contain ions selected for MS/MS analysis.

Acknowledgments

This work is supported by grants from the Canadian Institute of Health Research (CIHR), the Canada Funds for Innovation (CFI), the Multiple Sclerosis Society of Canada (MSSC), the CIHR sponsored Neuroinflammation training program and Funds for Research in Science in Quebec (FRSQ). A.P. is a Research Scholar of the FRSQ and is also the recipient of the Donald Paty Career Development Award from the MSSC. R.C. and A. D. D. are recipients of a MSSC studentship.

References

1. Abbott NJ, Ronnback L, Hansson E (2006) Astrocyte-endothelial interactions at the blood-brain barrier. Nat Rev Neurosci 7:41–53

2. Pachter JS, de Vries HE, Fabry Z (2003) The blood-brain barrier and its role in immune privilege in the central nervous system. J Neuropathol Exp Neurol 62:593–604

3. Prat A, Biernacki K, Wosik K, Antel JP (2001) Glial cell influence on the human blood-brain barrier. Glia 36:145–155

4. Lampugnani MG, Dejana E (2007) The control of endothelial cell functions by adherens junctions. Novartis Found Symp 283:4–13

5. Tsukita S, Furuse M, Itoh M. Molecular dissection of tight junctions (1996) Cell Struct Funct 21:381–385

6. Wallez Y, Huber P (2008) Endothelial adherens and tight junctions in vascular homeostasis, inflammation and angiogenesis. Biochim Biophys Acta 1778:794–809

7. Deli MA, Abraham CS, Kataoka Y, Niwa M (2005) Permeability studies on in vitro blood-brain barrier models: physiology, pathology, and pharmacology. Cell Mol Neurobiol 25:59–127

8. Ghazanfari FA, Stewart RR (2001) Characteristics of endothelial cells derived from the blood-brain barrier and of astrocytes in culture. Brain Res 890:49–65

9. Jacobson K, Mouritsen OG, Anderson RG (2007) Lipid rafts: at a crossroad between cell biology and physics. Nat Cell Biol 9:7–14

10. Foster LJ, Chan QW (2007) Lipid raft proteomics: more than just detergent-resistant membranes. Subcell Biochem 43:35–47

11. Manes S, Viola A (2006) Lipid rafts in lymphocyte activation and migration. Mol Membr Biol 23:59–69

12. Munro S (2003) Lipid rafts: elusive or illusive? Cell 115:377–388

13. Wosik K, Cayrol R, Dodelet-Devillers A, Berthelet F, Bernard M, Moumdjian R et al (2007) Angiotensin II controls occludin function and is required for blood brain barrier maintenance: relevance to multiple sclerosis. J Neurosci 27:9032–9042

14. Chamberlain LH (2004) Detergents as tools for the purification and classification of lipid rafts. FEBS Lett 559:1–5

15. McCaffrey G, Staatz WD, Quigley CA, Nametz N, Seelbach MJ, Campos CR et al (2007) Tight junctions contain oligomeric protein assembly critical for maintaining blood-brain barrier integrity in vivo. J Neurochem 103:2540–2555

16. Cayrol R, Wosik K, Berard JL, Dodelet-Devillers A, Ifergan I, Kebir H et al (2008) Activated leukocyte cell adhesion molecule promotes leukocyte trafficking into the central nervous system. Nat Immunol 9:137–145

17. Haqqani AS, Kelly JF, Stanimirovic DB. Quantitative protein profiling by mass spectrometry using label-free proteomics. Methods Mol Biol 2008; 439:241–256

18. Haqqani AS, Kelly JF, Stanimirovic DB (2008) Quantitative protein profiling by mass spectrometry using isotope-coded affinity tags. Methods Mol Biol **439**:225–240

19. Robinson WH, Steinman L, Utz PJ (2002) Proteomics technologies for the study of autoimmune disease. Arthritis Rheum **46**:885–893

20. Jack CS, Arbour N, Manusow J, Montgrain V, Blain M, McCrea E et al (2005) TLR signaling tailors innate immune responses in human microglia and astrocytes. J Immunol **175**:4320–4330

21. Calabria AR, Weidenfeller C, Jones AR, de Vries HE, Shusta EV (2006) Puromycin-purified rat brain microvascular endothelial cell cultures exhibit improved barrier properties in response to glucocorticoid induction. J Neurochem **97**:922–933

22. Ge S, Pachter JS (2006) Isolation and culture of microvascular endothelial cells from murine spinal cord. J Neuroimmunol **177**:209–214

23. Ifergan I, Kebir H, Bernard M, Wosik K, Dodelet-Devillers A, Cayrol R et al (2008) The blood-brain barrier induces differentiation of migrating monocytes into Th17-polarizing dendritic cells. Brain **131**:785–799

Chapter 14

Analysis of Mouse Brain Microvascular Endothelium Using Laser Capture Microdissection Coupled with Proteomics

Nivetha Murugesan, Jennifer A. Macdonald, Qiaozhan Lu, Shiaw-Lin Wu, William S. Hancock, and Joel S. Pachter

Abstract

The blood–brain barrier (BBB) has been well studied in terms of its pharmacological properties. However, for a better understanding of the molecular mechanisms regulating these activities, means to thoroughly investigate the BBB at the genomic and proteomic levels are essential. Global gene expression analysis platforms have, in fact, provided a venue for cataloguing the BBB transcriptome. By comparison, and largely because of technical issues, there have been few comprehensive studies of the cerebral microvasculature at the protein level. Recent advances in both microdissection techniques and proteomic analytical tools have nonetheless circumvented many of these obstacles, allowing for isolation of relatively pure cell populations from complex tissues in situ and profiling of cellular proteomes. For example, immunohistochemistry-guided laser capture microdissection (immuno-LCM) provides the unique opportunity to selectively remove brain microvascular endothelial cells from the surrounding cell populations at the BBB, while supporting downstream proteomic analysis. In this chapter, we describe the use of immuno-LCM coupled with a sensitive, high resolution, hybrid linear ion trap coupled with Fourier transform mass spectrometry (FTMS) for proteomic profiling of mouse brain microvascular endothelium, a crucial cellular component of the BBB. We provide details of the quick double-immunostaining protocol for immuno-LCM, laser capture process, sample pooling, and protein recovery followed by in-gel digestion of protein sample, mass spectrometric analysis, and protein identification. Using such an approach to obtain comprehensive protein expression profiles of the cerebral endothelium in situ will enable detailed understanding of the crucial mediators of brain microvascular signaling and BBB function in both normal and pathophysiological conditions.

Key words: Laser capture microdissection, Immuno-laser capture microdissection, Mass spectrometry, Brain microvascular endothelial cells.

1. Introduction

The monolayer of endothelial cells that line cerebral microvessels is a major constituent of the blood–brain barrier (BBB), which critically regulates the passage of soluble substances between the peripheral

Sukriti Nag (ed.), *The Blood-Brain and Other Neural Barriers: Reviews and Protocols*, Methods in Molecular Biology, vol. 686, DOI 10.1007/978-1-60761-938-3_14, © Springer Science+Business Media, LLC 2011

circulation and the neural environment. These microvascular endothelial cells are in constant molecular interplay with other cell types such as pericytes, astrocytes and neuronal processes, and basement membrane components, the collective ensemble comprising a functional syncytium termed the neurovascular unit (NVU) (1–3). Such dynamic multicellular interactions within the central nervous system microenvironment dictate and regulate overall BBB function. Hence, efforts to best gauge molecular mechanisms governing BBB activity need to assess the microvascular endothelium within its in situ context. It is further important to keep in mind that the cerebral endothelium accounts for only one-thousandth of total brain volume (4, 5); therefore, genomic or proteomic analyses of whole brain homogenates could easily result in subtle vascular changes being obscured. These caveats demand a methodology that enables the cerebral endothelium to be "plucked" from its native environment, so that the cells might retain an accurate impression of their physiological or pathophysiological state. In this regard, laser capture microdissection (LCM) offers such an opportunity. Haqqani et al. (6) provided the first glimpse of the potential for coupling LCM to proteomic analysis for BBB studies. Specifically, they linked lectin-guided LCM to an isotope-coded affinity tag-based nano-liquid chromatography-tandem mass spectrometry (MS) platform to study a select group of cerebral microvascular protein expression patterns during cerebral ischemia – though a detailed study of the extent of the microvascular proteome was not performed (6).

The LCM process involves placing of a "cap" coated with a thin, thermoplastic (ethylene vinyl acetate) polymeric film over the tissue section, in the path of a low-power infrared laser beam aligned with the optical axis of the microscope. The laser beam, the spot size of which can be adjusted, is directed over specific cellular regions of interest within the tissue section. When the laser is pulsed, the polymeric film melts and cellular material below the pulsed spot adheres to the cap, leaving behind the rest of the heterogeneous tissue specimen (7, 8). In order to facilitate identification of the cell types targeted for retrieval from within a complex heterogeneous tissue specimen, dye-based histological stains and lectins have been employed in LCM applications coupled to downstream proteomic analyses (6, 9, 10). However, these staining methods either suffer from their inadequate cell type selectivity and/or variability in reactivity among subpopulations of a given cell type. These problems are particularly evident with regard to staining brain microvascular endothelial cells (BMEC) (11, 12).

By contrast, immunohistochemistry-guided LCM (immuno-LCM) utilizes highly selective antibodies that recognize cell-specific markers, in order to selectively label cell populations for capture. The use of combinations of antibodies to distinguish closely apposed cells affords increased ability to carefully remove designated cell types within a complex tissue microenvironment. Here we describe the use of a double-label immuno-LCM method compatible with downstream proteomic applications. In preparation for laser capture, tissue sections are generally fixed with precipitating fixatives (such as ethanol or acetone), which, because they avoid protein cross-linking, do not adversely affect downstream proteomic profiling (13, 14). Fixation of tissue samples with 75% ethanol, prior to immunostaining for CD31, allows for better protein recovery compared to using 100% acetone fixation (Fig. 1). Hence, all our tissue sections are fixed with 75% ethanol, prior to the LCM process. Apart from the choice of fixation agent for effective coupling of LCM with proteomic applications, the selection of immunohistochemical staining method is also important. A previous study has reported on the suitability of different immunostaining techniques when coupling LCM to two-dimensional gel electrophoresis (2D)-based proteomic analysis (9). According to their observations, immunofluorescent staining provides no reduction in protein recovery in comparison to unstained tissue samples, whereas immunohistochemical methods adopting avidin–biotin complexes and enzymatic detection led to poor protein yields on 2D gels (9, 13). However, we have observed that it is the *choice of enzyme/enzyme substrate* pair (which leads to formation of the colored precipitant on the immunostained tissue) that is the critical determinant affecting protein recovery postimmunostaining. In our experience, while avidin-biotinylated enzyme complex (ABC) conjugated with horseradish peroxidase (HRP) and using DAB (3,3′-diaminobenzidine) as

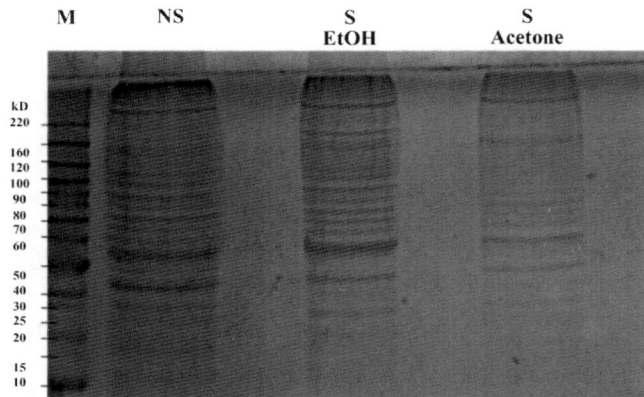

Fig. 1. Fixation and immunostaining of tissue for LCM. Consecutive frozen brain sections (7 μm) were either immunostained for CD31 with ABC-AP/NBT-BCIP (S) or not (NS), after fixation with either ethanol (EtOH) or acetone. Tissue scrapes were then collected into lysis buffer (1% SDS, 10% glycerol, 6.3 mM Tris base, pH 6.8) and equal volumes of extract resolved by SDS-PAGE and Coomassie Blue staining. M: molecular weight standards.

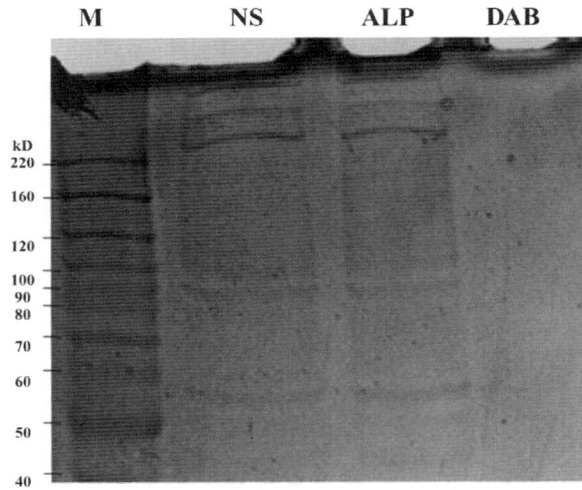

Fig. 2. Immunostaining for CD31 with ABC-AP/NBT-BCIP vs ABC-HRP/DAB. Cultures of b.End3 cells (a capillary endothelial cell line), grown in 12-well cluster plates, were fixed with EtOH, and either immunostained for CD31 using ABC-AP/NBT-BCIP (ALP) or ABC-HRP/DAB (DAB), or not stained (NS). Samples were collected into lysis buffer, and equal volumes of extract (representing 8,000 cellular equivalents) were resolved by SDS-PAGE and Coomassie Blue staining. M: molecular weight standards.

substrate results in a drastic loss in protein yield, colorimetric staining with analogous ABC-conjugated alkaline phosphatase (AP) and NBT/BCIP (Nitro blue tetrazolium/5-Bromo-4-chloro-3-indolyl phosphate) substrate results in comparatively good protein recovery (see Fig. 2).

In this chapter we describe an immuno-LCM protocol that can be effectively coupled to a proteomic platform. It involves double-immunostaining, combining bright-field and fluorescence microscopy, to identify BMECs and clearly resolve them from perivascular astrocytes (15). We have previously reported immuno-LCM coupled to downstream mRNA expression studies (16, 17), adopting a similar double-immunostaining method developed in our laboratory. In these previous studies, DAB was used for the enzymatic colorimetric detection as it does not interfere with post-LCM RNA isolation and/or analysis by RT-PCR or qRT-PCR. But, due to the aforementioned limitation of using DAB for downstream proteomic studies, our double immunostaining method had to be further optimized for use with alkaline phosphatase substrate NBT/BCIP-based detection. Details of the double-label immuno-LCM method for retrieval of BMECs from mouse brain tissue are outlined in Fig. 3.

One of the earliest reports of the application of proteomics technology combined with LCM was a report in 2002 by Jones et al. (18) which described a 2D gel study of microdissected ovarian tumors. In subsequent years, LCM was applied to brain

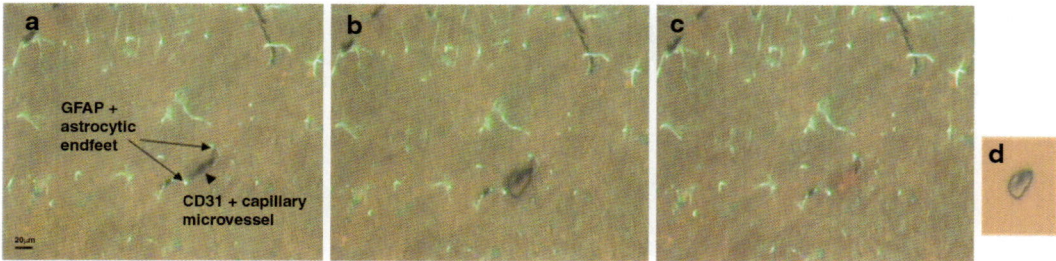

Fig. 3. Selective retrieval of brain microvascular endothelial cells using Immuno-LCM. A 7 μm thick frozen coronal section of mouse brain double-immunostained with antibodies against CD31 and GFAP is shown. CD31 immunoreactivity was visualized by alkaline phosphatase and NBT-BCIP substrate, and GFAP was detected by immunofluorescence. (a) *Prelift:* Tissue stained with anti-CD31 (*dark purple*) and anti-GFAP (*green*), viewed under bright-field and epifluorescence optics simultaneously, prior to LCM. *Arrows* indicate the CD31+ stained endothelial cells of the microvessel (*dark purple*) and the perivascular distribution of astrocytic endfeet (*green fluorescence*). (b) *Tissue during-lift:* Tissue section is shown during the laser-melting, with the LCM cap placed over the CD31+ capillary microvessel. (c) *Tissue postlift:* Tissue after LCM shows that the entire endothelial layer was removed and the fluorescent distribution of astrocytic end-feet was not disturbed. (d) *Cap:* Tissue transferred to cap after LCM, showing intact endothelium, was retrieved, with no detection of astrocytic end-feet. Copyright Wiley-VCH Verlag GmbH & Co. KGaA. Reproduced with permission from Lu et al. (15).

tissue samples, gastric mucosal metaplasa, pancreatic ductal adenocarcinoma, hepatocellular carcinoma, lung cancer, and nasopharyngeal cancer with 2D gel-based proteomic studies (19–24). An early study coupling LCM with the more sensitive reversed phase high performance liquid chromatography (HPLC) MS analysis of tryptic digests was performed on a breast cancer cell line (SKBR-3) and detected cancer markers such as Her-2 (25). Subsequent LCM-based liquid chromatography-proteomic studies include bladder cancer and squamous lung cancer (26, 27). The HPLC proteomics approach was further advanced for LCM samples by the incorporation of differential quantitation of hepatocellular carcinoma samples with an isotope-coded affinity tag (28), with $^{16}O/^{18}O$ isotopic labeling for breast cancer samples (29), and development of a differential multiplex radioactive prototype method (30). Further advances in methodology have enabled application of high resolution mass spectrometers (FTMS) to LCM samples prepared from breast cancer cells and to cervical cytological samples (31, 32). Most recently, such a mass spectrometry platform has been coupled to immuno-LCM to enable proteomic analysis of the brain microvasculature (15). From this analysis, 881 proteins were identified, out of which 181 were resolved by two or more peptide hits. The BBB-enriched and other endothelial-associated proteins identified in this study are listed in Table 1.

The opportunity to combine the exquisite in situ dissecting capability of LCM with increasingly sensitive proteomic platforms offers unparalleled promise of defining tissue, as well as cellular proteomes in even the most complex anatomical settings. By so

Table 1
Data from LC-MS analysis of immuno-LCM samples. Copyright Wiley-VCH Verlag GmbH & Co. KGaA

Protein identified	Category
BBB enriched proteins	
Tight junction protein 1 (ZO1)	Junctional proteins
Gap junction membrane channel protein alpha 7 (cx45)	
Junction plakoglobin	
Claudin 11	
ATP-binding cassette, subfamily A (ABC1), member 3 (MRP1)	Transporters
Glutamate/neutral amino acid transporter, Slc1a4 (ASCT1)	
Mitochondrial carrier, phosphate carrier, Slc25a3	
Solute carrier family 26, member 8 (TAT1)	
Solute carrier family 4 (anion exchanger), member 1	
Similar to solute carrier family 22, member 16 isoform 4	
UDP glucuronosyltransferase S	Metabolic enzymes
CytochromeP450, family 19, subfamily a, CPY19a1	
CytochromeP450, family 2, subfamily g, CYP2g1	
CytochromeP450, family 2, subfamily c, CYP2c37	
CytochromeP450, family 3, subfamily a, CYP3a44	
Glutathione S transferase, mu 1	
Plasma membrane calcium ATPase2 isoform 2	
Na, K-ATPase, ATP1a3	
Endothelial-associated proteins	
Apolipoprotein E	
Lactotransferrin	
Vascular endothelial zinc finger 1	
Endoglin	
Vimentin	
Cerebral vascular amyloid peptide (APP)	
Neural cell adhesion molecule 1	
Cell adhesion molecule with homology to L1CAM	
CD9 antigen	
Purinergic receptor, P2Y1	
Myelin basic protein (isoforms 1 and 2)	

Modified and reproduced with permission from Lu et al. (15)

doing, such technology will uniquely provide a window into the intricate processes regulating both cerebrovascular function in general and BBB activity in particular. Applications abound for using LCM/proteomics to study vascular properties in health and disease within and outside the nervous system.

2. Materials

2.1. Brain Tissue Procurement and Preparation of Frozen Tissue Sections

1. Compressed CO_2 in glass cylinders is used for euthanasia, as this allows the influx of gas to the induction chamber in a controlled manner. Cages are placed in the chamber, the chamber lid closed, and 100% CO_2 introduced at a rate of 10–20% of the chamber volume per minute. This rate of CO_2 introduction minimizes animal distress. After the animals become unconscious, the flow rate is increased to minimize the time to death. This euthanasia procedure is in accordance with measures stipulated by the Animal Care and Use Guidelines of the University of Connecticut Health Center (Animal Welfare Assurance #A3471-01).

2. Mouse brains.

3. Sterile surgical instruments: scalpel, forceps, scissors.

4. Isopentane.

5. Shandon Cryomatrix embedding medium (Thermo Fisher Scientific, Waltham, MA).

6. Microtome cryostat (Thermo Electron Corporation, San Jose, CA).

7. Shandon MX35 disposable microtome blades (Thermo Fisher Scientific).

8. RNase Away (Invitrogen Inc., Carlsbad, CA).

9. Uncoated glass slides.

2.2. Fixation of Frozen Tissue Sections and Quick Immunostaining for Laser Capture Microdissection

1. Ethanol (75, 95, and 100%).

2. Phosphate-buffered saline (PBS), pH 7.6.

3. Antibody diluting solution: All antibodies are diluted in 1× PBS containing 0.2% Tween-20 (see Note 1).

4. Pap-pen hydrophobic marker pen.

5. Rat anti-mouse CD31 antibody, 1:10 dilution (BD Biosciences, San Jose, CA).

6. Biotinylated rabbit anti-rat IgG antibody, 1:250 dilution (Vector Labs, Burlingame, CA).

7. Alexa 488 glial fibrillary acidic protein (GFAP) antibody, 1:5 dilution (Invitrogen Inc.).

8. Alkaline phosphatase avidin-biotinylated enzyme ABC kit (Vector Labs).

   ABC Complex: Add one drop of reagent A to one drop of reagent B. Add 2.5 mL of 1× PBS. Mix well and incubate for 30 min at RT prior to use. Prewarm an aliquot to 37°C before use.

9. Alkaline phosphatase substrate (NBT/BCIP) kit IV, (Vector Labs).

   NBT/BCIP substrate solution: Add one drop of reagent 1 (BCIP) to one drop of reagent 2 (NBT). Add one drop of reagent 3 ($MgCl_2$) and 1 mL of 0.1 M Tris-HCl, pH 9.5. Mix well.

10. Protease inhibitor cocktail: Add to all staining reagents in 1:100 dilution. (Sigma-Aldrich, St. Louis, MO).

2.3. Dehydration of Tissue Sections Prior to Immuno-LCM

1. Distilled water (DNAse, RNAse free)
2. Xylene

2.4. Laser Capture Microdissection

1. Pixcell IIe LCM microscope equipped with epifluorescent optics (Molecular Devices, MDS Analytical Technologies, Sunnyvale, CA).
2. Capsure HS caps (Molecular Devices).
3. Sterile 0.5 mL microfuge tubes.

2.5. Protein Isolation: Pooling Together Tissue Material from LCM Caps

1. Sterile scalpel blades.
2. Lysis buffer: 2% SDS, 50mM NH_4HCO_3 in water, pH 7.8.
3. Sonication bath.

2.6. SDS-PAGE and Coomassie Stain of LCM Retrieved Brain Microvessels Protein Sample

1. NuPAGE Novex 4–12% Bis-Tris gradient gel (Invitrogen Inc.).
2. SimplyBlue SafeStain (Invitrogen Inc.).

2.7 In-Gel Digestion

1. Acetonitrile (ACN).
2. 0.1 M NH_4HCO_3.
3. 10 mM DTT in 0.1 M NH_4HCO_3.
4. 55 mM iodoacetamide in 0.1 M NH_4HCO_3.
5. 5 mM $CaCl_2$, 8 ng/μL trypsin in 50 mM NH_4HCO_3.
6. 5% formic acid.
7. Digestion buffer: 5 mM $CaCl_2$, 8 ng/μL trypsin in 50 mM NH_4HCO_3, pH 8.0.
8. CentriVap® speed vacuum (Labconco Corporation, Kansas City, MO).

9. Shaker/Incubator.

10. Vortex mixer.

2.8. LTQ-FT Mass Spectrometry and Protein Identification

1. Solution B: 0.1% formic acid in ACN

2. LTQ-FT MS (Thermo Electron Corporation)

3. Dionex nano-LC (Ultimate 3000, Sunnyvale, CA)

4. 75 μm id×15 cm C-18 capillary column packed with Magic C18 (3 μm, 200 Å pore size) (Michrom Bioresources, Auburn, CA)

5. BioWorks 3.3.1 imbedded with Sequest (Thermo Electron Corporation) from a mouse database (ncbi.nih.gov/blast.db/FASTA)

6. Protein Prophet probability software (Institute of Systems Biology, Seattle, WA)

3. Methods

3.1. Brain Tissue Procurement and Preparation of Tissue Sections

1. Sacrifice the animal by euthanasia and spray the head with 75% ethanol.

2. Carefully remove the whole brain from the cranium using sterile surgical instruments.

3. Immediately immerse the brain into isopentane freezing medium (precooled on dry ice) for 1–2 min.

4. Remove the frozen brain from isopentane with a clean spatula.

5. Embed the frozen specimen in embedding medium, on dry ice.

6. Store the embedded tissue at −80°C until ready for sectioning

7. Cut 7 μm coronal sections of frozen brain using a cryostat.

8. Affix sections onto uncoated clean glass slides and keep on dry ice until cutting session is completed.

9. Place the tissue sections in a clean slide box precleaned with Rnase Away.

10. Store at −80°C (see Note 2).

3.2. Fixation of frozen Tissue Sections and Quick Immunostaining for LCM

1. Remove frozen 7 μm tissue section from −80°C storage and quickly thaw to room temperature.

2. Fix the tissue section in 75% ethanol for 3 min.

3. Briefly air-dry the tissue and draw a water repellent circle around the tissue section with a Pap-pen hydrophobic marker pen.

4. Incubate with monoclonal rat anti-mouse CD31 antibody for 3 min at room temperature (RT).

5. Wash briefly by dipping slide in 1× PBS for 5 s.

6. Incubate with biotinylated rabbit anti-rat antibody for 2 min at RT.

7. Wash briefly by dipping slide in 1× PBS for 5 s.

8. Add prepared ABC, prewarmed to 37°C, to the tissue for 3 min (see Notes 3, 4).

9. Wash briefly by dipping slide in 1× PBS for 5 s.

10. Add prepared NBT/BCIP substrate solution and incubate until purple color develops (~5–6 min).

11. Wash briefly by dipping slide in 1× PBS for 5 s.

12. Add Alexa 488 GFAP antibody for 5 min.

13. Wash briefly by dipping slide in 1× PBS for 5 s.

3.3. Dehydration of Tissue Sections Prior to Immuno-LCM

1. Dip the immunostained slide in 75% ethanol for 10 s.

2. Transfer to 95% ethanol for 30 s.

3. Immerse in first 100% ethanol for 60 s.

4. Immerse in second 100% ethanol for 90 s.

5. Transfer to first xylene wash for 2 min.

6. Transfer to second xylene wash for 3 min.

7. Air-dry the slide for 5 min after the final xylene wash (see Note 5).

3.4. Laser Capture Microdissection

1. The dehydrated double-immunostained slide is placed on the microscope stage.

2. A vacuum seal is used to keep the slide in place.

3. A HS Capsure LCM cap is placed in position, in the path of the laser beam, above the tissue section.

4. Both bright field and epifluorescence optics are used in order to visualize the CD31 positive microvessels (purple) and the green fluorescent GFAP in apposing astrocytic processes (see Fig. 3).

5. A 7.5 μm laser spot size is used at a power range of 65–80 mW and pulse duration of 550–800 μs. This combination of parameters allows for efficient retrieval of microvessel sample and limits contamination from perivascular cells.

6. Microvessels typically <10 μm, representing mainly capillaries, are targeted.

7. The cap containing LCM captured material is fitted onto a clean 0.5 mL microfuge tube and stored at −80°C until further analysis (see Note 6).

3.5. Protein Isolation: Pooling Together Tissue Material from LCM Caps

1. Take a cap containing LCM-captured material from −80°C.

2. With a sterile scalpel carefully remove the polymer membrane outside the black ring on the cap and discard. All the LCM-

captured tissue material is contained within the area enclosed by the black ring.

3. Transfer the remaining membrane (containing the LCM-retrieved tissue) to a clean microfuge tube containing 20 μL of lysis buffer.

4. Keep on ice for 20 s.

5. Sonicate for 20 s in ice cold water.

6. Repeat steps 4 and 5, five times for each cap's membrane.

7. Then briefly spin down the microfuge tube.

8. Carefully remove the membrane to the side of the tube (above the liquid level) and spin again for 1 min.

9. Remove the dry membrane with sterile forceps and discard.

10. Take the next cap from –80°C and repeat **step 2**.

11. Transfer the cut out membrane to the lysate containing tissue material from cap 1 and repeat **steps 4–9**.

12. Pool tissue from all LCM caps into the same 20 μL lysate volume to collect total protein sample for downstream proteomic analysis.

3.6. SDS-PAGE and Coomassie Stain of LCM-Retrieved Brain Microvessels Protein Sample

1. Load the entire extract (~15 μL) on a 4–12% gradient gel to separate proteins by molecular weight.

2. Stain the gel with Coomassie blue.

3.7. In-Gel Digestion

1. Cut the gel into three individual sections with molecular mass ranges below 20 kDa (section 3), between 20 and 50 kDa (section 2), and above 50 kDa (section 1).

2. Further mince each section into very small pieces (approximately 0.5 mm^2).

3. Wash the gel slices with 500–600 μL water for 15 min, spin down, and remove the liquid.

4. Add 400 μL of ACN and shake for 15 min. Spin down and remove the liquid.

5. Dry the gel slices in a speed vacuum.

6. Rehydrate the gel slices in 300 μL 0.1 M NH_4HCO_3 for 10–15 min. Then add an equal amount of CAN and vortex for 15–20 min. Spin down and remove liquid. Repeat up to three times if necessary, or until no visible Coomassie stain remains.

7. Swell the gel slices in 250 μL of 10 mM DTT and incubate for 30 min at 56°C.

8. Spin down and remove liquid.

9. Add 400 μL of ACN to shrink the gel slices.

10. Replace ACN with 250 µL of iodoacetamide and incubate for 60 min at RT in the dark.

11. Remove liquid, wash the gel slices with 400–500 µL of 0.1 M NH_4HCO_3, shaking for 15 min.

12. Spin down, remove liquid, shrink the gel slices with 400 µL ACN, remove the liquid, and dry in a speed vacuum.

13. Rehydrate the gel in 250 µL digestion buffer at 4°C for 45–50 min.

14. Remove the remaining supernatant. Add 50–100 µL 50 mM NH_4HCO_3 to cover gel pieces. Incubate overnight at 37°C to elute peptides from the gel slices.

15. Remove and save the supernatant.

16. Add 100 µL of 5% formic acid to further extract the gel slices at 37°C for 10 min and sonicate the solution for 1 h.

17. Combine the formic acid solution with the previous supernatant and concentrate to 15–20 µL.

3.8. LTQ-FT Mass Spectrometry

1. Inject the in-gel digested peptides using a 2 µL sample loop (see Note 7). The online LC using a linear IT coupled to a Fourier transfer mass spectrometer with a Dionex nano-LC instrument and a 75 mm id × 15 cm C-18 capillary column packed with Magic C18.

2. Operate the LTQ-FTmass spectrometer in the data-dependent mode to switch automatically between MS and MS2 acquisition. Acquire Survey full-scan MS spectra (m/z 400–2,000) in the Fourier transform ion cyclotron resonance cell with a mass resolution of 100,000 at m/z 400 (after accumulation to a target value of 2×10^6 ions in the linear IT, followed by ten sequential LTQ-MS/MS scans throughout the 120 min separation).

3. Utilize dynamic exclusion with exclusion duration of 60 s and no repeat counts.

4. Carry out the analytical separation using a three-step linear gradient, starting from 2 to 40% of Solution B over 60 min. Increase to 60% Solution B over the next 40 min, and then to 80% Solution B over 20 min.

5. Maintain the column flow rate at 200 nL/min.

3.9. Protein Identification

1. Identify peptide sequences first utilizing BioWorks 3.3.1 imbedded with Sequest from a mouse database which identifies peptides with Xcorr scores above the following thresholds: ≥3.8 for 3+ and higher charge state ions, ≥2.2 for 2+ ions, and ≥1.9 for 1+ ions, with full trypsin specificity and up to two internal missed cleavages.

2. Further confirm the identified peptides by the high mass accuracy (≤5 ppm).

3. Then utilize Protein Prophet probability software to identify proteins based upon corresponding peptide sequences with ≥95% confidence.

4. Notes

1. No serum or blocking agents are used in the immunostaining protocol in order to avoid introduction of exogenous proteins which could hinder proper mass spectrometric analysis of the sample.

2. Tissue sections are used within 1 week of sectioning in order to minimize degradation of proteins which can occur with an increase in storage time.

3. Choice of immunostaining method: Alkaline phosphatase-based colorimetric detection was selected rather than HRP, as the peroxidase-based substrate deposition diaminobenzidine causes drastic loss in protein recovery after the immunostaining steps.

4. Prewarm the ABC reagent to 37°C, prior to use. This enables quicker colorimetric development of alkaline phosphatase-NBT/BCIP reaction product and hence shorter incubation time.

5. Proper dehydration of the slides is critical for good LCM capture efficiency. Alcohol dilutions and xylene washes should be fresh and changed frequently after several dehydrations.

6. Any loosely adhered nonspecific tissue material, if present, can be removed from the LCM cap using an adhesive cleaning pad such as clean Post-it™ notes, prior to storage.

7. To avoid potential sample loss, larger loops with higher contact surface should be avoided for LC-MS analysis. We recommend using a 2 μL sample loop. If large volumes are encountered (e.g., 10 μL), multiple injections for each sample (e.g., 5 injections for 10 μL sample) can be conducted, as needed. In this large volume case, the LC gradient will not start until the total volume of the sample is loaded and desalted properly.

Acknowledgements

This work was supported by National Institutes of Health grants RO-1-MH54718 and R21-NS057241 to J.S. Pachter. The authors thank Dr. Barry Karger of Northeastern University for application development of LCM with MS.

References

1. Abbott N J, Ronnback L, Hansson E (2006) Astrocyte-endothelial interactions at the blood-brain barrier. Nat Rev Neurosci 7, 41–53

2. Lok J, Gupta P, Guo S, Kim WJ, Whalen M J, van Leyen K, Lo EH (2007) Cell-cell signaling in the neurovascular unit. Neurochem Res 32, 2032–2045

3. Hawkins BT, Davis T P (2005) The blood-brain barrier/neurovascular unit in health and disease. Pharmacol Rev 57, 173–185

4. Pardridge WM (1998) CNS drug design based on principles of blood-brain barrier transport. J Neurochem 70, 1781–1792

5. Enerson BE, Drewes L R (2006) The rat blood-brain barrier transcriptome. J Cereb Blood Flow Metab 26, 959–973

6. Haqqani AS, Nesic M, Preston E, Baumann E, Kelly J, Stanimirovic D (2005) Characterization of vascular protein expression patterns in cerebral ischemia/reperfusion using laser capture microdissection and ICAT-nanoLC-MS/MS. FASEB J 19, 1809–1821

7. Espina V, Wulfkuhle JD, Calvert V S, VanMeter A, Zhou W, Coukos G, Geho DH, Petricoin EF, III, Liotta LA (2006) Laser-capture microdissection. Nat Protoc 1, 586–603

8. Espina V, Heiby M, Pierobon M, Liotta LA (2007) Laser capture microdissection technology. Expert Rev Mol Diagn 7, 647–657

9. Gutstein HB, Morris JS (2007) Laser capture sampling and analytical issues in proteomics, Expert Rev Proteomics 4, 627–637

10. Mustafa D, Kros JM, Luider T (2008) Combining laser capture microdissection and proteomics techniques. Methods Mol Biol 428, 159–178

11. Harrison JK, Luo D, Streit WJ (2003) In situ hybridization analysis of chemokines and chemokine receptors in the central nervous system. Methods 29, 312–318

12. Vorbrodt AW (1988) Ultrastructural cytochemistry of blood-brain barrier endothelia. Prog Histochem Cytochem 18, 1–99

13. Mouledous L, Hunt S, Harcourt R, Harry JL, Williams KL, Gutstein HB (2003) Proteomic analysis of immunostained, laser-capture microdissected brain samples. Electrophoresis 24, 296–302

14. Ahram M, Flaig MJ, Gillespie JW, Duray PH, Linehan WM, Ornstein DK, Niu S, Zhao Y, Petricoin EF, III, Emmert-Buck MR (2003) Evaluation of ethanol-fixed, paraffin-embedded tissues for proteomic applications. Proteomics 3, 413–421

15. Lu Q, Murugesan N, Macdonald JA, Wu SL, Pachter JS, Hancock WS (2008) Analysis of mouse brain microvascular endothelium using immuno-laser capture microdissection coupled to a hybrid linear ion trap with Fourier transform-mass spectrometry proteomics platform. Electrophoresis 29, 2689–2695

16. Kinnecom K, Pachter JS (2005) Selective capture of endothelial and perivascular cells from brain microvessels using laser capture microdissection. Brain Res Brain Res Protoc 16, 1–9

17. Macdonald JA, Murugesan N, Pachter JS (2008) Validation of immuno-laser capture microdissection coupled with quantitative RT-PCR to probe blood-brain barrier gene expression in situ. J Neurosci Methods 174, 219–226

18. Jones MB, Krutzsch H, Shu H, Zhao Y, Liotta LA, Kohn EC, Petricoin EF, III (2002) Proteomic analysis and identification of new biomarkers and therapeutic targets for invasive ovarian cancer. Proteomics 2, 76–84

19. Mouledous L, Hunt S, Harcourt R, Harry J, Williams KL, Gutstein HB (2003) Navigated laser capture microdissection as an alternative to direct histological staining for proteomic analysis of brain samples. Proteomics 3, 610–615

20. Lee JR, Baxter TM, Yamaguchi H, Wang TC, Goldenring JR, Anderson MG (2003) Differential protein analysis of spasmolytic polypeptide expressing metaplasia using laser capture microdissection and two-dimensional difference gel electrophoresis. Appl Immunohistochem Mol Morphol 11, 188–193

21. Shekouh AR, Thompson CC, Prime W, Campbell F, Hamlett J, Herrington CS, Lemoine NR, Crnogorac-Jurcevic T, Buechler MW, Friess H, Neoptolemos JP, Pennington SR, Costello E (2003) Application of laser capture microdissection combined with two-dimensional electrophoresis for the discovery of differentially regulated proteins in pancreatic ductal adenocarcinoma, Proteomics 3, 1988–2001

22. Ai J, Tan Y, Ying W, Hong Y, Liu S, Wu M, Qian X, Wang H (2006) Proteome analysis of hepatocellular carcinoma by laser capture microdissection. Proteomics 6, 538–546

23. Yao H, Zhang Z, Xiao Z, Chen Y, Li C, Zhang P, Li M, Liu Y, Guan Y, Yu Y, Chen, Z (2008)

Identification of metastasis associated proteins in human lung squamous carcinoma using two-dimensional difference gel electrophoresis and laser capture microdissection. Lung Cancer doi:10.1016/j.lungcan.2008.10.024

24. Li MX, Xiao ZQ, Liu YF, Chen YH, Li C, Zhang PF, Li MY, Li F, Peng F, Duan CJ, Yi H, Yao HX, Chen ZC (2009) Quantitative proteomic analysis of differential proteins in the stroma of nasopharyngeal carcinoma and normal nasopharyngeal epithelial tissue. J Cell Biochem 106, 570–579

25. Wu SL, Hancock WS, Goodrich GG, Kunitake ST (2003) An approach to the proteomic analysis of a breast cancer cell line (SKBR-3). Proteomics 3, 1037–1046

26. Niu HT, Zhang YB, Jiang HP, Cheng B, Sun G, Wang YEYJ, Pang DQ, Chang JW (2008) Differences in shotgun protein expression profile between superficial bladder transitional cell carcinoma and normal urothelium. Urol Oncol doi:10.1016/j.urolonc.2008.07.007

27. Nan Y, Yang S, Tian Y, Zhang W, Zhou B, Bu L, Huo S (2008) Analysis of the expression protein profiles of lung squamous carcinoma cell using shot-gun proteomics strategy. Med Oncol doi: 10.1007/s12032-008-9109-4

28. Li C, Hong Y, Tan YX, Zhou H, Ai JH, Li SJ, Zhang L, Xia QC, Wu JR, Wang HY, Zeng R (2004) Accurate qualitative and quantitative proteomic analysis of clinical hepatocellular carcinoma using laser capture microdissection coupled with isotope-coded affinity tag and two-dimensional liquid chromatography mass spectrometry. Mol Cell Proteomics 3, 399–409

29. Zang L, Palmer Toy D, Hancock WS, Sgroi DC, Karger BL (2004) Proteomic analysis of ductal carcinoma of the breast using laser capture microdissection, LC-MS, and 16O/18O isotopic labeling. J Proteome Res 3, 604–612

30. Neubauer H, Clare SE, Kurek R, Fehm T, Wallwiener D, Sotlar K, Nordheim A, Wozny W, Schwall GP, Poznanovic S, Sastri C, Hunzinger C, Stegmann W, Schrattenholz A, Cahill MA (2006) Breast cancer proteomics by laser capture microdissection, sample pooling, 54-cm IPG IEF, and differential iodine radioisotope detection. Electrophoresis 27, 1840–1852

31. Umar A, Luider TM, Foekens JA, Pasa-Tolic L (2007) NanoLC-FT-ICR MS improves proteome coverage attainable for approximately 3000 laser-microdissected breast carcinoma cells. Proteomics 3, 323–329

32. Gu Y, Wu SL, Meyer JL, Hancock WS, Burg LJ, Linder J, Hanlon DW, Karger BL (2007) Proteomic analysis of high-grade dysplastic cervical cells obtained from ThinPrep slides using laser capture microdissection and mass spectrometry. J Proteome Res 6, 4256–4268

Chapter 15

Molecular and Functional Characterization of P-Glycoprotein In Vitro

Gary N. Y. Chan and Reina Bendayan

Abstract

The blood-brain barrier (BBB) physically and metabolically functions as a neurovascular interface between the brain parenchyma and the systemic circulation, and regulates the permeability of several endogenous substrates and xenobiotics in and out of the central nervous system. Several membrane-associated transport proteins, such as P-glycoprotein (P-gp), multidrug resistance-associated proteins, breast cancer resistance protein, and organic anion transporting polypeptides, have been characterized at the BBB and identified to play a major role in regulating the brain bioavailability of several pharmacological agents. This chapter reviews several well-established techniques for the study of the molecular expression, cellular localization, and functional activity of transport proteins in primary and immortalized cell culture systems of the BBB. In particular, we describe the molecular characterization of P-gp/MDR1 at the transcript level using semiquantitative polymerase chain reaction (PCR), at the protein level using immunoblotting, and at the cellular level using immunofluorescence. In addition, the uptake/efflux and transepithelial flux studies, which characterize P-gp transport activity, are described.

Key words: Astrocytes, ATP-binding cassette membrane transporter, Blood-brain barrier, Brain microvessel endothelial cells, Cell culture, Drug accumulation studies, Immunoblotting, P-glycoprotein, Polymerase chain reaction (PCR), Transepithelial flux, Transport activity

1. Introduction

The blood-brain barrier (BBB) constitutes a remarkable physical and biochemical barrier between brain and systemic circulation. Structurally, it is primarily composed of nonfenestrated endothelial cells characterized by the presence of tight junctions. These junctions form a continuous almost impermeable cellular barrier that limits paracellular flux. The high transendothelial electrical resistance further restricts the free flow of water and solutes. It is now well accepted that the functional unit of the BBB includes

Sukriti Nag (ed.), *The Blood-Brain and Other Neural Barriers: Reviews and Protocols*, Methods in Molecular Biology, vol. 686, DOI 10.1007/978-1-60761-938-3_15, © Springer Science+Business Media, LLC 2011

more than just capillary endothelial cells. Several other cell types, in particular pericytes and perivascular astrocytes, are in constant and intimate contact with the endothelium and maintenance of the brain capillary phenotype seems to be critically dependent on interactions with these other cells (1, 2). In addition to the physical barrier, there is a selective metabolism-driven barrier that largely reflects expression and function of several receptors, ion channels, metabolic enzymes, and influx/efflux transport proteins expressed prominently at the BBB. In particular, ATP-binding cassette membrane transporters such as P-glycoprotein (P-gp), multidrug resistance-associated proteins (MRP), and breast cancer resistance proteins (BCRP) play a significant role in restricting the permeability of several pharmacological agents including anticancer and anti-HIV agents (3–6). These transporters are primarily expressed at the luminal membrane of the brain capillary and serve as efflux pumps to extrude substrates from the brain tissue back into the circulation (5–11). In contrast, influx transporters such as members of the organic anion transporting polypeptide family can facilitate substrate delivery into the brain. Together, these influx and efflux transport proteins are believed to ultimately regulate the overall pharmacokinetic and pharmacodynamic profile of xenobiotics in the brain (3, 6, 9, 11, 12). Functionally, brain transport proteins are similar to well-characterized systems in other tissues, although the capacity and rate of transport can vary widely (13).

Since the early 1990s, cell culture models have proven to be a useful approach to study the membrane permeability of several substrates including therapeutic agents at several epithelial barriers. In this chapter, we focus on the molecular characterization of the P-gp/MDR1 transcript level, protein expression, localization, and transport activity using several cell lines derived from brain, such as the human cerebral microvessel endothelial cells/D3 (hCMEC/D3), rat brain endothelial cells (RBE4), and primary cultures of rat and human astrocytes. These cell systems are known to retain several properties of cellular compartments, which are part of the neurovascular unit in vivo, including the functional expression of many transport proteins necessary for the permeability of nutrients and pharmacological agents, adhesion molecules, tight junction molecules, and some of their respective regulatory pathways (1, 2, 7, 8, 14–21).

In vitro, in order to improve transepithelial electrical resistance (TEER) and reduce paracellular permeability, astrocytes can be cocultured with brain microvessel endothelial cells in a transwell culture system in the presence of a cyclic adenosine monophosphate analog and hydrocortisone (22, 23). Coculture of astrocytes and brain microvessel endothelial cells provides a powerful in vitro tool, which mimics the complexity of the in vivo BBB, for the study of bidirectional flux (apical to basolateral and basolateral to apical) of substrates across brain vasculature.

2. Materials

2.1. Cell Culture Models of Brain Microvessel Endothelial Cells and Astrocytes

1. Reagents and culture medium were optimized for the hCMEC/D3 cell system, an immortalized human brain microvessel endothelial cell line originally developed by Dr. P.O. Couraud's group (7, 15) and for the RBE4 cell system, an immortalized rat brain microvessel endothelial cell line developed by Dr. F. Roux's group (8, 16, 24).

2. Reagents and culture medium were optimized for primary cultures of rat and human astrocytes implemented in our laboratory (18–20, 25).

2.2. Characterization of P-gp at the Transcript Level

2.2.1. Isolation of Total Cellular mRNA

1. TRIZOL reagent (Invitrogen Inc, Grand Island, NY, USA).

2. Sterile Dulbecco's phosphate buffered saline (DPBS) (Invitrogen Inc.).

3. Solvents: Chloroform, Isoamyl alcohol, Isopropanol (Sigma-Aldrich, St. Louis, MO, USA), 75 % Ethanol (prepared with DNAase and RNAase free water, Invitrogen Inc).

4. Tris–HCl buffer (Sigma-Aldrich).

2.2.2. Reverse Transcription Assay and Semiquantitative PCR

1. Reaction Mix (Invitrogen Inc): PCR grade Oligo(dt) primer, 10× PCR buffer, 10 mM dNTP mix, 25 mM $MgCl_2$, 0.1 M DTT and 1 U/µL DNase I per 2 µg RNA, SuperScript II reverse transcriptase.

2. UV Spectrophotometer (Beckman Coulter Canada Inc., Mississauga, ON, Canada).

3. Primer pairs used in semiquantitative PCR (see Table 1).

4. Agarose, electrophoresis purity grade (Sigma-Aldrich).

5. Ethidium bromide: 1 mg/mL in DNase-RNase-free water.

6. 1× Tris Borate EDTA (TBE) buffer (Invitrogen Inc).

7. 100 bp DNA ladder (Invitrogen Inc).

8. PCR Mix Kit (Invitrogen Inc.): 2.5 µL of 10× PCR buffer, 1.5 mM of 25 mM $MgCl_2$, 400 µM for each dNTP's, 1.0 µM of forward primer, 1.0 µM of reversed primer, 2.5 U of Platinum Taq DNA polymerase, top volume up to 20 µL using DEPC water or DNAase and RNAase free water.

9. GeneAmp 2400 Thermocycler (Perkin-Elmer, Mississauga, ON, Canada).

10. ImageQuant 5.2 densitometric software (Molecular Dynamics, Sunnyvale, CA, USA).

11. Bio-Rad Wide Mini-Sub® Cell GT (Bio-Rad Laboratories Canada Ltd., Mississauga, ON, Canada).

12. 6× Gel loading dye: 0.25% Bromophenol blue, 40% w/v sucrose in ddH_2O.

Table 1
Reverse transcription PCR conditions and primers used

Genes	Species	Accession no.	Primer sequence (5'→ 3')	Annealing temperature (°C)	Fragment length (bp)
MDR1	Human	NM 000927.3	FP: GGT-GCT-GGT-TGC-TGC-TTA-CA RP: TGG-CCA-AAA-TCA-CAA-GGG-T	55	291
mdr1a	Rodent	NM 133401	FP: GGA-CAG-AAA-CAG-AGG-ATC-GC RP: CCC-GTC-TTG-ATC-ATG-TGG-CC	55	440
mdr1b	Rodent	NM 012623	FP: GGA-CAG-AAA-CAG-AGG-ATC-GC RP: TCA-GAG-GCA-CCA-GTG-TCA-CT	55	355
β-Actin	human	NM 001101.2	FP: GAC-TAT-GAC-TTA-GTT-GCG-TTA RP:GCC-TTC-ATA-CAT-CTC-AAG-TTG	55	504

FP forward primer; *RP* reverse primer

2.3. Characterization of P-gp at the Protein Level

2.3.1. Preparation of Whole Cell and Membrane Lysates from Adherent Cells

1. 0.25% Trypsin/EDTA (Invitrogen Inc).
2. Whole cell lysis buffer: 150 mM NaCl, 20 mM Tris-HCl pH 7.5, 5 mM EDTA, 1% NP-40, 2 μL of protease inhibitor cocktail/mL of buffer (Sigma-Aldrich), 1 mM of Phenylmethylsulphonyl fluoride (PMSF) (Sigma-Aldrich).
3. Bradford protein assay (Bio-Rad Laboratories Canada Ltd.).
4. Crude-membrane isolation buffer: 250 mM sucrose buffer containing 1.0 mM EDTA and 0.1% (v/v) protease inhibitor cocktail (Sigma-Aldrich).

2.3.2. Immunoblotting of P-gp

1. Power-Pac™ HC (Bio-Rad Laboratories Canada Ltd.).
2. Mini – Protean Tetra Cell (Bio-Rad Laboratories Canada Ltd.).
3. Mini-trans Blot Electrophoretic Transfer Cell (Bio-Rad Laboratories Canada Ltd.).
4. 100% Methanol.
5. SDS-PAGE Gel: Stacking: 3% (w/v) Acrylamide – Bisacrylamide (36:1) (Sigma-Aldrich), 0.25% (v/v) N,N,N,N-Tetramethyl ethylenediamine (TEMED) (Sigma-Aldrich), 0.1% (v/v) SDS (Sigma-Aldrich), 25% (v/v) Tris-Cl pH 6.8, 0.05% (v/v) Ammonium sulfate (Sigma-Aldrich). Running

Gel: 12% (w/v) Acrylamide – Bisacrylamide (36:1) (Sigma-Aldrich), 0.15% (v/v) *N,N,N,N*-Tetramethyl ethylenediamine (TEMED) (Sigma-Aldrich), 0.1% (v/v) SDS (Sigma-Aldrich), 25% (v/v) Tris-Cl pH 6.8, 0.05% (v/v) Ammonium per sulfate (Sigma-Aldrich), 25% (v/v) Tris-Cl pH 8.8 (see Note 1).

6. Hybond-P PVDF membrane (GE Healthcare Bio-Sciences, Piscataway, NJ, USA).

7. Gel holder cassette with sponge set (Bio-Rad Laboratories Canada Ltd.).

8. Filter paper (Bio-Rad Laboratories Canada Ltd.).

9. Western blotting Enhanced Chemiluminescence (ECL) system (GE Healthcare Bio-Sciences).

10. Hyper film ECL (GE Healthcare Bio-Sciences).

11. Developer and Fixer (Kodak, Rochester, NY, USA): Both are diluted 1:4 with ddH_2O.

12. Antibodies used (see Table 2).

13. Tris–HCl, pH 8.8: 15 mM Trizma-base (Sigma-Aldrich) in ddH_2O, pH 8.8.

14. Laemmli buffer: 350 mM sodium dodecylsulphate (SDS, Sigma-Aldrich), 17% (v/v) ddH_2O, 40% (v/v) Tris-Cl pH 6.8, 33% (v/v) glycerol (GE Healthcare Bio-Sciences), 0.006%(w/v) bromophenol blue (Sigma-Aldrich) and 10% 2-Mercaptoethanol (Sigma-Aldrich).

15. Electrophoresis running buffer: 25 mM Trizma-base, 200 mM glycine (Sigma-Aldrich), 3.5 mM SDS (Sigma-Aldrich) in ddH_2O.

16. Transfer buffer: 25 mM Tris-base, 200 mM glycine, 20% (v/v) methanol, pH 8 in ddH_2O.

17. Ponceau-S solution: 0.1% (w/v) Ponceau S (Sigma-Aldrich) in 5% (v/v) acetic acid in ddH_2O.

18. Tris Buffered Saline pH 7.6 w/Tween (TBS-T): 20 mM Trizma-base, 140 mM NaCl, 0.05% (v/v) Tween-20 (Sigma-Aldrich), pH 7.6.

19. Blocking reagent: 5% (w/v) low fat milk powder in TBS-T.

20. Primary and secondary antibody staining solution: 2.5–5% (w/v) low fat milk powder in TBS-T.

2.3.3. Reprobing of PVDF Membrane and Deglycosylation of Transport Proteins

1. Western blot stripping solution (Pierce, Thermo Fisher Scientific Inc., Waltham, MA, USA).

2. Peptide *N*-glycosidase F (PNGase F) (Biolabs Inc., Lawrenceville, GA, USA).

3. Endoglycosidase H (Endo-H) (Biolabs Inc.).

Table 2
Antibodies used for immunoblotting analysis of transport proteins

Protein of Interest	Antibody	Epitope region	Species specificity	Dilution range	Secondary antibody dilution	Source
P-gp (MDR1, mdr1a/b)	C219	Intracellular	Human, mouse and rat	WB 1:500 IF 1:10	Antimouse 1:3,000 (WB) Alexa Fluor antimouse 1:100 (IF)	ID Labs Biotechnology Inc. (London, ON, Canada)
P-gp (MDR1, mdr1a/b)	JSB-1	Intracellular	Human, mouse and rat	WB 1:50–1:200	Antimouse 1:3,000 (WB)	Santa Cruz Biotechnology Inc.(Santa Cruz, CA, USA)
β-Actin	Actin	Intracellular	Human, mouse and rat	WB 1:500	Antimouse 1:3,000 (WB)	ID Labs Biotechnology Inc

IF immunofluorescence studies; *WB* Western blot)

2.4. Cellular Localization of Transport Proteins by Immunofluorescence

2.4.1. Preparation of Adherent Cultured Cells and Fixation

1. Cover slips (22×22 mm) thickness #1 (Fisher Scientific, Nepean, ON, Canada).

2. Sterile petri dish or 6-well plate (Sarstedt, Newton, NC, USA).

3. 1% glutaraldehyde solution: Dilute 50% glutaraldehyde, EM grade (Canemco Inc., St. Laurent, PQ, USA) using 0.1 M DPBS, pH 7.2 (see Note 2).

2.4.2. Immunostaining and Mounting of Adherent Cells

1. Blocking buffer and Antibody Diluting Solution: 5% goat serum (Invitrogen Inc.), 0.2% Triton X-100 dissolved in DPBS.

2. Vectashield hard set mounting solution (Vector Laboratories Inc, Burlington, ON, Canada).

3. Microscope slide (Frosted on both sides at one end, 76×26 mm) (VWR, West Chester, PA, USA).

4. Fluorescence microscope.

2.5. Functional Activity of P-gp

2.5.1. Drug Accumulation Studies Using Radioactive Labeled and Fluorescent Substrates

1. Transport buffer: 10 mM of HEPES (Sigma-Aldrich), 0.01% of BSA (Sigma-Aldrich) dissolved in Hanks balance solution (Invitrogen Inc.), pH 7.4.

2. 1% Triton-X solution: Mix 5 mL of Triton-X solution in 500 mL of distilled water to create 1% (v/v).

3. 24-well or 48-well cell culture plate (Sarstedt).

4. Scintillation counting fluid: Picofluor 40 (Perkin Elmer Life Sciences, Boston, MA, USA).

5. Bio-Rad Dc protein assay kit (Bio-Rad Laboratories Canada Ltd.).

6. P-gp inhibitors:

   (a) PSC833, 1–10 µM (kindly donated by Novartis Pharmaceuticals Canada Inc, Dorval, QC, Canada).

   (b) GF120918, 10 µM (BCRP and P-gp inhibitor, kindly donated by GlaxoSmithKline, Research Triangle Park, NC, USA).

   (c) Cyclosporine A 10–50 µM (Sigma-Aldrich) (see Table 3).

7. Radiolabelled P-gp substrate: 100 nM [^{3}H] Digoxin (Perkin Elmer Life Sciences, see Note 3 and Table 3).

8. Fluorescent P-gp substrates: 1 µM Rhodamine-6G, 5 µM Rhodamine-123 (Invitrogen Inc., see Table 3).

2.5.2. Transepithelial Flux Studies

1. Costar®12-well tissue culture plate with 6.5 mm diameter and 0.4 µm pore size transwell insert (Corning Inc., Corning, NY, USA).

Table 3
Selective P-gp inhibitors, P-gp fluorescence substrates and P-gp radiolabeled substrates

Compounds	Concentration used to characterize P-gp	Source
Radiolabeled P-gp substrates		
[³H] Digoxin (37Ci/mmol)	100 nM	Perkin Elmer Life Sciences, Boston, MA, USA
Fluorescence P-gp substrates		
Rhodamine123 (R-123) (Excitation λ: 485 nm, emission λ: 520)	5 μM	Invitrogen Inc., Grand Island, NY, USA
Rhodamine-6G (R-6G) (Excitation λ: 530 nm, emission λ: 560)	1 μM	Invitrogen Inc
P-gp Inhibitors		
PSC 833	1–10 μM	Novartis Pharmaceuticals Canada Inc., Dorval, Québec, Canada
Cyclosporine A	10–50 μM	Sigma-Aldrich, St. Louis, MO, USA
GF120918	10 μM	GlaxoSmithKline, Research Triangle Park, NC, USA

2. Liquid scintillation counter (Beckman Coulter Canada Inc).

3. Fluorescence Plate Reader: SpectraMax Gemini XPS (Molecular Devices, Sunnyvale, CA, USA).

4. 70% Ethanol.

5. Electrical resistance reader (Millicell-D) (Millipore, Billerica, MA, USA).

6. [¹⁴C] d-Mannitol (Perkin Elmer Life Sciences) (see Note 4).

3. Methods

3.1. Cell Culture Models of Brain Microvessel Endothelial Cells and Astrocytes

1. The protocols for cell culture, subculture, cryostorage, and culture thawing are well described for hCMEC/D3 and RBE4 cell culture system (7, 8, 15, 16, 24).

2. The protocols for the isolation and maintenance of primary cultures of rat and human astrocytes are well established in our laboratory (18–20, 25).

3.2. Characterization of P-gp at the Transcript Level

Reverse transcription-polymerase chain reaction (RT-PCR) has been well-utilized to study changes in mRNA or transcript expression. The semiquantitative PCR technique that is described in this chapter can be used to detect mRNA expression of transport proteins.

3.2.1. Isolation of Cellular mRNA

1. Add 1 mL of TRIZOL to lyse approximately two million cells and transfer to a sterile microcentrifuge (see Note 5).

2. Incubate sample for 5 min at room temperature to permit complete dissociation.

3. Add 200 μL of chloroform-isoamyl alcohol solution for every mL of TRIZOL used.

4. Vortex vigorously for ~1–3 min and incubate at room temperature for ~2–3 min.

5. Centrifuge at 4°C for 15 min at $10,500 \times g$.

6. Transfer as much of the aqueous phase as possible to a clean, sterile microcentrifuge tube without touching the organic interphase.

7. Add 0.5 mL ice-cold isopropanol to the aqueous phase for every mL of TRIZOL used.

8. Mix vigorously and chill on wet ice for 5 min.

9. Centrifuge at $10,500 \times g$ for 10 min at 4°C to pellet RNA.

10. Remove supernatant carefully and wash RNA pellet by adding 1 mL ice-cold 75% ethanol for every mL of TRIZOL used.

11. Centrifuge at $6,600 \times g$ for 5 min at 4°C and remove nearly every drop of 75% ethanol.

12. Air dry RNA pellet in fumehood for ~10–15 min.

13. Resuspend pellet in 30–50 μL DNAse- and RNAse-free water (see Note 6).

14. Record exact volume of water added and store sample at −80°C.

3.2.2. Reverse Transcription Assay

1. Quantitation of mRNA should be performed using Tris–HCl buffer. Determine absorbance at 260 nm and 280 nm and the 260/280 nm ratio using an UV spectrometer.

2. Mix 6 μL DNAase and RNAase free water with every 2 μg RNA and 1 μL of oligo(dt) primer in a sterile microcentrifuge tube.

3. Incubate at 65°C for 10 min and quickly chill on ice.

4. Prepare a reaction mix: 2 μL of 10× PCR buffer, 1 μL of 10 mM dNTP mix, 4 μL of 25 mM $MgCl_2$, 1 μL of 0.1 M DTT, and 2 μL of 1 U/μL DNase I per 2 μg RNA per sample.

5. Mix gently and pulse centrifuge.

6. Incubate at 37°C for 30 min, and then incubate at 75°C for 5 min to denature DNase I.

7. Cool mixture on ice and add 1 μL of Superscript II, mix gently and pulse centrifuge.

8. Incubate at 42°C for 20 min and store at −20°C until PCR.

3.2.3. Semiquantitative PCR

1. Prepare 20 μL of PCR mix solution for each sample (see Subheading 2.2.2, item 8).

2. Add 20 μL of PCR master mix to 5 μL of template cDNA in each tube in triplicate.

3. Start thermocycling using the following guideline for human MDR1 (P-gp): 95°C for 5 min, (94°C for 1 min, 55°C for 1 min, 72°C for 1.5 min) 72°C for 10 min (see Note 7 and Table 1).

4. If necessary, forward and reverse primers for a loading control such as β-actin can be added to the reaction mix after 10–15 cycles.

5. Prepare 1.7% (w/v) of agarose using 1× TBE buffer with 1 mg/mL of ethidium bromide.

6. Heat agarose solution in a microwave oven to dissolve agarose completely.

7. Pour gel solution into gel tray avoiding entrapment of air bubbles.

8. Gently place comb into gel and allow it to solidify for at least 30 min.

9. Load 13 μL of 100 bp DNA ladder and samples-loading dye mixture to respective wells.

10. Run gel at 75 V for 2 h in 1× TBE buffer.

11. An example of RT-PCR analysis of MDR1 mRNA from hCMEC/D3 cells is shown in Fig. 1.

12. Densitometric analysis of ethidium bromide stained gels is performed using ImageQuant 5.2 densitometric software.

13. The ratio between the target mRNA and the appropriate housekeeping gene is calculated to obtain the relative mRNA expression of the particular gene of interest.

3.3. Characterization of P-gp at the Protein Level

Expression of transport proteins is commonly quantified using immunoblotting analysis of whole cell lysates or crude membrane lysates. The antibodies used and their working dilution as described in Table 2 serve as a guideline, since the sensitivity of antibodies can vary depending on the manufacturer and lot. Furthermore, a positive control consists of a purified transporter protein or lysate of cells overexpressing the protein of interest and should always be included in the experiment. To further confirm the identity of a band, transport proteins, which are heavily glycosylated with carbohydrates, such as P-gp can be deglycosylated

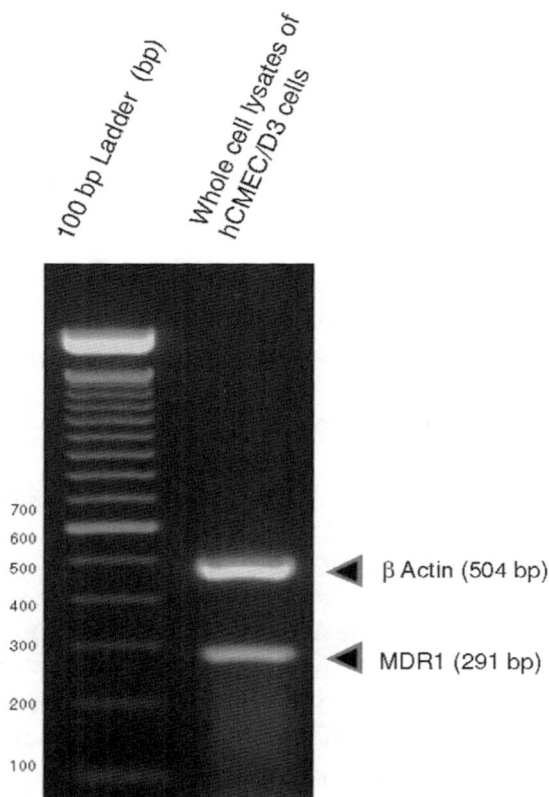

Fig. 1. RT-PCR analysis of MDR1 mRNA in the immortalized human brain microvessel endothelial cell line, hCMEC/D3. A representative ethidium bromide stained gel shows amplification of specific bands for MDR1 (291 bp) and loading control β actin (504 bp) from whole cell lysates of hCMEC/D3 cells.

(25). This method may reduce the apparent molecular weight of the transporter protein determined by immunoblotting analysis and produce a shift in the location of the band.

3.3.1. Preparation of Whole-Cell Lysates from Adherent Cells

1. Add 0.25% trypsin/EDTA (i.e., 2 mL for a T-75 tissue culture flask) for approximately 3 min (see Note 8).

2. Incubate longer if cells do not detach from plate surface.

3. Spin cell suspension for 3 min at $1,000 \times g$ at 4°C.

4. Remove trypsin/EDTA.

5. Add ice-cold whole cell lysis buffer to the cell pellet (see Note 9) and homogenize with a Dounce homogenizer at 10,000 rev/min for 0.10 min.

6. Allow cells to lyse at 4°C for 10–15 min.

7. Centrifuge at $20,000 \times g$ for 5 min at 4°C and collect supernatant (whole cell lysate).

8. Determine protein concentration using the Bradford protein assay kit according to the manufacturers' protocol.

3.3.2. Preparation of Crude Membrane Lysates from Adherent Cells

1. Harvest cells as described in Subheading 3.3.1, steps 1–4.

2. Incubate cell pellet for 30 min at 4°C in crude-membrane isolation buffer.

3. Homogenize cell suspension with a Dounce homogenizer at 10,000 rev/min for three cycles of 10 s each.

4. Centrifuge homogenates at $3,000 \times g$ for 10 min to eliminate cellular debris.

5. Collect and centrifuge supernatant at $10,000 \times g$ for 1 h at 4°C.

6. Resuspend pellet in 10 mM Tris buffer, pH 8.8, and keep frozen at −20°C until required.

7. Protein concentration of the crude membrane preparation is determined using the Bradford protein assay kit according to the manufacturers' protocol.

3.3.3. Immunoblotting to Demonstrate P-gp Protein Expression

1. Aliquot the appropriate volume which contains 50 μg of protein samples.

2. Add 4 μL of Laemmli buffer (5×) and top sample volume up to 20 μL with 10 mM Tris buffer, pH 8.8.

3. Mix by vortexing and load the entire sample mix into each lane of a 10–15%, SDS-PAGE gel (see Note 1).

4. Run the electrophoresis at 60 V for 30 min to stack sample on top of the resolving gel.

5. Increase voltage to 150 V and run gel until front dye reaches bottom.

6. Cut a piece of PVDF membrane having the same size as the gel and immerse membrane in 100% methanol for 15 s and rinse in water before use.

7. Equilibrate membrane in transfer buffer for at least 10–15 min.

8. Sandwich gel and PVDF membrane together with pieces of sponge and filter papers in a Gel holder cassette (see Note 10).

9. Transfer proteins to membrane at 4°C at 200 mA for 2 or 3 h for two gels in transfer buffer.

10. Remove membrane from the cassette using forceps.

11. Immediately pour Ponceau-S solution over the membrane and incubate for 1 min at room temperature (see Note 11).

12. Discard solution and wash membrane with water once.

13. Check for protein bands on membrane (unwanted lanes can be cut at this point).

14. Wash membrane with TBS-T for 10 min and incubate membrane in blocking reagent overnight at 4°C.

15. Incubate membrane with primary antibody for 3–4 h at room temperature with constant rocking (see Table 2).

16. Wash membrane with TBS-T, 3 times for 10-min each (see Note 12).

17. Incubate membrane with secondary antibody solution for 1.5 h at room temperature with constant rocking (see Table 2).

18. Repeat step 16 (see Note 12).

19. Incubate membrane in ECL solution for 4 min.

20. In a dark room, place film over the membrane and close the cassette.

21. At the desired time, immerse film in developer until bands appear (see Note 13).

22. Wash film in water to remove excess developer and transfer into a tray containing fixer and wait until the film becomes transparent.

23. Wash film with water and dry it.

24. Save membrane for stripping and reprobing or dry it for future reference.

25. P-gp bands derived from RBE4, hCMEC/D3, and primary culture of rat astrocytes are shown in Fig. 2.

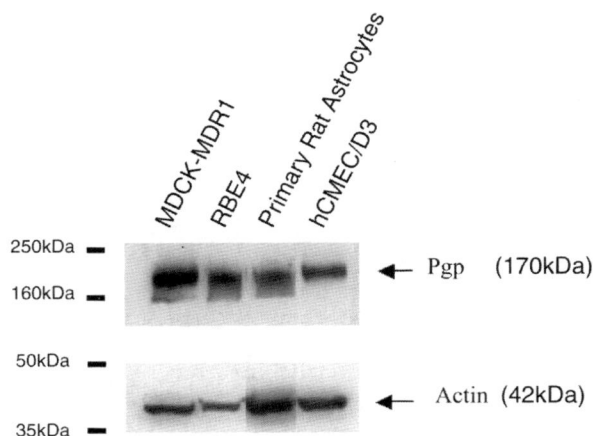

Fig. 2. Western blot analysis of P-gp in rat brain microvessel endothelial cells (RBE4), hCMEC/D3, primary culture of rat astrocytes and MDCK-MDR1 cells. Whole cell lysates were resolved in a 12% sodium dodecyl sulfate-polyacrylamide gel and transferred to a Polyvinylidene Difluoride (PVDF) membrane. P-gp (~170 kDa) was detected using the monoclonal C219 antibody (1:500 dilution) and secondary antimouse antibody (1:3,000 dilution). The MDCK-MDR1 cells served as the positive control.

3.3.4. Reprobing
of PVDF Membrane

1. Incubate membrane in western blot stripping solution with gentle rocking for 5–10 min at room temperature.

2. Wash membrane in TBS-T three times for 10-min each.

3. Membrane can now be probed with other antibody.

3.3.5. Deglycosylation
of P-gp

1. The cell lysate is denatured in 0.5% SDS and 1% β-mercapto-ethanol at 100°C for 10 min.

2. Incubate suspension with either 5 μL of PNGase F or Endo-H for 60 min at 37°C.

3. Proteins can then be resolved by SDS-PAGE and probed with the appropriate antibody.

4. An example of bands of deglycosylated P-gp is shown in Fig. 3.

3.4. Cellular
Localization
of P-gp by
Immunofluorescence

The cellular localization of P-gp can be detected by immunofluorescence. This technique requires that cells be attached on a surface. The glutaraldehyde fixation method we describe and other alternatives generally cause proteins to cross link in a meshwork, preserving the protein mass and cell morphology. The fixation step is required to expose antigenic sites of the transport proteins allowing the binding of primary antibodies and subsequent binding of fluorochrome-conjugated secondary antibodies. However, concentrations of the fixation solution and fixation period can affect the antigen recognition by antibodies.

In this chapter, we describe the glutaraldehyde fixation method, which forms a looser matrix for antibodies to diffuse compared to other methods such as ethanol fixation (26). Not all primary antibodies are suitable for immunofluorescence, and it is necessary to perform a careful selection to optimize antigen recognition. As well, if two or more fluorochrome-conjugated antibodies are used, one must avoid overlapping of emission or absorbance spectra between different fluorochrome-conjugated

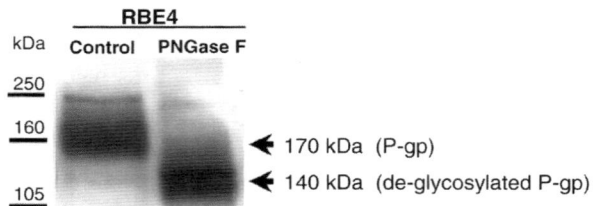

Fig. 3. Western blot analysis of P-gp expression in RBE4 cells. Crude membrane preparations (50 μg protein) from RBE4 cells were treated with or without PNGase F endoglycosidase, resolved on a 10% sodium dodecyl sulfate-polyacrylamide gel and transferred to a Polyvinylidene Difluoride (PVDF) membrane. P-gp (~170 kDa) was detected using the monoclonal C219 antibody (1:500 dilution) and secondary antimouse antibody (1:3,000 dilution) (*Lane 1*). P-gp was deglycosylated to ~140 kDa (*Lane 2*) (reproduced with permission from the University of Toronto).

antibodies. Spectra overlapping can produce high background making detection of transport proteins difficult.

3.4.1. Preparation of Adherent Cultured Cells and Fixation

1. Grow cells according to appropriate conditions on cover slips placed in a tissue culture petri dish or 6-well plate.

2. Remove medium and gently rinse cells twice with ice-cold PBS.

3. Fix with 1% glutaraldehyde for 1 h at room temperature (see Note 14).

4. Remove glutaraldehyde or other fixing solution and rinse cells gently twice with DPBS.

3.4.2. Immunostaining and Mounting of Adherent Cells

1. Cover cells with blocking buffer for 30 min at room temperature.

2. Shake off the blocking buffer and cover cells with the primary antibody staining solution for 1 h at room temperature (see Table 2).

3. Shake off the primary antibody staining solution and wash cells with fresh DPBS at room temperature 3 times with gentle rocking for 15 min.

4. Shake off the DPBS and cover cells with Alexa Fluor fluorochrome-conjugated antibody for 1 h in the dark to avoid photo bleaching (see Table 2).

5. Repeat washes as given in step 3.

6. Allow the cells to dry at room temperature in the dark.

7. Add one drop of Vectashield mounting solution on a new microscope slide.

8. Invert the cover slip on the mounting medium avoiding bubbles so that the cell growing side of the coverslip comes in contact with the mounting medium.

9. Let the mounting solution dry at room temperature for at least 1 h before viewing under a fluorescence microscope.

10. An example of P-gp localization in RBE4 cells is shown in Fig. 4.

3.5. Functional Activity of P-gp

The detection of both intracellular mRNA and protein expression does not necessarily predict the functional activity of the transport proteins. For example, we have previously observed that BCRP expressed in primary human and rat brain microvessel endothelial cells has a low transport activity in a cell culture system (27).

In order to measure activity of transporter proteins, several drug uptake/accumulation and flux studies can be performed using a combination of substrates and inhibitors selective for the

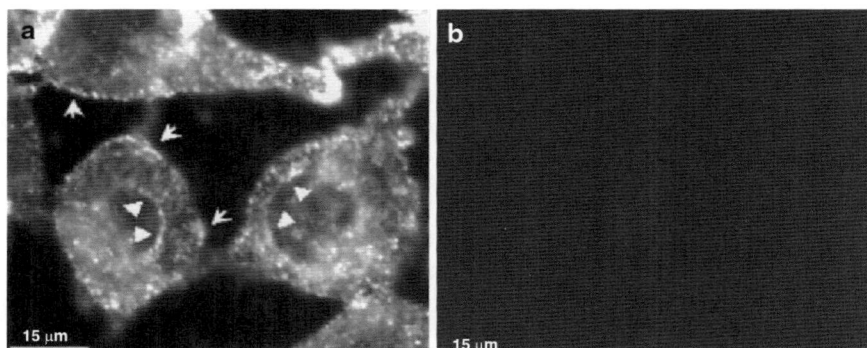

Fig. 4. Immunocytochemical localization of P-gp in fixed RBE4 cells. (**a**) P-gp protein was localized with the monoclonal C219 antibody (1:10 dilution) and Alexa-Fluor 594-conjugated secondary antibody (1:100 dilution). Labeling is present at the nuclear (*solid triangles*) and plasma (*arrowheads*) membranes. (**b**) Images of cells incubated with only secondary antibody (negative control) show no immunostaining. Original magnification ×100 (reproduced with permission from the University of Toronto).

transport protein of interest. Radioactive or fluorescence labeled substrates are often used to characterize the functional activity of the transporter protein of interest. To specifically investigate the directional flux across the endothelial cells, transepithelial flux assays using transwell tissue culture plates can be done to measure the basolateral to apical flux and apical to basolateral flux of substrates in the presence or absence of inhibitors.

3.5.1. Drug Accumulation Studies Using Radioactive Substrates

1. Culture cells in 24-well or 48-well plates at appropriate cell density (see Note 15).

2. Prepare nonradiolabeled (cold) substrate buffer (at the required concentration) in the transport buffer, and "spike" it with radiolabeled (hot) substrate solution (see Notes 3 and 16).

3. Use the substrate (experimental stock) buffer prepared from step 2 to prepare the substrate buffers needed for the specific experiment (i.e., buffer 1: substrate only, buffer 2: substrate with inhibitor A, buffer 3: substrate with inhibitor B) to avoid variability.

4. Pipette three samples of 50 μL of the experimental stock buffer into three scintillation vials to determine the specific activity of the radiolabeled compound for each experiment: the ratio between the concentration for hot substrate (mole) added and the disintegrations per minute (DPM) count value obtained.

5. Warm buffers to 37°C and place the tissue culture plate on a heating plate set at 37°C.

6. Replace medium in all the wells with blank transport buffer.

7. Let the system equilibrate for 15 min.

8. Replace blank transport buffer with the desired transport buffer prepared in step 3.

9. Reserve some wells (three wells for 24-well plate) for the protein determination assay (see Note 17).

10. At the desired time, aspirate transport solution and replace with ice-cold DPBS.

11. Rinse cells twice with fresh ice-cold DPBS and aspirate all DPBS before moving to the next well.

12. Reserve wells empty until the entire plate is used.

13. To harvest cells, add 250 µL of 1% Triton-X solution for 30 min at 37°C with gentle agitation.

14. Collect the Triton-X solution.

15. Transfer into separate labeled scintillation vials for scintillation counting and use fresh 1% Triton-X solution for measurement of background signal.

16. Measure radioactivity of all samples using scintillation counting.

17. Harvest wells designated for protein assay with 1 N NaOH.

18. Determine protein content using the Bio-Rad Dc Protein Assay Kit.

19. Substrate accumulation per well is expressed as mol of substrate (by using the specific activity value generated from step 4) per mg of protein (generated from step 19).

20. An example illustrating PSC-833-mediated enhancement of 100 nM [^{3}H] Digoxin uptake in primary cultures of rat astrocytes is shown in Fig. 5.

3.5.2. Drug Accumulation Studies Using Fluorescence Substrates

1. Follow steps 1–13 from Subheading 3.5.1.

2. The volume of 1% Triton-X solution used to harvest cells should be adjusted according to the amount of the fluorescence substrate (i.e., 200 µL 1% Triton-X solution/well for a 24-well plate in the case of 1 µM Rhodamine-6G (R-6G) used in the uptake study) (see Note 18).

3. Transfer each sample into three wells in the fluorescence reading plate to minimize variability in pipetting.

4. Use 1 N NaOH to harvest cells in wells and determine protein content using the Bio-Rad Dc Protein Assay Kit.

5. An example illustrating PSC-833-mediated enhancement of 1 µM R-6G uptake by hCMEC/D3 cells is shown in Fig. 6.

3.5.3. Transepithelial Flux Studies

3.5.3.1. Cell Adhesion to the Outside of the Transwell Insert

1. Perform collagen coating of the insert to prepare for steps in Subheading 3.5.3.2.

2. Invert transwell insert (outside of inserts are facing up) and place it on a glass petri dish.

3. Plate 45,000 astrocytes/well on day 1.

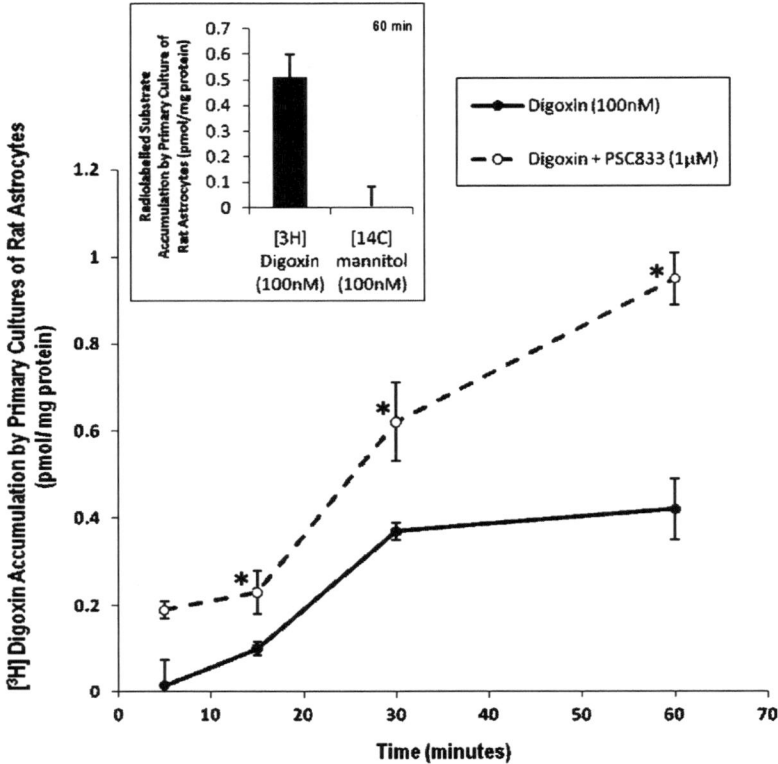

Fig. 5. Functional activity of P-gp in primary cultures of rat astrocytes is shown. Cellular accumulation of [³H] Digoxin (100 nM) was examined in the presence or absence of 1 μM PSC-833 over 60 min. Inset: Cellular accumulation of [³H] Digoxin (100 nM) and [¹⁴C] Mannitol (100 nM) at 60 min. Results are expressed as mean ± SD of three separate experiments, with each data point in an individual experiment representing quadruplicate measurements. Significant difference was observed between control and the PSC group (*$p < 0.05$) (reproduced with permission from the University of Toronto).

4. Pour cell suspension onto insert. Make sure enough medium is added to cover insert.

5. Allow astrocytes to proliferate and adhere for 1–2 days.

3.5.3.2. Cell Adhesion to the Inside of Transwell Insert

1. Plate collagen-coated upper chamber with 15,000 per/cm² endothelial cells in 250 μL of medium and allow cells to proliferate and adhere for 3–7 days.

2. Monitor cell growth and change medium in upper and lower chambers every other day (see Note 19).

3.5.3.3. Transepithelial Flux Assay (Apical-to-Basolateral Transport)

1. If an inhibitor for P-gp is used, first prepare stock transport buffer with the inhibitor concentration of interest, then divide the buffer into two; one transport buffer without substrate (buffer A) and one buffer with substrate (buffer B). Add to buffer B the required concentration of cold substrate and "spike" with radiolabelled substrate solution in transport buffer (see Subheading 3.5.1, step 2).

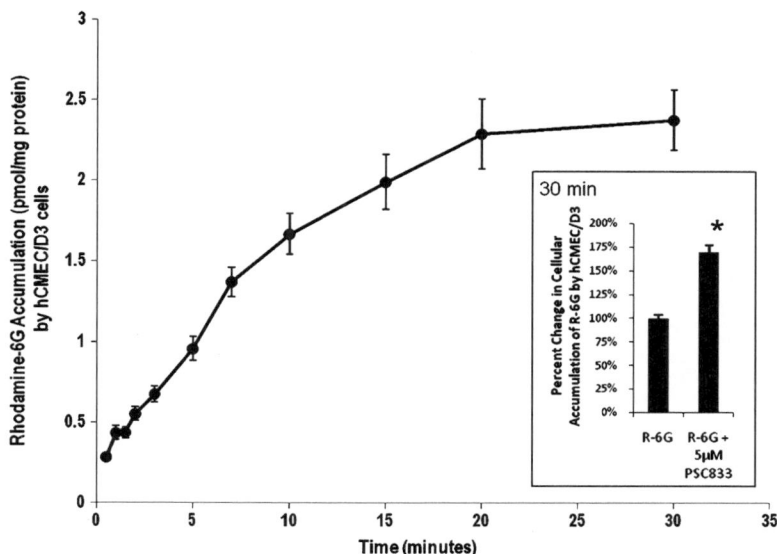

Fig. 6. Functional activity of P-gp in hCMEC/D3 cells is shown. Cellular accumulation of 1 μM Rhodamine-6G (R-6G) (pmol/mg protein) was examined over 30 min. *Insert:* Percent change in R-6G cellular accumulation in the presence of 5 μM PSC-833 at 30 min. Excitation and emission wavelengths for R-6G were 530 nm and 560 nm respectively. Results are expressed as mean ± SEM of three separate experiments, with each data point in an individual experiment representing triplicate measurements (* $p < 0.05$).

2. Warm all solutions to 37°C and place transwell plate on a heating plate at 37°C.

3. Replace medium with buffer A into upper (0.25 mL) and lower (1 mL) compartments.

4. Let the system equilibrate for 15 min and replace buffer A in the top compartment with 0.25 mL of buffer B.

5. At the desired time, transfer entire insert into an empty well with cold DPBS.

6. Remove solution in upper compartment and rinse the upper compartment with ice-cold DPBS twice.

7. Save buffer B in the lower compartment for radioactive scintillation counting (this allows quantification of substrate flux from basolateral to apical side by using the specific activity value generated from Subheading 3.5.1, step 4.)

8. To quantify substrate intracellular accumulation, cut membrane of insert and transfer membrane into scintillation vials containing 1–2 mL scintillation fluid.

9. Cut membrane of blank insert (only incubated with blank buffer) and use 1 N NaOH to harvest cells and determine protein content using the Bio-Rad Dc Protein Assay Kit.

10. Intracellular substrate accumulation per insert is expressed as mol of substrate (by using specific activity generated from Subheading 3.5.1, step 4) per mg of protein (generated from step 9 above).

3.5.3.4. Transepithelial Flux Assay (Basolateral-to-Apical Transport)

1. Follow steps 1–3 in Subheading 3.5.3.3.

2. Replace buffer A from bottom compartment with 1 mL of buffer B.

3. At the desired time, transfer entire insert into an empty well with cold DPBS.

4. Save buffer A in top compartment for radioactive scintillation counting (this allows quantification of substrate flux from basolateral to apical side by using specific activity generated from Subheading 3.5.1, step 4).

5. Quantify substrate intracellular accumulation (see Subheading 3.5.3.3, steps 9–11).

3.5.3.5. Transepithelial Electrical Resistance Measurement

To assess the integrity of a monolayer prior to and after an experiment, the TEER measurement, and permeation of the paracellular marker, [^{14}C] d-Mannitol should be performed in each transepithelial flux assay in order to ensure that the cell monolayer is not leaky (see Note 4).

1. Clean the electrical probe using 70% ethanol and rinse with sterile PBS.

2. Place probe in medium and let it equilibrate for 10–15 min.

3. Turn on electrical reader.

4. Recalibrate resistance to zero using fresh medium.

5. Insert probe into upper chamber of transwell insert.

6. One of the ends should be in contact with medium in upper chamber, while the other end should be in contact with medium in the lower chamber without touching the sides.

7. Record TEER value.

8. Interpretation: TEER value is expressed as the product of the resistance (Ω) and the surface area of the endothelial monolayer (cm^2). Therefore, indication of the insert surface area in published papers is important to allow for a comparison of TEER values between cell culture models. Both the hCEMC/D3 and RBE4 cell culture systems exhibit TEER value of at least 200 Ω cm^2 when cultured on transwell plates.

9. The techniques described in this chapter can be adapted to study other transport proteins such as the MRPs and BCRP.

4. Notes

1. 10–15% SDS-PAGE gel can be used to detect P-gp. Low gel density is recommended for a better resolution of bands detected at 170 kDa and above.

2. Glutaraldehyde solution must contain monomers and low polymers (oligomers) with molecules small enough to penetrate membranes fairly quickly. Electron Microscopy grade glutaraldehyde (25 or 50% solution) must be used and not "technical" grade, which consists largely of polymer molecules too large to fit between macromolecules.

3. Radioactive waste includes surplus radioisotope material in any form, material that has come into direct contact with radioactive exposed minor equipment, material used for radioactive decontamination (e.g., towels), and material that has come into incidental contact with other radioactive material (e.g., bench top covering material, gloves). Contaminated equipment used during radioisotope handling procedures that cannot be easily cleaned (e.g., centrifuges) must be segregated, stored or decontaminated according to isotope physical state (liquid or solid), half-time of the isotope, and regulations set by the radiation protection authority. Wash contaminated objects with water and mild detergent. Perform contamination swap test routinely and keep records.

4. Flux studies using [^{14}C] Mannitol or [^{3}H] inulin of the same concentration of the substrate can provide detailed information on paracellular permeability or the tightness of confluent cell monolayer grown on the transwell culture plate. The aim is to obtain permeability for the paracellular marker of less than 5%.

5. All RT-PCR steps should be performed using filter pipette tips.

6. If necessary, heat sample for 10 min at 55–60°C to completely dissolve the mRNA pellet.

7. Each gene of interest may require unique experimental conditions. Modifications can be made based on the guideline for human MDR1 (P-gp).

8. Trypsin can decrease the yield of transport protein. An alternative method is to scrape off cells from the surface using a cell scraper and centrifuge at $1,000 \times g$ for 3 min to pellet cells.

9. Typically 40 μL of whole cell lysis buffer is used in a cell pellet containing approximately four million cells.

10. Gas bubbles will restrict protein movement from gel to membrane. Roll with a glass test tube over the membrane to remove gas bubbles between membrane and gel.

11. We keep PVDF membrane wet throughout the entire immunoblotting procedure. Other protocols involve drying the membrane completely and then reactivating it with 100% methanol for 1 min before the blocking step. This strengthens the binding of protein to the membrane.

12. The number and the time of T-BST rinses after each primary and secondary antibody incubation should be optimized to yield best intensity and minimal background. Longer and increased number of rinses can remove nonspecific background; however, it will yield fainter signals.

13. Time of film exposure depends on the amount of antibody binding to the membrane, enzymatic activity of the secondary antibody and strength of ECL. Exposure time can range from 15 s to 10 min, but typically it is 1.5–4 min for C219 anti-P-gp antibody and the antimouse secondary antibody. In order to detect changes in intensity between lanes, one should always avoid over-exposure.

14. Other cell fixation methods such as BD Cytofix fixation solution (Becton Dickinson, Franklin Lakes, NJ) can be used according to the manufacturers' instructions.

15. Cell culture wells on the edge of the plate can be left blank to avoid "edge effect" due to condition differences, e.g., humidity. Ensure that an equal number of cells are transferred into each well. Differences in cell numbers between wells can introduce significant variability. RBE4 seeding density is 12,500 cells/cm^2 and cells reach confluence in 4 days. hCMEC/D3 seeding density is 25,000 cells/cm^2 and cells reach confluence in 4 days. The seeding density of primary cultures of human and rat astrocytes is 5,000 cells/cm^2. These cells reach confluence in 4 days.

16. Hot: cold substrate ratio is approximately 1: 1,000. Given that the radiolabeled stock of [^{3}H] Digoxin has a specificity of 37Ci/mmol, one should spike every 10 mL of 100 nM cold digoxin with 3 μL of the hot [^{3}H] Digoxin and adjust according to the scintillation counts obtained.

17. Wells for protein assay are used to estimate the amount of protein in a typical well in a given plate. It is important to seed each well with identical initial cell density.

18. One should be cautious in the selection of the cell harvesting buffer to minimize interference with the fluorescence substrate used. As well, pH, temperature and albumin concentration in the transport buffer can affect physical and chemical properties of substrates, such as charge and protein binding.

19. It is difficult to observe cell proliferation either inside or outside the insert under phase contrast microscopy. As a reference one should culture cells separately in blank wells on the same plate to avoid the formation of multicell layering.

References

1. Abbott NJ, Rönnbäck L, Hansson E (2006) Astrocyte-endothelial interactions at the blood-brain barrier. Nat Rev Neurosci 7:41–53

2. Hawkins BT, Davis TP (2005) The blood-brain barrier/neurovascular unit in health and disease. Pharmacol Rev 57(2):173–185

3. Deeken JF, Löscher W (2007) The blood-brain barrier and cancer: transporters, treatment, and trojan horses. Clin Cancer Res 13(6):1663–1674.

4. Potschka H, Löscher W (2007) Handbook of neurochemistry and molecular neurobiology. In: Lajtha, A., Reith, M.E.A. (eds.) Chapter 23. Efflux transporter in the Brain. pg 461–483, Plenum, New York

5. Bendayan R, Ronaldson PT, Gingras D, Bendayan M (2006) In situ localization of P-glycoprotein (ABCB1) in human and rat brain. J Histochem Cytochem 54(10):1159–1167

6. Ronaldson PT, Babakhanian K, Bendayan R (2007) Drug transporters: molecular characterization and role in drug disposition. In: You G. and Morris M.E.(eds) Chapter 14. Drug transport in the brain, pg 411–461. Wiley, New Jersey

7. Zastre J, Chan GNY, Ronaldson PT, Ramaswamy M, Couraud P, Romero IA, Weksler B, Bendayan R (2008) Up-regulation of P-glycoprotein by HIV protease inhibitors in a human brain microvessel endothelial cell line. J Neurosci Res 87(4): 1023–1036

8. Bendayan R, Lee G, Bendayan M (2002) Functional expression and localization of P-glycoprotein at the blood brain barrier. Microsc Res Tech 57:365–380

9. Dallas S, Miller DS, Bendayan R (2006) Multidrug resistance-associated proteins: Expression and function in the central nervous system. Pharmacol Rev 58: 140–161

10. Kusuhara H, Sugiyama Y (2001) Efflux transport systems for drugs at the blood-brain barrier and blood-cerebrospinal fluid barrier (Part 1). Drug Discov Today 6(3):150–156

11. Lee G, Bendayan R (2004) Functional expression and localization of P-glycoprotein in the central nervous system: Relevance to the pathogenesis and treatment of neurological disorders. Pharm Res 21(8):1313–1330

12. Kusuhara H, Sugiyama Y (2001) Efflux transport systems for drugs at the blood-brain barrier and blood-cerebrospinal fluid barrier (Part 2). Drug Discov Today 6(4):206–212

13. Choudhuri S and Klaassen C (2006) Structure, function, expression, genomic organization, and single nucleotide polymorphisms of human ABCB1 (MDR1), ABCC (MRP), and ABCG2 (BCRP) efflux transporters. Int J Toxicol 25(4):231–259

14. Cucullo L, Couraud P, Weksler B, Romero IA, Hossain M, Rapp E, Janigro D (2008) Immortalized human brain endothelial cells and flow-based vascular modeling: A marriage of convenience for rational neurovascular studies. J Cereb Blood Flow Metab 28:312–328

15. Weksler BB, Subileau EA, Perriere N, Charneau P, Holloway K, Leveque M, Tricoire-Leignel H, Nicotra A, Bourdoulous S, Turowski P, Male DK, Roux F, Greenwood J, Romero IA, Couraud PO (2005) Blood-brain barrier-specific properties of a human adult brain endothelial cell line. FASEB J 19(13): 1872–1874

16. Roux F, Durieu-Trautmann O, Chaverot N, Claire M, Mailly P, Bourre J, Strosberg AD, Couraud P (1994) Regulation of gamma-glutamyl transpeptidase and alkaline phosphatase activities in immortalized rat brain microvessel endothelial cells. J Cell Physiol 159:101–113

17. Walz W (2000) Controversy surrounding the existence of discrete functional classes of astrocytes in adult gray matter. Glia 31:95–103

18. Ronaldson PT, Bendayan R (2008) HIV-1 viral envelope glycoprotein gp120 produces oxidative stress and regulates the functional expression of multidrug resistance protein-1 (Mrp1) in glial cells. J Neurochem 106(3):1298–1313

19. Ronaldson PT, Bendayan R (2006) HIV-1 viral envelope glycoprotein gp120 triggers an inflammatory response in cultured rat astrocytes and regulates the functional expression of P-glycoprotein. Mol Pharmacol 70(3):1087–1098

20. Ronaldson PT, Bendayan M, Gingras D, Piquette-Miller M, R Bendayan (2004) Cellular localization and functional expression of P-glycoprotein in rat astrocyte cultures. J Neurochem 89:788–800

21. Ronaldson PT, Lee G, Dallas S, Bendayan R (2004) Involvement of P-glycoprotein in the transport of saquinavir and indinavir in rat brain microvessel endothelial and microglia cell lines. Pharm Res 21:811–818

22. Colgan OC, Collins NT, Ferguson G, Murphy RP, Birney YA, Cahill PA, Cummins PM (2008) Influence of basolateral condition on the regulation of brain microvascular endothelial tight junction properties and barrier function. Brain Res 1193:84–92

23. Perrière N, Yousif S, Cazaubon S, Chaverot N, Bourasset F, Cisternino S, Declèves X, Hori S, Terasaki T, Deli M, Scherrmann J, Temsamani J, Roux F, Couraud P (2007) A functional in vitro model of rat blood-brain barrier for molecular analysis of efflux transporters. Brain Res 1150:1–13

24. Babakhanian K, Bendayan M, Bendayan R (2007) Localization of P-glycoprotein at the nuclear envelope of rat brain cells. Biochem Biophys Res Commun 361(2):301–306

25. Richert ND, Aldwin L, Nitecki D, Gottesman MM, Pastan I (1988) Stability and covalent modification of P-glycoprotein in multidrug-resistant KB cells. Biochemistry 27:7607–7613

26. Kiernan JA (2000) Formaldehyde, formalin, paraformaldehyde and glutaraldehyde: What they are and what they do. Micros Today 00-1:8–12

27. Lee G, Babakhanian K, Ramaswamy M, Prat A, Wosik K, Bendayan R (2007) Expression of the ATP-binding cassette membrane transporter, ABCG2, in human and rodent brain microvessel endothelial and glial cell culture systems. Pharm Res 24(7):1262–1274

Chapter 16

Methods to Study Glycoproteins at the Blood-Brain Barrier Using Mass Spectrometry

Arsalan S. Haqqani, Jennifer J. Hill, James Mullen, and Danica B. Stanimirovic

Abstract

Glycosylation is the most common posttranslational modification of proteins in mammalian cells and is limited mainly to membrane and secreted proteins. Glycoproteins play several key roles in the physiology and pathophysiology of the blood-brain barrier (BBB) and are attractive as diagnostic markers and therapeutic targets for many neurological diseases. However, large-scale glycoproteomic studies of the BBB have been lacking, largely due to the complexity of analyzing glycoproteins and a lack of available tools for this analysis. Recent development of the hydrazide capture method and significant advances in mass spectrometry (MS)-based proteomics over the last few years have enabled selective enrichment of glycoproteins from complex biological samples and their quantitative comparisons in multiple conditions. In this chapter, we describe methods for: (1) isolating membrane and secreted proteins from BEC and other cells of the neurovascular unit, (2) enriching glycoproteins using hydrazide capture, and (3) performing label-free quantitative proteomics to identify differential glycoprotein expression in various biological conditions. Hydrazide capture, when coupled with label-free quantitative proteomics, is a reproducible and sensitive method that allows for quantitative profiling of a large number of glycoproteins from biological samples for the purposes of differential expression measurements and biomarker discovery.

Key words: Hydrazide, Proteomics, Neurovascular unit, Blood-brain barrier, Glycoprotein

1. Introduction

Glycosylation is the most common posttranslational modification of proteins in mammalian cells. It has an impact on a wide range of biological functions of proteins, including protein stability, solubility, protease resistance, antigenicity, and immune reactions, as well as protein roles in cell–cell interactions and signaling (1). Protein glycosylation is particularly prevalent on proteins destined for the extracellular environment; most of the known human

Sukriti Nag (ed.), *The Blood-Brain and Other Neural Barriers: Reviews and Protocols*, Methods in Molecular Biology, vol. 686, DOI 10.1007/978-1-60761-938-3_16, © Springer Science+Business Media, LLC 2011

glycoproteins are either cell membrane proteins or secreted proteins present in the extracellular space, CSF, and other body fluids.

The blood-brain barrier (BBB) physiological functions are determined by a specialized phenotype of brain endothelial cells (BEC), which includes tightly sealed, nonpermissive intercellular contacts, polarized expression of transporters and carriers, and low pinocytic activity (2). BEC also display an exceptionally thick glycocalyx, the surface coating being mainly comprised of oligosaccharide moieties of plasmalemmal glycoproteins, particularly enriched in sialic acid residues (3). These proteins include many glycosylated transporters polarized to the luminal endothelial membrane, including ABC transporters such as P-glycoprotein, some multidrug resistance-associated proteins (MRP), breast cancer resistance protein (BCRP) and solute carrier family transporters such as glucose and amino-acid transporters. Some of these transporters such as glucose transporter-1 (GLUT-1) are selectively glycosylated in BEC compared to endothelium of other vascular beds (4). In addition to modulating activities of specific proteins, BEC surface-displayed sugars (glycocalyx) have important physiological roles in BBB function. According to the "fiber-matrix" model (5), the endothelial glycocalyx acts as a molecular sieve, which differentiates molecules based on their size and charge and plays a role in selective permeability of the endothelial barrier. Furthermore, a paracellular matrix, formed by arrays of glycosylated junctional adhesion molecules such as VE-cadherin, JAMs, and PECAM-1 may similarly influence molecular size and charge restriction through the paracellular route. Experiments in which the glycocalyx was disrupted or reduced by mild heparinase treatments demonstrated increase in endothelial permeability, cerebral blood flow and perfusion (6, 7). Endothelial glycocalyx has also been suggested to play a role as a mechano-sensor and transducer of flow shear-stress affecting endothelial biochemical responses and cytoskeletal rearrangements (8).

In pathological conditions, modifications of the endothelial glycocalyx play a role in angiogenesis (9) and neuroinflammatory conditions, such as those caused by ischemic stroke and multiple sclerosis, where cell–cell interactions are modified by a set of glycoproteins involved in cell adhesion and migration including selectins, intercellular adhesion molecule 1 (ICAM-1), activated leukocyte cell adhesion molecule (ALCAM), integrins, and others (10, 11). The endothelial glycocalyx also serves as a pool of angiogenic and growth factors bound to heparan-sulfate proteoglycans, which can be released during injury and inflammation (7).

Glycoproteins secreted in the extracellular space surrounding brain microvessels also play significant biological roles. Secreted factors such as cytokines, chemokines, and growth factors involved in *inter*cellular signaling especially among BEC, astrocytes, and leukocytes are mainly glycoproteins. In addition, most of the molecules present in the basal lamina of brain vessels, including extracellular matrix structural proteins such as collagens and laminins and regulatory proteins such as agrin, SPARC and dystroglycans, are known glycosylated proteins (12, 13).

The endothelial cell glycocalyx has traditionally been studied by analyzing the affinity and specificity for various lectins, which bind to terminal and internal sugar residues on cell membrane carbohydrates (14). Large-scale glycoproteomic studies of the BBB have been lacking, largely due to the complexity of analyzing glycoproteins and a lack of available tools for this analysis. Therefore, the diversity of glycoproteins and oligosaccharide structures displayed by BECs and their roles in various pathological states has not been fully understood.

Significant advancements in high-performance liquid chromatography (HPLC) and mass spectrometry (MS) instrumentation over the past 5 years and a recent development of the hydrazide capture technology (15, 16) have enabled selective enrichment of glycoproteins from cells and tissues for large-scale identification using LC-MS based quantitative proteomics. In this chapter, hydrazide capture, label-free LC-MS, and bioinformatics are described to selectively enrich and identify mainly N-linked glycosylated proteins, from luminal membranes and the secreted fraction of BEC.

2. Materials

2.1. Biological Samples

1. Samples for membrane isolation may include cells such as BEC, glia, and neurons, isolated brain vessels, whole brain tissue or laser capture microdissected samples, under different biological states. This protocol was optimized for HCMEC/ D3, an immortalized human BEC cell line created by the Couraud group (17).

2. For secreted proteins, samples may include proteins secreted into the culture medium by BEC, glia, and neurons in response to various treatments or under different biological states. Proteins in body fluids such as serum and CSF of diseased patients can also be analyzed. This protocol was optimized for proteins secreted by glioblastoma U87MG cells (18) (see Note 1).

2.2. Protein Isolation

1. Pre-chill acetone by refrigerating at −20°C for 20 min.

2. Buffer-exchange/concentrating columns: such as Centriprep or Amicon Ultra columns (Millipore, QC, Canada) with a nominal molecular cut-off of 5,000 Da.

3. Eppendorf Centrifuge 5415D (Brinkmann Instruments, Westbury, NY).

4. 550 Sonic Dismembrator (Fisher Scientific, Canada).

5. 10 mM Tris-HCl buffer, pH 7.4

6. Sucrose solution: 38% (w/v) sucrose in 10 mM Tris-HCl, pH 7.4.

7. 0.22 μm Millex®-GV syringe filters (Millipore).

8. Tris-HCl buffer 1: 250 mM sucrose in 10 mM Tris-HCl, pH 7.4

9. Tris-HCl buffer 2: 38% (w/v) sucrose in 10 mM Tris-HCl, pH 7.4

10. 13×51 mm, Thickwall polycarbonate ultracentrifugation tubes (Beckman Coulter, Mississauga, Canada).

11. Optima TLX ultracentrifuge with SW 50.1 rotor (Beckman Coulter).

2.3. Hydrazide Capture of Glycopeptides and Elution of N-Linked Deglycosylated Peptides

1. ACN-AA solution: 50% acetonitrile (ACN) and 5% acetic acid.

2. ACN-TFA solution: 60% ACN and 0.1% trifluoroacetic acid (TFA).

3. AMBIC solution: 50 mM ammonium bicarbonate.

4. AMBIC-ACN solution: 50 mM AMBIC solution and 15% ACN.

5. Coupling buffer: 100 mM sodium acetate, pH 5.5, 150 mM NaCl.

6. 100 mM of dithiothreitol (DTT), freshly prepared in 50 mM AMBIC solution.

7. Affi-Gel Hz Hydrazide gel bead suspension containing ~50% beads by volume (catalog #153-6047, BioRad, Hercules, CA).

8. IAA solution: 250 mM of iodoacetamide, freshly prepared in 50 mM AMBIC solution.

9. 100% methanol.

10. MilliQ water: Milli-Q® Ultrapure MS-grade water (Millipore).

11. Minispin columns: Pierce Centrifuge Empty Columns (#89868, Pierce, Rockford, IL) to fit in 1.5 mL collector tubes, with top and bottom caps.

12. 300 mM of sodium sulfite (Na_2SO3).

13. 1.5 M sodium chloride.

14. 150 mM of freshly prepared sodium *meta*-periodate ($NaIO_4$).

15. PNGase F: N-glycosidase F, recombinant, reconstituted as recommended by the manufacturer (Roche, Mississauga, Canada, Cat. #11365185001).

16. Protease Inhibitor Cocktail (Sigma, catalog #P8340) or equivalent.

17. Protein concentration assay kit: Bradford (BioRad) or Lowry (Pierce).

18. Rotator: capable of rotating 1.5 mL conical tubes end-over-end.

19. SpeedVac® Concentrator SPD111V (Thermo Scientific, Waltham, MA).

20. Trypsin Gold, Mass Spectrometry Grade (Cat. # V5280, Promega, Madison, WI).

21. Urea-AMBIC solution: 8 M Urea and 0.4 M AMBIC solution.

2.4. Mass Spectrometry and Bioinformatics Software

1. An online nanoflow liquid chromatography (nanoLC) system with a reverse phase nanoLC column such as the highly reproducible nanoAcquity UPLC system (Waters, Millford, MA) with a 100 μm I.D. 10 cm × 1.7 μm BEH130C18 column (Waters).

2. A high-resolution MS instrument is connected to the nanoLC system that is capable of performing electrospray ionization (ESI) directly on the eluting peptides followed by full MS scans on precursor ions with high mass accuracy such as a hybrid quadrupole time-of-flight Q-TOF™ Ultima (Waters).

3. A tandem MS instrument capable of performing ESI directly on the eluting peptides followed by high-throughput MS/MS analysis such as the LTQ XL™ (Thermo) or the QTOF™ Ultima (Waters). The LTQ XL™ is the instrument of choice due to its ability to do high-throughput sequencing. In our laboratory, the LTQ XL™ can obtain 7–10 times more MS/MS spectra than QTOF Ultima.

4. LC-MS and LC-MS/MS data may be converted into a number of data formats that are compatible for analysis by software such as MSight, Mascot or MatchRx. The software that comes with the MS instrument or a number of available tools from http://tools.proteomecenter.org, such as Mass Wolf can also be used.

   (a) MSight software is a visualization tool (19) available free of charge from http://www.expasy.ch/MSight. It allows graphical representation of the LC-MS and LC-MS/MS data. MSight version 1.0 was used here.

   (b) Mascot® software (Matrix Science Ltd., London, UK) is a probability-based search engine for identifying peptide sequences from the nanoLC-MS/MS data using protein database searching (20). Mascot® version 2.2.0 was used here.

   (c) MatchRx software (from NRC-IBS) extracts peptide abundance values from LC-MS data and allows quantitative comparison of peptide levels in two or more samples (21). MatchRx also overlays the quantitative differences of peptides and LC-MS/MS identification results (from Mascot) into MSight images. MatchRx version 4.0 was used here.

3. Methods

Methods for extracting proteins from membrane or secreted fractions of BEC or other cells in the neurovascular unit, a detailed method for the enrichment of N-linked glycoproteins using hydrazide capture (Fig. 1) (15, 16), and their subsequent MS

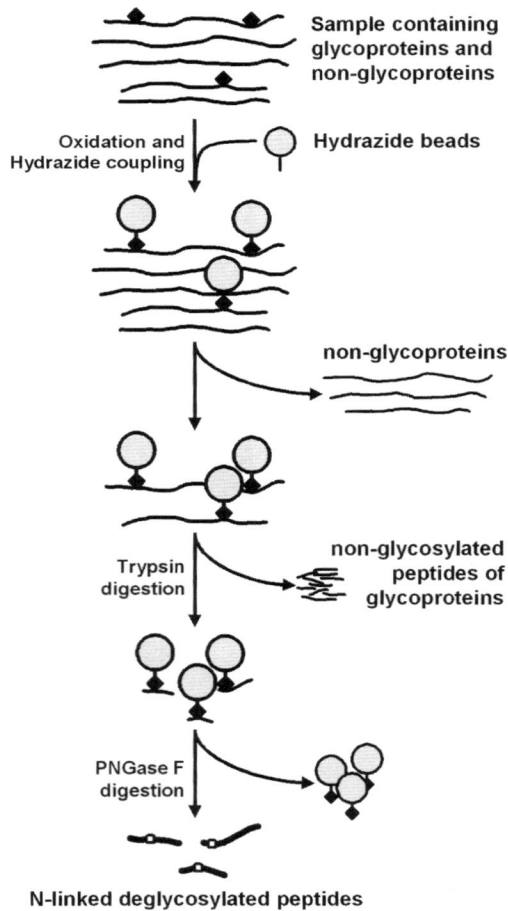

Fig. 1. Isolation of N-linked glycopeptides using hydrazide capture.

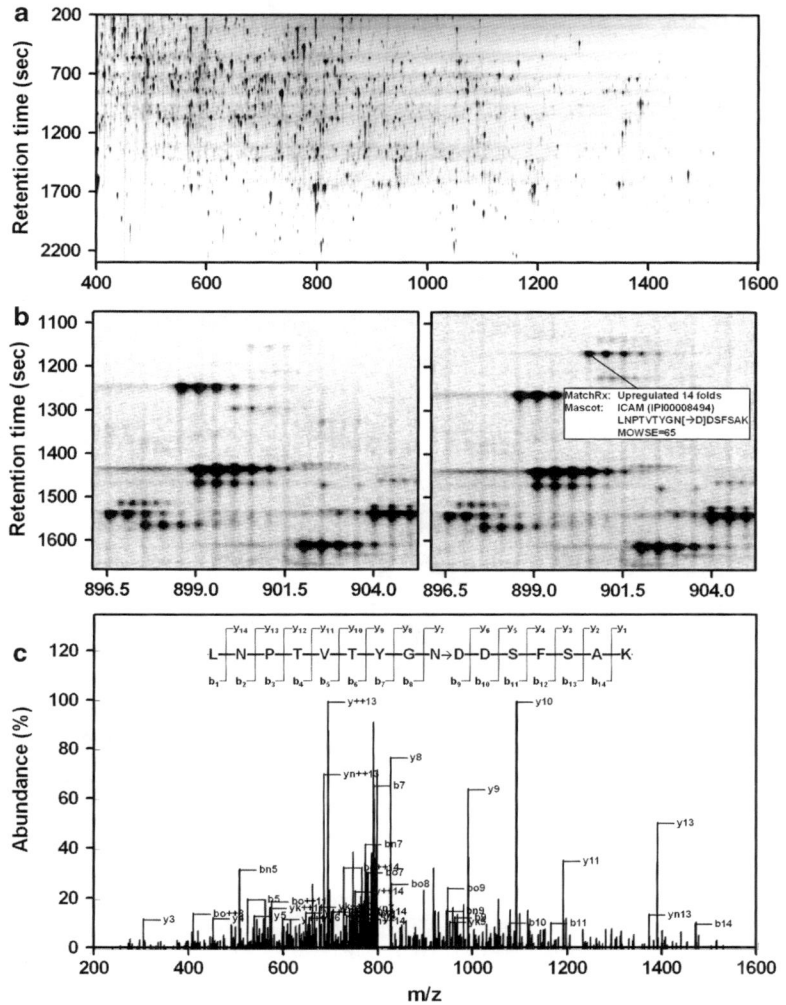

Fig. 2. NanoLC-MS and nanoLC-MS/MS analysis of hydrazide-captured glycopeptides from BBB total membranes. (**a**) An image representing the nanoLC-MS data. Each spot represents a deglycosylated peptide isolated by the capture. (**b**) A portion of nanoLC-MS image under two biological conditions (*left*: control, *right*: interleukin-1beta-treated). Label indicates a glycopeptide showing >14-fold overexpression (as determined by MatchRx) in response to interleukin-1beta. The peptide was sequenced by Mascot to be LNPTVTYGN[→D]DSFSAK and belongs to ICAM-1 protein. (**c**) NanoLC-MS/MS spectrum of the peptide LNPTVTYGN[→D]DSFSAK. Indicated are b- and y-fragment ions of the peptide. Note that deglycosylation using PNGase F results in N→D conversion as shown in (**b**, **c**).

analysis using label-free proteomics are described (Fig. 2) (21). The membrane isolation method can be used to either examine total membranes or further separate the membranes into luminal and abluminal fractions. In addition, the method for isolation of detergent-resistant membrane microdomains (lipid rafts) from BEC can also be used as described previously (11) and in Chap. 13.

3.1. Protein Isolation

3.1.1. Total, Luminal, and Abluminal Membrane Proteins from Human BEC

1. Harvest at least 1×10^6 human BEC in PBS by scraping from culture plates or flasks.

2. Collect cells and centrifuge at $500 \times g$ for 5 min.

3. Aspirate off the supernatant.

4. Resuspend the pellet in 0.5 mL of Tris-HCl buffer.

5. To lyse the cells, sonicate the suspension for 4–5 cycles of 15 s each at a low setting of three, followed by 20 s on ice.

6. Centrifuge at $10,000 \times g$ for 10 min at 4°C to pellet the cell debris and nuclei.

7. Transfer supernatant to a fresh ultracentrifugation tube and dilute to 3.5 mL using Tris-HCl buffer.

8. Ultracentrifuge at $100,000 \times g$ for 1 h at 4°C.

9. Remove the supernatant containing cytoplasmic proteins.

10. The pellet contains total BEC membranes and, if desired, can be used for hydrazide capture.

11. Redissolve the membrane pellet in 1 mL of Tris-HCl buffer 1 and sonicate as described in **step 5**.

12. Carefully layer the membrane sample on top of 2.5 mL Tris-HCl Buffer 2 in a fresh ultracentrifugation tube.

13. Ultracentrifuge at $165,000 \times g$ for 2 h at 4°C.

14. Collect the turbid layer at the interface and dilute to 3.5 mL with Tris-HCl Buffer 1. The pellet at the bottom is the abluminal BEC membrane fraction and, if desired, can be either stored at –80°C for up to 1 month or used for hydrazide capture.

15. Ultracentrifuge at $100,000 \times g$ for 0.5 h at 4°C.

16. The pellet contains luminal BEC membranes and can be used for hydrazide capturing or stored at –80°C for up to 1 month. It may be necessary to validate enrichment of luminal proteins (see Subheading 3.5).

3.1.2. Cell Secreted Proteins from Culture Medium of BEC or Astrocytes

1. Grow at least 1×10^6 cells in serum-free conditions for 24 h or more to allow enough proteins to accumulate in the culture medium (see Note 1).

2. Collect the culture medium and filter through a 0.22-μm syringe filter to remove any floating cells.

3. Concentrate and desalt the medium to less than 0.5 mL using buffer-exchange/concentrating columns using the manufacturer's protocol.

4. Precipitate the proteins by adding ten-volumes of ice-chilled acetone and incubating at –20°C for at least 1 h.

5. Centrifuge at $10,000 \times g$ for 10 min to pellet the proteins.

6. Decant the supernatant.

7. The pellet contains secreted proteins and can be used for hydrazide capturing or stored at −80°C for up to 1 month.

3.2. Hydrazide Capture of Glycopeptides and Elution of N-Linked Deglycosylated Peptides (see Fig. 1)

1. The protein pellet, containing either the total or luminal membranes (see Subheading 3.1.1) or secreted proteins (see Subheading 3.1.2), is dissolved in 100 μL of coupling buffer. Measure the protein concentration using the Bradford, Lowry or another protein assay. Transfer 150–250 μg of protein to a fresh tube and adjust the concentration to about 1.0 mg/mL using the coupling buffer. Add the protease inhibitor cocktail.

2. Add $NaIO_4$ from a 10× stock solution to a final concentration of 15 mM. Incubate the sample at room temperature in the dark for 1 h to oxidize the glycans of the glycoproteins.

3. Quench the oxidation reaction by adding Na_2SO_3 from a 10× stock solution to a final concentration of 30 mM. Incubate the sample at room temperature for 10 min to complete the quenching action.

4. Prepare an oxidized and quenched protein sample by diluting the sample to a final volume of 500 μL using coupling buffer. If desired, remove 5% of the sample for analysis by SDS-PAGE ("input" sample, see Note 2).

5. Transfer 300 μL of hydrazide beads (containing 50% hydrazide beads slurry) to a fresh 1.5-mL tube. Centrifuge at $3,000 \times g$ for 1 min to pellet the beads (see Note 3). Aspirate the supernatant. Resuspend the beads in 1 mL of coupling buffer. Repeat the wash two additional times for a total of three washes. The beads can stay in the final coupling buffer until ready to capture glycopeptides.

6. Couple glycoproteins to the hydrazide beads by removing the coupling buffer from the prewashed hydrazide beads by centrifuging at $3,000 \times g$ for 1 min and aspirating the supernatant. Add the oxidized and quenched protein sample from step 4 to the beads and mix by flicking the tube gently. Incubate at room temperature overnight in an end-over-end rotator to covalently couple the glycoproteins to the beads.

7. Collect unbound nonglycosylated proteins by centrifuging the sample at $3,000 \times g$ for 1.5 min. Carefully remove the supernatant containing nonglycosylated proteins to a fresh 1.5-mL tube. If desired, this fraction can be analyzed by SDS-PAGE ("unbound" sample, see Note 2).

8. Add 1 mL of Urea-AMBIC solution to the beads from step 7 and mix gently by flicking. Prewash a minispin column by adding 200 μL of the Urea-AMBIC solution and centrifuging at $10,000 \times g$ for 10 s to remove the wash. Transfer the

resuspended beads to the prewashed minispin column. Centrifuge the minispin column at $100 \times g$ for 1 min to remove liquid. Repeat, if necessary, to ensure complete transfer of beads.

9. Wash nonspecifically bound proteins from the hydrazide beads by washing the beads on the column 6 times with 500 μL of Urea-AMBIC solution, 3 times with 500 μL of AMBIC solution, and 3 times with 500 μL of AMBIC-ACN solution. For each wash, centrifuge at $100 \times g$ for 1 min to remove the wash. The flow-through may be discarded.

10. Cap the bottom of a minispin column and wrap in parafilm to ensure that there are no leaks. Add 450 μL of AMBIC-ACN solution to beads. Add 30 μL of trypsin, cap and parafilm the top of the minispin column and mix gently by flicking. Incubate at 37°C overnight in an end-over-end rotator to digest the hydrazide-coupled glycoproteins.

11. To reduce and alkylate cysteine residues add DTT from a freshly prepared 10× stock to a final concentration of 10 mM. Recap columns and incubate at 56°C for 1 h with occasional gentle mixing. Add IAA solution from a freshly prepared 10× stock solution to a final concentration of 25 mM. Incubate at room temperature in the dark for 1 h with occasional gentle mixing.

12. To collect released tryptic peptides (nonglycosylated), place each column in a clean 1.5 mL tube and centrifuge at $100 \times g$ for 1.5 min. Save the flow-through fraction, which contains nonglycopeptides released from glycoproteins by trypsin digestion.

13. Wash nonspecifically bound tryptic peptides from the hydrazide beads by washing the beads, 3 times with 500 μL of NaCl, 3 times with 500 μL of ACN-TFA solution, 3 times with 500 μL of MeOH, and 6 times with 500 μL of AMBIC solution. Perform the washes as described in **step 9**. After the final wash in the AMBIC solution, allow the extra wash solution to flow-through by gravity, but do not centrifuge, so that the beads remain fully hydrated.

14. To cleave N-glycopeptides, cap and parafilm the bottom of the minispin column. Add 100 μL of AMBIC solution and 7 μL of PNGase F to the sample and cap the top of the column. Incubate at 37°C overnight with rotation to enzymatically cleave N-linked glycopeptides from their hydrazide-coupled glycans.

15. Collect N-linked deglycosylated peptides by placing the column in a new 1.5-mL tube and centrifuge at $100 \times g$ for 1.5 min. Save the flow-through and place the column into a fresh 1.5 mL tube. Add 100 μL of 50 mM AMBIC solution

to the column and centrifuge as before. Repeat this step once. Combine the three flow-through fractions (total volume about 300 µL). Add 100 µL of ACN-AA solution to the spin column, centrifuge, and repeat. Combine this with the others, resulting in a total volume of ~500 µL. Dry the sample in a SpeedVac® until the volume is less than 10 µL. Add MilliQ water to a final volume of 30 µL. This sample can be stored at -80°C.

3.3. Relative Quantification of Peptides Derived from Glycoproteins Using NanoLC-MS

1. NanoLC-MS analysis: Separately inject 2–5 µL of N-linked deglycosylated peptides (from Subheading 3.2, step 15) and 10–20 µL of nonglycopeptides from glycoproteins (from Subheading 3.2, step 12) into a nanoLC system setup online to an ESI-MS instrument. Separate peptides on the nanoAcquity system by gradient elution (1–95% ACN, 0.2% formic acid) over 60 min at a flow rate of 400 nL/min. Acquire MS data on ions with mass/charge (m/z) values between 400 and 2,000 with 1.0 s scan duration and 0.1 s interscan interval (see Note 4).

2. Convert the LC-MS data for each run into mzXML format using public domain software. Mass Wolf was used to convert the QTOF Ultima data.

3. To visually examine the number of peptides and quality of each LC-MS run, use the MSight software to convert each mzXML file into a "2-D gel like" image. An example is shown in Fig. 2a (see Note 5).

4. A quantitative comparison of peptide levels in multiple samples can be done by using the MatchRx software and the following automated steps:

   (a) Identify isotopic distribution pattern, charge state, and quantitative abundance of peptides in each LC-MS run.

   (b) Align the multiple LC-MS runs.

   (c) Quantitatively compare the levels of each peptide in the runs to identify differentially expressed peptides.

   (d) Overlay the results on MSight images (see step 3) for visual verification.

   (e) Export the coordinates of the verified peptides (m/z and elution time) to an "include list" for sequence identification using LC-MS/MS (see Fig. 2b).

3.4. Identification of Peptides Derived from Glycoproteins Using NanoLC-MS/MS

1. LC-MS/MS analysis: Inject 2–5 µL of N-linked deglycosylated peptides (from Subheading 3.2, step 15) and 10–20 µL of nonglycopeptides from glycoproteins (from Subheading 3.2, step 12) separately into a nanoLC system setup online to an ESI-MS/MS instrument. Separate peptides by LC-MS/MS as described in Subheading 3.3. Acquire MS/MS data on 2+, 3+, and 4+ charged precursor ions with m/z values

between 400 and 2,000 using either data-dependent MS/MS mode (a full-scan MS is followed by MS/MS on most abundant peaks) or an "include list"-dependent MS/MS mode (a full-scan MS is followed by MS/MS on only the m/z values and elution times specified in the list) (see Note 6).

2. If required, convert the LC-MS/MS data into an appropriate format such as PKL, mgf, DAT or mzXML, using public domain software or software provided by the MS/MS instrument.

3. Search the acquired MS/MS spectra against a human protein database using Mascot® or another search engine. The following steps are used:

   (a) Specify trypsin enzymatic cleavage with one possible missed cleavage.

   (b) Allow variable modification of oxidation (+15.99 Da) at the Met residues.

   (c) Allow fixed modification of carbamidomethyl (iodoacetamide derivative; +57.02 Da) at the Cys residues.

   (d) For the sample containing N-linked deglycosylated peptides only, allow variable modification of deamidation (+0.9840 Da) for the conversion of glycosylated Asn to Asp residues upon deglycosylation with PGNase F.

   (e) Set parent ion tolerance at 0.5 Da.

   (f) Set fragment ion tolerance at 0.2 Da. An example of an N-linked deglycosylated peptide identified by Mascot® is shown in Fig. 2c.

4. Using MatchRx software, reformat the list of differentially expressed proteins identified by LC-MS/MS and import into MSight the LC-MS images to overlay the identified proteins. This gives a visual validation that the identified peptide is in fact the originally chosen ion (see Fig. 2b).

3.5. Validation Studies   It is recommended that the enrichment of luminal proteins in the membrane fractions and some of the results from the hydrazide capture protocol be validated using alternative molecular techniques.

1. It may be necessary to validate the enrichment of luminal proteins in these samples compared to total or abluminal membranes. This can be done by determining a known list of luminal and abluminal membrane markers using either western blotting or label-free quantitative proteomics (Table 1). This table shows results from label-free quantitative proteomics showing enrichment of luminal proteins and depletion of abluminal and contaminating proteins in a luminal fraction compared to a total membrane fraction.

Table 1
List of known luminal and abluminal membrane proteins in human brain endothelial cells and their relative ratios in luminal vs. total membrane fractions[a]

Proteins	Symbol(s)	Mean ratio[b]
Luminal		
Platelet/endothelial cell adhesion molecule	PECAM1/CD31	4.6
Activated leukocyte cell adhesion molecule	ALCAM/CD166	4.3
Multidrug resistance-associated protein 4	ABCC4/MRP4	3.8
P-glycoprotein 1	ABCB1/PGP	3.3
Monocarboxylic acid transporter 1	SLC16A1/MCT1	2.6
Multidrug resistance-associated protein 5	ABCC5/MRP5	1.7
Glucose transporter 1	SLC2A1/GLUT1	1.5
Intercellular adhesion molecule 1	ICAM1	[c]
Abluminal		
Multiple drug resistance protein 1	ABCC1/MRP1	<0.1
Organic anion transporter 2	SLCO2A1/OATP2	<0.1
Cell surface glycoprotein CD44	CD44	<0.1
Na+/K+ ATPase (various polypeptides)	ATP1A1, ATP1A2, ATP1B1, ATP1B3	0.1–0.5
Organic anion transporter 3	SLC22A8/OAT3	0.56
Various integrins		0.2–0.7
Contaminating proteins		
Various ER proteins		<0.2
Various golgi-associated proteins		<0.3
Various mitochondrial proteins		<0.6

Peptide ratio = (corrected abundance in luminal fraction)/(corrected abundance in total fraction).
Peptide abundance was calculated using MatchRx software and corrected for the amount of protein injected for nanoLC-MS analysis. Thus, mean ratios demonstrate the enrichment of luminal proteins and depletion of abluminal and contaminating proteins in the luminal fraction relative to the total fraction.
[a]Proteins isolated from luminal and total membrane fractions were trypsin digested and analyzed by label-free quantitative proteomics using nanoLC-MS and MS/MS (21).
[b]Each mean ratio is an average of the ratios of peptides originating from the protein in the two fractions
[c]Only detectable under inflammatory conditions.

2. To validate whether a protein identified from the hydrazide capture protocol is indeed glycosylated, a combination of immunoprecipitation, glycostaining, PNGase F treatment, and/or western blotting can be carried out. This can be combined with mutagenesis studies to confirm the site(s) of the N-glycosylation.

3. To validate the changes in expression of glycoproteins in response to treatment, western blotting, ELISA and/or immunochemistry can be used, provided that the antibodies used in these techniques recognize the glycosylated form of the protein.

3.6. Conclusions

The procedures described here have been successfully used to isolate membrane or secreted proteins from BECs and other cells of the neurovascular unit, enrich for glycoproteins using hydrazide chemistry and perform label-free quantitative proteomics to identify differential glycoprotein expression in multiple biological conditions ((18), unpublished data). We recently captured >150 glycoproteins from human BECs and found that >25 showed similar differential expressions in three biological replicates in response to an inflammatory stimulus (unpublished data). In addition, we have identified >140 glycoproteins in the secreted fraction of glioblastoma cells and found that 35 glycoproteins showed reproducible differential expression in response to cAMP analog treatment (18). We have also demonstrated that the hydrazide chemistry is efficient in capturing glycoproteins from complex biological samples, as confirmed by the absence of staining by the glycosylation-specific stain in the unbound fraction (18). In the Uniprot human database, 21% of the proteins are known to be glycosylated (Fig. 3a). Using hydrazide capture of glycoproteins from total membranes of human BEC, we observed that >90% of proteins are glycoproteins (Fig. 3b). Although most of these were previously known to be glycosylated, we were able to identify 23 novel glycoproteins that were not reported in the Uniprot database (Fig. 3b). Thus, the hydrazide method efficiently captures glycoproteins from complex biological samples, allowing significant glycoprotein enrichment as well as detection of novel glycoproteins.

In conclusion, hydrazide capture is a reproducible method that, when combined with sensitive label-free nanoLC-MS techniques and bioinformatics tools, enables quantitative profiling of a large number of glycoproteins from biological samples for the purposes of differential expression measurements and biomarker discovery.

4. Notes

1. For examining culture medium, it is important that the cells be incubated for the appropriate amount of time to allow enough proteins to accumulate in the medium. It may be necessary to measure the protein content at the end of the incubation to ensure that there is enough protein (>150 μg) for hydrazide capture. Usually one 10-mm plate of confluent

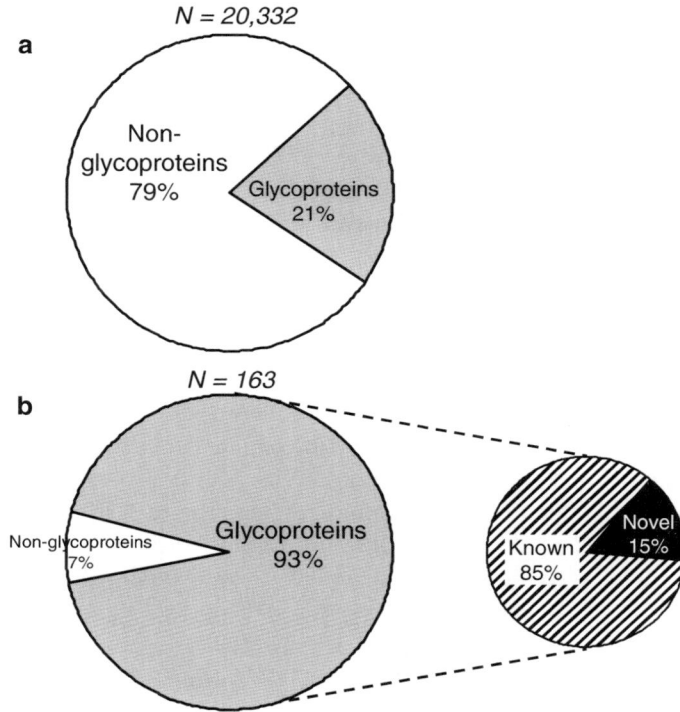

Fig. 3. Pie graphs showing the percentages of glycoproteins in (**a**) human protein database and (**b**) hydrazide captured proteins in human BECs. In (**a**), the glycoproteins represent the number of proteins that are known to be glycosylated as determined from the Uniprot human knowledgebase containing 20,332 entries (www.Uniprot.org). In (**b**), nonglycoproteins represent proteins that bind nonspecifically to the hydrazide beads. Although most of the glycoproteins identified by the hydrazide capture are known N-linked glycoproteins, 15% of these are novel glycoproteins, which are not reported in the Uniprot knowledgebase.

cells yields about 100–500 μg of protein, and multiple plates may be pooled if necessary. It is also very important that the medium is free of any serum or protein supplements. We usually grow cells for 1–3 days in serum-free and protein-free medium. To ensure that the medium is free of serum and supplements, cells should be washed at least 3 times with a buffered saline solution and then incubated in serum-free medium such as DMEM. For examining body fluids, depletion of abundant proteins such as albumin and IgGs may be necessary to allow the detection of lower abundant proteins (*22*).

2. If desired, the efficiency of the hydrazide capture can be monitored by SDS-PAGE. Collect the "input" and "unbound" samples as described in the protocol. Take an equal volume of both samples, add an SDS loading buffer, and separate the proteins by SDS-PAGE using standard methods. Next, visualize the glycoproteins by using a glycoprotein specific stain, such as ProQ Emerald (Pierce). If the capture efficiency is high,

glycoproteins that are visible in the input sample should have been removed in the unbound sample. Next, stain the same SDS-PAGE gel with a total protein stain, such as Sypro Ruby. This will allow for visualization of nonglycoproteins which will be present in both the input and unbound samples.

3. To maintain consistency between samples, it is important to handle the hydrazide beads with care to avoid damaging or losing the beads. Before any manipulation, the bead suspension should be carefully mixed by inversion. Avoid vortexing, high speed centrifugations, or rigorous mixing that may damage the soft agarose beads. When pipetting the bead suspension, it is best to cut the end of the pipette tip to produce a larger sized opening. Carefully pipette the bead solution up and down a few times before transferring the beads to ensure good mixing and consistent bead volumes in each sample. To minimize bead losses, it is helpful to set up a vacuum aspirator that has a gel loading tip or a blunt-ended syringe needle at the end for washes that take place in conical tubes.

4. Injecting at least one blank run in-between samples to clean out the column and prevent cross-contamination between samples is recommended.

5. If the file size is large, it may be necessary to split the files into two m/z ranges: 400–1,200 and 1,200–2,000.

6. The "include" list usually contains a list of m/z values, charge states, and retention times in a format accepted by the MS instrument. Using a m/z tolerance of 0.12 for QTOF and 0.5 for LTQ instruments is recommended. In addition, using a retention time tolerance of 100 s for nanoAcquity and >300 s for CapLC instruments is recommended.

Acknowledgments

Thanks are expressed to Dr. P.O. Couraud and his group (Institut Cochin, Université Paris Descartes, Paris, France) for providing the immortalized HBEC cells (hCMEC/D3) used in this study.

References

1. Varki A, Cummings RD, Esko JD, Freeze HH, Stanley P, Bertozzi CR, Hart GW, Etzler ME (2008) Essentials of Glycobiology. Cold Spring Harbor Laboratory Press, Plainview

2. Begley DJ, MW Brightman (2003) Structural and functional aspects of the blood-brain barrier. Prog Drug Res 61, 39–78

3. Lawrenson JG, Cassella JP, Hayes AJ, Firth JA, Allt G (2000) Endothelial glycoconjugates: a comparative lectin study of the brain, retina and myocardium. J Anat 196 (Pt 1), 55–60

4. Kumagai AK, Dwyer KJ, Pardridge WM (1994) Differential glycosylation of the

GLUT1 glucose transporter in brain capillaries and choroid plexus. Biochim Biophys Acta **1193**, 24–30

5. Michel CC (1996) Transport of macromolecules through microvascular walls. Cardiovasc Res **32**, 644–653

6. Vogel J, Sperandio M, Pries AR, Linderkamp O, Gaehtgens P, Kuschinsky W (2000) Influence of the endothelial glycocalyx on cerebral blood flow in mice. J Cereb Blood Flow Metab **20**, 1571–1578

7. Weinbaum S, Tarbell JM, Damiano ER (2007) The structure and function of the endothelial glycocalyx layer. Annu Rev Biomed Eng **9**, 121–167

8. Weinbaum S, Zhang X, Han Y, Vink H, Cowin SC (2003) Mechanotransduction and flow across the endothelial glycocalyx. Proc Natl Acad Sci U S A **100**, 7988–7995

9. Krum JM, More NS, Rosenstein JM (1991) Brain angiogenesis: variations in vascular basement membrane glycoprotein immunoreactivity. Exp Neurol **111**, 152–165

10. Man S, Ubogu EE, Ransohoff RM (2007) Inflammatory cell migration into the central nervous system: a few new twists on an old tale. Brain Pathol **17**, 243–250

11. Cayrol R, Wosik K, Berard JL, Dodelet-Devillers A, Ifergan I, Kebir H, Haqqani AS, Kreymborg K, Krug S, Moumdjian R, Bouthillier A, Becher B, Arbour N, David S, Stanimirovic D, Prat A (2008) Activated leukocyte cell adhesion molecule promotes leukocyte trafficking into the central nervous system. Nat Immunol **9**, 137–145

12. Zamze S, Harvey DJ, Pesheva P, Mattu TS, Schachner M, Dwek RA, Wing DR (1999) Glycosylation of a CNS-specific extracellular matrix glycoprotein, tenascin-R, is dominated by O-linked sialylated glycans and "brain-type" neutral N-glycans. Glycobiology **9**, 823–831

13. Hughes RC (1992) Role of glycosylation in cell interactions with extracellular matrix. Biochem Soc Trans **20**, 279–284

14. Vorbrodt AW, Dobrogowska DH, Lossinsky AS, Wisniewski HM (1986) Ultrastructural localization of lectin receptors on the luminal and abluminal aspects of brain micro-blood vessels. J Histochem Cytochem **34**, 251–261

15. Zhang H, Li XJ, Martin DB, Aebersold R (2003) Identification and quantification of N-linked glycoproteins using hydrazide chemistry, stable isotope labeling and mass spectrometry. Nat Biotechnol **21**, 660–666

16. Sun B, Ranish JA, Utleg AG, White JT, Yan X, Lin B, Hood L (2007) Shotgun glycopeptide capture approach coupled with mass spectrometry for comprehensive glycoproteomics. Mol Cell Proteomics **6**, 141–149

17. Weksler BB, Subileau EA, Perriere N, Charneau P, Holloway K, Leveque M, Tricoire-Leignel H, Nicotra A, Bourdoulous S, Turowski P, Male DK, Roux F, Greenwood J, Romero IA, Couraud PO (2005) Blood-brain barrier-specific properties of a human adult brain endothelial cell line. FASEB J **19**, 1872–1874

18. Hill JJ, Moreno MJ, Lam JC, Haqqani AS, Kelly JF (2009) Identification of secreted proteins regulated by cAMP in glioblastoma cells using glycopeptide capture and label-free quantification. Proteomics **9**, 535–549

19. Palagi PM, Walther D, Quadroni M, Catherinet S, Burgess J, Zimmermann-Ivol CG, Sanchez JC, Binz PA, Hochstrasser DF, Appel RD (2005) MSight: an image analysis software for liquid chromatography-mass spectrometry. Proteomics **5**, 2381–2384

20. Hirosawa M, Hoshida M, Ishikawa M, Toya T (1993) MASCOT: multiple alignment system for protein sequences based on three-way dynamic programming. Comput Appl Biosci **9**, 161–167

21. Haqqani AS, Kelly JF, Stanimirovic DB (2008) Quantitative protein profiling by mass spectrometry using label-free proteomics. Methods Mol Biol **439**, 241–256

22. Haqqani AS, Hutchison JS, Ward R, Stanimirovic DB (2007) Biomarkers and diagnosis; protein biomarkers in serum of pediatric patients with severe traumatic brain injury identified by ICAT-LC-MS/MS. J Neurotrauma **24**, 54–74

Part IV

Models to Study the Barriers

Chapter 17

Novel Models for Studying the Blood-Brain and Blood-Eye Barriers in Drosophila

Robert L. Pinsonneault, Nasima Mayer, Fahima Mayer, Nebiyu Tegegn, and Roland J. Bainton

Abstract

In species as varied as humans and flies, humoral/central nervous system barrier structures are a major obstacle to the passive penetration of small molecules including endogenous compounds, environmental toxins, and drugs. In vivo measurement of blood-brain physiologic function in vertebrate animal models is difficult and current ex vivo models for more rapid experimentation using, for example, cultured brain endothelial cells, only partially reconstitute the anatomy and physiology of a fully intact blood-brain barrier (BBB). To address these problems, we and others continue to develop in vivo assays for studying the complex physiologic function of central nervous system (CNS) barriers using the fruit fly *Drosophila melanogaster (Dm)*. These methods involve the introduction of small molecule reporters of BBB physiology into the fly humoral compartment by direct injection. Since these reporters must cross the *Dm* BBB in order to be visible in the eye, we can directly assess genetic or chemical modulators of BBB function by monitoring retinal fluorescence. This assay has the advantage of utilizing a physiologically intact BBB in a model organism that is economical and highly amenable to genetic manipulation. In combination with other approaches outlined here, such as brain dissection and behavioral assessment, one can produce a fuller picture of BBB biology and physiology. In this chapter, we provide detailed methods for examining BBB biology in the fly, including a *Dm* visual assay to screen for novel modulators of the BBB.

Key words: Subperineural glia, Blood-brain barrier, Xenobiotic, ABC transport, Chemoprotection, Live assays, Drug partition, Hemolymph, Retinal chemo-exclusion, Drosophila

1. Introduction

Considerable effort is being spent on the development and optimization of methods to allow drug transit across the blood-brain barrier (BBB). However, because the BBB is such a complex and intricately regulated structure, producing a comprehensive in vitro facsimile of the barrier using cultured cells has, at least to date,

Sukriti Nag (ed.), *The Blood-Brain and Other Neural Barriers: Reviews and Protocols*, Methods in Molecular Biology, vol. 686, DOI 10.1007/978-1-60761-938-3_17, © Springer Science+Business Media, LLC 2011

been problematic. The role of the BBB is tied to CNS function. Robust responsiveness (i.e., preserving the state of readiness) is critical to neuronal performance, as free-living organisms must be ready to respond to constantly changing environmental circumstances such as predation or food sources (1). All organisms must guard against the chemical world, since many naturally occurring toxins would cause rapid deterioration of CNS function if allowed to accumulate in neurons (2). This strong selective pressure favored the origin and elaboration of BBB-based chemoprotection in the majority of higher organisms (3). Unfortunately, even though molecular targets, toxicologic sensitivities and complex nervous system functions are conserved from worms and flies through man, little is known about the evolutionary origin of chemoprotective physiologies of the CNS.

In vivo BBB animal models, primarily in rodents, have confirmed the importance of the two components mentioned above and are the leading systems for brain-specific drug partition studies in the pharmaceutical literature (4, 5). Nevertheless, the anatomically compact nature of the vertebrate BBB makes live pharmacokinetic assays technically challenging and cumbersome for forward chemical or genetic screens. Furthermore, analysis of the multiplicity of individual components involved in chemical protection physiology is not practical because vertebrate life spans are long and the number of testable individual gene reduction permutations greatly limited. To fill this void, a number of in vitro models using polarized cells are available for testing the transport efficiency of compounds across barriers, but each system has substantial limitations (6). Madin–Darby canine kidney cells and composite glia/endothelium double cell layers are amenable to gene knockdowns and high throughput screens, but neither system fully recapitulates the elaborate physiologic constraints of in vivo BBB function. These systems also lack CNS or humoral inputs that might be essential for establishment and maintenance of barrier physiology. Hence, there is a pressing need for a system that can combine molecular genetic, genomic, chemical biology and integrative physiology tools to probe how CNS-specific chemoprotective physiologies and their control paradigms are made manifest. For this, we turned to the fruit fly *Drosophila melanogaster* (*Dm*) and asked what aspects of BBB physiology can be modeled in an invertebrate.

While substantial physiologic differences exist between flies and vertebrates in the content of their humoral space and the cell types and structure of the *Dm* CNS barrier, we will refer to this structure as the *Dm* BBB for simplicity. *Dm* has an open circulatory system in which hemolymph is pumped around the body and bathes organs externally. Here, the CNS is surrounded and separated from the humoral space by a thin layer of glial-derived epithelial cells (7–10), making the Dm humoral/CNS interface topologically much simpler than the vertebrate BBB. However on a cellular level, the vertebrate and insect BBBs share many common features. In particular, one specific cell layer of the *Dm* BBB, the subperineural glia (SPG), possesses elaborate laterally-localized homotypic junctional complexes, or pleated septate junctions, that create a tight barrier to paracellular diffusion (11, 12). The *Dm* proteins that make up the pleated septate junctions are nearly identical to the vertebrate proteins that compose the tight junctions (13, 14). Furthermore, disruption of the pleated septate junctions leads to defects in *Dm* BBB function (9, 15). Furthermore, *Dm* possess a full array of xenobiotic transporters from both the ATP-binding cassette (ABC) and solute carrier families that participate in active drug flux between biologic compartments (16–18). Together with tight paracellular borders, these transport mechanisms provide two means of protecting the CNS from xenobiotic and other threats (Fig. 1).

We have developed several assays for movement of xenobiotics across the BBB using *Dm*. These include real-time live imaging of the fly retina to assess blood retinal barrier (BRB) integrity in two modes (paracellular and transcellular routes of entry), quantitative measure of dye penetration into the brain parenchyma, confocal imaging of barrier layers and finally, quantitative measures of hemolymph drug concentration through hemolymph extraction. The retinal assay in particular takes advantage of using a barrier in which normal cell–cell interactions and regulatory systems are entirely intact. And by coinjecting a fluorescent effluxable substrate such as Rhodamine123 with a noneffluxable

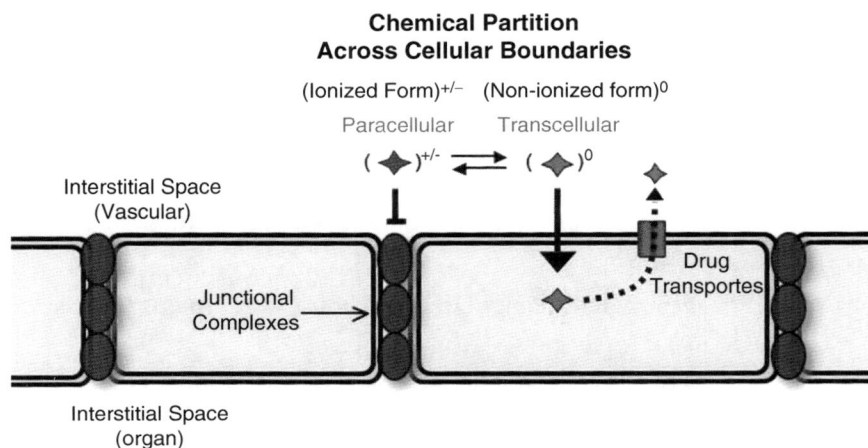

Chemical Partition Across Cellular Boundaries

Fig. 1. To form a barrier against xenobiotics, the BBB employs two main modes of exclusion: (1) tight paracellular junctions that exclude charged molecules and (2) a wide array of drug transporters that efflux charge neutral, lipophilic compounds back into the vascular space. A charge neutral dye is not excluded from WT or mutant animals as it can take a transcellular route for entry across the BBB.

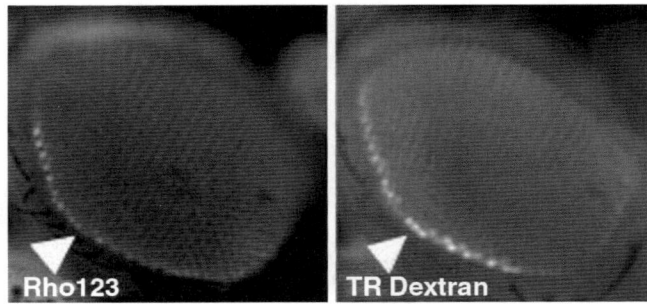

Fig. 2. Simultaneous assessment of transport and diffusion barriers is done by hemo-lymph coinjection of Rhodamine 123 (Rho123) and Texas Red Dextran (TRD). Images are taken live under CO_2 anesthesia 15 min post-injection using green and red filter sets respectively. Hemolymph exclusion from the CNS is seen as a bright line around the retina (*white triangles*). We call this the hemolymph exclusion line (HEL).

molecule such as 10 kDa Texas Red-dextrans, we can simultaneously assess both modes of exclusion at the barrier (Fig. 2).

Taken as a whole, *Dm* is a highly tractable to chemical and genetic manipulations. The ability to (1) screen for mutants defective in specific aspects of drug transport function, (2) express genes involved in drug transport in specific cell types in vivo, and (3) perform live assays of BBB function make *Dm* an ideal system to elucidate the physiology of the BBB in a robust and evolutionarily conserved model.

2. Materials

2.1. Intrahemolymph Dye Dosing and Retinal Image Acquisition

1. White minus (w⁻) *Dm* flies (Heberlein Lab, San Francisco, CA).
2. Cornmeal (Iowa Corn Processors Lc, Glidden, IA or Woolco Foods Inc, Jersey City, NJ).
3. Molasses agar medium (Moorhead, Rocklin, CA).
4. CO_2 gas.
5. Micromanipulator (Technical Instrument Co, San Francisco, CA).
6. Borosilicate needle (Sutter Instruments, Novato, CA).
7. Permeable pad (see Note 1).
8. Polystyrene food vial 28.5 mm wide × 95 mm high (Fisher Scientific, Houston, TX).
9. Round bottom bottles, 64 mm diameter × 103 mm height, 6 oz (Fisher Scientific).

10. 3 kDa Fluorescein isothiocyanate (FITC) dextran (D-3306, Invitrogen Inc, Carlsbad, CA, USA).

11. 0.48 Da Rhodamine B (RhoB) fluorescent dye (R6626, Sigma-Aldrich, St Louis, MO, USA).

12. 0.38 kDa Rhodamine 123 (Rho123) fluorescent dye (Sigma-Aldrich).

13. 0.33 kDa Fluorescein isothiocyanate sodium salt (FITC) fluorescent dye (F6377, Sigma-Aldrich).

14. Dissecting instruments including sharpened, fine point forceps.

15. 1 mL pipette with feather affixed to the tip (Fisher Scientific).

16. 0.1% SDS.

17. 1× phosphate buffered saline (PBS).

18. Plastic petri dish, 5 1/2 in. diameter for dissection (Fisher Scientific).

19. Dissecting light microscope (Zeiss, Santa Clara, CA)

20. Dissecting fluorescent microscope with two color/channel camera (Zeiss).

21. Costar Special Optics 96-well plates, black-walled (Fisher Scientific).

2.2. Brain Dissection and Quantitative Measure of Brain Specific Chemical Retention

1. Tecan SpectraFluorPlus Reader (MTX Lab Systems, Inc, Vienna, Virginia).

2.3. Assessing BBB Physiology In Situ

1. Cyclosporine A dissolved in H_2O (C1832, Sigma-Aldrich).

2. Fixative: 3.7% paraformaldehyde (Fisher Scientific) in PBS.

3. Blocking buffer: PBS, 5% goat serum, 4% Tween 20.

4. C219 monoclonal antibody which detects a 200 kDa protein: 1:50 (Abcam Inc, Cambridge, MA).

5. Secondary antibody: Goat AntiMouse FITC, 1:200 dilution (Invitrogen Inc).

6. DakoCytomation Fluorescent Mounting Medium (Dako, Glostrup, Denmark).

7. 10 kDa Tetramethyl-Rhodamine Dextran (TMRD) (D-1817, Invitrogen Inc).

8. 10 kDa Texas Red Dextran (TRD) (D-1863, Invitrogen Inc).

9. Zeiss LSM-510 laser scanning confocal microscope.

10. Glass coverslips (#286518, Corning, NY).

11. Glass slides (#12-550-33, Fisher Scientific).

12. Nail polish.

2.4. Drug Pharmacokinetics in Dm

1. Fine point needle (26G 3/8 in.).

2. 0.5 mL microfuge tubes (perforated at the very bottom).

3. 1.5 mL microfuge tubes.

3. Methods

3.1. Intrahemolymph Dye Dosing and Retinal Image Acquisition

Live imaging is a useful and highly reproducible means of probing the functional properties of the BBB. The goal of this method is to assess the integrity of the BBB in flies without invasive procedures. Various dyes are chosen for their size or charge and injected into a remote site of the fly. The presence or absence of a hemolymph exclusion line (HEL, see Note 2) is a real-time measure of BBB chemical permeability. This approach is remarkably handy for screening through various genetic backgrounds to find changes in barrier permeability to test dyes and assessing the chemical nature of small molecules, putative chemical modulators and genetic contributors to BBB function.

3.1.1. Intrahemolymph Injection

1. Prepare a micromanipulator fitted with a thin borosilicate needle that has been preloaded with 5 μL dye(s) of choice (FITC-D and/or RhoB).

2. Using the dissecting microscope, break the very end of the injection needle to induce flow of dye to tip of the needle.

3. Using a permeable pad (see Note 1), anesthetize a sufficient number of flies for the experiment (5–10 flies per experimental condition) using CO_2.

4. To immobilize the flies for accurate and reproducible injection, suction is applied through a small hole in a paraffin-covered vacuum box and the fly is held with forceps at the base of the wings.

5. Insert the microinjection needle between the posterior abdominal wall body segments of anesthetized flies under a dissecting microscope (Fig. 3, bright line seen in 0.5 s image).

6. Apply positive pressure to the needle under direct visualization over 1–2 s (just until the abdominal segments begin to separate) to deliver an average volume of 100 ± 25 nL dye per injection (range 70 130 nL) (see Note 3).

7. Allow animals to recover from injection in food vials at 25°C.

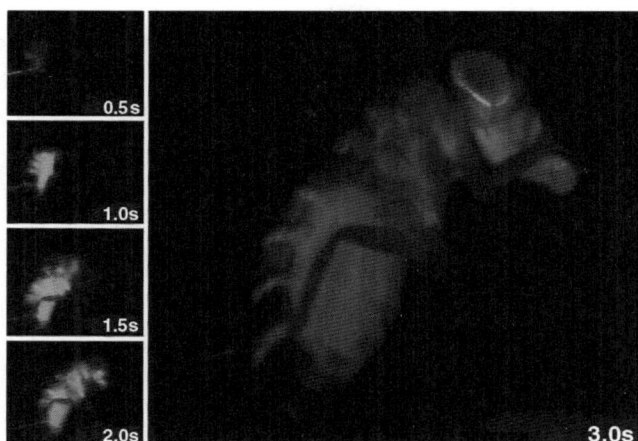

Fig. 3. Live whole animal dye injection and chemical partition in the hemolymph compartment. A series of images taken on a fluorescent dissecting microscope during a live injection of 3 kDa FITC dextran are shown. Note the rapid delivery of hemolymph around the entire animal and specific hemolymph partition at the retina/cuticle interface.

Table 1
Size of the tracers, their concentration and their excitation, and emission spectra are given

Tracer	Size (kDa)	Concentration(mg/mL)	Excitation (nm)	Emission (nm)
Fluorescein isothiocya-nate sodium salt	0.33	5	485	535
Fluorescein isothiocya-nate dextran	3.0	25	485	535
Rhodamine 123	0.38	1.25	485	535
Rhodamine B	0.48	1.25	535	595
Tetramethyl-Rhodamine Dextran	10.0	25	535	595
Texas Red Dextran	3/10	25	535	595

3.1.2. Visual Assessment of Drug Partition

1. Two hours following the dye injection, flies are CO_2 anesthetized and placed on a CO_2 pad at low flow to maintain narcosis.

2. Using the feather probe, gently orient the flies so that they are on their sides with the compound eye facing up toward the objective of the fluorescent dissecting scope.

3. Adjust the excitation (ex) and emission (em) of the scope in accordance with the dye that is being detected (see Table 1).

4. The retinal dye penetration phenotype is then scored qualitatively by assessment of sharpness and intensity of the HEL (see Fig. 4 for mutant extreme mutant phenotype).

WT   Moody null

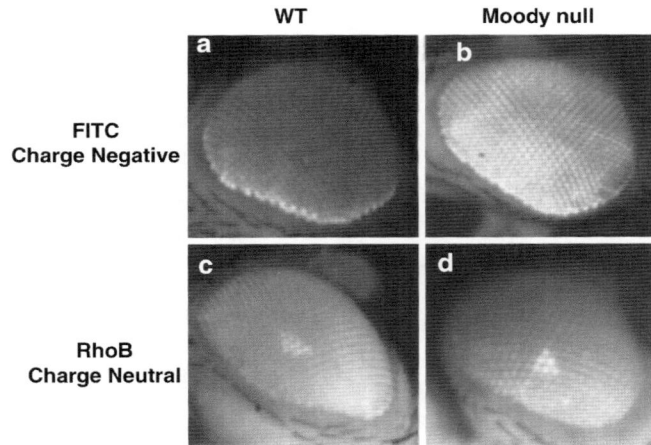

FITC
Charge Negative

RhoB
Charge Neutral

Fig. 4. Small molecule fluors have different properties in different mutant backgrounds. In *Dm*, FITC demonstrates exclusion from the retina of a white minus (W−) wild type (WT) animal. Note the distinct hemolymph exclusion line and lack of dye in the center of the retina (**a**). A moody null animal allows dye into the retina and demonstrates loss of the hemolymph exclusion line (**b**). Visual-retinal leakiness of the moody null BBB to small molecule dyes is corroborated by quantitative brain dissection assays (see Subheading 3.2). A charge neutral dye like RhoB has equal access to WT (**c**) and mutant flies (**d**) as it takes a transcellular route into the CNS by diffusing across lipid bilayers. Thus, choosing the appropriate fluorescent reporter is essential for the type of transport physiology to be tested.

3.2. Brain Dissection and Quantitative Measure of Brain Specific Chemical Retention

Once differences in retinal phenotypes have been observed, we can then look at how much or how little dye actually transits the BBB into the brain parenchyma. This quantitative fluorimetry approach affords this opportunity (see Note 4).

1. Rhodamine B and 3 kDa FITC dextran are dissolved in H_2O at 1.25 mg/mL and 25 mg/mL, respectively, and brought to neutral pH (see Notes 4 and 5).

2. Sibling animals of flies used for retinal assay are decapitated, and brains rapidly dissected (<90 s) from the cuticle, trachea and fat body in PBS (see Note 6).

3. With single forceps tip, brains are washed once in PBS and placed in the fluorimeter plate well (1 brain/well) containing 50 µL 0.1% SDS.

4. Incubate in SDS for 5–10 min at room temperature to allow the dye to dissociate from the brain tissue.

5. Dye released from brain samples is measured using a Tecan Genios fluorimeter in both the red (RhoB) and green channels (FITC) (see Table 1).

6. Average values of dye retained in CNS tissue are determined after subtracting the fluor counts or units obtained from

brains of noninjected flies (or SDS alone) from brains of injected flies (see Note 7).

3.3. Assessing BBB Physiology and Anatomy In Situ

To directly visualize the amount of dye leaking into the brain parenchyma, brains are dissected from the fly head and examined by confocal microscopy. Using a fluorescent dye such as tetramethylrhodamine dextran (TMRD) that will only pass through the barrier if it is compromised, one can, in combination with the above methods, produce a full picture of the integrity of the *Dm* BBB.

1. Animals to be immunostained with C219 antibody are injected with 200 μM cyclosporine A in H_2O (see Note 8) and 25 mg/mL, 3 kDa Texas Red dextran and allowed to recover overnight.

2. The next day, flies are anesthetized with CO_2, decapitated, and the proboscis removed with forceps.

3. Whole heads are placed in fixative solution for 15 min at room temperature.

4. Central brains are removed from the cuticle in PBS, carefully preserving brain surface structures, and washed with PBS.

5. Isolated brains are incubated in blocking buffer for 1 h and then probed in the C219 antibody overnight at 4°C.

6. Brains are washed 3 times for 30 min in PBS and probed with goat antimouse FITC for 45 min at room temperature.

7. Brains are washed 3 times for 45 min in PBS and mounted on glass slides using mounting medium.

8. Brains are placed on glass slides with ~40 μm posts (see Note 9) then covered with a coverslip and sealed with nail polish.

9. Laser and detector gain settings for fluorescent background noise are defined using brains with no primary antibody exposure.

10. At the coverslip interface, the brain is slightly pressed against the glass providing a flat brain interface with widths of 10–20 μm. This preparation allows for proper anatomic identification of dorsal-ventral brain orientation, and overall quality of the brain preparation.

11. Since the BBB is a continuous surface around the *Dm* CNS, the depth of confocality is changed to find a cross-sectional image.

12. To observe a tangential section, follow the edge of the brain to its greatest extent laterally. This provides the highest resolution of the apical-basal polarization of the BBB epithelia (see Fig. 5).

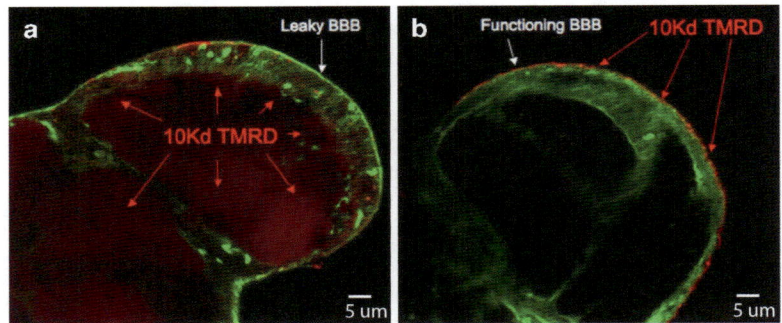

Fig. 5. Merged confocal images of the physiologic barrier to drug transport. Cross-sections of a *Dm* brain at the lobular plate hemolymph injected with 10 kDa tetramethylrhodamine dextran (TMRD). Both brains are marked with a pan glial driver (REPO-GAL4) crossed to a transgene expressing green fluorescent protein (GFP, *green channel*) using the UAS/GAL4 system of Brand and Perrimon (24). (**a**) The *moody* mutant shows a strong dextran signal (red) penetrating the brain. (**b**) The *moody* mutant rescued with Moody-GFP shows wild type BBB function. Here the dextran is fixed to the surface tissue outside the brain (red).

3.4. Drug Pharmacokinetics in Dm

To assess dye whole animal pharmacokinetics (half-life), we collect hemolymph injected whole flies at 0, 1, 2, 4, 8, and 16 h postinjection.

3.4.1. Dye Assessment in Whole Flies

1. All flies are injected with 100 nL mixture of RhoB and, FITC-D.

2. Put the 20 flies in a 1.5 mL centrifuge tube and add 200 µL of 0.1% SDS.

3. Crush the flies by hand for 1 min using a plastic micropestle.

4. Grind mechanically for another 1 min using a pestle motor and inspect the slurry to be sure flies have been completely homogenized.

5. Vortex for 30 s.

6. Centrifuge for 5 min at maximum speed ~18,590×g.

7. Repeat crush and spin once.

8. Samples are diluted 10× in 0.1% SDS, vortexed, centrifuged and 100 uL placed in a 96 well fluorimeter plate wells.

9. Fluorescent units are determined using a TECAN fluorimeter for FITC or RhoB (see Table 1). All samples are measured in the linear range of a standard curve for each fluor.

3.4.2. Hemolymph Drug/ Xenobiotic Concentration: Hemolymph Extraction Methods

This method is employed to compare transbarrier partition efficiency by assessing the chemical content of the humoral compartment. Hemolymph can be recovered postinjection from flies and drug half-life can be established for the humoral space. From this humoral/CNS partition, ratios can be established similar in principle to Lung and Wolfner (19).

1. Inject 25 flies with ~100 nL of a specific fluorescent dye (such as RhoB).

2. Select 20 of the injected flies and, using a fine needle, perforate the dorsal aspect of the fly at midthorax.

3. Collect these flies in porous filtered 0.5 mL tubes placed inside a larger 1.5 mL solid microcentrifuge tube.

4. Centrifuge this at 6,000×g for 15 min. Typically recovery is ~1–2 μL of hemolymph from the 20 flies.

5. Add 0.5 μL of the collected hemolymph in 50 μL 0.1% SDS and measure the fluorescence in a fluorimeter with excitation and emission settings for RhoB (see Table 1).

6. To calculate the hemolymph dye content or concentration:

$$(FI_h) = FI_h / V_m$$

Where (FI_h) is the hemolymph fluor concentration, FI_h is the measured fluor units and V_m is the volume of hemolymph measured.

4. Notes

1. The CO_2 pad is constructed from a 5 in. by 3 in. plastic box, the top of which is covered with a porous polyethylene sheet. One end of the box has a nozzle which is connected to a CO_2 regulator that will allow low but free flow of gas to flies placed on the surface of the pad.

2. The distinct HEL, combined with a lack of dye in the center of the retina, demonstrate that the BBB is physiologically intact. BBB vulnerability is demonstrated through observing a *moody* null animal, which has increased drug transport into the retina and loss of the hemolymph exclusion line. The *moody* null lacks the G-protein coupled receptor moody, a protein that has been demonstrated to affect the full integrity of the *Dm* BBB (20).

3. Fly hemolymph volume is estimated to be between 150 and 250 nL. With practice, very reproducible injection quantities can be introduced into the humoral compartment without significant injury to the animal.

4. Brain dissection of *Dm* is relatively easy and quite reproducible. Since the flies are grown under standard conditions and are genetically identical, it can be assumed that they possess nearly identical brain size. Fluorimetric analysis of

chemical content of the brain is easily correlated with live imaging described above. However, this method could be applied easily to radiolabeled drugs using a scintillation counter for quantitation. Furthermore, very small quantities of experimental compounds can be used, thus allowing chemical libraries to be tested as BBB modulators in a whole animal. Finally, individual dye chemistry can determine the transport pathway into and out of the brain and thus different physiologic properties of the BBB can be tested by choosing specific chemical properties of your reporter. RhoB, a charge neutral dye, will penetrate the brain and retina quickly and efflux slowly and is best for quantitative measures of efflux. Other dyes like fluorescein penetrate poorly in the wild type (WT) flies and are only found in the brain because of genetic or chemical perturbations and are thus best for live functional assessment of chemical partition.

5. FITC, a classic small molecule substrate for assessing paracellular barrier transport in vertebrates, is highly charge negative and shows exclusion from the retina of a white minus (W–) WT fly and is not a substrate for efflux transport in vertebrates (21). RhoB, another BBB effluxable dye, was used instead of Rho123, as RhoB produces better fluorescent signal in a quantitative assay. Dextrans are stable over very long periods of time in hemolymph and poorly penetrate the brain of WT flies (20).

6. For a very helpful demonstration on how to dissect a brain from *Drosophila*, you can direct your web browser to: http://jfly.iam.u-tokyo.ac.jp/html/movie/

7. Isolated brains from fluor uninjected animals contain insignificant intrinsic fluorescent signal.

8. Cyclosporine A holds ABC B1 in an open conformation and improves C219 antibody staining in situ (22, 23).

9. In order to allow sufficient space between the slide and coverslip so as to not crush the delicate brain, four posts are fashioned on the slide from drops of common nail polish, one for each corner of the coverslip.

Acknowledgements

This work was done with support of the NIH (GM081863) and the UCSF Department of Anesthesia and Perioperative Care, San Francisco, CA.

References

1. Davis GW (2006) Homeostatic control of neural activity: from phenomenology to molecular design. Annu Rev Neurosci 29:307–23

2. Sarkadi B, Homolya L, Szakacs G, Varadi A (2006) Human multidrug resistance ABCB and ABCG transporters: participation in a chemoimmunity defense system. Physiol Rev 86:1179–236

3. Abbott NJ (2005) Dynamics of CNS barriers: evolution, differentiation, and modulation. Cell Mol Neurobiol 25:5–23

4. Schinkel AH, Mayer U, Wagenaar E, Mol CA, van Deemter L, Smit JJ, van der Valk MA, Voordouw AC, Spits H, van Tellingen O, Zijlmans JM, Fibbe WE, Borst, P (1997) Normal viability and altered pharmacokinetics in mice lacking mdr1-type (drug-transporting) P-glycoproteins. Proc Natl Acad Sci U S A 94:4028–33

5. Nitta T, Hata M, Gotoh S, Seo Y, Sasaki H, Hashimoto N, Furuse M, Tsukita S (2003) Size-selective loosening of the blood-brain barrier in claudin-5-deficient mice. J Cell Biol 161:653–60

6. Garberg P, Ball M, Borg N, Cecchelli R, Fenart L, Hurst RD, Lindmark T, Mabondzo A, Nilsson JE, Raub TJ, Stanimirovic D, Terasaki T, Oberg JO, Osterberg T (2005) In vitro models for the blood-brain barrier. Toxicol In Vitro 19:299–334

7. Treherne JE, Pinchon Y (1972) The insect blood-brain barrier. In: Advances in Insect Physiology, JE Treherne, MJ Berridge, VB Wigglesworth, eds. London: Academic

8. Carlson SD, Juang JL, Hilgers SL, Garment MB (2000) Blood barriers of the insect. Annu Rev Entomol 45:151–74

9. Stork T, Engelen D, Krudewig A, Silies M, Bainton RJ, Klambt C (2008) Organization and function of the blood-brain barrier in Drosophila. J Neurosci 28:587–97

10. GA Kerkut, LI Gilbert, ed. (1985) Comprehensive Insect Physiology, Biochemistry, and Pharmacology, 5:115–37. London: Pergamon

11. Edwards JS, Swales LS, Bate M (1993) The differentiation between neuroglia and connective tissue sheath in insect ganglia revisited: the neural lamella and perineurial sheath cells are absent in a mesodermless mutant of Drosophila. J Comp Neurol 333:301–8

12. Tepass U, Hartenstein V (1994) The development of cellular junctions in the Drosophila embryo. Dev Biol 161:563–96

13. Banerjee S, Sousa AD, Bhat MA (2006) Organization and function of septate junctions: an evolutionary perspective. Cell Biochem Biophys 46:65–77

14. Wu VM, Beitel G J (2004) A junctional problem of apical proportions: epithelial tube-size control by septate junctions in the Drosophila tracheal system. Curr Opin Cell Biol 16:493–9

15. Schwabe T, Bainton RJ, Fetter RD, Heberlein U, Gaul U (2005) GPCR signaling is required for blood-brain barrier formation in drosophila. Cell 123:133–44

16. Mayer F, Mayer N, Chinn L, Pinsonneault RL, Kroetz D, Bainton RJ (2009) Evolutionary conservation of vertebrate blood-brain barrier chemoprotective mechanisms in Drosophila. Journal of Neuroscience 29(11):3538–50

17. Dean M, Annilo T (2005) Evolution of the ATP-binding cassette (ABC) transporter superfamily in vertebrates. Annu Rev Genomics Hum Genet 6:123–42

18. Torrie LS, Radford JC, Southall TD, Kean L, Dinsmore AJ, Davies SA, Dow JA (2004) Resolution of the insect ouabain paradox. Proc Natl Acad Sci U S A 101:13689–93

19. Lung O, Wolfner M F (1999) Drosophila seminal fluid proteins enter the circulatory system of the mated female fly by crossing the posterior vaginal wall. Insect Biochem Mol Biol 29:1043–52

20. Bainton RJ, Tsai LT, Schwabe T, DeSalvo M, Gaul U, Heberlein U (2005) Moody encodes two GPCRs that regulate cocaine behaviors and blood-brain barrier permeability in Drosophila. Cell 123:145–56

21. Nag S (2003) The Blood-Brain Barrier: Biology and research protocols. Humana, Totowa, NJ

22. van Den Elsen JM, Kuntz DA, Hoedemaeker FJ, Rose DR (1999) Antibody C219 recognizes an alpha-helical epitope on P-glycoprotein. Proc Natl Acad Sci U S A 96:13679–84

23. Demeule M, Vachon V, Delisle MC, Beaulieu E, Averill-Bates D, Murphy GF, Beliveau R (1995) Molecular study of P-glycoprotein in multidrug resistance using surface plasmon resonance. Anal Biochem 230:239–47

24. Brand AH, Perrimon N (1993) Targeted gene expression as a means of altering cell fates and generating dominant phenotypes. Development 118:401–15

Chapter 18

Zebrafish Model of the Blood-Brain Barrier: Morphological and Permeability Studies

Brian P. Eliceiri, Ana Maria Gonzalez, and Andrew Baird

Abstract

The blood-brain barrier (BBB) is a monolayer of endothelial cells that is regulated by the proximity of a unique basement membrane and a tightly controlled molecular interaction between specialized subsets of cells including pericytes, astrocytes, and neurons. Working together, these cells form a neurovascular unit (NVU) that is dedicated to the local regulation of vascular function in the brain and BBB integrity. Accordingly, the intrinsic complexity of the cell–matrix–cell interactions of the NVU has made analyzing gene function in cell culture, tissue explants, and even animal models difficult and the inability to study gene function in the BBB *in vivo* has been a critical hurdle to advancing BBB research.

Zebrafish has emerged as a premier vertebrate organism to model and analyze complex cellular interactions *in vivo* and genetic mechanisms of embryonic development. To this end, we provide a technical overview of the procedures that can be used in Zebrafish to analyze BBB integrity with a focus on the cerebrovasculature of adult fish where the BBB is now defined. The techniques that are used to measure the functional integrity, the cell biology, and the ultrastructure of the BBB include permeability assays, fluorescent imaging of reporter genes, and electron microscopy, respectively. Each can be applied to the functional analysis of mutant fish in ways that characterize the molecular sequelae to pathological insults that compromise BBB integrity. Due to the highly conserved nature of both the genetics and cell biology of zebrafish when compared with higher vertebrates, drug discovery techniques can be used in zebrafish models to complement drug development studies in other model systems.

Key words: Blood-brain barrier, Blood vessel, Vascular permeability, Drug screening, Genetic analyses, Zebrafish

1. Introduction

The blood-brain barrier (BBB) is a highly regulated multicellular partition separating blood from the brain parenchyma. Because of intrinsic difficulties in studying complex cellular interactions *in vivo*, preclinical studies of the BBB have focused on its pathophysiology, whereas cell biology techniques have been the

Sukriti Nag (ed.), *The Blood-Brain and Other Neural Barriers: Reviews and Protocols*, Methods in Molecular Biology, vol. 686, DOI 10.1007/978-1-60761-938-3_18, © Springer Science+Business Media, LLC 2011

primary tool to study the control of drug delivery in the central nervous system. Unfortunately, the *in vitro* cell culture techniques that have been developed to date have severe limitations that preclude a full understanding of the molecular control of barrier function let alone its physiological homeostasis. The development of multicellular coculture methods to mimic barriers has significantly increased our understanding of cell–cell interactions, but strategies to analyze the BBB in intact animal models have been limited.

In recent years, zebrafish have emerged as an ideal animal model system to study complex biological processes as, for example, the biology of development and growth. Specifically, the power of using forward and reverse genetics in zebrafish combined with the ease of generating large numbers of optically transparent embryos has resulted in the generation of large collections of mutant and transgenic reporter lines of fish that can be exploited to dissect the functional roles of specific cell types *in vivo*. As the smallest vertebrate model with a functional BBB and an endothelial cell-based vasculature, the range of genetic tools available to study Zebrafish are comparable to those available to study mice and make zebrafish a nearly ideal model to examine the cell–cell interactions *in vivo*.

An analysis of brain blood vessels in vertebrate reveals that there is a widely distributed and evolutionarily conserved requirement for a functional barrier between the lumen of blood vessels and the parenchyma of the brain. For example, in the case of teleost fish, the presence of tight junctions was inferred by electron microscopy (EM) and established by the restricted permeability of their BBB to classical markers of BBB function like horseradish peroxidase and sulfosuccinimidyl-biotin, which are consistent with the presence of tight junctions between endothelial cells. In addition, immunohistochemical markers localize specific molecular components of functional BBBs in the Zebrafish brain microvasculature including the tight junction proteins zonula occludens-1 and claudin 5 and association of astrocyte markers with endothelium (1). In addition to these histological and ultrastructural similarities between the BBB of zebrafish, rodents, and man, the methods discussed in this chapter describe an application of classic BBB tracers such as Evans blue dye to demonstrate the restricted permeability of brain blood vessels when compared to the fenestrated endothelium of zebrafish gills. Taken together, the findings are all consistent with the original observations of Paul Ehrlich (2) who first suggested the existence of a BBB that protects the brain of vertebrates.

While immunohistochemistry and transgenic reporter models are available to monitor the function of the vasculature in zebrafish embryos, there are simpler techniques that can be applied such as the endogenous alkaline phosphatase (AP) activity that is present in the endothelium and electron microscopy. While this staining technique has less resolution than required for most applications, it is inexpensive, straightforward, and rapid making it useful for screens, large scale applications, and teaching labs. Although non-endothelial cells also have endogenous AP activity, it was first described in human arterial endothelial cells and amphibian blood vessels (3, 4) and using a modified protocol (5–7), it provides a useful approach for examining developing vasculature.

2. Materials

2.1. Ex Vivo Tracking of Permeability Tracers in Zebrafish Adult Tissues

1. Zebrafish (Zebrafish International Resource Center, Eugene, OR; http://zebrafish.org/zirc).
2. 1 L Tank.
3. Tricaine: 0.4% in water (3-aminobenzoic acid ethyl ester, Cat No. 5040, Sigma-Aldrich, St. Louis, MO) with 1% $Na_2HPO_4 \bullet 2H_20$, pH 7.2. 6 mL of 0.4% solution is used per 100 mL of swimming water.
4. Evans blue dye: 1% in water prepared fresh daily (Cat No. E2129, Sigma-Aldrich).

2.2. In Vivo Tracking of Permeability Tracers in Live Zebrafish Embryos

1. Depression Slide (Cat No. 48339-009, VWR, West Chester, PA).
2. Olympus SZX 10 stereoscope (Lehigh Valley, PA) with appropriate filter cubes.
3. Olympus BX61 fluorescence microscope equipped with the Fluoview 1000 laser scanning confocal imaging system.

2.3. Endothelial Cell Staining

1. 50× stock 1-phenyl-2-thiourea (PTU) 0.01 M in water.
2. Fixatve solution: 3.7 % paraformaldehyde in PBS.
3. Permeabilization buffer: 100% methanol at -20°C.
4. Nitro blue tetrazolium (NBT)/5-Bromo-4-chloro-3-indoyl-phosphate (BCIP) stock solution: 18.75 mg/mL NBT + 9.4 mg/mL BCIP in 67% dimethyl sulfoxide (Cat No. 11 681 451 001 Boehringer Mannheim, Mannheim Germany).
5. Developing buffer: 0.1 M Tris-HCl, pH 9.5, 0.1 M NaCl and 0.05 M $MgCl_2$.
6. NBT/BCIP substrate solution: 200 µL of NBT/BCIP stock solution in 9.8 mL developing buffer.

2.4. Electron Microscopy

1. EM fixative: 2% paraformaldehyde, 2% glutaraldehyde in 0.1 M sodium cacodylate, pH 7.4.
2. Osmium tetroxide solution: 1% osmium tetroxide and 0.5 % potassium ferrocyanide.
3. Ethanol: 70, 95, and 100%.

4. Propylene oxide.

5. Embed 812 Resin (Electron Microscopy Sciences, Hatfield, PA)

6. 2% Uranyl acetate (Ted Pella, Redding, CA)

7. Reynolds Lead citrate: 2.66 g lead citrate, 3.52 g sodium citrate to 150 mL boiled water followed by addition of 0.8 g sodium hydroxide (Electron Microscopy Sciences, Hatfield, PA)

8. Zeiss 10C Electron Microscope, 100 kV (Zeiss, Peabody, MA).

3. Methods

3.1. Ex Vivo Tracking of Permeability Tracers in Zebrafish Adult Tissues

In several instances, it is useful to characterize the BBB in transgenic fish or in animals where function has been genetically modified. In these instances, we use permeability tracers that can be injected into anesthetized zebrafish and the tissues are analyzed *ex vivo*. This protocol is amenable to modifications with other fluorescent tracers (see Notes 1 and 2); however, here we focus on a classical marker of BBB breakdown due to the capacity of Evans blue dye to bind to serum albumin.

1. Anesthetize adult zebrafish in tricaine added to the swimming water to immobilize them.

2. Positioning the zebrafish to expose the ventral aspect, inject 50 μL of 1% Evans blue dye into the sinus venosus with a 30-gauge needle attached to a 0.5 mL syringe.

3. Return the zebrafish to the swimming water to recover for 10 min.

4. Anesthetize adult zebrafish in tricaine by immersion in a dish for 5 min, dissect anterior, and mount head and gill arches, remove the gill plate with care, and place it on a glass slide.

5. Cover the tissue with a 24×50 mm cover slip supported on each end by a stack of 22×22 mm cover slips to match the height of the specimen.

6. Add PBS as needed to the edges of the cover slip with a transfer pipette to prevent drying of the tissue.

7. Image fluorescence of the Evans blue dye with a 568 nm or equivalent filter set using a fluorescence stereoscope, upright fluorescence compound microscope or upright confocal microscope.

3.2. In Vivo Tracking of Permeability Tracers in Live Zebrafish Embryos

The capacity to analyze large numbers of zebrafish embryos combined with markers of BBB integrity is a novel approach focused on the identification of BBB components based on forward or reverse genetic screens.

1. Transfer 48–168 h embryos to 3–4 wells of a 24 well dish.

2. Add Evans blue dye to the swimming water at a range of recommended concentrations (0.01, 0.1 and 1%), dependent on age and desired intensity of stain and incubate for 20 min.

3. Carefully remove the dye-labeled water with a transfer pipette, leaving embryos in the well.

4. Wash 5–10 times with swimming water to remove excess dye until the swimming water is clear.

5. Anesthetize to immobilize for imaging by the addition of tricaine to swimming water.

6. Transfer embryos to a depression slide for imaging on a fluorescence stereoscope, microscope or CCD imaging system (see Notes 3 and 4).

7. Embryos may be returned to normal swimming water for at least 48 h further if imaging time is kept to a minimum and the dose of tricaine minimized to avoid overdose.

3.3. Endothelial Cell Staining

In cases where it is necessary to identify general vascular structure in intact fish, the following staining protocol is a useful and practical technique without having to rely on immunohistochemistry or transgenic models.

1. Treat embryos at 24 h postfertilization with PTU to prevent melanization and facilitate imaging by addition to swimming water

2. Fix in fixative solution overnight at 4°C.

3. Place in permeabilization buffer for at least 1 h at −20°C. Embryos may be stored at this point in 100% methanol at −20°C if needed.

4. Wash in Developing buffer twice for 10 min each.

5. Incubate in NBT/BCIP substrate solution for 15–30 min, monitoring color for development.

6. Before the sample overdevelops, wash three times in Developing buffer, 5 min each.

7. Fix in fixative solution for at least 30 min and store at 4°C.

8. Mount on depression slide and image using a stereoscope (Fig. 1).

3.4. Electron Microscopy

To characterize the ultrastructure of capillaries and associated basal lamina and astrocytes, electron microscopy is necessary (Fig. 2).

1. Anesthetize intact zebrafish in tricaine for 5 min.

2. For embryos, fix intact embryos in EM fixative, and for adults, remove the brains first and then fix for 12–16 h.

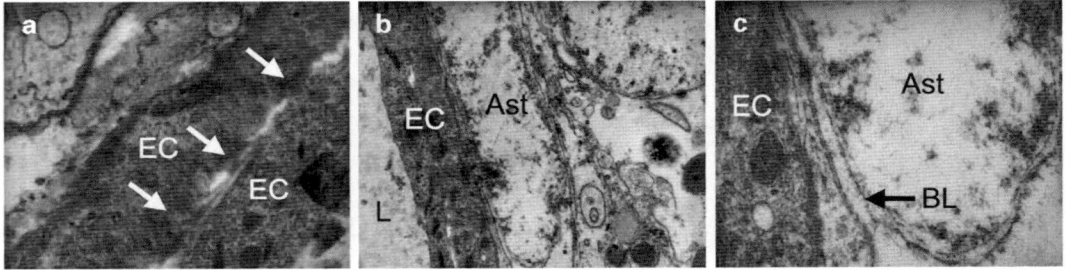

Fig. 1. Electron micrographs show the blood-brain barrier in zebrafish. Adult zebrafish brains were fixed, embedded, and sectioned using standard techniques and imaged using a transmission electron microscope. (a) The junctions formed between overlapping edges of brain endothelial cells (EC) are shown (*arrows*). (b) The presence of an astrocyte (Ast) endfoot adjacent to an endothelial cell and the lumen (L) are indicated. (c) The basal lamina (BL) constituting the extracellular matrix component of the blood-brain barrier found in higher vertebrates is shown between the astrocyte endfoot and endothelial cell. (a) ×25,000; (b) ×12,000; (c) ×50,000.

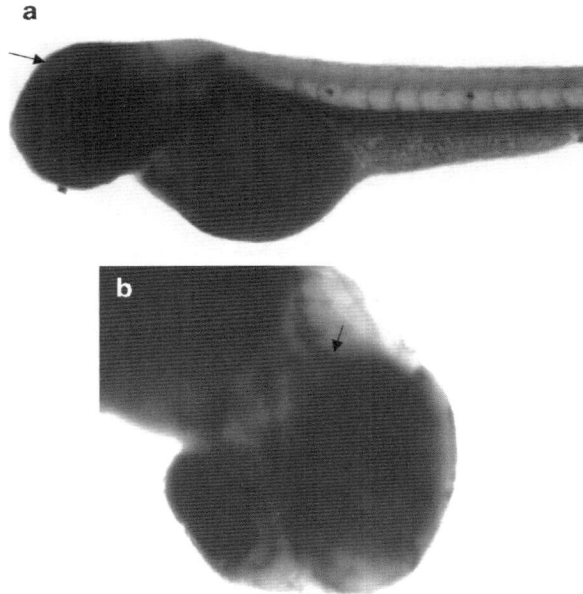

Fig. 2. Alkaline phosphatase staining of blood vessels in intact zebrafish embryos is shown. Seventy-two hour zebrafish embryos were fixed and subjected to incubation with NBT/BCIP alkaline phosphatase substrate to localize blood vessels. The image was acquired with an Olympus SZX10 stereoscope and CCD camera. Lateral (a) and ventral (b) images of cranial blood vessels (*arrows*) are shown.

3. Rinse embryos in PBS three times, and fix in osmium tetroxide buffer for 1 h followed by alcohol dehydration 10 min each with 70% ethanol, 95% ethanol, and three changes of 100% ethanol.

4. Incubate dehydrated embryos in two changes of 100% propylene oxide for 15 min each.

5. Infiltrate tissue in a 50% mixture of resin in 50% propylene oxide for 18 h, followed by 1 h in 100% resin, and a final 100% resin incubation. Polymerize for 18 h in a 60°C oven.

6. The block is ready for 60 nm thin sectioning.

7. Stain ultrathin sections with uranyl acetate for 30 min and Reynold's lead citrate for 15 min and examine using a transmission electron microscope.

4. Notes

1. For the imaging of fluorescent tracers, a wide array of various wavelengths and of various molecular weights are commercially available (Invitrogen, Carlsbad, CA). See also the following URL: http://www.invitrogen.com/site/us/en/home/References/Molecular-Probes-The-Handbook/Fluorescent-Tracers-of-Cell-Morphology-and-Fluid-Flow/Fluorescent-and-Biotinylated-Dextrans.html#head6.

2. The choice of fluorescent tracer is dependent on the hardware limitations of the available instrumentation. Selection of different molecular weights will affect the rate and bulk transfer of a given tracer in a specific vascular bed. Using the technique described in Subheading 3.2, in a 24-well dish format various tracers can be optimized based on molecular weight, concentration, and incubation time.

3. The methods described in this chapter are compatible with multiphoton confocal systems and spinning disk confocal, each generally optimized for superior imaging depth/resolution and faster kinetics, respectively.

4. Imaging with a deep cooled CCD imaging system was performed with a Lumina imaging system (Caliper Life Sciences, Hopkinton, MA) equipped with appropriate fluorescence filter cubes, background subtraction, and image integration software. This system is especially compatible with plate format screening of fluorescent and luminescent tracers.

Acknowledgments

This work was supported by grants from the NIH. The authors thank the contribution of Alexandra Borboa and Montha Pao toward several of these protocols.

References

1. Jeong JY, Kwon HB, Ahn JC, Kang D, Kwon SH, Park JA, Kim KW (2008) Functional and developmental analysis of the blood-brain barrier in zebrafish. Brain Res Bull 75:619–628

2. Ehrlich P (1904) Ueber die beziehungen von chemischer constitution, verteilung und pharmakologischer wirkung, in *Gesammelte Arbeiten zur Immunitaetsforschung*, Hirschwald, Berlin

3. Bannister RG, Romanul FC (1963) The localization of alkaline phosphatase activity in cerebral blood vessels. J Neurol Neurosurg Psychiatry 26:333–340

4. Stolk A (1963) Localized areas of high alkaline phosphatase activity in the endothelium of arteries in the axolotl. Experientia 19:21

5. Childs S, Chen JN, Garrity DM, Fishman MC (2002) Patterning of angiogenesis in the zebrafish embryo. Development 129:973–982

6. Schulte-Merker S (2002) Looking at Embryos, Oxford University Press, New York

7. Kamei M, Isogai S, Weinstein, BM (2004) Imaging blood vessels in the zebrafish. Methods Cell Biol 76:51–74

Chapter 19

Methods to Assess Pericyte-Endothelial Cell Interactions in a Coculture Model

Gokulan Thanabalasundaram, Jehad El-Gindi, Mira Lischper, and Hans-Joachim Galla

Abstract

The blood-brain barrier (BBB) comprises the microvascular endothelial cells, pericytes, and astrocytes, which are connected by the extracellular matrix (ECM). Current BBB models focus solely on the microvascular endothelial cells which constitute a physical barrier by formation of tight junctions (TJs), while the impact of pericytes on barrier regulation is poorly understood. We established a coculture model from primary porcine brain capillary endothelial cells (PBCECs) and pericytes (PBCPs) to approach the in vivo situation. This model allows the examination of pericyte impact on pharmacological, transport, migration, and metabolic activity of the BBB. In vivo the interaction between pericytes and endothelial cells is partly controlled by the ECM which is remodeled by matrix metalloproteinases (MMPs). Both endothelial cells and pericytes secrete MMPs which are important not only for ECM remodeling but also for TJ cleavage. In this chapter, current methods to study the interactions of these cell types by ECM signaling as well as MMP secretion are described.

Key words: Blood-brain barrier, Coculture model, Electric cell-substrate impedance sensing, Endothelial cells, Extracellular matrix, Matrix metalloproteinases, Pericytes, Transendothelial electrical resistance, Zymography

1. Introduction

The blood-brain barrier (BBB) is a unique, selective barrier formed mainly by the brain capillary endothelial cells (BCECs), which together with pericytes, astrocytes, and neurons constitute the neurovascular unit. BCECs in the cerebrum have one of the tightest barriers against para- and transcellular diffusion found in the body. Pericytes are able to regulate endothelial proliferation, differentiation, influence capillary blood flow, and synthesize structural constituents of the extracellular matrix (ECM) (1, 2).

Sukriti Nag (ed.), *The Blood-Brain and Other Neural Barriers: Reviews and Protocols*, Methods in Molecular Biology, vol. 686, DOI 10.1007/978-1-60761-938-3_19, © Springer Science+Business Media, LLC 2011

Furthermore, they are involved in specific microvascular diseases such as diabetic retinopathy (3) and were recently discovered to play a critical role in tumor angiogenesis (4).

An important part of the cell–cell interaction is mediated by the ECM which is composed of two layers called *lamina lucida* and *lamina densa* (5, 6) which together form a 30–40-nm thick ECM called basement membrane (7). The ECM consists of a variety of proteins and glycoproteins such as collagens, laminin, fibronectin, nidogen, and perlecan. These proteins are secreted locally and form an organized network which is in close association to the surface of the ECM-producing cells. The ECM exerts essential functions in integrin or growth factor mediated key signaling events during angiogenesis, cell adhesion, migration, proliferation, and differentiation (8, 9). This matrix is modulated among others by matrix metalloproteinases (MMPs) which are secreted by endothelial cells and pericytes (10). MMPs regulate the structure and function of the ECM by proteolytic modification of the ECM compounds during steady states and pathological conditions (11). Currently, 23 MMPs are known to be expressed in humans. Most of them are expressed as precursor zymogens and remain inactive until they are activated proteolytically. Active MMPs can be inhibited by tissue inhibitors of metalloproteinases (TIMPs) by forming MMP-TIMP complexes. The activity of MMPs depends on the balance between MMPs and TIMPs (12). MMPs are not only capable of ECM protein cleavage but may also be able to degrade tight junction proteins such as occludin and claudin-5 (13). Thus, BBB integrity could be dependent on MMP-expression by cells of the neurovascular unit.

Models of the BBB are important in increasing our understanding of how brain integrity is maintained and how the influx and efflux of endogenous and therapeutic substances are controlled. Improving in vitro models leads to more reliable tools for pharmacological studies and for improved understanding of BBB regulation and function under various conditions.

This chapter describes a BBB coculture model consisting of endothelial cells and pericytes which have been characterized previously (14–17). Methods for the measurement of the transendothelial electrical resistance (TEER), permeability, electric cell-substrate impedance sensing (ECIS), and methods to analyze MMP secretion and activity are described.

2. Materials

2.1. Culture Techniques

2.1.1. Isolation and Culture of Primary Porcine Brain Capillary Endothelial Cells

1. Ethanol (70%).
2. Phosphate buffered saline (PBS): 140 mM NaCl, 2.7 mM KCl, 8.1 mM Na_2HPO_4, 1.5 mM KH_2PO_4.
3. PBS Plus: PBS, 0.5 mM $MgCl_2$, 0.9 mM $CaCl_2$.

4. Preparation medium: Earle's medium M199, 0.7 mM l-glutamine, 1% (v/v) penicillin (10,000 U/mL)/streptomycin (10 mg/mL), 1% (v/v) gentamycin (10 mg/mL).

5. Plating medium: Earle's medium M199, 0.7 mM l-glutamine, 1% (v/v) penicillin (10,000 U/mL)/streptomycin (10 mg/mL), 1% (v/v) gentamycin (10 mg/mL), 10% (v/v) newborn calf serum (NCS).

6. Culture medium: Earle's medium M199, 0.7 mM l-glutamine, 1% (v/v) penicillin (10,000 U/mL)/streptomycin (10 mg/mL), 10% (v/v) NCS.

7. Serum-free medium (SFM): DMEM/Ham's F12 Medium, 4.1 mM l-glutamine, 1% (v/v) penicillin (10,000 U/mL)/streptomycin (10 mg/mL), 1% (v/v) gentamycin (10 mg/mL).

8. Dextran solution: 200 mL Earle's buffer (10x) containing phenol red, 4.4 g $NaHCO_3$, 360 g dextran (~160 kDa/mol), 2 L ddH_2O.

9. Percoll solution I ($\rho = 1.03$ g/cm^3): 400 mL PBS, 90 mL percoll, 10 mL Earle's medium M199 (10x).

10. Percoll solution II ($\rho = 1.07$ g/cm^3): 200 mL PBS, 270 mL percoll, 30 mL Earle's medium M199 (10x).

11. Protease/Dispase II (Becton Dickinson, Heidelberg, Germany).

12. Collagenase/Dispase II (Roche, Basel, Switzerland).

13. Puromycin (Sigma-Aldrich, Munich, Germany).

14. Collagen G (Biochrom, Berlin, Germany), 133 µg/mL in ddH_2O, 1 mL/75 cm^2.

15. Trypsin solution (0.25% w/v).

16. Dimethyl sulfoxide (DMSO) in 10% NCS (Sigma-Aldrich).

17. Nylon mesh, 180 µm pore size (Heidland, Harsewinkel, Germany).

18. Freezing medium: NCS containing 10% DMSO v/v.

2.1.2. Isolation and Culture of Primary Porcine Brain Capillary Pericytes

1. Pericyte medium: DMEM/Ham's F12 Medium, 2 mM l-glutamine, 1% (v/v) penicillin (10,000 U/mL)/streptomycin (10 mg/mL), 1% (v/v) gentamycin (10 mg/mL), 10% (v/v) NCS.

2.1.3. Coculture Model

1. Collagen IV is extracted from rat tails or acquired from Becton Dickinson Biosciences.

2. Trypsin/ethylenediaminetetraacetic acid (EDTA): 0.25% (w/v) trypsin in PBS, 1 mM Na-EDTA.

3. Hydrocortisone, 550 µM in 100% ethanol (Sigma-Aldrich, Munich, Germany).

4. Polycarbonate Membrane Transwell® Inserts, pore size 0.4 µm, membrane area 1.12 cm² (Corning Lifesience Schiphol-Rijk, Netherlands).

5. Jars, wide mouth, with screw cap (Cat. No. 216-8224, VWR International, Darmstadt, Germany).

6. Pincers.

2.2. TEER Measurement

1. CellZscope® – Automated Cell Monitoring System, nanoAnalytics, Münster, Germany.

2. Autoclaved ultrapure water.

2.3. Permeability Measurement

1. ^{14}C-sucrose (Amersham Biosciences, Buckinghamshire, UK).

2. Scintillation cocktail aqua safe 500 plus (Zinsser Analytic, Frankfurt, Germany).

3. Vial, Poly-Q (Beckmann Coulter Fullerton, CA, USA).

4. Liquid scintillation counter LC 6000 (Beckmann Coulter).

2.4. Electrical Cell-Substrate Impedance Sensing (ECIS)

1. ECIS™ Model 1600R (Applied BioPhysics Inc, Troy, USA).

2. ECIS 8W10E* Arrays (Applied Biophysics Inc).

3. Argon plasma cleaner (Harrick Plasma, New York, USA).

4. l-cystein: 10 mM, pH 7.

5. Collagen G (90% collagen I, 10% collagen III): 4 mg/mL in 12 mM HCl, pH 1.5.

6. ECM Isolation Solution 1: 1% (v/v) Nonidet P-40 (NP-40) in ultrapure water.

7. ECM Isolation Solution 2: 1% (w/v) desoxycholat and 1% (v/v) Nonidet P-40 (NP-40) in ultrapure water.

2.5. MMP Activity

2.5.1. Fluorogenic Substrate Based MMP-2/-7 Activity Assay

1. Fluorescence spectrometer: LS 50 B (Perkin Elmer, Überlingen, Germany).

2. Fluorescence cuvette: Type No.105.250 quartz, three windows, volume 100 µL (Hellma, Müllheim, Germany).

3. MMP-2/-7 fluorogenic substrate: MCA-Pro-Leu-Gly~Leu-Dpa-Ala-Arg-NH$_2$ ·TFA (Calbiochem, Merck Chemicals, Nottingham, UK). Prepare a 1-mM stock in methanol. Protect from light and store at –20°C.

4. MMP-2/-7 activity buffer: 200 mM NaCl, 50 mM Tris/HCl, 5 mM CaCl$_2$, 20 µM ZnSO$_4$, 0,05% Brij-35 (v/v), pH 7.4.

2.5.2. Gelatin Zymography

1. Prestained protein marker (New England Biolabs, Ipswich, USA).

2. 2× zymography-sample buffer: 20% (v/v) glycerin, 4% (w/v) sodium dodecyl sulfate (SDS), 2 mM EDTA, 125 mM Tris/HCl, 0.01% (w/v) bromophenol blue, pH 5.8.

3. Stacking gel buffer: 0.5 M Tris/HCl, 0.4% (w/v) SDS, pH 6.8.

4. Separating gel buffer: 1.5 M Tris/HCl, 0.4% (w/v) SDS, pH 8.8.

5. Acrylamide solution: 30% (w/v) acrylamide, 0.8% bisacrylamide in ddH$_2$O.

6. Gelatin from porcine skin (Sigma-Aldrich, Munich, Germany). Prepare a 1% (w/v) solution in ddH$_2$O. It can be stored at −20°C.

7. N,N,N′,N′-tetramethylethylenediamine (TEMED).

8. APS solution: 10% (w/v) ammonium peroxodisulfate in ddH$_2$O.

9. Running buffer: 25 mM Tris, 0.2 M glycine, 1% (w/v) SDS.

10. Developing buffer: 50 mM Tris/HCl, 200 mM NaCl; 5 mM CaCl$_2$, 20 μM ZnSO$_4$, 0.02% (v/v) Brij-35.

11. Renaturating buffer: 2.7% (v/v) Triton-X 100 in distilled water.

12. Destaining buffer 1: 50% (v/v) methanol, 10% acetic acid (v/v), 40% ddH$_2$O.

13. Destaining buffer 2: 10% (v/v) methanol, 10% acetic acid (v/v), 80% ddH$_2$O.

14. Staining solution: 0.25% (w/v) coomassie brilliant blue, 25% (v/v) isopropanol, 10% (v/v) acetic acid.

15. Gel drying frame (Kem-en-Tech, Copenhagen, Denmark).

16. Gel dry solution: 20% methanol, 2% glycerol, 78% ddH$_2$O.

3. Methods

3.1. Culture Techniques

The isolation and cultivation of primary porcine PBCECs which is one of the critical components needed for the coculture model is described.

3.1.1. Isolation and Culture of Primary Porcine Brain Capillary Endothelial Cells

1. Take porcine brains from freshly slaughtered 6 month old pigs and collect them in ice-cold 70% ethanol.

2. For the transport replace the ethanol by ice-cold PBS with 1% (v/v) penicillin/streptomycin (10 mg/mL).

3. Replace with fresh PBS containing streptomycin (200 μg/mL) and penicillin (200 U/mL).

4. After briefly flaming the hemispheres, remove the meninges and larger vessels carefully but completely.

5. Isolate the cortex from other brain areas such as cerebellum and brain stem. Then mechanically homogenize the cortex.

6. Supplement the homogenate with preparation medium to a final volume of 100 mL per brain.

7. Add 6.5 mL protease/dispase II and stir the suspension at 37°C for 2 h.

8. Mix 100 mL of the digested homogenate with 150 mL of 18% (w/v) dextran solution (4°C, MW ~160 kDa) and centrifuge at $6,800 \times g$ for 10 min at 4°C, for blood vessel isolation.

9. Discard the supernatant and resuspend the pellet containing the capillaries in 10 mL preparation medium per brain.

10. Filter this suspension through a nylon mesh (180 μm pore size) to remove larger blood vessels.

11. Triturate the capillaries up and down 5 times in a 10-mL sized pipette with the pipette tip placed directly onto the bottom of a petri dish to singularize the capillaries mechanically.

12. Digest with a mix of collagenase/dispase II (1 mg/mL) at 37°C for 30 min in a flask with a hanging stirrer (see Note 1).

13. Collect the isolated cells by centrifugation for 10 min at $110 \times g$ and resuspend in 10 mL preparation medium per percoll gradient. This discontinuous gradient should contain 15 mL percoll solution I ($\rho = 1.07$ g/mL) and 20 mL percoll solution II ($\rho = 1.03$ g/mL). Place 10 mL of the cell suspension carefully onto the top of the gradient and centrifuge (without brake) at $1,300 \times g$ for 10 min at 4°C. The cell load of each gradient is adjusted to correspond to 1–1.5 brains.

14. After centrifugation, isolate the gradient interface containing the endothelial cells and dilute in plating medium.

15. Centrifuge at $110 \times g$ for 10 min again and resuspend the cell pellet in plating medium containing 2 μg/mL puromycin.

16. Plate the cell yield of one brain on a total of 6×75 cm^2 collagen-G coated cell culture flasks.

17. After 2 days in culture, wash the cells twice with PBS and detach them with trypsin solution (0.25% w/v) at room temperature. Control the detachment of endothelial cells by light microscopy. Stop the digestion by addition of 8 mL plating medium as soon as most of the cells have lifted-off the culture flask.

18. Centrifuge the cell suspension at $110 \times g$ for 10 min and then resuspend in plating medium for direct use or in freezing medium for long-term storage. For direct use, seed the cells on a substrate and start the experiments after the cells become confluent. PBCECs can be identified by their typical spindle-shaped morphology (Fig. 1) and expression of von Willebrand factor (15). For storage, freeze the cells first at –70°C and store finally in liquid nitrogen at –196°C.

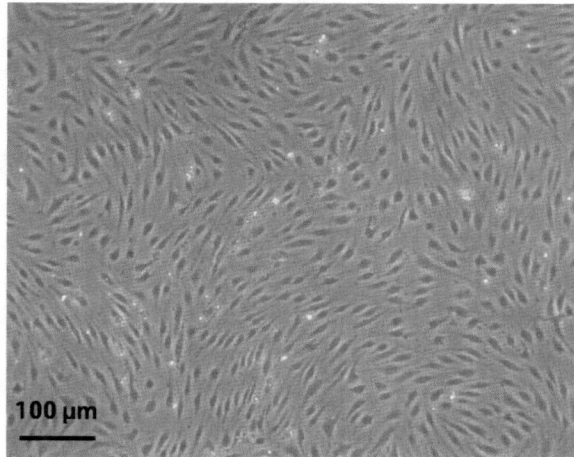

Fig. 1. The appearance of porcine brain capillary endothelial cells (PBCECs) by phase contrast microscopy is shown.

3.1.2. Isolation and Culture of Primary Porcine Brain Capillary Pericytes

The isolation and cultivation of primary porcine brain capillary pericytes (PBCPs) which is the second critical component needed for the coculture model is described.

1. The preparative steps are identical to the protocol described in Subheading 3.1.1 up to step 15.

2. Then resuspend the cell pellet in plating medium without puromycin.

3. Plate the cell yield from one brain on a total of 6×75 cm^2 collagen-G coated cell culture flasks.

4. After 2 days in culture, wash the cells twice with PBS and detach the endothelial cells with trypsin solution (0.25% w/v) at room temperature. Control the detachment by light microscopy. Stop the digestion by addition of plating medium as soon as ~90% of the cells have lifted-off the culture flask.

5. Remove the cell suspension containing the endothelial cells and add pericyte-medium into the flask.

6. Exchange the pericyte medium every 3 days and culture the pericytes for 14 days until they reach 80–90% confluence. Brain capillary pericytes can be identified by their typical stellate morphology with long extended processes (Fig. 2) and expression of α-smooth muscle actin (18).

7. Wash the cells twice with PBS and detach them with trypsin solution (0.25% w/v) at room temperature. Control the detachment of pericytes by light microscopy. Stop the digestion by addition of pericyte-medium as soon as most of the cells had lifted-off the culture flask.

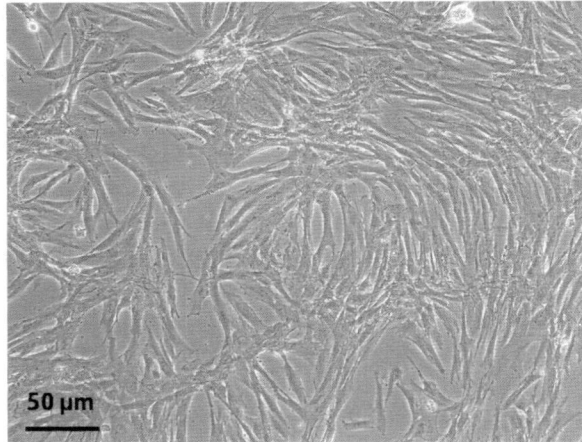

Fig. 2. The appearance of cultured porcine brain capillary pericytes by phase contrast microscopy is shown.

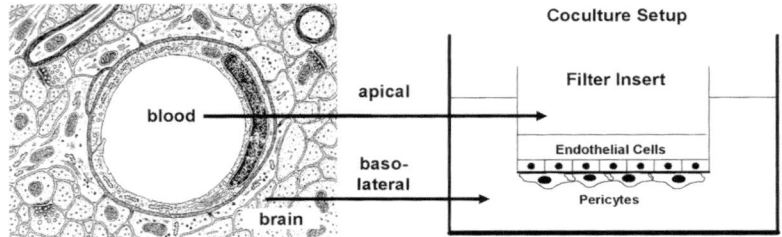

Fig. 3. Coculture setup based on filter inserts. Barrier building endothelial cells cover the apical side of the porous filter membrane while pericytes cover the basolateral side.

8. Centrifuge the cell suspension at $110 \times g$ for 10 min and then resuspend in freezing medium and store as indicated in Subheading 3.1.1, step 18.

3.1.3. Coculture Model

Primary porcine PBCECs are seeded at in vitro day (IVD) 2 on the apical (upper) compartment (representing the blood side) of Transwell® filter inserts. The porous filter membrane with well-defined pore sizes enables the exchange of soluble factors and allows the contact between cells of the apical and basolateral (lower, representing the brain side) compartment. By modifying the experimental setup with primary PBCPs on the basolateral side, the BBB-model is optimized and represents a more in vivo-like situation (Fig. 3). In order to induce their differentiation and retain their in vivo phenotype, the cells are cultured after IVD 4 under serum-free conditions and the natural glucocorticoid-hydrocortisone is added to the medium (19). Isolation of mRNA or total protein can be performed at IVD 6 when the endothelial cells are completely differentiated and the barrier reaches its maximal tightness. After isolation of each cell type separately, it is possible to research various constituents such as the ECM components and MMPs at the mRNA level by real time PCR and at the protein level by Western Blot analysis (18).

1. Turn the Transwell® filter inserts upside down with a sterilized pincer and place them in the jars.

2. Coat the basolateral side with 112 μL collagen IV and incubate overnight at 37°C (see Note 2).

3. Turn the filter inserts again and place them into the 12-well plate.

4. Coat the apical side with 112 μL collagen IV and incubate over night at 37°C.

5. Air-dry the filter under a clean bench.

6. Wash the cultivated pericytes twice with PBS.

7. Use the Trypsin/EDTA-solution and incubate for 2 min to detach cells from the culture flask.

8. Add Plating-medium containing serum to stop the trypsin digest and centrifuge at 800×g for 10 min.

9. Remove the medium, add 1 mL of fresh plating medium and count the cells.

10. Turn the Transwell® filter inserts upside down with a sterilized pincer and place them in the jars.

11. Seed 50,000 cells in 150 μL plating medium on the top of each turned filter insert (see Note 3).

12. Incubate the filters inside the jars for 6 h at 37°C.

13. Add plating-medium, 1.5 mL/well in a 12-well plate.

14. After the incubation, invert the filter inserts back into the 12-well plate.

15. Add 250 μL Plating-medium to the apical side and store the plate in the incubator. Avoid agitation of the plate.

16. Defrost cryopreserved endothelial cells, add Plating-medium and centrifuge the cells at 800×g for 10 min (see Note 4).

17. Seed 250,000 endothelial cells (IVD 2) in 250 μL Plating-medium on the apical side of each of the filter inserts.

18. Remove the medium after 2 days (IVD 4) and replace it with equal volume SFM with 550 nM hydrocortisone supplement (see Note 5). Therefore, dilute hydrocortisone stock solution 1:1,000 in SFM.

19. After 1 day (IVD 5), the endothelial cells are differentiated and ready to use for experiments.

3.2. TEER Measurement

This technique determines the barrier integrity of endothelial cells since the permeability of a cell layer correlates with its electric resistance. By using the Transwell® filter inserts it is possible to measure the TEER with the cellZscope® device (20). This setup allows the measurement of barrier properties of cell layers under various conditions and it is suitable for time-resolved quantification of barrier integrity. It is used in particular to study the influence of substances such as drugs, cytokines, and toxins or to determine the impact of cells like pericytes on endothelial barrier function.

1. The seeding and preparative experimental setup is identical to the protocol described in Subheading 3.1.3.

2. Then disinfect the cellZscope® device with ethanol and wash it 3× with autoclaved ultrapure water.

3. Store the device in an incubator for 30 min until the chambers are equilibrated to 37°C.

4. Fill the chambers with 1 mL of medium (warmed to 37°C) having the same composition as in the plates.

5. Transfer the filter inserts from the 12-well plate into the cellZscope® device with a pincer.

6. Place the device into an incubator and start the measurement (see Note 6).

7. After the measurement transfer the cells back to the 12-well plate. Wash the device 3× with autoclaved ultrapure water and finally with ethanol.

3.2.1. Results

Using this technique the TEER of the porcine endothelium is reported to be 2,000–1,400 $\Omega \bullet cm^2$ and the TEER in a coculture system is reported to be 800–400 $\Omega \bullet cm^2$ (18) at IVD 6. The absolute TEER values and the influence of pericytes varies between the in vitro models used (e.g., rat primary cocultures display TEER values of 380 $\Omega \bullet cm^2$ at IVD 5, (21)). The sample measurement displayed in Fig. 4. reveals that in contrast to endothelial cells, pericytes themselves do not build up a barrier and therefore present TEER values of about 30 $\Omega \bullet cm^2$. Endothelial cells solely show constant values between 1,353 and 1,487 $\Omega \bullet cm^2$ after IVD 5. Cocultured endothelial cells (CoEC) display decreasing TEER values from 747 $\Omega \bullet cm^2$ at IVD 5 to 62 $\Omega \bullet cm^2$ at IVD 9. These results indicate that porcine pericytes decrease the barrier integrity of porcine endothelial cells.

3.3. Permeability Measurement

Another common technique to analyze barrier properties is the measurement of the permeability for substances such as sucrose, fluorescein, and FITC-labeled dextran through the endothelial cell layer. Radiolabeled ^{14}C-sucrose is often used in addition to TEER to confirm the quality of in vitro BBB models (22). Sucrose is not taken up by endothelial cells either by active or facilitated transport, so the permeability for sucrose solely depends on the paracellular barrier tightness. In this technique, radiolabeled ^{14}C-sucrose is added to the apical side of the Transwell® filter.

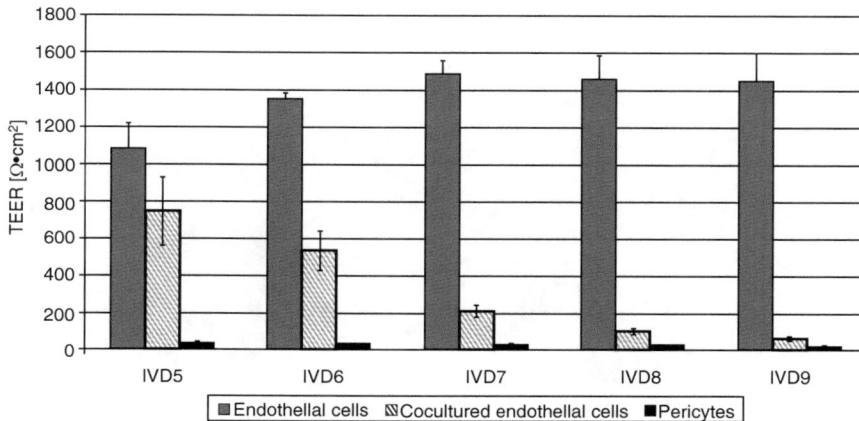

Fig. 4. TEER measurements showing the absolute values for endothelial cells (*dark grey*, 1,085–1,487 Ω•cm²), cocultured endothelial cells (CoEC) (*light grey*, 747–762 Ω•cm²), and pericytes (*black*, 27–38 Ω•cm²). These results indicate that coculture with pericytes decreases the barrier property of endothelial cells.

By measuring the time-dependent amount of ^{14}C-sucrose which passes to the basolateral side, the permeability can be calculated.

1. The seeding and preparative steps are identical to the steps described in Subheading 3.1.3.

2. Check barrier integrity by TEER measurement as described in Subheading 3.2 (see Note 7).

3. Remove the basolateral medium and replace it with fresh SFM containing 550 nM hydrocortisone supplement (see Note 8).

4. Prepare vials with 1 mL scintillation cocktail for each sample.

5. Add 1 µCi ^{14}C-sucrose to the apical compartment of each well.

6. Determine the radioactivity of the apical compartment (c_{api}).

7. Remove 50 µL from the basolateral compartment twice at 10, 20, 30, 40, 60, 80 min and add them to the vials.

8. Vortex all vials and measure the radioactivity with the liquid scintillation counter.

9. Calculate the counts per min for the rate of radioactive labeled ^{14}C-sucrose which passes through the endothelial cell layer.

10. The restriction of the paracellular transport of ^{14}C-sucrose is determined based on the measurement of the apparent permeability of a cell layer (P_{app}). The P_{app} is calculated by the amount of the initial apical concentration of ^{14}C-sucrose (c_{api}), the surface area (A) and the basolaterally accumulated substance (n_{baso}) after the time (t) (1).

$$P_{app} = \frac{n_{baso}}{c_{api} \, A \, t}.$$ (1)

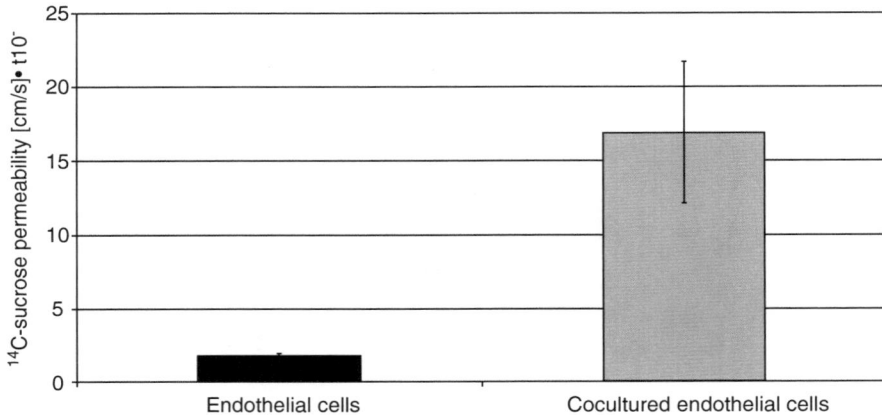

Fig. 5. Permeability measurement for ^{14}C-sucrose shows the absolute values of endothelial cells (*black*, 1.87•10^{-7} cm/s) and the CoEC (*grey*, 16.9•10^{-7} cm/s). Pericytes alone do not restrict the permeability (data not shown). The permeability values indicate that pericytes decrease endothelial barrier integrity.

11. To calculate the absolute permeability of the cell layer (P_c), the apparent permeability (P_{app}) and the permeability of an empty filter (P_f) must be measured and the values applied to (2).

$$\frac{1}{P_c} = \frac{1}{P_{app}} - \frac{1}{P_f}. \tag{2}$$

3.3.1. Results

Using this technique, the permeability of the porcine endothelium for ^{14}C-sucrose is reported to be about $1.87\,10^{-7}$ cm/s (**22**) and the permeability in the coculture system is reported to be about 16.9×10^{-7} cm/s (Fig. 5, unpublished data). The permeability values in conformity with the TEER measurement show a barrier integrity decreasing influence of pericytes.

3.4. Electrical Cell-Substrate Impedance Sensing (ECIS)

In order to study the impact of pericyte derived ECM on the endothelial cell monolayer, we took advantage of the electric cell-substrate impedance sensing (ECIS) technique which measures the electrical resistance of cell layers grown on ECM coated electrode arrays (**23, 24**). In the first step, pericytes are cultured in ECIS dishes with small goldfilm electrodes deposited on the bottom, where they produce and secrete basement membrane components (Fig. 6). The ECM is then isolated by removing the overlaying cells by hypotonic lysis leaving the ECM on the array surface. In the second step, the barrier building endothelial cells are seeded on the array. The time course of the resistance at the altered current frequency of 400 Hz serves as a quantitative parameter for barrier tightness. Using this technique, ECM derived from pericytes has an inductive impact on barrier function of cerebral microvascular endothelial cells in vitro (**25**).

a   Seeding of extracellular matrix producing cell type (e.g.pericytes)

b   Production of endogenous extracellular matrix

c   ECM deliverance via hypotonic lysis & detergent treatment

d   Seeding of barrier building cells (e.g. PBCEC ' s)

e   Electrical Cell Impedence sensing (ECIS) measurement

time (n)

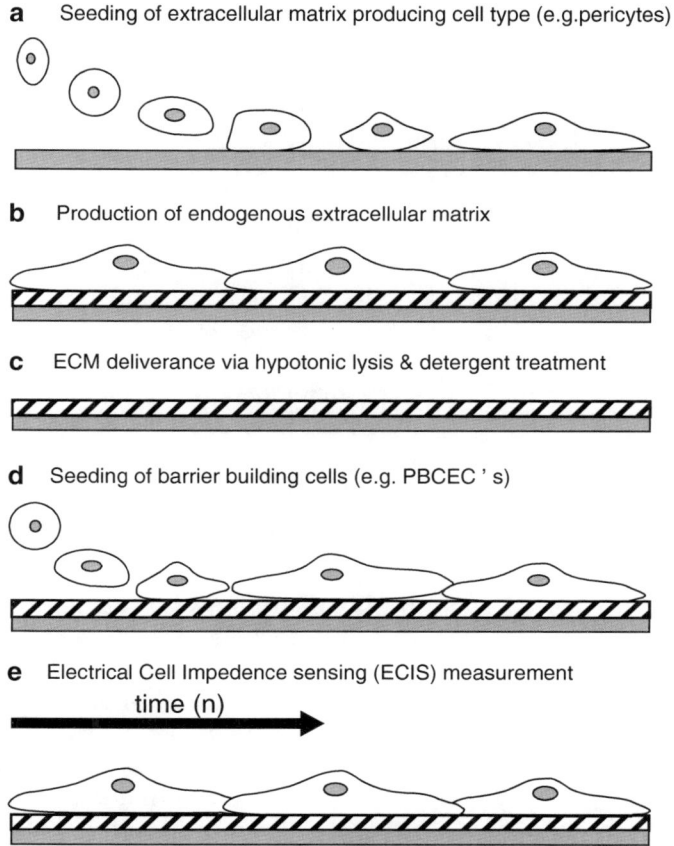

Fig. 6. **(a-e)** Schematic of the electrical cell-substrate impedance sensing (ECIS) setup is shown. It determines the impact of pericytic endogenous ECM on barrier formation of PBCECs.

3.4.1. Seeding of Pericytes

1. Clean and sterilize the ECIS 8W10E* Array for 1 min in an argon plasma cleaner.

2. Cover each well with 200 μL l-cysteine for 15 min, then wash the wells 3× with ultrapure water (see Note 9).

3. Coat the gold electrodes with 200 μL of collagen G solution diluted 1/30 (133.3 μg/mL) in ultrapure water for 2 h at room temperature (see Note 10).

4. Remove the collagen G solution and wash each well once carefully with 200 μL SFM supplemented with 550 nM hydrocortisone.

5. Thaw the pericytes and endothelial cells as described in Subheading 3.1.3 and 1 mL of SFM supplemented with 550 nM hydrocortisone to each pellet.

6. Count the cells and dilute the cell suspension to obtain 400,000 cells in 1.6 mL medium.

7. Seed 400,000 pericytes into four of the eight wells so that each well has 100,000 cells in 100 µL. Seed the endothelial cells in the same way into the last four wells. These serve as an internal control (see Note 11).

8. Incubate the array for 48 h at 37°C, with 5% CO_2 and water vapor-saturated atmosphere.

9. Wash each well 3× with 400 µL ultrapure water.

10. Incubate each well for 2 h at 4°C with ultrapure water. Wash each well with ultrapure water 3–5 times. To remove remaining cells and cell membrane constituents, incubate each well with 400 µL of ECM Isolation Solution 1 for 3–5 min at room temperature. To remove all detergent, wash 5–10 times with ultrapure water (see Note 12).

11. Check if all cells have been removed by microscopy. Incubate the wells with ECM Isolation Solution 2 for 1–3 min to remove any remaining cells and then wash 5–10 times with ultrapure water.

12. Fill 200 µL SFM supplemented with 550 nM hydrocortisone into each well and start the ECIS measurement (see Note 13).

3.4.2. Seeding of the Endothelial Cells

1. Thaw the PBCECs and count the cells as described before (see Subheading 3.4.1, steps 5 and 6) and seed 200,000 cells into each well in 200 µL SFM supplemented with 550 nM hydrocortisone during the running measurement.

3.4.3. Results

Using this technique the resistance of the porcine endothelium on endothelial derived ECM is reported to be about 12.8 kΩ and the resistance on pericyte derived ECM is reported to be about 18.0 kΩ displaying a barrier tightening influence of the pericytic ECM (Fig. 7) (25).

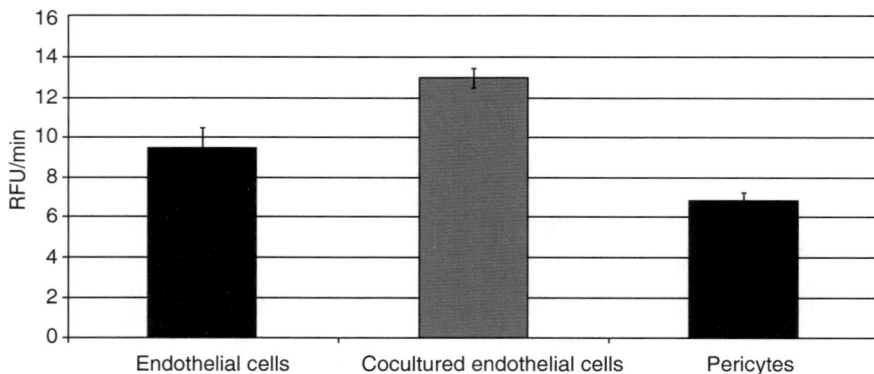

Fig. 7. Electrical cell-substrate impedance sensing measurements showing absolute resistance values for endothelial cells on endothelial derived ECM (*dark grey*, 12.8 kΩ) and pericyte derived ECM (*light grey*, 18.0 kΩ). These results show the tightening influence of pericyte derived ECM on the endothelial barrier.

3.5. MMP Activity Assay

MMPs are Zn^{2+}-containing and Ca^{2+}-dependent endopeptidases with a wide substrate specificity (26). Their proteolytic potential is strictly controlled not only on transcriptional and translational level but also on the level of secretion, proteolytic activation of zymogens and TIMP inhibition. It is possible to analyze the MMP activity in both apical and basolateral compartments of the coculture dish independently from each other. MMP activity in this pericyte-endothelial cell coculture model can be studied by two complementary techniques as described below.

3.5.1. Fluorescent Substrate Based MMP Activity Assay

In fluorogenic substrate based activity assays the level of MMP activity in cell culture supernatants can be determined by kinetic measurements on commercially available substrates. The substrate specificity correlates with the substrate spectrum of the analyzed MMPs. An assay using a MMP-2/-7 substrate which is also sensitive to detect other MMPs such as MMP-9 will be described. The substrate MCA-Pro-Leu-Gly~Leu-DPA-Ala-Arg-NH$_2$ ·TFA consists of an oligopeptide that is coupled to the fluorophor MCA ((7-methoxycoumarin-4-yl)acetyl) and possesses an incorporated quencher DPA(N-3-(2,4-dinitrophenyl)-l-2,3-diaminopropionyl). MMPs cause substrate cleavage to take place between the amino acids glycine and leucine, thereby removing the quencher from the fluorophor. The increase of the fluorescence signal in a certain period is displayed in the slope of the signal line and describes the MMP activity in relative fluorescence units per min (RFU/min). However, due to the wide substrate specificity of MMPs the assay cannot definitely identify the activity of a specific MMP in many cases. Furthermore, it is not possible to trace back a low MMP activity to the control level responsible. One cannot distinguish whether no MMPs are present in the culture supernatant, or that MMP zymogens are not activated, or active MMPs are inhibited by TIMPs. The assay is carried out according to Knight et al. (27) with minor modifications (28).

1. Isolate a minimum volume of 60 µL cell culture supernatant (see Note 14).

2. Freshly prepare a 10-µM substrate solution in MMP-2/-7 activity buffer from the 1 mM substrate stock solution.

3. Heat the cuvette holder to 37°C.

4. Pipette 60 µL of cell culture supernatant into a tube.

5. Add 60 µL of the substrate solution.

6. Quickly transfer the mixture into the cuvette.

7. Start the time driven measurement (excitation wavelength 325 nm, emission wavelength 393 nm.)

8. Measure the signal increase over 15 min (see Note 15).

9. Calculate the MMP activity in relative fluorescence units per min (RFU/min) from the slope of the signal line.

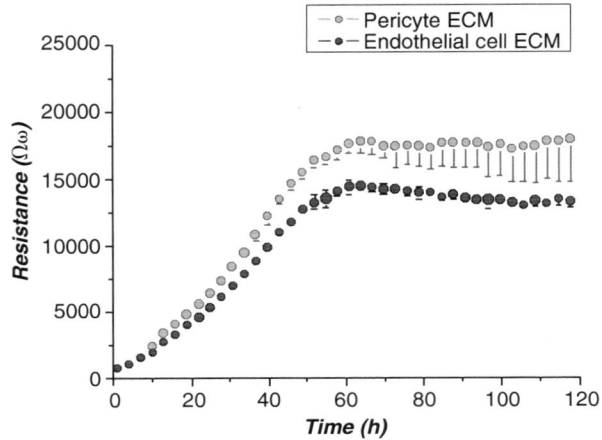

Fig. 8. MMP-2/-7 activity assay showing the activity of endothelial cells (*black*, 9.5 RFU/min), CoEC (*light grey*, 12.9 RFU/min), and pericytes (*dark grey*, 6.8 RFU/min). The coculture presents higher MMP activity in comparison to endothelial cells or pericytes cultured alone.

3.5.2. Results

The results of an MMP activity assay for endothelial cells, cocultured cells, and pericytes are shown in Fig. 8 (18). Supernatants from pericyte cultures display a MMP-2/-7 activity of 6.8 RFU/min while the activity in endothelial cell culture supernatants is about 9.5 RFU/min. In the supernatant of the coculture, the MMP-2/-7 activity is 12.9 RFU/min which reveals that MMP activity increases when endothelial cells are cocultured with pericytes.

3.6. Gelatin Zymography

This method is based on the widely used sodium dodecyl sulfate polyacrylamide gel electrophoresis (SDS-PAGE), which allows the distinction of different MMPs, their zymogens, and active species due to their different molecular weight. Several variations of this method have been described using different MMP substrates such as gelatin or casein, and even reverse zymography to detect TIMPs has been developed (29). The protocol for gelatin zymography which is the standard technique for detection of MMP-2 and MMP-9 is described in this section. Cell culture supernatants are diluted in a nonreducing, SDS containing sample buffer that destroys MMP/TIMP complexes. Gelatin is incorporated into the separating gel during the casting procedure of an SDS-PAGE. The separation of the proteins contained in the cell culture supernatant takes place according to the molecular weights by electrophoresis. After the SDS-PAGE, the separating gel is washed in a Triton-X 100 solution to renature MMP proteins by elimination of SDS. In this step, not only active MMP species are recovered but the zymogens are activated, probably due to partial misfolding and autocatalytic cleavage in the acidic Triton-X 100 solution. After renaturation, the gel is exposed to the proteolytic substrate digest in developing buffer. For MMP detection a coomassie staining is performed. The coomassie dye stains the undigested substrate and leads to a blue coloration of the gel, while white bands appear in spots where MMPs are located due to gelatin degradation. The MMPs are identified by their apparent molecular weight in reference to the protein marker and/or commercially available MMP controls.

1. Prepare a 1% gelatin solution in distilled water and heat to 60°C until the gelatin melts.

2. Cast a 10% SDS-PAGE containing gelatin: For a 10×10 cm^2 minigel constitute stacking and separating gel according to the following instructions: For the 10% polyacrylamide separating gel, mix 1.64 mL ddH$_2$O, 0.6 mL 1% gelatin solution, 1.76 mL separating gel buffer, 2.0 mL acrylamide solution, and 15 μL TEMED.

3. Start polymerization with 15 μL APS, cast the gel, cover with a layer of ethanol (70%) and wait until the gel has polymerized.

4. Meanwhile mix the 4% polyacrylamide stacking gel: 1.81 mL ddH$_2$O, 0.78 mL stacking gel buffer, 0.4 mL acrylamide solution, and 15 μL TEMED.

5. When the separating gel has polymerized, remove ethanol and start polymerization of the stacking gel by adding 15 μL of the APS solution. Cast the gel upon the stacking gel and insert the sample spacer. Wait until gel is polymerized.

6. Dilute cell culture supernatants with an equal volume of sample buffer and load it onto the stacking gel.

7. Also load a lane with prestained protein marker and, if at hand, commercially available MMP controls onto the gel (see Note 17).

8. Run the electrophoresis at 80 V for the stacking gel (approximately 15 min) and at 130 V for the separating gel (approximately 90 min).

9. Adequate tracking can be assessed by the use of a prestained protein marker.

10. When electrophoresis is complete, remove the stacking gel.

11. Transfer the separating gel to a sealable box/container.

12. Put the gel container on a shaker. All further washing steps are performed in a generous buffer volume under gentle shaking.

13. Wash the gel twice for 30 min in renaturating buffer.

14. Remove the detergent by washing the gel 5× in ddH$_2$O.

15. Equilibrate the gel for 5 min in developing buffer.

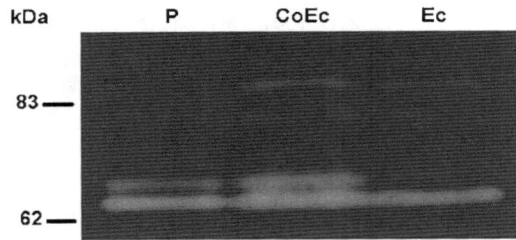

Fig. 9. Zymography analysis displaying the MMP-2 (active: 62 kDa, intermediate form: 65 kDa) and MMP-9 (active: 84 kDa) expression of pericytes (P), CoEc, and endothelial cells (Ec). The expression levels of the active and intermediate form of MMP-2 as well as the active form of MMP-9 are higher in the supernatant when endothelial cells are cocultivated with pericytes in contrast to the supernatants of cells cultured alone.

16. Exchange the buffer for a fresh volume.

17. Seal the gel container.

18. Incubate the gel at 37°C for 20–24 h in developing buffer with gentle shaking.

19. Stop gelatin degradation by shaking in destainer 1 solution for 30 min.

20. Apply the staining solution for 30 min.

21. Destain the gel in destainer 1 solution for 30 min (see Note 18).

22. Apply destainer 2 until white bands appear in the blue gel (see Note 19).

23. Place the gel between two transparent plastic sheets and scan the image for further analysis.

24. For long-term storage dry the gel between two cellophane sheets (see Note 20).

3.6.1. Results

The results of an MMP zymography assay for endothelial cells, CoEC, and pericytes for MMP-2 (active: 62 kDa, intermediate form: 65 kDa) and MMP-9 (active: 84 kDa) are shown in Fig. 9. The expression levels of active and latent MMP-2 as well as active MMP-9 are higher in the supernatant when endothelial cells are cocultured with pericytes rather than when cultured alone (18).

4. Notes

1. By this procedure, the mainly collagen containing basement membrane is removed and the endothelial cells are released. The time for the digest depends on the batch of the collagenase/dispase II and should be checked and varied for every new batch. The hanging stirrer is essential to avoid any cell damage.

2. Add 1 μL coating material for each 0.01 cm² surface area on the filter. Further coating materials are collagen G, various

other collagens, and other ECM components such as laminin or fibronectin. Do not touch the membrane with the pipette.

3. It is possible to check after 6 h to determine whether the cells have attached by performing nonspecific cell staining (e.g., with carbolfuchsin). The seeding ratio of 1:5 (50,000/250,000) is chosen because it is similar to the ratio in vivo.

4. If endothelial cells in culture are used it is possible to seed them like pericytes as given in Subheading 3.1.3, steps 6–9.

5. First, remove the medium on the basolateral side and then the medium from the apical side. If the apical side medium is removed first the cell layer will be damaged by the influx of medium from the basolateral side. Leave a small amount of medium at the apical side to prevent the cells from drying.

6. Avoid agitation after the measurement has started. If the measurement is commenced directly after a medium exchange, the cells might be stressed. In this case, leave the device containing the filters in the incubator for 2 h and then start the measurement. It is also possible to add substances during the measurement through vents to avoid further stress.

7. If the TEER does not reach typical values, the cells have not yet fully differentiated. It is possible to wait 1 more day and to start the permeability experiment on IVD 7 after checking the TEER again. Pericyte and endothelial monocultures should be compared to the coculture as controls in the permeability measurement.

8. The volume of medium must be exactly 500 µL in the apical compartment and 1,500 µL on the basolateral side.

9. This step is important to form a monomolecular self-assembly layer on the electrode surface and for stabilization of the electrode/electrolyte interface impedance.

10. It is necessary to promote adhesion of the pericytes so that they secrete their own ECM on the electrode.

11. To prevent a high deviation between the sensitive measurements, it is important to have four parallel determinations per condition. Endothelial cells are seeded as an internal control for every measurement.

12. If the cell removal is insufficient, it is possible to incubate the cells with ECM Isolation Solution 1 at 37°C or up to 10 min at room temperature. It might be helpful to exert some mechanical force (e.g., gentle shaking) to remove the cells during the incubation with the Isolation Solution 1.

13. Check the contact of the array with the "electrode check" tool before starting the measurement. The electrode capacitance without cells should not be less than 50–60 nF.

14. Serum usually contains MMP-2 and MMP-9 in considerable amounts. To determine the activity of MMP secreted by

endothelial cells and pericytes, cultivate cells in SFM as suggested for our coculture model.

15. Perform an assay with culture medium/substrate solution 1:1 as a control. If working with phenol red free culture medium, the curve will decrease slightly due to substrate bleaching. Subtract the negative slope of the control from your sample slope to calculate the absolute MMP activity. If working with phenol red containing culture supernatant, the fluorescence signal initially strongly decreases due to phenol red bleaching. In this case determine the period until the initially hyperbolic phenol red bleaching effect devolves to linear curve decay (approximately 2 min in our assay). Take only time points after this period into the slope calculation and subtract the medium control slope from your sample slope to calculate the absolute MMP activity.

16. Qualitative MMP-controls containing different MMPs are commercially available from Calbiochem, Merck Chemicals, Nottingham, UK.

17. Destaining solutions can be recovered from coomassie contamination by filtering with activated carbon.

18. White bands should appear after approximately 15 min. The contrast is increased if you leave the gel in destainer 2 overnight.

19. For long-time storage, incubate the gel in gel dry solution for at least 15 min. Immerse two sheets of cellophane (Bio-Rad, Munich, Germany) with the same solution, arrange the gel on one cellophane sheet and cover with the second cellophane. Smooth to eliminate bubbles and clamp the sandwich in a gel drying frame. Air-dry the gel for at least 2 days in a flue.

Acknowledgments

The authors are grateful to Sabine Hüwel for his comments on this work, technical assistance and helpful discussions. This work was supported by a fellowship awarded to Gokulan Thanabalasundaram by the International Graduate School of Chemistry, Münster.

References

1. Allt G, Lawrenson JG (2001) Pericytes: cell biology and pathology. Cells Tissues Organs 169(1):1–11

2. Balabanov R, Dore-Duffy P (1998) Role of the CNS microvascular pericyte in the blood-brain barrier. J Neurosci Res 53(6):637–644

3. Cai J, Boulton M (2002) The pathogenesis of diabetic retinopathy: old concepts and new questions. Eye 16(3):242–260

4. Bergers G, Benjamin LE (2003) Tumorigenesis and the angiogenic switch. Nat Rev Cancer 3(6):401–410

5. Merker HJ (1994) Morphology of the basement membrane. Microsc Res Tech 28(2): 95–124

6. Osawa T, Feng XY, Yamamoto M, Nozaka M, Nozaka Y (2003) Development of the basement membrane and formation of collagen fibrils below the placodes in the head of anuran larvae. J Morphol 255(2):244–252

7. Hawkins BT, Davis TP (2005) The blood-brain barrier/neurovascular unit in health and disease. Pharmacol Rev 57(2):173–185

8. Farkas E, Luiten PG (2001) Cerebral microvascular pathology in aging and Alzheimer's disease. Prog Neurobiol 64(6):575–611

9. Davis GE, Senger DR (2005) Endothelial extracellular matrix: biosynthesis, remodeling, and functions during vascular morphogenesis and neovessel stabilization. Circ Res 97(11):1093–1107

10. Rosenberg GA (2002) Matrix metalloproteinases in neuroinflammation. Glia 39(3): 279–291

11. Nagase H, Visse R, Murphy G (2006) Structure and function of matrix metalloproteinases and TIMPs. Cardiovasc Res 69(3):562–573

12. Visse R, Nagase H (2003) Matrix metalloproteinases and tissue inhibitors of metalloproteinases: structure, function, and biochemistry. Circ Res 92(8):827–839

13. Rosenberg GA, Yang Y (2007) Vasogenic edema due to tight junction disruption by matrix metalloproteinases in cerebral ischemia. Neurosurg Focus 22(5):E4

14. Franke H, Galla HJ, Beuckmann CT (1999) An improved low-permeability in vitro-model of the blood-brain barrier: transport studies on retinoids, sucrose, haloperidol, caffeine and mannitol. Brain Res 818(1): 65–71

15. Franke H, Galla H, Beuckmann CT (2000) Primary cultures of brain microvessel endothelial cells: a valid and flexible model to study drug transport through the blood-brain barrier in vitro. Brain Res Brain Res Protoc 5(3):248–256

16. Mischeck U, Meyer J, Galla HJ (1989) Characterization of gamma-glutamyl transpeptidase activity of cultured endothelial cells from porcine brain capillaries. Cell Tissue Res 256(1):221–226

17. Bowman PD, Ennis SR, Rarey KE, Betz AL, Goldstein GW (1983) Brain microvessel endothelial cells in tissue culture: a model for study of blood-brain barrier permeability. Ann Neurol 14(4):396–402

18. Zozulya A, Weidenfeller C, Galla HJ (2008) Pericyte-endothelial cell interaction increases MMP-9 secretion at the blood-brain barrier in vitro. Brain Res 1189:1–11

19. Hoheisel D, Nitz T, Franke H, et al (1998) Hydrocortisone reinforces the blood-brain properties in a serum free cell culture system. Biochem Biophys Res Commun 247(2):312–315

20. Wegener J, Abrams D, Willenbrink W, Galla HJ, Janshoff A (2004) Automated multi-well device to measure transepithelial electrical resistances under physiological conditions. Biotechniques 37(4):590; 2–4; 6–7

21. Nakagawa S, Deli MA, Nakao S, et al (2007) Pericytes from brain microvessels strengthen the barrier integrity in primary cultures of rat brain endothelial cells. Cell Mol Neurobiol 27(6):687–694

22. Lohmann C, Huwel S, Galla HJ (2002) Predicting blood-brain barrier permeability of drugs: evaluation of different in vitro assays. J Drug Target 10(4):263–276

23. Giaever I, Keese CR (1993) A morphological biosensor for mammalian cells. Nature 366(6455):591–592

24. Wegener J, Keese CR, Giaever I (2000) Electric cell-substrate impedance sensing (ECIS) as a noninvasive means to monitor the kinetics of cell spreading to artificial surfaces. Exp Cell Res 259(1):158–166

25. Hartmann C, Zozulya A, Wegener J, Galla HJ (2007) The impact of glia-derived extracellular matrices on the barrier function of cerebral endothelial cells: an in vitro study. Exp Cell Res 313(7):1318–1325

26. Sternlicht MD, Werb Z (2001) How matrix metalloproteinases regulate cell behavior. Annu Rev Cell Dev Biol 17:463–516

27. Knight CG, Willenbrock F, Murphy G (1992) A novel coumarin-labelled peptide for sensitive continuous assays of the matrix metalloproteinases. FEBS Lett 296(3):263–266

28. Lohmann C, Krischke M, Wegener J, Galla HJ (2004) Tyrosine phosphatase inhibition induces loss of blood-brain barrier integrity by matrix metalloproteinase-dependent and -independent pathways. Brain Res 995(2):184–196

29. Clark IM, ed. Matrix Metalloproteinase Protocols. Humana Press, Totowa, 2001

Chapter 20

Isolation and Properties of an In Vitro Human Outer Blood-Retinal Barrier Model

Robin D. Hamilton and Lopa Leach

Abstract

The outer blood–retinal barrier is composed of a monolayer of retinal pigment epithelium (RPE), Bruch's membrane, and the choriocapillaris, which is fenestrated. An in vitro model that includes all these layers within a 3-D architecture confers a clear advantage over traditional monolayer cultures. Cells here, whether endothelial or epithelial, reside in conditions resembling that in vivo and can participate in cell–cell and cell–matrix cross talk. This chapter describes how a human trilayer culture model was generated with RPE (ARPE-19) cells cultured on the epithelial surface of amniotic membrane and with human umbilical vein derived endothelial cells (HUVEC) on the interstitial surface. This model resembles the outer retinal barrier both in restricting transport of small molecules (<4 kDa), possession of occludin-rich tight junctions in the RPE and fenestrated endothelial cells. Techniques used to test the generated trilayer properties are also described and these include imaging of structure and molecular occupancy of tight and adherens junctions, estimation of the barrier efficiency of trilayer by measurement of fluorescein and fluorescein-conjugated tracers under flow, measurement of secreted vascular endothelial growth factor-A and ultrastructural studies, which allow analyses of the fine structure of the tight junctions in the RPE, and the endothelial fenestra.

Key words: 3D culture, Fenestrated endothelium, HUVEC, In vitro permeability, Outer retinal barrier, Retinal pigment epithelium, VEGF-A

1. Introduction

Reductionist approaches into mechanisms underlying diseases of the outer retinal barrier, such as age related macular degeneration (AMD), have been hampered by the lack of optimal in vitro models utilising human cells to provide the 3-D architecture and allow expression of the in vivo phenotype for both the retinal pigment epithelial cells and the endothelium. There is good evidence from in vivo studies that the choriocapillaris is dependant upon the

Sukriti Nag (ed.), *The Blood-Brain and Other Neural Barriers: Reviews and Protocols*, Methods in Molecular Biology, vol. 686, DOI 10.1007/978-1-60761-938-3_20, © Springer Science+Business Media, LLC 2011

overlying retinal pigment epithelium (RPE) as it atrophies when the RPE is experimentally removed (1). The morphological phenotype of the choriocapillaris appears to be dictated by trophic factors secreted across the basement membrane by the RPE (2, 3). In active neo-vascular AMD, breakdown of the outer retinal barrier and endothelial cell (EC) proliferation are key processes that lead to retinal damage and involve factors such as vascular endothelial growth factor –A (VEGF-A) (3–5).

In this chapter, we describe techniques employed to generate a viable human in vitro model for the outer blood–retinal barrier consisting of the RPE, amnion, and human umbilical vein endothelial cells (HUVEC) (2). This model possesses some of the key properties of the outer barrier and can be experimentally manipulated for mechanistic studies while allowing cross talk between neighbouring tissue layers.

The RPE and the choriocapillaris are separated by Bruch's membrane in vivo. The amniotic membrane can be compared to Bruch's membrane, as it consists of an inner layer of epithelial cells resting on a basement membrane (BM), which is composed of mainly type IV collagen, laminin, elastin, and heparan sulphate (6). This BM interfaces with the vascular stroma, which consists mainly of collagen I, III, IV, fibronectin and laminin, forming an ideal extracellular matrix for cultivation of ECs. We and others have shown that the human amniotic membrane is suitable for the growth of epithelial and endothelial monolayers (2, 7, 8). This chapter describes how placing RPE cells first on the epithelial side of the amnion and then seeding the opposite surface with ECs (see below) results in formation of a polarised RPE layer and a fenestrated endothelial monolayer. Co-cultures of RPE and bovine or human choroidal EC on transwell inserts have been attempted before (5, 9), but the phenotypic endothelial modifications we have reported were not seen on the synthetic membranes.

Primary HUVEC is used in the trilayer since these cells have been extensively used for vascular research. They are experimentally pliable and show phenotypic plasticity ranging from formation of quiescent continuous monolayers with VE-cadherin rich adherens junctions to angiogenesis/tubulogenesis in 3-D culture (10, 11). Moreover, isolated HUVEC continue to express signalling molecules, which respond to inflammatory mediators (12), hypoxia, and angiogenic growth factors such as VEGF-A (11). In our generated trilayer, HUVEC show loss of fidelity of origin by becoming fenestrated (2). Moreover, they still possess paracellular clefts with defined junctional regions, which include presence of VE-cadherin, zonula occludens (ZO)-1, and occludin (see Fig. 1).

In polarised epithelial monolayers, tight junctions are the key entities that seal the paracellular pathway between adjacent cells.

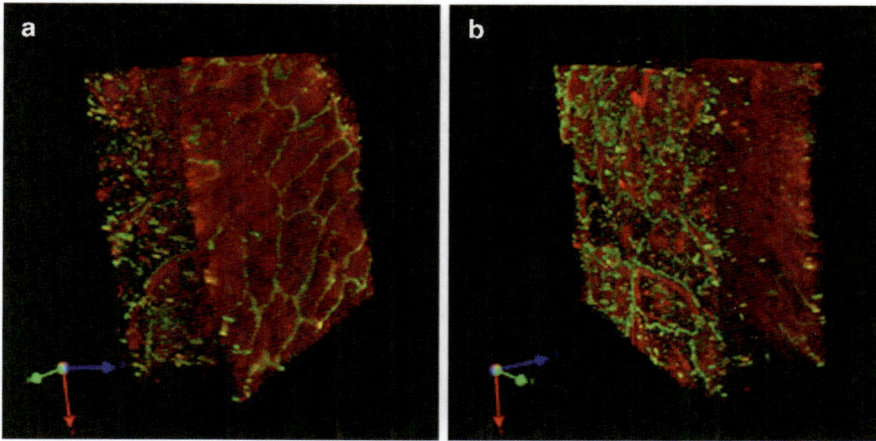

Fig. 1. Laser scanning confocal images of optical sections of the generated outer retinal barrier tilted around its axis showing the RPE cell surface (**a**) and endothelial cell surface (**b**). Monolayers of both surfaces have been immunostained with ZO-1 (*green*) and occludin (*red*).

This prevents diffusion of solutes and differentiation of the apical and basolateral membrane domains to allow active transport across the monolayer (13). The presence of tight junctions must therefore be established in any barrier model, alongside measurements of transepithelial flux of solutes. This chapter describes some of the standard protocols used for localisation of the key structural proteins such as occludin and the cytoplasmic linking molecule ZO-1. Localization can be performed with immunocytochemistry; however, confocal laser scanning microscopes should be used for imaging the precise locations of these molecules (Fig. 1). Tight junction (TJ) proteins such as the claudins, occludin, and junctional adhesion molecule (JAM) dictate the complexity of tight junctional strands and paracellular permeability and their expression should be ascertained (14–16).

The tracer leakage studies are performed using a commercially available MINUCELL system, which allows the upper and lower chambers to be independently perfused with different media, under chosen flow rates. This compartmentalisation also allows frequent sampling of different compartments without disturbance of the monolayers and eliminates the unstirred layer, which can complicate tracer flux studies in static culture. Within this system, the generated model, after full confluence of both layers, excluded 95% of the transfer of 4 kDa dextran and 90% of the smaller sodium fluorescein from the upper (HUVEC chamber) to the lower chamber (RPE chamber) over a 2-h period of sampling at flow rates that do not cause shear stress (17). The system also allows collection of culture medium from the two different compartments over the duration of culture, for later measurement of secreted growth factors by ELISA. In this in vitro

3-D model of the outer retina, VEGF-A was found to be secreted with highest measured levels being reached by 72 h in culture.

Electron microscopy studies are used to detail the TJs present between RPE in the generated model. Despite the high restrictivity to or negligent transfer of small molecules measured, ultrastructural analyses show that the TJs in the RPE do not occupy and occlude the entire length of the paracellular clefts, but are present as discrete points of apposition, more reminiscent of non-CNS endothelial tight junctions seen in vivo (18). This may be an in vitro artefact or perhaps the addition of neural factors may lead to the formation of the archetypical tight junctional strands. Ultrastructural studies also revealed the fenestrated nature of the endothelial layer in our 3-D model. The resolution of the light microscope would not have allowed this observation, leading one to advocate, whenever possible, the addition of an ultrastructural analytical step.

In conclusion, this is a reproducible and near physiological model that mimics, with some limitations, the different layers of the back of the eye both in structure and barrier function and may be used to study the interactions between the RPE and vascular endothelium and for mechanistic investigations into development, functioning, and pathology of the outer retinal barrier.

2. Materials

2.1. Cell Culture

2.1.1. Chemicals and Media

1. Phosphate Buffered saline (PBS), calcium and magnesium free (Oxoid Ltd, Basingstoke, UK).

2. Phosphate-buffered saline (PBS): 2.69 mM KCl, 137 mM NaCl, 1.47 mM KH_2PO_4 8.1 mM, Na_2HPO_4, pH 7.6 (Oxoid Ltd).

3. Dulbeccos' modified eagle medium (DMEM), HAM's F12 medium, Hanks Balanced Salt Solution (HBSS); amphotericin B; l-glutamine; and trypsin (0.05%) with ethylene diamine tetra-acetic acid (EDTA) disodium salt (0.02%) in PBS (trypsin/EDTA) (Life Technologies, Paisley,UK) (see Note 1).

4. For HUVEC, supplement Medium 199 with l-glutamine (2 mM), penicillin (100 U/mL), Streptomycin (100 µg/mL) and amphotericin B (0.2 µg/mL). Name this medium M0. Supplement this with 20% (v/v) Hyclone FCS, ECGS (30 µg/mL) and Heparin (90 µg/mL) before use for cell culture and name medium M20.

5. Medium for corneal epithelial cells: Epilife medium (Cascade Biologics, Portland, USA) with calcium chloride (120 µM) and FCS (10%).

6. Foetal calf serum (FCS) (Autogen Bio Clear UK Ltd, Wiltshire, UK).

7. Tracers: FITC-labelled 70 kDa (FD70), 40 kDa (FD40), 4 kDa (FD40), Fluorescein Sodium salt (Sigma-Aldrich).

8. Collagenase type II (Sigma-Aldrich).

9. Chemicals: Dimethylsulphoxide (DMSO), Triton-X-100, paraformaldehyde, phorbol 12-myristate 13-acetate (PMA), thermolysin, sodium acetate, calcium acetate, (Sigma-Aldrich).

10. Bovine serum albumin (BSA) (Sigma-Aldrich).

11. Endothelial cell growth supplement (ECGS) (First Link, Briery Hill, UK).

12. Sodium chloride (Baxter Healthcare Ltd, Thetford, UK).

13. Trigene™ (Scientific Laboratory Supplies Ltd, Nottingham, UK).

14. 1% (w/v) gelatin in sterile water, autoclaved and stored at 4°C.

2.1.2. ARPE-19 Cells

1. RPE Cells (ARPE-19, ATCC CRL-2302) at a passage of 10 (American Type Culture Collection (ATCC), Manassas, Virginia, USA).

2. Medium for RPE cells: Supplement a 1:1 mixture of HAM's F12 medium and Dulbecco's modified eagle's medium with l-glutamine (2.5 mM), and 1.5 g/L sodium bicarbonate with penicillin (100 Units/mL), Streptomycin (100 µg/mL), and amphotericin B (0.2 µg/mL). Call Medium D0. Supplement this was with 10% (v/v) Hyclone FCS prior to use with cell culture and name medium D10.

2.1.3. Tissue Culture Plastics and Accessories

1. NUNC™ tissue culture flasks 25 cm² and 75 cm², maxisorp wells, plates, chamber slides, and 0.2 mM and 0.4 mM inserts and filters (Life Technologies).

2. Cryopreservation tubes (1 mL) (Sarstedt, Leicester, UK).

3. 3-way stopcocks (Vygon UK Ltd., Gloucester, UK).

4. Suture (Davies and Geck, Gosport, UK).

5. Container to incubate umbilical cord and steel trays (Jencons-PLS, Leighton Buzzard, UK).

6. Cover slips (13 mm, Life Technologies).

2.1.4. Antibodies for Immunocyto-chemistry

1. Antibodies used, their source and dilution are given in Table 1.

2. Confocal laser scanning microscopy: Leica TCS-4d or Leica SP2.

3. Software: LSM 5-image analysis software.

Table 1
The source of the antibodies used and their dilution are shown

Antibody	Catalogue number	Source	Dilution
Cellular retinoic acid binding protein 1 (CRALBP-1)	(Ascitic fluid 1:250, MA3-813)	Affinity Bioreagents via Cambridge Bioscience	1 µg/mL
Anti-Occludin	ab31721	Affinity Bioreagents	4 µg/mL
Anti-ZO-1	ab59720	Affinity Bioreagents	1 µg/mL or dilute1 in 50–100
Mouse monoclonal anti-human VE-cadherin	Clone 55-7H1, Pharmingen,	Cambio, Cambridge, UK	5 µg/mL
Mouse monoclonal anti E-cadherin	FAB7481P	R&D systems	20 µg/mL
TRITC conjugated goat anti-mouse	T 7782	Sigma-Aldrich	Dilute 1 in 50–100
FITC conjugated goat anti-rabbit	F 0382	Sigma-Aldrich	Dilute 1 in 50–100
Rabbit polyclonal anti-occludin	Clone ZMD.467	Zymed Laboratories, USA	Dispense into 10 µL aliquots 50 µg/mL and stored at −80˚CFinal dilution 10 µg/mL
Rabbit polyclonal anti-ZO-1	Clone 61-7300	Zymed Laboratories, USA	Dispense into 10 µL aliquots 20 µg/mL and stored at −80°C.Final dilution 2.5 µg/mL

2.2. MINUCELLS Cells and MINUTISSUE System

The 13 mm MINUSHEETs (MINUCELLS and MINUTISSUE, Bad Abbach, Germany) consist of a black base ring, in which the selected support material, i.e. amnion is placed. A white tension ring is used to hold the support in place on the base ring (Fig. 2). The MINUSHEETS optimum sterilisation method is autoclaving at 105°C and 0.3 bar for 20 min or at a maximum of 135°C and 2 bar for 10 min.

2.3. ELISA for VEGF-A

1. Anti-human VEGF-A antibody (Cat no: af-293-na, R&D Systems, UK). To 100 µg add 1 mL of sterile PBS to produce a concentration of 0.1 mg/mL. Store in 50 µL aliquots (1:250 = 0.4 µg/mL).

2. Recombinant VEGF-A (Cat no: 293-ve, R&D systems). Ten micrograms in vial, add 1 mL of 0.1% PBS/BSA to produce a concentration of 10 µg/mL. Store in 10 µL aliquots in the

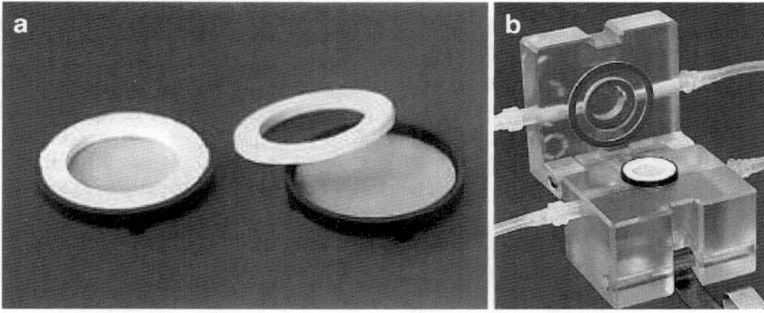

Fig. 2. The MINUCELL SYSTEM (**a**) The amniotic membrane is placed within the black base ring and the white tension ring, prior to seeding with cells. (**b**) The culture system allows independent perfusions of HUVEC (*upper chamber*) and RPE (*lower chamber*) and efficient sampling of tracers from either chamber.

freezer –4°C. Dilute at 1:100 = 100 ng/mL, dilute again for 2,000 pg/mL (first step of standard curve).

3. Biotinylated anti-human VEGF-A Antibody (Cat no: baf293, R&D systems). Fifty micrograms in vial, add 1 mL of tris buffered saline + 0.1% BSA to produce a concentration of 50 µg/mL. Aliquot in 50 µL and freeze. 44 µL in 11 mL is 1:250 dilution to leave 200 ng/mL.

4. Biotiyinylated-streptavidin-HRP kit (DY998, R&D systems).

5. TMB kit 50-76-00 www.kpl.com (add substrates at 1:1 less than 15 min before use).

6. Tris buffered saline: To 10 mL of 0.1% BSA, add 24.2 mg of Tris and 67.6 mg of NaCl.

7. 6% H_2SO_4: 6 mL of 100% H_2SO_4 and 94 mL of dH_2O (see Note 2).

3. Methods

3.1. Coating Tissue Culture Plastic and Cover Slips with Extracellular Matrix Protein

1. Coat the culture flasks with sterile gelatin 1% (w/v) in water.

2. Use 2 mL for the 25 cm^2 culture flasks, 5 mL for the 75 cm^2 culture flasks, and 1 mL for each cover slip in an 8-well plate.

3. Incubate at 37°C with 5% (v/v) CO_2 for 30 min.

4. Remove gelatin and air-dry in a laminar flow hood under sterile conditions.

3.2. Preparation of Tissues and Cells

3.2.1. Preparation of Amnion

1. Immediately after clinical inspection of the placenta and umbilical cord by the midwife, transfer the amniotic membrane with chorion still attached, but removed from the placenta, to a 500-mL sterile bottle containing 200 IU/mL penicillin and 200 µg/mL streptomycin (SAL/AB). Store the bottle with the amnion at 4°C for up to 24 h.

2. Handle the amnion and umbilical cord in a class two laminar flow hood vented to the outside, and decontaminate all surfaces that have come into contact with the tissue, cells or blood with Trigene™ as a precaution against the potential infectious elements in human tissue.

3. Place the amnion in a bowl for manipulation. Under sterile conditions, process the amnion as previously described by Tseng and co-workers (19) and described below.

4. The chorion interstitial side is easily distinguishable from the epithelial surface because it is more irregular and messy. Peel the chorion carefully off the amnion with blunt dissection, making sure to remove all the interstitial material with it and leave the amnion alone behind.

5. During this process, clip one of the minusheets onto the amnion with the black side to the epithelial surface to maintain the amnion polarity. Completely remove and dispose of the chorion (see Note 3).

6. Wash the amnion repeatedly in further SAL/AB to remove all blood and protein remnants, and clip the rest of the MINUSHEETs onto the amnion.

7. Once all the clips are attached, perform careful dissection of the clips with curved scissors to produce a clean edge.

8. Place each of the clips into a sterile 12 well plate and suspend in 2 mL of SAL/AB and store overnight at 4°C.

9. After 24 h, remove any further chorionic material that is visible by light microscopy in the laminar flow hood under sterile conditions.

10. The clips can remain in the sterile 12 well plate suspended in 2 mL of SAL/AB and stored at 4°C for up to 1 month (see Note 4).

3.2.2. Isolation of Human Umbilical Vein Endothelial Cells (HUVEC)

1. Obtain umbilical cord and amnion (see Subheading 3.2.1).

2. Isolate HUVEC based on the method by Jaffe (20).

3. Inspect the cord for any damage that could affect the isolation process or the endothelium of the umbilical vein; e.g. needle stick holes or clamp marks.

4. Blot the ends of the cord on sterile absorbable paper to dry and remove blood residues. Tease open vein at either end with forceps.

5. Cannulate the vein at either end using the 3-way stopcocks, and tie securely to prevent any leak.

6. Flush the vein three times with SAL/AB to remove blood cells and proteins. Fill vein completely to full turgor to ensure no leak or puncture marks are present and then empty.

7. Fill with a 0.5-mg/mL solution of collagenase type II in Medium M0 until the vein is under pressure and close the second stopcock.

8. Incubate the cord in a water bath for 10 min at 37°C. Using two syringes attached at the stopcocks at either end of the vein, agitate the collagenase solution back and forth through the vein to ensure that all the endothelial cells have been removed from the vessel wall.

9. Transfer the cell suspension from the syringe into a 25-mL sterile tube and centrifuge at $10 \times g$ for 5 min.

10. Siphon off the supernatant and re-suspend the pellet in 5 mL of full medium M20 with 50 μg/mL ECGS. Seed this onto a gelatin coated flask. Incubate at 37°C with 5% (v/v) CO_2 in air, and re-feed every 2–3 days with fresh medium.

11. Cells can be subcultured when they reached 90% confluence.

3.2.3. HUVEC Subculture

1. Aspirate medium from 25cm² culture flasks or 75cm² culture flasks and wash the cells with calcium and magnesium free PBS (pre-warmed to 37°C).

2. Incubate these cultures at 37°C, 5% (v/v) CO_2 with 2 mL trypsin/EDTA (0.05/0.02%) for 5 min. Observe cells rounding up and assist detachment by tapping the flask gently.

3. Once the cells have all detached, add medium M20 to inhibit the action of the trypsin/EDTA, and transfer the suspension to a sterile 25 mL tube.

4. Centrifuge at $10 \times g$ for 5 min and re-suspend pellet in 5 mL of the M20 medium.

5. Seed cells on to 75 cm² culture flasks at a split ratio of approximately 1:4 or 1:5. Add medium to flasks to give a final volume of 15 mL and add 50 μg/mL ECGS.

6. Maintain cultures at 37°C with 5% (v/v) CO_2 in the air, and re-feed with fresh medium every 48–72 h (see Note 5).

3.2.4. Culture of Retinal Pigment Epithelium (RPE)

1. Culture RPE cells routinely in the D10 medium.

2. Maintain cultures at 37°C with 5% (v/v) CO_2 in air, and re-feed with fresh medium every 48–72 h.

3. Antibodies to cellular retinoic acid binding protein 1 (ascitic fluid 1:250) should be used periodically to prove the cells retain their phenotype.

3.2.5. Subculture of RPE

1. Culture RPE cells in 75 cm² culture flasks and subculture at approximately 90% confluence. Aspirate medium and wash the cells with calcium and magnesium free PBS (pre-warmed to 37°C).

2. Incubate cultures at 37°C, 5% (v/v) CO_2 with 2 mL trypsin/EDTA for 5 min.

3. Observe cells rounding up and assist detachment by tapping the flask gently as previously described for HUVEC. Add Medium D10 to inhibit the action of the trypsin/EDTA, and transfer the suspension to a sterile 25 mL tube.

4. Centrifuge at $125 \times g$ for 10 min and re-suspend the pellet in 5 mL of the D10 medium. Seed cells on to 75 cm² culture flasks at a split ratio of approximately 1:4 or 1:5.

5. Add Medium D10 to give a final volume of 15 mL. Cultures are maintained at 37°C with 5% (v/v) CO_2 in air, and re-fed with fresh medium every 48–72 h (see Note 6).

3.3. Culture on Amnion

3.3.1. Culture of RPE on Amnion

1. The native epithelium of the amnion can be neutralised using distilled water or remove using 4% deoxycholate for two h at 20°C (21), or 0.25 M NH_4OH for 1–2 h at RT (22), or by Thermolysin 125 µg/mL buffered with sodium acetate 5 mM and calcium acetate 10 mM for 16 h at 4°C (22–24) (see Note 7).

2. Gently wash the membrane three times with sterile SAL/AB and remove any cells remaining on the amniotic membrane by gently rubbing with a sterile cotton bud.

3. After treatment as above, transfer each clip to a fresh 24 well plate containing 2 mL of Medium D0 per well. Store at 4°C until use.

4. Carefully seed the RPE cells at 1×10^5/mL (with 2 mL per well) onto the epithelial surface of the amnion. Wash carefully after 24 h to remove excess unattached RPE cells.

5. Results: The RPE cells exhibit normal epithelial morphology when grown on the amnionitic membrane. As mono-cultures, RPE cells became confluent 24 h after seeding onto the epithelial surface of the amniotic membrane only. They show a typical columnar appearance, with no multiple layering. Cell–cell contact regions show immunoreactivity to the TJ markers, occludin, and ZO-1 by 24 h but are negative for the endothelial junctional marker, VE-cadherin.

3.3.2. HUVEC Growth on Amnion

1. In a mirrored experiment to the above, the amniotic membrane in the MINUSHEET is placed into a 24-well plate (Fig. 2).

2. Carefully seed HUVEC cells at 1×10^5/mL (with 2 mL per well).

3. Results: HUVEC exhibit the typical cobble-stone appearance and reach confluence 48 h after seeding onto the interstitial surface of the amniotic membrane only. They fail to grow on

the other surface of the membrane. Immunocytochemistry confirms full expression of the adherens junction molecule VE-cadherin at cell–cell contacts. The TJ marker ZO-1, but not occludin, is also localised to the paracellular clefts.

3.3.3. Production of the Trilayer

1. Seed RPE cells onto amnion (as described in Subheading 3.3.1).

2. After 24 h, wash the membranes to remove unattached RPE cells and turn over the minusheet (with RPE attached).

3. The cells remain bathed in epithelial medium, up to the level of the minusheet. Seed the HUVECs at 1×10^5/mL onto the other (interstitial) surface and place endothelial media on to the minusheet.

4. Wash the membrane again 24 h later to remove unattached HUVECs.

5. Once confluence has been reached (~48 h), the trilayer can be placed into a single tissue gradient carrier in the dual perfusion chamber system (Minucells and Minutissue), endothelial side up, with endothelial medium (in upper chamber) nourishing the HUVECs, and RPE medium (in lower chamber) nourishing the RPE.

6. Connect the chambers to separate reservoirs to form a closed circuit containing 25 mL media for both epithelial and endothelial circuits. Replace Medium on a daily basis. This system is amenable to both static and flow culture (see Note 8).

7. Results: RPE and HUVEC form continuous monolayers on their preferred sides of the amnion. HUVEC grow to confluence within 24 h of co-culture with RPE and amnion.

3.4. Immunocytochemistry

1. Transfer the trilayers contained within the MINUSHEETS from the tissue carriers to 12 well plates.

2. Aspirate the culture media, wash cells and fix with 1% paraformaldehyde at room temperature (RT) for 10 min.

3. Wash cells 3 times with normal saline.

4. Permeabilise with 0.5% Triton in PBS.

5. Block with 5% BSA in PBS for 30 min at RT.

6. Incubate with the primary antibody for 1 h at 37°C or for 24 h at 4°C.

7. Wash with PBS.

8. Incubate with the secondary antibodies for 1 h at 37°C or for 24 h at 4°C.

9. Overlay with PBS/glycerol mixed in a ratio of 1:1 (see Note 9).

10. Add Vectashield with or without Propidium iodide (if desirable to discern cell nuclei) on the slide under the trilayer and over the trilayer under the cover slip.

11. Optical images can be compiled and tilted on the *x* and *z* axes to obtain information on the extent of paracellular clefts occupied by JAMs (2).

12. Results: The co-culture conditions do not disturb the molecular phenotype of the endothelial adherens junctions since VE-cadherin is still found at cell–cell contacts throughout this period. At 24 h, HUVEC retain localisation of the TJ marker, ZO-1, with the surface expression of occludin remaining cytoplasmic. However, increased paracellular localisation of occludin is observed by 48 h, and after 72 h co-culture with RPE most of the occludin is localised to cell–cell contacts in HUVEC.

3.5. Trilayer Culture Under Flow

1. The trilayer can be placed into a flow culture system (Minucells and Minutissue), 48 h after seeding the HUVEC at known concentrations.

2. Flow above and below the membrane can be set to the desired flow rate. Ensure that endothelial medium is nourishing the HUVECs, and epithelial medium is nourishing the RPE.

3. Medium should be replaced on a daily basis.

3.6. VEGF-A Secretion of the Trilayer

1. After creating the trilayer in flow (see Subheading 3.5), 2 mL samples can be taken at the desired time points.

2. Samples should be snap-frozen and stored at $-20°C$ until further analysis.

3. Fresh medium should only replace the sample volume (2 mL) taken at each time point.

4. Coat Maxisorp wells with capture mAb (Anti-human VEGF-A antibody, from the 50 µL aliquots) at a dilution of 1:250 in PBS and leave at 4°C overnight. Use 100 µL per well for a 96-well plate.

5. Wash with PBS.

6. Block with 2% BSA for 1 h at RT.

7. Wash with PBS.

8. Prepare a standard curve of rVEGF-A from aliquots. Dilute in 0.1% PBS/BSA at 1:100 = 100 ng/mL, dilute again 1:50 for 2,000 pg/mL (first step of standard curve). Standard curve will be from 2,000 to 31 pg/mL (maybe further) in 1:2 dilutions. Dilute samples if necessary and leave standards and samples for at least 2 h at RT or preferably at 4°C overnight.

9. Wash with PBS.

10. Dilute with 0.5% PBS/BSA the detecting biotinylated anti-hVEGF to 200 ng/mL (44 µL of aliquoted Ab in 11 mL is 1:250 dilution) and leave for at least 1 h at RT.

11. Wash with PBS.

12. Add Biotinylated-streptavidin-HRP at 1: 2,000 in 1% PBS/BSA and leave for 1 h at RT.

13. Wash with PBS.

14. Add TMB and stop with acid after 10 min or whenever the colour changes.

15. Results: Secreted VEGF levels in the trilayer were seen to increase throughout the duration of co-culture and reached 7 ng/mL (+ 0.849 s.d.) by 72 h. Sampling at 24, 48, and 72 h revealed a sharp increase (doubling) in VEGF production between 48 and 72 h.

3.7. Measurement of Permeability or Tracer Flux

1. Place the trilayer in the MINUSHEET in the single tissue gradient carrier, endothelial side up.

2. Dissolve known concentrations of fluorescein and fluorescein-conjugated dextrans (4, 20 and 70 kDa) singly in the endothelial medium flowing through the upper chamber (see Note 10).

3. The lower chamber is bathed in the epithelial medium (as above) without tracer.

4. The chambers were connected to separate reservoirs and a closed circuit containing 25 mL media is obtained for both epithelial and endothelial circuits.

5. Samples of 200 mL can be taken at known time intervals from the lower chamber.

6. Examine fluorescence of the sample using an appropriate spectrometer, for example the Hitachi F-2000 Fluorescence Spectrophotometer.

7. The concentration of fluorescein and fluorescein-dextran in test samples are calculated by comparison against a standard concentration curve.

8. Results: The Amniotic membrane alone is freely permeable to sodium fluorescein (NaF) and 4 kDa FITC-conjugated dextran. Permeability of confluent monolayers of HUVEC to 4 kDa tracers was similar or slightly reduced to that measured for amnion alone. A further reduction in the rate of tracer leakage of 4 kDa tracers was seen for RPE monocultures on amnion. These leakage values of monolayers grown on amnion show no statistical significance when compared to amnion alone. Under flow and 72 h after establishment of the trilayer, permeability to 4 kDa tracer is abolished with less

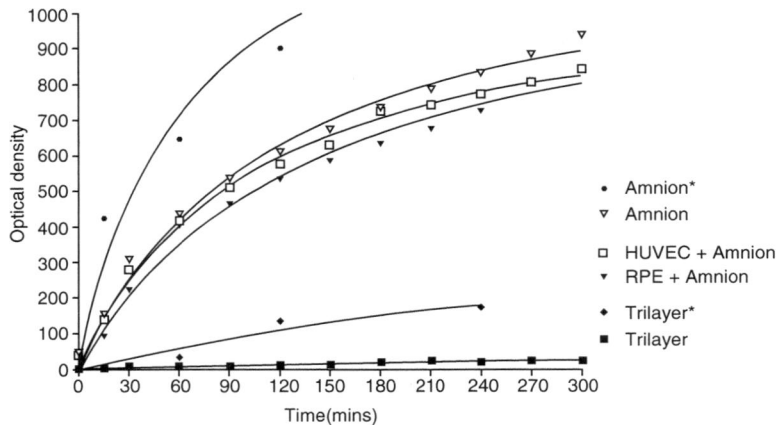

Fig. 3. Graph plotting leakage of 4 kDa fluorescein dextan tracer across amnion alone, amnion + HUVEC, amnion + RPE, amnion + HUVEC co-culture, and the trilayer (RPE + HUVEC co-culture) (*asterisks* shows sodium fluorescein as a comparison). Severe restriction of tracers can be seen in the trilayer plots.

than 5% of 4 kDa dextran being transferred. Moreover, transfer of NaF was seen to be severely restricted in the trilayer (Fig. 3).

3.8. Transmission Electron Microscopy (TEM)

1. To fix the trilayers and monolayers of RPE and HUVEC on amnion, as well as HCE and HUVEC on amnion, use modified Karnovsky's fixative (with 2.5% paraformaldehye, 2% glutaraldehdye in 0.1 M sodium cacodylate buffer, pH 7.4). Embed in resin and process for TEM, using standard protocols.

2. Thin (70 nm) sections should be viewed with an available TEM microscope (JOEL 1010 TEM microscope in our studies).

3. Results: In our study, the appearance of fenestrae in HUVEC after 3 days of co-culture with RPE was a novel exciting finding. In vivo and in monocultures HUVEC show a continuous non-fenestrated phenotype. Thus fidelity of origin appears to be lost when co-cultured with RPE in the trilayer architecture. This was not simply a consequence of co-culture with another epithelial cell type, since tri-layering with corneal epithelial cells did not induce such a change, suggesting this is an RPE-specific cross talk (2). The ultrastructural data highlights the usefulness of high resolution microscopy and strengthens the hypothesis that the EC phenotype (and indeed epithelial phenotype) is plastic and that its immediate environment is a more important determinant than the site of origin.

4. Notes

1. Store all media at 4°C and use within 4 weeks.

2. To make up the acid, goggles must be worn because of the exothermic nature of the reaction.

3. Maintaining the epithelial surface on the side of the black clip prevents confusion in later experiments.

4. Cryopreservation of amnion was not used (23) since freezing the amnion may produce breaks in otherwise healthy tissue that would then affect further studies, particularly the permeability studies.

5. For all RPE assays, it is recommended that the cells be used prior to passage 19 in order to maintain integrity of the phenotype. The cells should be regularly checked for phenotype by using antibodies to CRALBP.

6. For HUVEC assays, cells are used from passages 2 to 3 and are routinely checked for their endothelial nature by using antibodies to vascular specific VE-cadherin.

7. The authors prefer the use of neutralisation of the native epithelium of the amnion rather than using Thermolysin.

8. The flow system can be used to measure permeability of the trilayer to known tracers (see below). The flow rates for upper and lower chamber can be altered as necessary but if used at 0.1 mL/min, a shear stress of less than 0.25 dyne/cm^2 is present. This produces negligible shear stress for cells (17).

9. All steps are 100 µL per well. Wash with PBS containing 0.01% Tween between each step. Use 1:250 dilution (44 µL in 11 mL).

10. The reason for flow being present is also to ensure absence of any unstirred boundary layers, which would lead to an underestimation of solute flux.

Acknowledgements

This study was supported by grants from the Wellcome Trust.

References

1. Del Priore LV, Kaplan HJ, Hornbeck R, Jones Z, Swinn M (1996) Retinal pigment epithelial debridement as a model for the pathogenesis and treatment of macular degeneration. Am J Ophthalmol 122:629–643

2. Hamilton RD, Foss AJ, Leach L (2007) Establishment of a human in vitro model of the outer blood-retinal barrier. J Anat 211:707–716

3. Glaser BM, Campochiaro PA, Davis JL, Jr., Jerdan JA (1987) Retinal pigment epithelial

cells release inhibitors of neovascularization. Ophthalmology 94:780–784

4. Campochiaro PA, Soloway P, Ryan SJ, Miller JW (1999) The pathogenesis of choroidal neovascularization in patients with age-related macular degeneration. Mol Vis 5:34

5. Hartnett ME, Lappas A, Darland D, McColm JR, Lovejoy S, D'Amore PA (2003) Retinal pigment epithelium and endothelial cell interaction causes retinal pigment epithelial barrier dysfunction via a soluble VEGF-dependent mechanism. Exp Eye Res 77(5):593–599

6. Malak TM, Ockleford CD, Bell SC, Dalgleish R, Bright N, Macvicar J (1993) Confocal immunofluorescence localization of collagen types I, III, IV, V and VI and their ultrastructural organization in term human fetal membranes. Placenta 14(4):385–406

7. Capeans C, Pineiro A, Pardo M et al (2003) Amniotic membrane as support for human retinal pigment epithelium (RPE) cell growth. Acta Ophthalmol Scand 81(3):271–277

8. Ohno-Matsui K, Ichinose S, Nakahama K et al (2005) The effects of amniotic membrane on retinal pigment epithelial cell differentiation. Mol Vis 11:1–10

9. Geisen P, McColm JR, Hartnett ME (2006) Choroidal endothelial cells transmigrate across the retinal pigment epithelium but do not proliferate in response to soluble vascular endothelial growth factor. Exp Eye Res 82(4):608–619

10. Dejana E (1996) Endothelial adherens junctions: implications in the control of vascular permeability and angiogenesis. J Clin Invest 98(9):1949–1953

11. Wright TJ, Leach L, Shaw PE, Jones P (2002) Dynamics of vascular endothelial-cadherin and beta-catenin localization by vascular endothelial growth factor-induced angiogenesis in human umbilical vein cells. Exp Cell Res 280(2):159–168

12. Esser S, Lampugnani MG, Corada M, Dejana E, Risau W (1998) Vascular endothelial growth factor induces VE-cadherin tyrosine phosphorylation in endothelial cells. J Cell Sci 111 (Pt 13):1853–1865

13. Anderson JM, Van Itallie CM (1995) Tight junctions and the molecular basis for regulation of paracellular permeability. Am J Physiol 269(4 Pt 1):G467–G475

14. Furuse M, Hirase T, Itoh M et al (1993) Occludin: a novel integral membrane protein localizing at tight junctions. J Cell Biol 123(6 Pt 2):1777–1788

15. Furuse M, Fujita K, Hiiragi T, Fujimoto K, Tsukita S (1998) Claudin-1 and -2: novel integral membrane proteins localizing at tight junctions with no sequence similarity to occludin. J Cell Biol 141(7):1539–1550

16. Martin-Padura I, Lostaglio S, Schneemann M et al (1998) Junctional adhesion molecule, a novel member of the immunoglobulin superfamily that distributes at intercellular junctions and modulates monocyte transmigration. J Cell Biol 142(1):117–127

17. Dewey CF, Jr., Bussolari SR, Gimbrone MA, Jr., Davies PF (1981) The dynamic response of vascular endothelial cells to fluid shear stress. J Biomech Eng 103(3):177–185

18. Leach L, Firth JA (1992) Fine structure of the paracellular junctions of terminal villous capillaries in the perfused human placenta. Cell Tissue Res 268(3):447–452

19. Tseng SC, Prabhasawat P, Lee SH (1997) Amniotic membrane transplantation for conjunctival surface reconstruction. Am J Ophthalmol 124(6):765–774

20. Jaffe EA, Nachman RL, Becker CG, Minick CR (1973) Culture of human endothelial cells derived from umbilical veins. Identification by morphologic and immunologic criteria. J Clin Invest 52(11):2745–2756

21. Liotta LA, Lee CW, Morakis DJ (1980) New method for preparing large surfaces of intact human basement membrane for tumor invasion studies. Cancer Lett 11(2):141–152

22. Dua HS, Gomes JA, King AJ, Maharajan VS (2004) The amniotic membrane in ophthalmology. Surv Ophthalmol 49(1):51–77

23. Glade CP, Seegers BA, Meulen EF, van Hooijdonk CA, van Erp PE, van de Kerkhof PC (1996) Multiparameter flow cytometric characterization of epidermal cell suspensions prepared from normal and hyperproliferative human skin using an optimized thermolysin-trypsin protocol. Arch Dermatol Res 288(4):203–210

24. Kruse FE, Joussen AM, Rohrschneider K et al (2000) Cryopreserved human amniotic membrane for ocular surface reconstruction. Graefes Arch Clin Exp Ophthalmol 238(1):68–75

Chapter 21

Isolation and Properties of Endothelial Cells Forming the Blood-Nerve Barrier

Yasuteru Sano and Takashi Kanda

Abstract

The blood–nerve barrier (BNB) is one of the functional barriers sheltering the nervous system from circulating blood. It is very important to understand the cellular properties of endothelial cells of endoneurial origin because these cells constitute the bulk of the BNB. This chapter describes a standard protocol for isolating the endothelial cells forming the BNB. In addition, methods for confirming some of the barrier properties of isolated endothelial cells are also described.

Key words: Blood–nerve barrier, Peripheral nerve microvascular endothelial cells, Endoneurium, Cloning, Tight junction, Transporter

1. Introduction

The blood–nerve barrier (BNB) is one of the functional barriers sheltering the nervous system from circulating blood (1). The BNB is composed of the endoneurial microvasculature and the innermost layers of the perineurium. Blood-borne substances can reach the endoneurial extracellular space by crossing either the endoneurial vascular endothelium or the perineurium. Many studies (2, 3) indicate that perineurial permeability is much lower than that of endoneurial microvessels against various substances. Hence, peripheral nerve microvascular endothelial cells (PnMECs) which constitute the bulk of the microvessels in the endoneurium can be considered to be the real interface between blood and the peripheral nerves. Because PnMECs are the structural basis of the BNB, an investigation of the cellular properties of PnMECs using cell culture technique may provide new insights into the pathogenesis of immune-mediated neuropathies.

Sukriti Nag (ed.), *The Blood-Brain and Other Neural Barriers: Reviews and Protocols*, Methods in Molecular Biology, vol. 686, DOI 10.1007/978-1-60761-938-3_21, © Springer Science+Business Media, LLC 2011

The standard methods for isolating rat PnMECs are described in this chapter. The method described in this chapter can be applied for isolating PnMECs of human and other species such as mouse, cow, and pig. In addition, the basic properties of PnMECs as barrier forming cells are also indicated. Some important transporters of the BBB such as glucose transporter 1 (GLUT1) and p-glycoprotein (p-gp) are present in endoneurial microvessels in the peripheral nerves (4, 5). Various tight junction proteins such as claudin-5 and ZO-1 were also detected on the endoneurial microvessels in the human sural nerve (6). These results have been verified by an in vitro analysis which showed that rat PnMECs express tight junction associated molecules and transporters at the mRNA level (7). Hence, the method for characterizing PnMECs as barrier forming cells by RT-PCR analyses is also described in this chapter.

2. Materials

2.1. Culture Media

1. Dissecting medium (DM): Dulbecco's modified Eagle's medium (DMEM; Sigma-Aldrich, St. Louis, MO) supplemented with 5% fetal bovine serum (FBS; Sigma-Aldrich), 100 U/ml penicillin (Sigma-Aldrich), 100 μl/ml streptomycin (Sigma-Aldrich) and 25 ng/ml amphotericin B (Invitrogen Inc, Grand Island, NY).

2. Endothelial cell (EC) growth medium: EBM-2 medium (Cambrex, Walkersville, MD) supplemented with 20% FBS, EGM-2 MV (Cambrex), 100 U/ml penicillin (Sigma), 100 μl/ml streptomycin (Sigma-Aldrich) and 25 ng/ml amphotericin B (Invitrogen Inc).

2.2. Isolation of Rat PnMECs Forming the BNB

1. Wister rats weighing 150–300 g.
2. CO_2 is used for euthanasia procedures.
3. Collagenase type I (Sigma-Aldrich).
4. 1× Hanks' balanced salt solution (HBSS; Invitrogen Inc).
5. Rat tail Type I collagen-coated Petri dishes 30 and 60 mm (AGC Techno Glass Co, Funabashi, Japan).
6. Cloning cup (Hiroshima Resin Co, Hiroshima, Japan).

2.3. Identification of PnMECs

1. PBS:150 mM sodium phosphate, 145 mM NaCl, pH 7.2.
2. 4% Paraformaldehyde (Wako, Osaka, Japan).
3. Triton X-100 (Sigma-Aldrich).
4. Fetal Bovine Serum (FBS) (Sigma-Aldrich).
5. Mouse antihuman von Willebrand factor antibody, 100 dilution (Dako, A/S, Denmark).

6. FITC-conjugated antimouse IgG, 1:100 dilution (Zymed, CA, USA).

7. 1,1'-dioctacecyl-3,3,3',3', tetramethyl indocarbocyanine perchlorate acetylated low-density lipoprotein (DiI-Ac-LDL): Biogenesis, Poole, England.

8. Fluorescence microscope (Olympus, Tokyo, Japan).

9. Phase contrast microscope (Olympus).

2.4. Characterization of PnMECs

1. RNeasy® Plus Mini kit (Qiagen, Hilden, Germany).

2. StrataScript® First Strand Synthesis System (Stratagene®, Cedar Greek, TX, USA).

3. TaKaRa Ex Taq® (TaKaRa, Otsu, Japan).

4. Oligo-dT primer (Stratagene®).

5. Ethidium bromide (10 mg/ml; Sigma-Aldrich).

6. NuSieve® GTG® Agarose (Lonsa, ME, USA).

7. GeneRuler™ 50bp DNA Ladder (Fermentas, Burlington, Canada).

8. PCR Thermal Cycler (TaKaRa).

9. Agarose (Lonza).

3. Methods

3.1. Isolation of the Rat PnMECs Forming the BNB

1. Rats are euthanized using CO_2.

2. Remove the sciatic nerve and rinse the nerve well with DM to avoid bacterial contamination.

3. Strip-off the epi- and perineuria carefully with fine forceps. This procedure mimics the teased fiber preparation used to study peripheral nerve pathology (see Note 1).

4. Mince the endoneurium finely with a razor blade and digest the minced material with 0.25% collagenase type I in 1× Hanks' balanced salt solution (HBSS) at 37°C for 2 h.

5. After centrifugation ($800 \times g$, 5 min), wash the pellet with DM several times.

6. Suspend the pellet in 3 ml EC growth medium.

7. Spread the suspended cells onto a type I collagen-coated 60-mm Petri dish (see Note 2).

8. Culture the cells at 37°C in a humidified atmosphere of 5% CO_2 and 95% air (see Note 3).

9. Scratch and remove non-PnMECs mechanically with a sterilized pointed rubber spatula (see Note 4).

10. ECs are recognized by their spindle-shaped morphology (see Fig. 1).

11. When the PnMEC colonies grow sufficiently for cloning (several weeks after the first dissemination), pick up the PnMEC colonies with 0.25% Trypsin/EDTA using a cloning cup (plastic or glass ring).

12. Aspirate suspended ECs using a pipette and harvest in another 60-mm Petri dish containing EC growth medium.

13. Repeat steps 9–12 until 100% of PnMECs are obtained (see Note 5).

3.2. Identification of PnMECs

1. Culture PnMECs on a rat tail collagen type I-coated 30-mm dish at 37°C.

2. Wash the cells three times with PBS.

3. Fix the cells in 4% paraformaldehyde for 15 min at room temperature.

3.2.1. Immunoreactivity Against von Willebrand Factor

4. Wash the cells three times with PBS and permeabilize them with 0.1% Triton X-100 for 10 min.

5. Wash the cells three times with PBS and block the cells with 1% FBS in PBS for 1 h at room temperature.

6. Wash the cells three times with PBS and incubate cells with the mouse anti-human von Willebrand factor antibody in blocking solution at 4°C overnight.

7. Wash the cells 3 times with PBS and incubate cells with FITC-conjugated anti-mouse IgG for 1 h at room temperature.

8. Detect the von Willebrand factor signal in the cell cytoplasm using a fluorescence microscope (see Fig. 2).

3.2.2. Staining of PnMECs with DiI-Ac-LDL

1. Culture PnMECs at 37°C in a collagen type I-coated 30-mm dish containing EC growth medium.

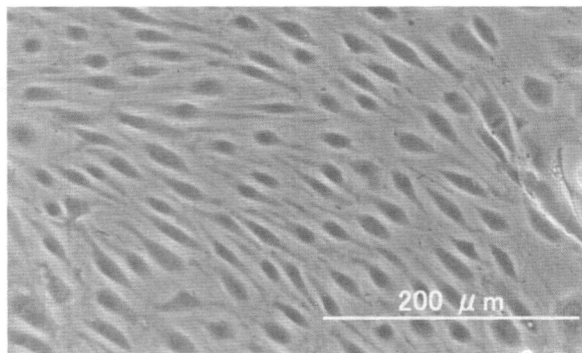

Fig. 1. Phase contrast micrograph of the isolated PnMECs. They are closely packed and show a spindle-shaped morphology.

Fig. 2. PnMECs are immunoreactive with anti-von Willebrand factor antibody confirming their endothelial nature.

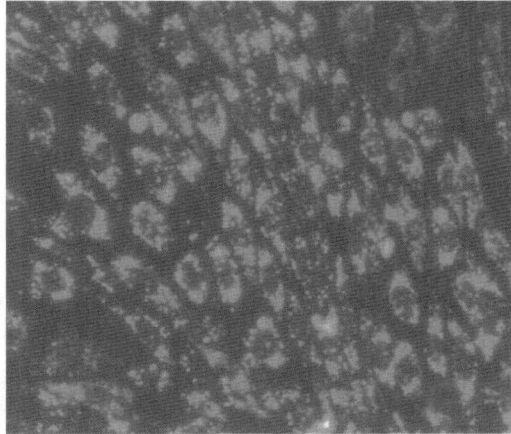

Fig. 3. Fluorescence micrograph of PnMECs showing cytoplasmic DiI-Ac-LDL uptake.

2. Incubate the cells with 10 µg/mL DiI-Ac-LDL at 37°C in culture medium overnight.

3. PnMECs incorporate DiI-Ac-LDL particles into their cytoplasm which appears bright when viewed with a fluorescence microscope (Fig. 3).

3.3. Characterization of PnMECs Derived From Rat BNB

1. Culture PnMECs in a rat tail collagen type I-coated 30-mm dish at 37°C until confluence is reached.

2. Wash the cells three times with PBS.

3. Prepare total RNA from PBS-washed cells or the rat cerebrum using an RNeasy® Plus Mini kit. Approximately $1 \times 10^5 \sim 10^7$ cells are used and approximately $5 \times 5 \times 5$ mm pieces of rat cerebrum are used.

4. Synthesize single-stranded cDNAs from 50–1,000 ng of total RNA using the StrataScript® First Strand Synthesis System with an oligo-dT primer.

5. Perform the sequential PCR with TaKaRa Ex Taq with PCR Thermal Cycler. Temperature cycling conditions for each primer consists of 5 min at 94°C followed by 35–40 cycles for 1 min at 94°C, 1 min at 55–65°C and 1 min at 72°C, with a final extension for 10 min at 72°C.

6. The primer pairs used and the expected size of the PCR fragments are shown in Table 1.

7. Visualize the PCR products by ethidium bromide staining following resolution on a 2% agarose gel. Estimate band size of each product by comparing them with a 50-bp ladder (see Fig. 4).

These techniques have been used to isolate and characterize BNB endothelial cells as reported previously (7).

4. Notes

1. The perineurial tissue is a hard and fibrous structure, and it comes into close contact with the relatively soft endoneurial tissue. Exclusion of vessels of nonendoneurial origin which do not have BNB properties is crucial in this isolation procedure, therefore, the hard perineurial tissue is completely discarded and only the endoneurial component is utilized in the next step.

2. A Petri dish must be used in this step. A cell culture flask is not appropriate for the PnMEC- purification steps which are essential for the final results.

3. The dish must not be moved for at least 12–24 h in order for the cells to attach to the dish.

4. The PnMECs are easily identified because of their spindle-shaped morphology (see Fig. 1). These cells also show contact inhibition and never pile up. Non-ECs such as pericytes, fibroblasts, and Schwann cells often pile up and definitely grow faster than PnMECs. Furthermore, these non-ECs often proliferate onto or beneath the PnMEC colonies and once these non-ECs invade the colony, the PnMEC purification procedure fails.

5. Generally, 100% of PnMECs should be obtained within 3–5 passages. PnMECs derived from BNB are closely packed and showed a spindle-shaped morphology that has been well recognized to be a characteristic of ECs constituting the barrier

Table 1
The primer pairs used and the expected size of the PCR fragments are listed

Molecules and fragment (gene accession no.)	Forward primers	Reverse primer size
claudin-5 (NM_031701) 230	5'-TTAAGGCACGGGTGGCACTCACG-3'	5'-TTAGACGTAGTTCTTCTTGTCG -3'
ZO-1 (NM_00110626) 81	5'-GCGAGGCATCGTTCCTAATAAG-3'	5'-TCGCCACCTGCTGTCTTTG-3'
GLUT-1 (NM_138827) 503	5'-GATGATGAACCTGTTGGCCT-3	5'-AGCGGAACAGCTCCAAGATG-3'
Mdr1a (NM_133401) 437	5'-ACAGAAACAGAGGATCGC-3'	5'-CGTCTTGATCATGTGGCC -3'
GAPDH (NM_017008) 240	5'-TGATGACATCAAGAAGGTGGTGAAG-3'	5'-TCCTTGGAGGCCATGTAGGCCAT-3'

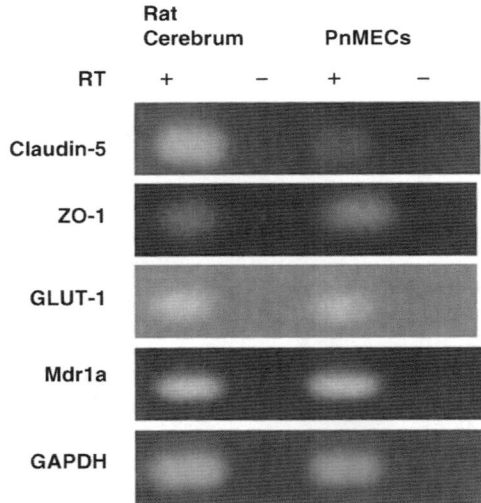

Fig. 4. mRNA expression of tight junction molecules and transporters in PnMECs. Expression of claudin-5, ZO-1, GLUT-1, and Mdr1a mRNAs in the rat cerebrum and PnMECs was investigated by RT-PCR analysis with a specific primer set for each gene. The reactions were performed against either total RNA with (+) or without (–) reverse transcription.

system ((4, 8); see Fig. 1). Some pericytes resemble PnMECs because of their elongated shape. Verification using DiI-Ac-LDL is very useful to distinguish pericytes from PnMECs because pericytes do not incorporate DiI-Ac-LDL.

Acknowledgments

This work was supported in part by a Neuroimmunological Disease Research Committee grant from the Ministry of Health, Labour and Welfare, Japan and also by a Grant-in-Aid for Scientific Research from the Ministry of Education, Science, Sports and Culture, Japan.

References

1. Poduslo JF, Curran GL, Berg CT (1994) Macromolecular permeability across the blood-nerve and blood-brain barriers. *Proc Natl Acad Sci USA* 91: 5705–5709

2. Soderfeldt B (1974) The perineurium as a diffusion barrier to protein tracers. Influence of histamine, serotonine and bradykinine. *Acta Neuropathol* 27: 55–60

3. Boddingius J (1984) Ultrastructural and histophysiological studies on the blood-nerve barrier and perineurial barrier in leprosy neuropathy. *Acta Neuropathol* 64: 282–296

4. Kanda T, Iwasaki T, Yamawaki M, Ikeda K (1997) Isolation and culture of bovine endothelial cells of endoneurial origin. *J Neurosci Res* 49: 769–777

5. Saito T, Zhang ZJ, Shibamori Y, Ohtsubo T, Noda I, Yamamoto T, Saito H (1997) P-glycoprotein expression in capillary

endothelial cells of the 7th and 8th nerves of guinea pig in relation to blood-nerve barrier sites. *Neurosci Lett* 232: 41–44

6. Kanda T, Numata Y, Mizusawa H (2004) Chronic inflammatory demyelinating poly-neuropathy: decreased claudin-5 and relocated ZO-1. *J Neurol Neurosurg Psychiatry* 75: 765–769

7. Sano Y, Shimizu F, Nakayama H, Abe M, Maeda T, Ohtsuki S, Terasaki T, Obinata M, Ueda M, Takahashi R, Kanda T (2007)

Endothelial cells constituting blood-nerve barrier have highly specialized characteristics as barrier-forming cells. *Cell Struct Funct* 32: 139–147

8. Hosoya K, Takashima T, Tetsuka K, Nagura T, Ohtsuki S, Takanaga H, Ueda M, Yanai N, Obinata M, Terasaki T (2000) mRNA expression and transport characterization of conditionally immortalized rat brain capillary endotherial cell lines; a new in vitro BBB model for drug targeting. *J Drug Target* 8: 357–370

Part V

Delivery of Therapeutic Agents Across the Barriers

Chapter 22

Treatment of Focal Brain Ischemia with Viral Vector-Mediated Gene Transfer

Hua Su and Guo-Yuan Yang

Abstract

Promoting functional recovery after ischemic brain injury has emerged as a potential approach for the treatment of ischemic stroke. An ideal restorative approach to enhance long-term functional recovery is to promote postischemic angiogenesis and neurogenesis. This chapter describes a system using adeno-associated viral (AAV) vector-mediated vascular endothelial growth factor (VEGF) gene transfer into the ischemic brain. The methods described here for construction, production, and purification of AAV vector expressing VEGF gene can also be applied to producing AAV vectors expressing other genes. This chapter also illustrates the methods to produce mouse middle cerebral artery occlusion (MCAO), injection of viral vector into the mouse brain, and standard assays for determining the success of brain ischemia and gene transfer.

Key words: Adeno-associated virus, Astrocyte, Capillary density, Endothelial cells, Gene transfer, Ischemia, Middle cerebral artery occlusion, Neurons, VEGF

1. Introduction

Stroke is the leading cause of serious and long-term disability in the United States. About 700,000 people suffer a new or recurrent stroke each year, about 80% of which is ischemic stroke. Although diagnosis is easy, available therapies for ischemic stroke are limited. Currently, only the FDA-approved drug for the treatment of clinical ischemic stroke, recombinant tissue plasminogen activator (tPA), shows therapeutic effect when given within 3 h of the onset of stroke in carefully selected patients (1). The disappointing results have encouraged scientists to explore other efficient strategies for the treatment of brain ischemia. It has been noted that except for short-term neuroprotection, a regenerative response, such as vascular remodeling, angiogenesis,

Sukriti Nag (ed.), *The Blood-Brain and Other Neural Barriers: Reviews and Protocols*, Methods in Molecular Biology, vol. 686, DOI 10.1007/978-1-60761-938-3_22, © Springer Science+Business Media, LLC 2011

and migration of neuroblasts from the subventricular zone (SVZ) to the ischemic zone, is triggered in the tissue adjacent to the ischemic core area. Ischemia-induced angiogenesis usually occurs in association with reactive astrocytes and blood vessels (2). The discovery of neurovascular coupling has led to the concept of a neurovascular unit, composed of functionally integrated cellular (including brain endothelial cells, astrocytes, pericytes, and smooth muscle cells) and acellular elements such as the basement membrane (3). The concept of "re-establishment of the neuro-vascular unit" suggests that cerebral vasculature-based therapy should be coupled with neuroprotection, a brand new direction in the development of new therapies for cerebral ischemia.

Adeno-associated virus (AAV) is currently a popular vector of choice for gene therapy and is widely used in various CNS disease models (4). AAV is a principal vector choice in experimental isch-emic stroke studies (5–10). The advantage of using AAV vector is its low immunogenecity and lack of association with human pathologies. In addition, the availability of multiple serotypes and their differential preferences to cells and tissues significantly broadens application of AAV vectors and improves their transduc-tion efficiency (11–17).

Vascular endothelial growth factor (VEGF), a 34–45-kDa dimeric glycosylated protein, is usually upregulated in ischemic tissues. VEGF exerts its biological effects through its receptors flt-1 and flk-1, both of which are expressed on endothelial cells. VEGF is explored as a therapeutic reagent in cerebral ischemia models because it not only induces cerebral angiogenesis, but also enhances neuroprotection and promotes neurogenesis after cere-bral ischemia. Pretreatment with VEGF through recombinant AAV-mediated gene transfer reduces infarct volume in the mouse transient middle cerebral artery occlusion (MCAO) model (6). Intraventricular administration of VEGF activates angiogenesis in the ischemic penumbra of the striatum and increases neuron sur-vival in the dentate gyrus and SVZ, consequently reducing infarct volume and improving neurological performance. Therefore, VEGF exerts its neuroprotective effect in the acute phase of cere-bral ischemia with the longer latency effects on angiogenesis and neoneuron survival (7).

Gene transfer technology has been applied to investigate the molecular mechanisms and therapeutic approaches to a spectrum of neurological diseases. Overexpression of beneficial genes by exogenous gene transfer to antagonize or prevent subsequent dis-ease progression underlies the central theme of gene therapy. Gene therapy targeting several neurological diseases, including Alzheimer's disease (18, 19), Canavan disease (20), and glioblas-toma multiforme (21–24), has progressed to clinical trials. However, gene therapy for cerebrovascular diseases, including ischemic stroke, is still in the experimental stage but shows promising results. The challenges in treating ischemic stroke with gene therapy technology are the acuity and the progression of this disease process. In addition, several other issues, such as vector preference for gene delivery, immunogenicity of vector, route of vector administration, selection of target gene, and examination of transduction efficacy, also influence the development of effective therapies (25–28).

The focal nature of ischemic stroke reinforces the significance of considering an efficient means of focal gene delivery. In contrast, more diffuse lesions in the brain require vector delivery that spans a larger, more generalized area of the brain. The possible routes of administration of vectors include injecting into cerebrospinal fluid (29–35), parenchyma (6, 36, 37), cerebral vasculature (26, 38), or peripheral site that has the potential to retroactively transport into the CNS (39). Each method offers different benefits and limitations that need to be taken into account when evaluating location, size, and mechanism of diseases in gene therapy studies.

Although gene therapy for cerebral ischemic injuries has yet to be tested in clinical trials, there have been many advances in the experimental stage since the first study (35). Animal studies in gene therapy have adapted models that mimic human cerebral ischemia, including transient MCAO (6, 36–38), permanent MCAO (5, 30, 40), four-vessel occlusion (41, 42), or transient bilateral carotid artery occlusions (8, 33). Using these models, research has manipulated various cellular pathways within cerebral ischemia pathophysiology to rescue brain tissue from ischemic injury. To test the efficacy of gene therapy in ischemia models, studies measuring infarct volume, cell proliferation, angiogenesis, neurogenesis, and neurobehavioral recovery in animals should be included (41–44). It is also useful to pay particular attention to the penumbral area of the infarction, as it is the remaining salvageable region after ischemic injury.

In many animal studies, gene transfer is performed before artery occlusion (37, 38, 45). It is more clinically relevant to further introduce promising genes postartery occlusion, as has been done in other animal studies (29, 36, 46–50). Some studies have even induced gene transfer days after onset of ischemia and demonstrated beneficial results (36, 51, 52). In patients, maximal functional recovery after a cerebral ischemic attack occurs within the first few months; therefore, the important step is to improve long-term recovery. Further, the time to protransduce genes into the disease model is primarily dependent on the vector utilized, and secondarily on the ability of transduced cells to express the gene.

Transduction efficiency is crucial in gene therapy research in vivo because of the limitation of diffusion after viral vector delivery, although the intraventricular or multiple parenchyma injection method effectively increases target protein expression in the experimental rat brain (35, 53). There is a vast size mismatch between the mouse brain and the human brain. Recently, new methodologies such as convection-enhanced AAV gene delivery have been developed, which enable widespread gene expression in the primate brain (54). It is important to apply gene therapy protocols to larger animals to better test for transduction efficiency. It is also important to recognize transduction efficiency within the microvasculature because speed of blood flow may hamper vector delivery into endothelial cells. Uncontrolled target gene expression can cause some unwanted side effects, such as hemangioma formation. Therefore, it is crucial to develop an efficient technique to control target gene expression (55).

2. Materials

2.1. AAV Virus Construction and Production

1. AAV Helper-Free Gene Delivery and Expression System (Stratagene Inc, Wilmington, DE).
2. Not 1 and T4 DNA ligase.
3. Agarose gel equipment.
4. Competent cells: *E. coli* strains XL1 blue, DH5α (Stratagene).
5. Dulbecco's modified Eagle's medium (DMEM) with 4.5 g/L glucose and l-glutamine (Cellgro, Herndon, VA).
6. Fetal bovine serum (FBS; Hyclone, Logan, UT).
7. HEK 293 cells (American Type Culture Collection, Gaithersburg, MD).
8. Quantikine human VEGF (R&D Systems, Minneapolis, MN).

2.1.1. Transfection, Purification, and Titer Determination

1. Sterile Glassware: 15 cm tissue culture dishes, 48-well tissue culture dishes, 50 mL Corning tubes, 1 mL pipets.
2. Chemicals: $CaCl_2$, Calcium phosphate, Cs Cl_2 1.4 g/mL (0.548 g/mL), EDTA, Tween-20, sorbital, polyethylene glycol (PEG).
3. 2×HBS:280 mM NaCl, 10 mM KCl, 1.5 mM $Na_2HPO4 \cdot 7H_2O$, 12 mM dextrose, and 50 mM HEPES.
4. Dialysis cassettes: Slide-A-Lyzer 7K (Pierce, Rockford, IL).
5. High speed centrifuge and ultra speed centrifuge.
6. Cell culture room equipped with CO_2 incubators and biological hood.
7. Hybond H+ membrane (Stratagene).
8. DIG High Prime DNA Labeling and Detection Kit II (Roche Diagnostics, Indianapolis, IN).
9. Taq DNA polymerase (Invitrogen Inc).
10. Random Primed DNA Labeling Kit (Roche).

11. 2× Standard Saline Citrate (SSC) solution:300 mM sodium chloride, 30 mM sodium citrate, pH 7.0, titrated with HCl pH 7.0.

2.2. Middle Cerebral Artery Occlusion (MCAO) Model and Viral Vector Injection

The optimal small animal surgery room needs to be kept quiet and pathogen-free, and the room temperature maintained at $25 \pm 3°C$. The items recommended for small animal surgery are shown in Fig. 1.

1. Sprague Dawley rats (Charles River, Davis, CA) weighing 260–300 g.
2. Isoflurane (Fisher Scientific, Pittsburgh, PA), ketamine (Sigma-Aldrich, St Louise MO), xylazine (Sigma-Aldrich).
3. PE-50 catheter (Fisher Scientific).
4. 4-0 nylon suture (Hospital Supply, Sherwood, AR).
5. Bipolar coagulator: Malis™ Precision-Control bipolar coagulator, CMC™-II-PC (Codman and Shurtleff Inc, Randolph, MA).
6. Exercise Physiological System (AD Instruments, PowerLab/4SP, Castle Hill, Australia).
7. Temperature controller: Homeothermic Blanket Control Unit (Harvard Apparatus, Cambridge, MA).
8. Laser Doppler Flowmeter: LASERFLO® Blood perfusion monitor BPM² (VASAMEDICS, St. Paul, MN).
9. Hot bead sterilizer (Fine Science Tools, Foster City, CA).
10. Matrix anesthesia machine: A vaporizer for Isofluorane (Summit Anesthesia Solutions, Bend, OR).
11. Stereotaxic frame: Model 900 (David Kopf Instruments, Tujunga, CA).

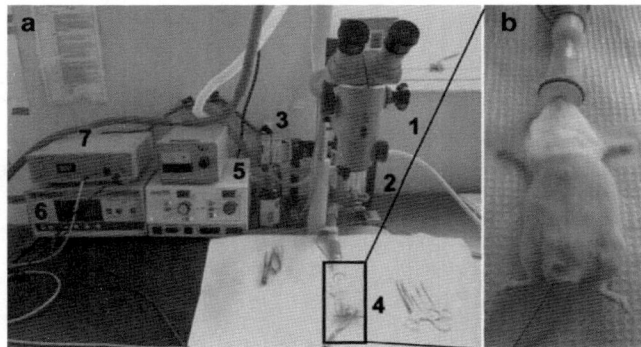

Fig. 1. (a) The animal surgical setup for the MCAO procedure includes: *1* a surgical microscope; *2* a vaporizer for anesthesia; *3* flowmeter for air gas control; *4* animal surgical table; *5* bipolar coagulator; *6* a laser Doppler flowmeter machine; and *7* a temperature controller. (b) A mouse under inhalant isoflurane anesthesia is shown.

12. pH/blood gas analyzer (Bayer, Radiolab 248, Tarrytown, NY).

13. High-speed micro drill (Fine Science Tools).

14. A set of small animal surgical equipments is shown in Fig. 2.

15. Hamilton syringe with replaceable Beveled Needle (World Precision Instruments, Sarasota, FL).

16. E type Thermocouples (Omega Engineering Inc, Stamford, CT).

2.3. Tissue Stains and Morphometry

1. 2% paraformaldehyde in 0.1 M PIPES pH 6.9.

2. 0.5% glutaraldehyde.

2.3.1. X-Gal Staining

3. Cryostat.

4. X-gal staining solution: 5 mM/L $K_3Fe (CN)_6$, 5 mM/L K4Fe (CN), 2 mM/L $MgCl_2$, 0.01% sodium deoxycholate, 0.02% Nonidet P-40, 1 mg/mL 5-bromo-4-chloro-3-indolyl-β-d-galactoside in PBS.

5. NIH Image 1.63 software (http://rsb.info.nih.gov/ij).

2.3.2. Cresyl Violet Staining

1. 0.1% Cresyl Violet acetate (Sigma-Aldrich).

2. Double distilled water.

3. 10% Acetic acid (Sigma-Aldrich).

4. Solvents: Xylene, graded ethanols (70, 90, 95, 100%).

5. Coverslips.

6. Permount.

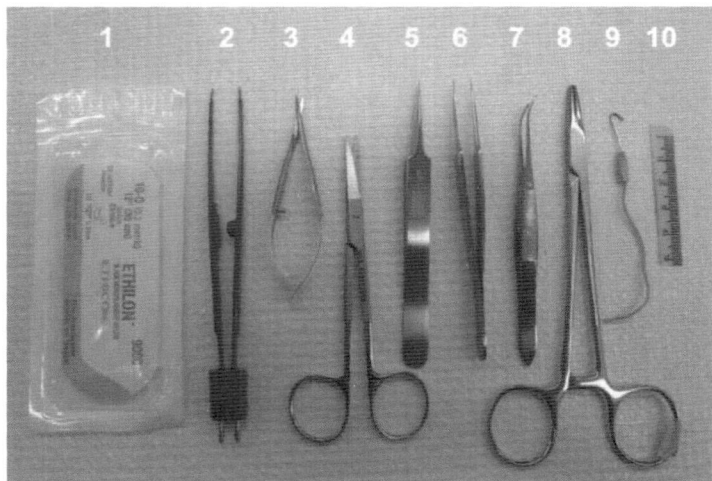

Fig. 2. The basic microsurgical equipment required for the MCAO model is shown. All items must be sterilized before use: *1* nylon suture; *2* bipolar forceps; *3* microscissors for small vessels; *4* surgical scissors for animal skin and tissue use; *5* and *6* microforceps; *7* tissue forceps; *8* needle holder; *9* skin hook for the exposure; *10* ruler.

2.3.3. Fluoro-Jade B Staining

1. 0.06% Potassium permanganate ($KMnO_4$) (Sigma-Aldrich).
2. A 0.001% fluoro-Jade staining solution (Chemicon, Temecula, CA).
3. 0.1% acetic acid (Sigma-Aldrich).
4. Light and fluorescence microscopes.

2.3.4. Lectin Staining for Capillary Density

1. Luorescein-lycopersicin esculentum lectin (Vector Lab, Burlingame, CA).

3. Methods

3.1. AAV Vector Construction

1. The pCMV-VEGF vector can be generated by inserting the human $VEGF_{165}$ cDNA to multiple cloning sites in pCMV-MCS vector (Fig. 3a). Plasmid AAV-LacZ (Fig. 3b) is included in the AAV Helper-Free Gene Delivery and Expression System.
2. The expression cassette includes CMV promoter, VEGF cDNA, and HGH polyadenylation signal (poly A) can be isolated from pCMV-VEGF vector by Not I digestion.
3. pAAV-VEGF vector can then be generated by replacing the LacZ expression cassette with VEGF expression cassette (Fig. 3c). VEGF expression mediated by pAAV-VEGF can be confirmed by infecting HEK 293 cells and analyzing VEGF concentration in the supernatant by ELISA using the Quantikine human VEGF Kit.

3.2. AAV Virus Production

3.2.1. Transfection

1. First, seed 4.5×10^6 HEK 293 cells in 15-cm tissue culture dishes 2 days before transfection. Distribute cells in one 90% confluent dish to eight dishes and maintain in 25 mL DMEM plus 10% FBS, and 25 mM HEPES.
2. Cotransfect the AAV vector containing the gene of interest with pAAV-RC and pHelper vectors into HEK 293 cells by using the calcium phosphate precipitation method as described below. pAAV-RC contains AAV rep and cap genes (see Note 1). pHelper has the adenoviral VA, E2A, and E4 regions that mediate AAV vector replication. To transfect one 15-cm dish, mix a total of 50 μg DNA (17 μg DNA of each plasmid) with 1 mL of 300 mM $CaCl_2$. A mixture for transfection of four dishes can be prepared in one 50 mL Corning tube.
3. Place a sterile pipet into the tube. Gently bubble air while 1 mL (4 mL for four plates) of 2× HBS is added to the tube drop by drop.
4. Distribute the transfection mix to each plate.
5. Incubate the plates at 37°C for 6 h.

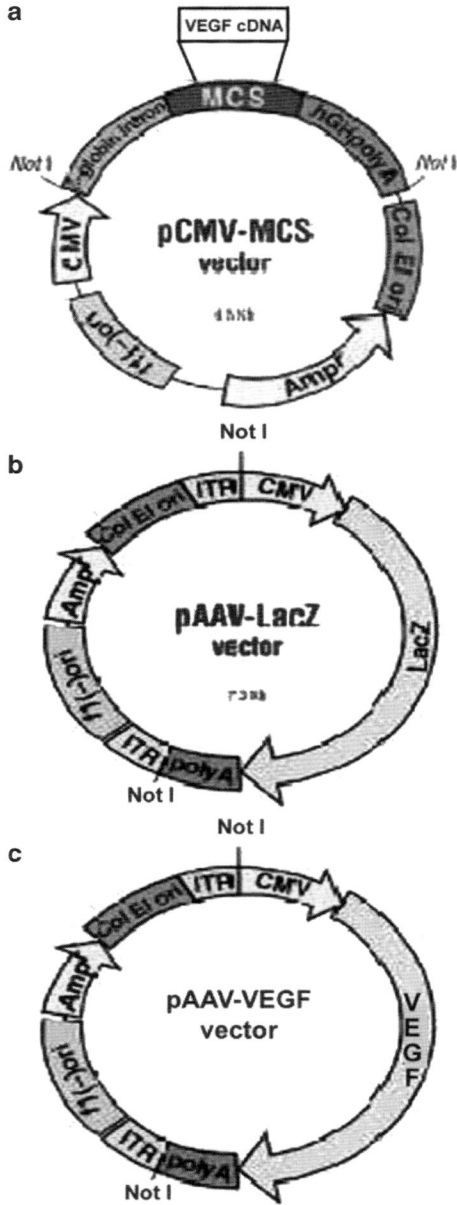

Fig. 3. Schematic drawing of pCMV-MCS, pAAV-LacZ, and pAAV-VEGF vectors. The expression cassette including the CMV promoter, LacZ or VEGF gene, and HGH polyadenylation signal in the vectors can be excised by *NotI* enzyme digestion. The expression cassette excised from the pCMV-VEGF-vector can be cloned between two *NotI* sites in pAAV-LacZ vector to replace LacZ expression cassette to generate pAAV-VEGF.

6. Change the medium DMEM containing 2% FBS and 25 mM HEPES. Culture cells at 37°C for 54 h.

3.2.2. Purification

1. Dislodge cells from the dishes by gentle pipeting and transfer to 50 mL Corning tubes.

2. Remove media by centrifugation ($1,000 \times g$ for 5 min at 4°C).

3. Resuspend cells in 100 mM Tris–HCl, 150 mM NaCl, pH 8.0 (1×10^7 cells/mL), and lyse using three freeze and thaw cycles (alternating between dry ice-ethanol and 37°C water baths).

4. Centrifute the lysates at $10,000 \times g$ for 15 min to remove the cell debris.

5. Precipitate the cleared supernatant with 25 mM CaCl$_2$ at 0°C for 1 h.

6. Remove precipitates by centrifugation at $10,000 \times g$ for 15 min at 4°C.

7. Add NaCl and PEG (8,000) to the supernatant to make the final concentration of 620 mM NaCl and 8% PEG.

8. Incubate the supernatants on ice for 3 h.

9. Collect AAV vector containing precipitate by centrifugation at $300 \times g$, for 30 min at 4°C, and resuspend in 5 mL of 50 mM HEPES, 150 mM NaCl, 25 mM EDTA, pH 8.0 (adjust pH to 8.0 with NaOH).

10. Remove insoluble materials by centrifugation at $10,000 \times g$ for 15 min at 4°C.

11. Purify the AAV vector by CsCl$_2$ gradient centrifugation. Add solid CsCl$_2$ to the supernatant to produce a density of 1.4 g/mL. Centrifuge samples at 15°C for 16 h at $223,000 \times g$.

12. Fractionate the gradient (0.5–1 mL/fraction) and assay by dot blot to detect the viral particles (see Subheading 3.2.3).

13. Pool the AAV vector-containing fractions, put in dialysis cassettes (Slide-A-Lyzer 7K), and dialyze against 1 L (1,000 times the volume of viral fraction) buffer containing 10 mM HEPES (pH 7.4), 140 mM NaCl, 0.1% Tween-80, and 5% sorbital, three times at 4°C. Change the dialysis buffer every 2 h.

14. The vectors can also be purified by other methods (see Note 2).

3.2.3. Titer Determination

Viral titers are determined by dot blot analysis of the DNA content (see Note 3).

1. First, add 2 μL of each fraction collected after CsCl$_2$ centrifugation or dialyzed viral stock to 20 μL of 500 mM NaOH, and dot on Hybond H$^+$ Membrane.

2. Then, serially dilute 1 ng of gene fragment (about 1 kb long) and dot on the same membrane as control.

3. Air-dry the membrane and neutralize in 500 mM Na$_2$HPO$_4$ buffer, pH 7.4, for 10 min, then wash in 2× SSC briefly and air-dry.

4. Hybridize the membrane with the isotope or digoxygenin-labeled probes for genes in the AAV vectors using the protocol provided in the DIG High Prime DNA Labeling and Detection Kit II or see Note 4.

5. Compare the density of each fraction or viral stock with that of the control. The relative copy number can be calculated. Copies of a 1-ng gene fragment can be calculated as follows (copies/mL): $1 \times 10-9/length\ of\ the\ fragment\ (bp) \times 660) \times 6 \times 1023 \times 2 \times 103$.

3.2.4. Toxicity

The toxicity of the viral stock is checked by infection of HEK293 cells.

1. Seed HEK 293 cells in a 48-well tissue culture dishes, 1×10^5 cells/well. Culture cells in 0.5 mL DMEM with 10% FBS for 24 h at 37°C.

2. Then, add 50 μL of testing viral stock to each well and culture with the cells for 24 h.

3. Change the medium 24 h later.

4. Monitor the status of the infected cells everyday for 3–5 days and compare with uninfected control cells. The viral stock is considered toxic if cell death or slower growth is observed.

5. More dialysis steps can be used to remove the toxic chemicals, if a viral stock is found to be toxic to cells.

3.3. Animal Protocol

3.3.1. Rat MCAO Model

1. This protocol is approved by the Animal Care Committee of the University of California, San Francisco.

2. Anesthesize adult rats using 1.5–2.0% isoflurane inhalation.

3. Insert a PE-50 catheter into the femoral artery for continuous monitoring of arterial blood pressure and obtaining blood for analysis of blood gases and blood pH.

4. Measure the temperature of both the temporal muscle and the rectum by E type thermocouples and maintain the temperature in the normal range.

5. Drill a burr hole in the pericranium 3.5 mm lateral to the sagittal suture and 1 mm posterior to the coronal suture (keep the inner layer intact).

6. Measure baseline cerebral blood flow using a Laser Doppler Flowmeter. Monitor and record surface blood flow before, during, and after occlusion.

7. Make a midline incision on the front of the neck and expose left common carotid artery (CCA).

8. Isolate left internal carotid artery (ICA) and coagulate other small branches.

9. Ligate the pterygopalatine artery from the bifurcation of the ICA and the pterygopalatine artery.

10. Insert a 3-cm length of 3-0 nylon suture with a slightly enlarged and rounded tip into the transected lumen of the

external carotid artery (ECA) and gently advance from the ICA across to the opening of the MCA. The distance from the tip of the suture to the bifurcation of CCA is 17–18 mm.

11. Measure surface blood flow to make sure the blood flow drops to below 20% of baseline.

12. For reperfusion, withdraw the inserted suture into the ECA to restore ICA-MCA blood flow.

13. Suture the skin and return the animal to the cage when it recovers from the anesthesia.

14. The control animal will undergo all the surgical procedures except for insertion of the suture.

15. For the procedure used in mice, see Note 5.

3.3.2. AAV-Mediated Gene Transfer

1. Anesthesize the rat with intraperitoneal injections of 50 mg/kg Ketamine and 10 mg/kg xylazine.

2. Place the rat in a stereotactic frame with a mouth holder.

3. Drill a burr hole in the pericranium 3.5 mm lateral to the sagittal suture and 1 mm posterior to the coronal suture.

4. Insert a 10-µL Hamilton syringe stereotactically 5.5 mm into the lateral caudate

5. Inject 10 µL AAV viral suspension containing 2×10^{10} genome copies of virus into the right caudate putamen at a rate of 0.2 µL per min (56). The volume and concentration can be adjusted depending on the experimental design. Withdraw the needle after 15 min of injection.

6. Seal the craniotomy site with bone wax and close the wound.

7. Return the rat to its home cage after it recovers from the anesthesia.

8. For the mouse procedure, see Note 6.

3.4. Tissue Stains and Morphometry

3.4.1. X-Gal Staining

1. Cut 20 µm coronal sections of the brain and fix in 0.5% glutaraldehyde for 10 min.

2. Incubate the sections in X-gal staining solution for 2 h, and photograph.

3. Calculate the transduction volumes by multiplying the transduction areas by the thickness of the sections using NIH Image 1.63 software.

3.4.2. Cresyl Violet Staining of Frozen Section for Infarct Volume

1. Place frozen sections in Xylene for 3 min.

2. Transfer sections to 100, 95, 90, and 70 ethanol for 3-min each and wash in distilled water for 3 min.

3. Place sections in Cresyl Violet solution for 5–10 min depending on the age of the stain.

4. Differentiate sections in 70, 90, and 100% ethanol sequentially, 2× for 3-min each. Incubation time can vary slightly according to staining density and section thickness.

5. Transfer sections to Xylene, 2× for 5 min each.

6. Mount sections with Permount.

7. The result of the staining should be: Nissl granules – violet; Nuclei – pale violet; and background – colorless (Fig. 4).

8. If sections are not differentiated completely after step 5, repeat steps 4-6. If the stain is too light, repeat steps 2-6.

3.4.3. Fluoro-Jade B Staining

1. Fix the frozen sections with 2% paraformaldehyde in 0.1 M PIPES pH (6.9) for 20 min.

2. After washing, stain the sections with 0.06% $KMnO_4$ for 30 min at room temperature.

3. Immerse the sections in a 0.001% fluoro-Jade staining solution for 20 min and then in 0.1% acetic acid for 20 min.

4. Evaluate results using a fluorescence microscope.

3.4.4. Determining Capillary Density Analysis

The method to determine capillary density has been described in our previous publications (57, 58).

1. Five minutes prior to euthanizing the animal, inject 100 μL of fluorescein-lycopersicin esculentum lectin (100 μg) intravenously.

2. Euthanize rats, remove their brains and fix 20 μm thick frozen sections with 100% ethanol at 20°C for 20 min.

3. Incubate with fluorescein-lycopersicin esculentum lectin 2 g/mL at 4°C overnight.

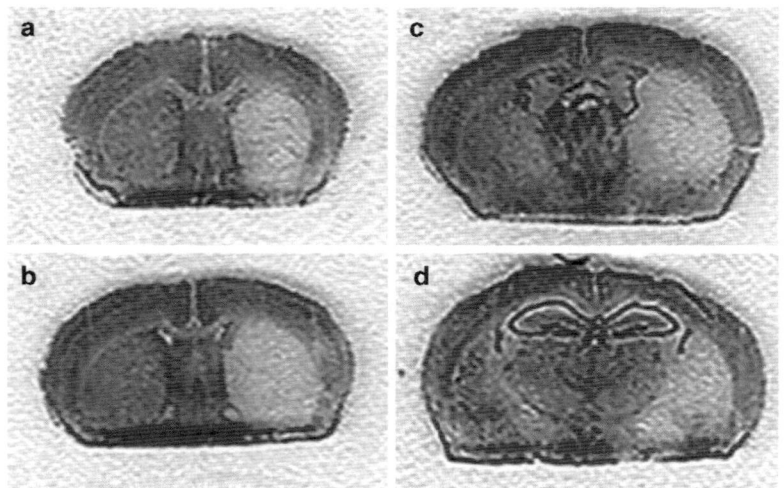

Fig. 4. A set of Cresyl violet-stained mouse brain coronal sections. The mouse underwent 45 min of MCAO, followed by 24 h of reperfusion. The infarct areas (*pale*) are usually limited to the basal ganglia region.

4. Choose two coronal sections from the lectin stained brain, 1 mm anterior and 1 mm posterior from the needle track. Quantify the microvessel density by counting the total number of vessel profiles in three microscopic fields ($10\times$ objective) to the left, right, and the bottom of the needle track per tissue section.

5. As a surrogate of vessel counting, pictures can be taken from the areas indicated above and the vessel density can be determined by measuring lectin optical density using NIH Image 1.63 software.

6. The vessel counts must be performed in a blinded manner with no knowledge of the experimental groups.

4. Notes

1. The AAV Helper-Free Gene Delivery and Expression System was generated based on AAV serotype 2 virus. Many new AAV serotypes have been cloned in recent years (11, 59–63). Recombinant cross-packaging of the AAV genome of one serotype into capsids of other AAV serotypes to achieve optimal tissue-specific gene transduction is now possible. Studies by us and others (64) have shown that the AAV serotype 1 results in more efficient transduction of genes in the murine and human adult heart compared with serotypes 2, 3, 4, and 5. AAV serotype 1 also mediates more efficient gene transduction in the brain than the AAV serotype 2 (6). The AAV packaged in serotypes other than serotype 2 can be made by replacing the serotype 2 CAP sequence in pAAV-RC with the CAP sequence of the corresponding serotype.

2. AAV titer can also be determined by quantitative PCR assay (65). Infectious particles can be determined by using serially diluted viral stocks to infect HEK 293 or other cells and quantitating the infected cells or transgene expression. DNase I digestion can be used to eliminate unpackaged DNA in viral stocks (66).

3. AAV can be purified by using other methods, such as non-ionic iodixanol gradients, ion exchange, or heparin affinity chromatography by either conventional or high-performance liquid chromatography columns (66, 67). Ion exchange or heparin affinity chromatography can also be used in combination with $CsCl_2$ gradients or nonionic iodixanol gradients.

4. The protocol for hybridization of the dot blot membrane with a radioactive probe is as follows: Membranes are prehybridized in a solution containing 0.5N Na_2HPO, 7%SDS, pH7.2 for 15 min at $65°C$ and then hybridized with the ^{32}P

labeled probe $(2 \times 10^6$ cpm/mL) in a mix containing 0.5 N Na$_2$HPO, 7% SDS, 10% dextran sulfate pH7.2, for 2 h at 65°C with gentle shaking. After hybridization, the membranes are washed in 2×SSC for 20 min at room temperature briefly and then washed in 2×SSC containing 1%SDS at 65°C for 1 h, and in 0.1×SSC containing 0.1%SDS for 20 min. After being rinsed with 2×SSC briefly, the membranes are place in a cassette. An X-ray film is placed on top of it in the dark room. Store the cassette in a –40° freezer. Signals for a typical genomic blot will show after a 30–60-min exposure.

5. The method for mouse MCAO is similar to the rat model except that the suture size is a 2-cm length of 5-0 nylon. The distance from the tip of the suture to the bifurcation of CCA is 10–11 mm (see Fig. 5).

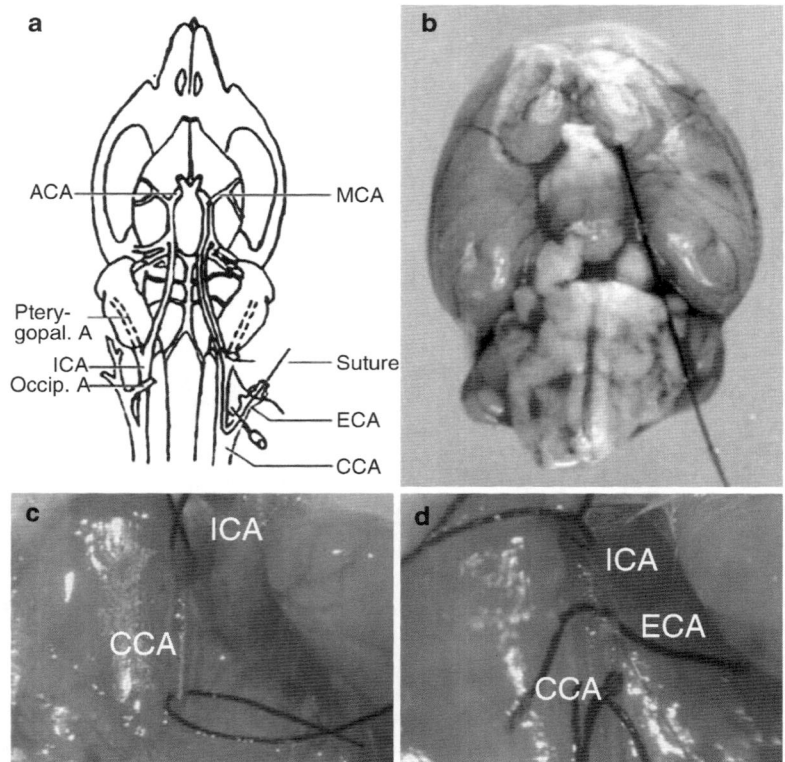

Fig. 5. (a) Diagrams of the cerebral anatomy of mouse brain illustrates extracranial and intracranial vascular relations. Anterior communicate artery (ACA), middle carotid artery (MCA), internal carotid artery (ICA), external carotid artery (ECA), common carotid artery (CCA), and their branches are shown. (b) Photograph of mouse brain at autopsy showing intravascular suture, inserted through ICA, within lumen of ECA. Suture occludes origin of MCA from ICA. (c) Photograph shows the anatomy of CCA and ICA in the neck; ICA is looped by a suture. (d) Photograph shows a right blue suture inserted into the opening of the ECA and advanced into ICA.

6. The method for mouse AAV-mediated gene transfer is similar to that used in the rat model except that, (1) the bone hole location is pericranium 2 mm lateral to the sagittal suture and 1 mm posterior to the coronal suture; and (2) the total injection volume of AAV vector is 2 μL.

References

1. Weintraub M I (2006) Thrombolysis (tissue plasminogen activator) in stroke: a medicolegal quagmire. Stroke 37, 1917–1922

2. Ohab JJ, Fleming S, Blesch A, Carmichael ST (2006) A neurovascular niche for neurogenesis after stroke. J Neurosci 26, 13007–13016

3. del Zoppo GJ (2006) Stroke and neurovascular protection. N Engl J Med 354, 553–555

4. McCown TJ (2005) Adeno-associated virus (AAV) vectors in the CNS. Curr Gene Ther 5, 333–338

5. Leker RR, Soldner F, Velasco I, Gavin DK, Androutsellis-Theotokis A, McKay RD (2007) Long-lasting regeneration after ischemia in the cerebral cortex. Stroke 38, 153–161

6. Shen F, Su H, Liu W, Kan YW, Young WL, Yang GY (2006) Recombinant adeno-associated viral vector encoding human VEGF165 induces neomicrovessel formation in the adult mouse brain. Front Biosci 11, 3190–3198

7. Sun Y, Jin K, Xie L, Childs J, Mao XO, Logvinova A, Greenberg DA (2003) VEGF-induced neuroprotection, neurogenesis, and angiogenesis after focal cerebral ischemia. J Clin Invest 111, 1843–1851

8. Bellomo M, Adamo EB, Deodato B, Catania MA, Mannucci C, Marini H, Marciano MC, Marini R, Sapienza S, Giacca M, Caputi AP, Squadrito F, Calapai G (2003) Enhancement of expression of vascular endothelial growth factor after adeno-associated virus gene transfer is associated with improvement of brain ischemia injury in the gerbil. Pharmacol Res 48, 309–317

9. Andsberg G, Kokaia Z, Klein RL, Muzyczka N, Lindvall O, Mandel R J (2002) Neuropathological and behavioral consequences of adeno-associated viral vector-mediated continuous intrastriatal neurotrophin delivery in a focal ischemia model in rats. Neurobiol Dis 9, 187–204

10. Tsai TH, Chen SL, Chiang YH, Lin SZ, Ma H I, Kuo SW, Tsao YP (2000) Recombinant adeno-associated virus vector expressing glial cell line-derived neurotrophic factor reduces ischemia-induced damage. Exp Neurol 166, 266–275

11. Xiao W, Chirmule N, Berta SC, McCullough B, Gao G, Wilson JM (1999) Gene therapy vectors based on adeno-associated virus type 1. J Virol 73, 3994–4003

12. Weber M, Rabinowitz J, Provost N, Conrath H, Folliot S, Briot D, Cherel Y, Chenuaud P, Samulski J, Moullier P, Rolling F (2003) Recombinant adeno-associated virus serotype 4 mediates unique and exclusive long-term transduction of retinal pigmented epithelium in rat, dog, and nonhuman primate after subretinal delivery. Mol Ther 7, 774–781

13. Zabner J, Seiler M, Walters R, Kotin RM, Fulgeras W, Davidson BL, Chiorini JA (2000) Adeno-associated virus type 5 (AAV5) but not AAV2 binds to the apical surfaces of airway epithelia and facilitates gene transfer. J Virol 74, 3852–3858

14. Davidson BL, Stein CS, Heth JA, Martins I, Kotin RM, Derksen TA, Zabner J, Ghodsi A, Chiorini JA (2000) Recombinant adeno-associated virus type 2, 4, and 5 vectors: transduction of variant cell types and regions in the mammalian central nervous system. Proc Natl Acad Sci U S A 97, 3428–3432

15. Chao H, Liu Y, Rabinowitz J, Li C, Samulski RJ, Walsh CE (2000) Several log increase in therapeutic transgene delivery by distinct adeno-associated viral serotype vectors. Mol Ther 2, 619–623

16. Chao H, Monahan PE, Liu Y, Samulski RJ, Walsh CE (2001) Sustained and complete phenotype correction of hemophilia B mice following intramuscular injection of AAV1 serotype vectors. Mol Ther 4, 217–222

17. Duan D, Yan Z, Yue Y, Ding W, Engelhardt JF (2001) Enhancement of muscle gene delivery with pseudotyped adeno-associated virus type 5 correlates with myoblast differentiation. J Virol 75, 7662–7671

18. Braddock M (2005) Safely slowing down the decline in Alzheimer's disease: gene therapy shows potential. Expert Opin Investig Drugs 14, 913–915

19. Tuszynski MH, Thal L, Pay M, Salmon DP, U HS, Bakay R, Patel P, Blesch A, Vahlsing HL, Ho G, Tong G, Potkin SG, Fallon J, Hansen L, Mufson EJ, Kordower JH, Gall C, Conner

J (2005) A phase 1 clinical trial of nerve growth factor gene therapy for Alzheimer disease. Nat Med 11, 551–555

20. McPhee SW, Janson CG, Li C, Samulski RJ, Camp AS, Francis J, Shera D, Lioutermann L, Feely M, Freese A, Leone P (2006) Immune responses to AAV in a phase I study for Canavan disease. J Gene Med 8, 577–588

21. Fulci G, Chiocca EA (2007) The status of gene therapy for brain tumors. Expert Opin Biol Ther 7, 197–208

22. Colombo F, Barzon L, Franchin E, Pacenti M, Pinna V, Danieli D, Zanusso M, Palu G (2005) Combined HSV-TK/IL-2 gene therapy in patients with recurrent glioblastoma multiforme: biological and clinical results. Cancer Gene Ther 12, 835–848

23. Ren H, Boulikas T, Lundstrom K, Soling A, Warnke PC, Rainov NG (2003) Immunogene therapy of recurrent glioblastoma multiforme with a liposomally encapsulated replication-incompetent Semliki forest virus vector carrying the human interleukin-12 gene–a phase I/II clinical protocol. J Neurooncol 64, 147–154

24. Rainov NG (2000) A phase III clinical evaluation of herpes simplex virus type 1 thymidine kinase and ganciclovir gene therapy as an adjuvant to surgical resection and radiation in adults with previously untreated glioblastoma multiforme. Hum Gene Ther 11, 2389–2401

25. Chu Y, Miller JD, Heistad DD (2007) Gene therapy for stroke: 2006 overview. Curr Hypertens Rep 9, 19–24

26. Heistad DD (2006) Gene therapy for vascular disease. Vascul Pharmacol 45, 331–333

27. Hsich G, Sena-Esteves M, Breakefield XO (2002) Critical issues in gene therapy for neurologic disease. Hum Gene Ther 13, 579–604

28. Yenari MA, Dumas TC, Sapolsky RM, Steinberg GK (2001) Gene therapy for treatment of cerebral ischemia using defective herpes simplex viral vectors. Ann N Y Acad Sci 939, 340–357

29. Watanabe T, Okuda Y, Nonoguchi N, Zhao M Z, Kajimoto Y, Furutama D, Yukawa H, Shibata MA, Otsuki Y, Kuroiwa T, Miyatake S (2004) Postischemic intraventricular administration of FGF-2 expressing adenoviral vectors improves neurologic outcome and reduces infarct volume after transient focal cerebral ischemia in rats. J Cereb Blood Flow Metab 24, 1205–1213

30. Shimamura M, Sato N, Oshima K, Aoki M, Kurinami H, Waguri S, Uchiyama, Y, Ogihara T, Kaneda Y, Morishita R (2004) Novel ther-

apeutic strategy to treat brain ischemia: overexpression of hepatocyte growth factor gene reduced ischemic injury without cerebral edema in rat model. Circulation 109, 424–431

31. Li H, Qian ZM (2002) Transferrin/transferrin receptor-mediated drug delivery. Med Res Rev 22, 225–250

32. Pang L, Ye W, Che XM, Roessler BJ, Betz A L, Yang GY (2001) Reduction of inflammatory response in the mouse brain with adenoviral- mediated transforming growth factor-ss1 expression. Stroke 32, 544–552

33. Hayashi K, Morishita R, Nakagami H, Yoshimura S, Hara A, Matsumoto K, Nakamura T, Ogihara T, Kaneda Y, Sakai N (2001) Gene therapy for preventing neuronal death using hepatocyte growth factor: in vivo gene transfer of HGF to subarachnoid space prevents delayed neuronal death in gerbil hippocampal CA1 neurons. Gene Ther 8, 1167–1173

34. Yang GY, Mao Y, Zhou LF, Ye W, Liu XH, Gong C, Betz AL (1999) Attenuation of temporary focal cerebral ischemic injury in the mouse following transfection with interleukin-1 receptor antagonist. Brain Res Mol Brain Res 72, 129–137

35. Betz AL, Yang GY, Davidson BL (1995) Attenuation of stroke size in rats using an adenoviral vector to induce overexpression of interleukin-1 receptor antagonist in brain. J Cereb Blood Flow Metab 15, 547–551

36. Davis AS, Zhao H, Sun GH, Sapolsky RM, Steinberg GK (2007) Gene therapy using SOD1 protects striatal neurons from experimental stroke. Neurosci Lett 411, 32–36

37. Badin RA, Lythgoe MF, van der Weerd L, Thomas DL, Gadian DG, Latchman DS (2006) Neuroprotective effects of virally delivered HSPs in experimental stroke. J Cereb Blood Flow Metab 26, 371–381

38. Baker AH, Sica V, Work LM, Williams-Ignarro S, de Nigris F, Lerman LO, Casamassimi A, Lanza A, Schiano C, Rienzo M, Ignarro LJ, Napoli C (2007) Brain protection using autologous bone marrow cell, metalloproteinase inhibitors, and metabolic treatment in cerebral ischemia. Proc Natl Acad Sci U S A 104, 3597–3602

39. Frampton AR Jr, Goins WF, Nakano K, Burton EA, Glorioso JC (2005) HSV trafficking and development of gene therapy vectors with applications in the nervous system. Gene Ther 12, 891–901

40. Wang C, Wang CM, Clark KR, Sferra TJ (2003) Recombinant AAV serotype 1 transduction efficiency and tropism in the murine brain. Gene Ther 10, 1528–1534

41. Xu DG, Crocker SJ, Doucet JP, St-Jean M, Tamai K, Hakim AM, Ikeda JE, Liston P, Thompson CS, Korneluk RG, MacKenzie A, Robertson GS (1997) Elevation of neuronal expression of NAIP reduces ischemic damage in the rat hippocampus. Nat Med 3, 997–1004

42. Pechan PA, Yoshida T, Panahian N, Moskowitz MA, Breakefield XO (1995) Genetically modified fibroblasts producing NGF protect hippocampal neurons after ischemia in the rat. Neuroreport 6, 669–672

43. Miao HS, Yu LY, Hui GZ, Guo LH (2005) Antiapoptotic effect both in vivo and in vitro of A20 gene when transfected into rat hippocampal neurons. Acta Pharmacol Sin 26, 33–38

44. Shirakura M, Fukumura M, Inoue M, Fujikawa S, Maeda M, Watabe K, Kyuwa S, Yoshikawa Y, Hasegawa M (2003) Sendai virus vector-mediated gene transfer of glial cell line-derived neurotrophic factor prevents delayed neuronal death after transient global ischemia in gerbils. Exp Anim 52, 119–127

45. Shen F, Su H, Fan Y-F, Zhu Y, Chen Y, Kan YW, Young WL, Yang GY (2006) Induction of focal angiogenesis through adeno-associated viral vector mediated VEGF165 gene transfer in the mature mouse brain (Abstract). Stroke 37, 685

46. Kurozumi K, Nakamura K, Tamiya T, Kawano Y, Ishii K, Kobune M, Hirai S, Uchida H, Sasaki K, Ito Y, Kato K, Honmou O, Houkin K, Date I, Hamada H (2005) Mesenchymal stem cells that produce neurotrophic factors reduce ischemic damage in the rat middle cerebral artery occlusion model. Mol Ther 11, 96–104

47. Gu W, Zhao H, Yenari MA, Sapolsky R M, Steinberg GK (2004) Catalase over-expression protects striatal neurons from transient focal cerebral ischemia. Neuroreport 15, 413–416

48. Hoehn B, Yenari MA, Sapolsky RM, Steinberg GK (2003) Glutathione peroxidase overexpression inhibits cytochrome C release and proapoptotic mediators to protect neurons from experimental stroke. Stroke 34, 2489–2494

49. Tsai TH, Chen SL, Xiao X, Chiang YH, Lin SZ, Kuo SW, Liu DW, Tsao YP (2003) Gene treatment of cerebral stroke by rAAV vector delivering IL-1ra in a rat model. Neuroreport 14, 803–807

50. Lawrence M S, McLaughlin JR, Sun GH, Ho DY, McIntosh L, Kunis DM, Sapolsky R M, Steinberg GK (1997) Herpes simplex viral vectors expressing Bcl-2 are neuroprotective when delivered after a stroke. J Cereb Blood Flow Metab 17, 740–744

51. Zhao MZ, Nonoguchi N, Ikeda N, Watanabe T, Furutama D, Miyazawa D, Funakoshi H, Kajimoto Y, Nakamura T, Dezawa M, Shibata MA, Otsuki Y, Coffin RS, Liu WD, Kuroiwa T, Miyatake S (2006) Novel therapeutic strategy for stroke in rats by bone marrow stromal cells and ex vivo HGF gene transfer with HSV-1 vector. J Cereb Blood Flow Metab 26, 1176–1188

52. Sugiura S, Kitagawa K, Tanaka S, Todo K, Omura-Matsuoka E, Sasaki T, Mabuchi T, Matsushita K, Yagita Y, Hori M (2005) Adenovirus-mediated gene transfer of heparin-binding epidermal growth factor-like growth factor enhances neurogenesis and angiogenesis after focal cerebral ischemia in rats. Stroke 36, 859–864

53. Sun Y, Jin K, Clark KR, Peel A, Mao XO, Chang Q, Simon RP, Greenberg DA (2003) Adeno-associated virus-mediated delivery of BCL-w gene improves outcome after transient focal cerebral ischemia. Gene Ther 10, 115–122

54. Hadaczek P, Kohutnicka M, Krauze MT, Bringas J, Pivirotto P, Cunningham J, Bankiewicz, K (2006) Convection-enhanced delivery of adeno-associated virus type 2 (AAV2) into the striatum and transport of AAV2 within monkey brain. Hum Gene Ther 17, 291–302

55. Vilaboa N, Voellmy R (2006) Regulatable gene expression systems for gene therapy. Curr Gene Ther 6, 421–438

56. Yang GY, Zhao Y, Davidson BL, Betz AL (1997) Overexpression of interleukin-1 receptor antagonist in the mouse brain reduces ischemic brain injury. Brain Res 751, 181–188

57. Yang GY, Xu B, Hashimoto T, Huey M, Chaly T, Jr, Wen R, Young WL (2003) Induction of focal angiogenesis through adenoviral vector mediated vascular endothelial cell growth factor gene transfer in the mature mouse brain. Angiogenesis 6, 151–158

58. Lee CZ, Xu B, Hashimoto T, McCulloch CE, Yang GY, Young WL (2004) Doxycycline suppresses cerebral matrix metalloproteinase-9 and angiogenesis induced by focal hyperstimulation of vascular endothelial growth factor in a mouse model. Stroke 35, 1715–1719

59. Gao G P, Alvira MR, Wang L, Calcedo R, Johnston J, Wilson JM (2002) Novel adeno-associated viruses from rhesus monkeys as vectors for human gene therapy. Proc Natl Acad Sci U S A 99, 11854–11859

60. Rutledge EA, Halbert CL, Russell DW (1998) Infectious clones and vectors derived from adeno-associated virus (AAV) serotypes other than AAV type 2. J Virol 72, 309–319

61. Muramatsu S, Mizukami H, Young NS, Brown K E (1996) Nucleotide sequencing and generation of an infectious clone of adeno-associated virus 3. Virology 221, 208–217

62. Chiorini JA, Yang L, Liu Y, Safer B, Kotin RM (1997) Cloning of adeno-associated virus type 4 (AAV4) and generation of recombinant AAV4 particles. J Virol 71, 6823–6833

63. Chiorini JA, Kim F, Yang L, Kotin RM (1999) Cloning and characterization of adeno-associated virus type 5. J Virol 73, 1309–1319

64. Du L, Sullivan CC, Chu D, Cho AJ, Kido M, Wolf PL, Yuan JX, Deutsch R, Jamieson S W, Thistlethwaite PA (2003) Signaling molecules in nonfamilial pulmonary hypertension. N Engl J Med 348, 500–509

65. Clark KR, Liu X, McGrath JP, Johnson PR (1999) Highly purified recombinant adeno-associated virus vectors are biologically active and free of detectable helper and wild-type viruses. Hum Gene Ther 10, 1031–1039

66. Zolotukhin S, Potter M, Zolotukhin I, Sakai Y, Loiler S, Fraites,TJ Jr, Chiodo VA, Phillipsberg T, Muzyczka N, Hauswirth WW, Flotte TR, Byrne BJ, Snyder RO (2002) Production and purification of serotype 1, 2, and 5 recombinant adeno-associated viral vectors. Methods 28, 158–167

67. Zolotukhin S, Byrne BJ, Mason E, Zolotukhin I, Potter M, Chesnut K, Summerford C, Samulski RJ, Muzyczka N (1999) Recombinant adeno-associated virus purification using novel methods improves infectious titer and yield. Gene Ther 6, 973–985

Chapter 23

Blood-Brain Barrier Disruption in the Treatment of Brain Tumors

Marie Blanchette and David Fortin

Abstract

Standard chemotherapy administered systemically has a limited efficacy in the treatment of brain tumors. One of the major obstacles in the treatment of brain neoplasias is the impediment to delivery across the intact blood-brain barrier (BBB). Many innovative approaches have been developed to circumvent this obstacle. One such strategy is BBB disruption (BBBD), which successfully increases the delivery of anti-neoplastic agents to the central nervous system (CNS). This chapter describes the application of the BBBD technique in rats. Different methods to evaluate and measure BBB permeability following hyperosmolar mannitol infusion including Evans blue staining, albumin immunohistochemistry, and dynamic magnetic resonance imaging are also described.

Key words: Brain tumors, Malignant gliomas, Chemotherapy delivery, Blood-brain barrier, Blood-brain barrier disruption

1. Introduction

The structural and physiological properties of the blood-brain barrier (BBB) present a formidable obstacle to drug delivery to the central nervous system (CNS), which limits the treatment of many CNS diseases. The therapeutic molecule must reach the target cell in sufficient concentration and in a suitable time frame for the treatment to be effective. Although many different CNS diseases are affected by this delivery impediment, in no other condition has it been as extensively documented as in malignant brain tumors which best exemplifies the problem of delivery across the BBB.

Chemotherapeutic drug trials for brain tumors have been conducted worldwide for more than 4 decades, and most investigators agree that little progress has been made since the introduction of the nitrosoureas, a class of chemotherapy agents (1).

Sukriti Nag (ed.), *The Blood-Brain and Other Neural Barriers: Reviews and Protocols*, Methods in Molecular Biology, vol. 686, DOI 10.1007/978-1-60761-938-3_23, © Springer Science+Business Media, LLC 2011

Moreover, the impact of these molecules on the natural course of malignant astrocytic neoplasms is rather limited (2). Limited therapeutic success in the treatment of CNS neoplasia with chemotherapy is generally attributed to two factors: natural or acquired resistance of tumor cells to chemotherapy, and difficulty in delivery related to the presence of the BBB (3).

The normal BBB prevents passage of ionized water-soluble compounds with a molecular weight greater than 180 Da (4). However, most currently available effective chemotherapeutic agents have a molecular weight between 200 and 1,200 Da. Although the integrity of the barrier is often compromised within the tumor, this alteration in permeability is variable and dependent on tumor type and size (4). Moreover, it is extremely heterogeneous in a given lesion. Although the BBB is frequently leaky in the center of malignant brain tumors, the brain adjacent to tumor (BAT) has been shown to have variable and complex barrier integrity (4). Therefore, by steeply reducing the concentration of intravenously administered chemotherapeutic agent at the periphery of the tumor, the phenomenon of sink effect is yet another mechanism that can contribute to failure of chemotherapy in the treatment of CNS neoplasms (5). Therefore, a strategy to increase dose intensity to the CNS must take into account the impediment posed by the BBB, and somehow, bypass it (5–7). Interestingly, despite the fact that this limitation in delivery imposed by the BBB is more and more acknowledged, this topic of delivery is infrequently discussed in the neuroscience field (7–9).

Different approaches to improve delivery across the BBB may be local, regional, or global in their ability to circumvent the BBB. This chapter focus on a technique which transiently increases BBB permeability causing BBB disruption (BBBD) by intracarotid infusion of hypertonic solutions. Although this is an invasive technique, it offers the potential of global delivery. There is now extensive animal and human clinical data on the use of this approach (3, 4, 6). The extensive vascular network supplying the brain makes global delivery using this vascular network possible. The importance of this network was reported by Bradbury and colleagues, who stated that the entire network covers an area of 12 m^2/g of cerebral parenchyma (10). The magnitude of this network is further exemplified by the fact that the brain receives 20% of the total cardiac output, even though its weight constitutes only 2% of the total body mass (11). Increasing BBB permeability in the BBBD model represents the ideal means for global delivery (Fig. 1) since it is transient, but of sufficient duration so as to allow the intra-arterial infusion of a therapeutic molecule.

Different methods to evaluate and measure BBB permeability following hyperosmolar mannitol infusion including Evans blue staining, albumin immunohistochemistry, and dynamic magnetic resonance imaging (MRI) will also be described.

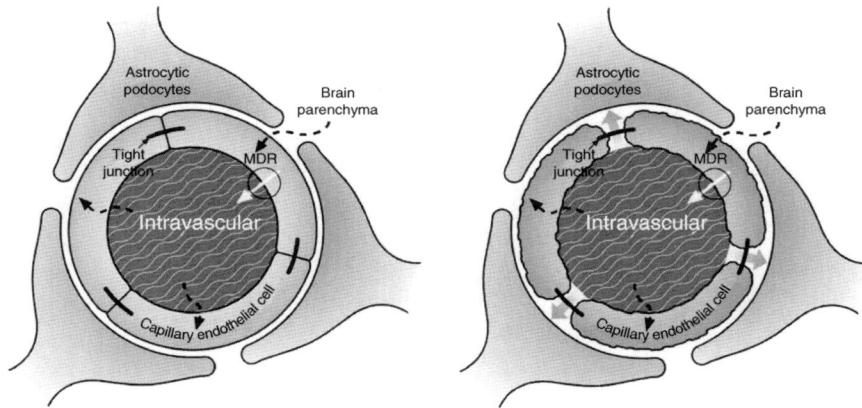

Fig. 1. Diagrams illustrate the principle underlying the osmotic BBB modification. (**a**) Brain endothelial tight junctions are shown with no anatomic space between the endothelial cells. The multidrug resistance (MDR) gene product, or p-glycoprotein efflux pump, is also illustrated as it is integral to the barrier mechanism. (**b**) The osmotic BBBD procedure induces shrinkage of endothelial cells resulting in a physical space between cells accompanied by a modification of the Ca^{2+} metabolism in the cell.

2. Materials

2.1. Animal Surgery and BBBD Procedure

1. Anesthetics: isoflurane (Abbott Laboratories Limited, Montreal, QC, Canada) and propofol, 10 mg/mL, (Novopharm Limited, Toronto, ON, Canada) (see Notes 1–3).

2. Male Wistar rats, 250–300 g body weight (Charles River Laboratories, St-Constant, QC, Canada) (see Notes 4 and 5).

3. Microsurgery instruments and supplies: forceps, Vannas scissors, fine scissors, artery clamps, wound clips, surgical blades, surgical nylon (World Precision Instruments, Sarasota, NY).

4. Endotracheal intubation: i.v. catheter 16G, 1.77 in (Becton Dickinson Medical, Sandy, Utah).

5. Intra-arterial catheter: 23G1-gauge needles for attachment to PE-50 intramedic Clay Adams polyethylene tubing (Becton Dickinson Diagnostics, Franklin Lakes, NJ).

6. 25% Mannitol, 50 mL, (Hospira, Lake Forest, IL, USA). The mannitol solution is heated at 37°C, prior to its infusion to prevent crystallization that occurs at room temperature (see Note 6).

7. Disposable 1, 10 and 60 mL syringes (Becton Dickinson Diagnostics).

8. Equipment:

    (a) Hot plate: Model Isotemp (Fisher Scientific, Ottawa, ON, Canada).

(b) Syringe perfusion pump: Model KDS210 (KD Scientific, Holliston, USA).

(c) Pressure-controlled ventilator (Kent Scientific, Torrington, USA).

(d) High temperature cautery (World Precision Instruments).

(e) Homeothermic blankets (Harvard Apparatus, Saint-Laurent, QC, Canada).

2.2. Evans Blue Extravasation

1. 2% Evans blue in saline, dose: 2 mL/kg body weight (Sigma-Aldrich, St-Louis, USA).

2. 4% paraformaldehyde solution: Add 216.2 mL of paraformaldehyde 37% to 2 mL of glutaraldehyde, adjust pH to 7.4, and add distilled water to obtain a final volume of 2 L.

2.3. Albumin Immunohistochemistry

1. Single-edged industrial razor blades No 9 (VWR, Mississauga, ON).

2. Brain matrix: RMBA 600-C (World Precision Instruments Inc).

3. Leica Model RM2125 Microtome (Nußloch, Germany).

4. Superfrost plus slides (VWR, Mississauga, ON, Canada).

5. 0.1 M Citrate buffer, pH 6.0.

6. Polyclonal antialbumin antibody 1:100 dilution, (MP Biomedicals, Solon, OH, USA).

7. Vectastain ABC Kit containing biotinylated goat antirabbit IgG 1:200 dilution (Vector Laboratories, Burlingame, CA, USA).

8. Vectastain conjugation solution (avidin-biotinylated peroxidase conjugate):

   Add 5 mL of PBS to 2 drops of Reagent A and 2 drops of Reagent B (from the Vectastain Kit). Prepare 30 min before use and keep at RT.

9. DAB substrate (Roche, Indianapolis, IN, USA).

10. 10 mM PBS, pH 7.5.

11. PBC: PBS containing 2% skim milk.

12. 1% Lithium carbonate.

13. Microscope cover glasses (VWR, Mississauga, ON, Canada).

14. Mounting medium: Cytoseal 60 (Richard-Allan Scientific, Kalamazoo, MI, USA).

15. Olympus BX41 Microscope (Olympus, Melville, USA).

16. QICAM FAST 1394 Microscope camera (QImaging, Surrey, BC, Canada).

17. Software used: QCapture Pro (QImaging, Surrey, BC, Canada), and Sigmascan Pro (Hearne Scientific Software, Chicago, IL, USA).

2.4. MRI Quantification of BBB Permeability

1. 7 T MRI scanner (Varian Inc, Palo Alto, CA, USA).

2. Intra-arterial catheter: intravenous catheter 24G, 0.75 in (Becton Dickinson Medical) for attachment to PE-50 intramedic Clay Adams polyethylene tubing (Becton Dickinson Diagnostics).

3. Intravenous catheter for the contrast agent administration : 30 G1-gauge needles for attachment to intramedic Clay Adams polyethylene tubing PE-10 (Becton Dickinson Diagnostics) inserted in an intravenous (i.v.) catheter 24G, 0.75 in (Becton Dickinson Medical).

4. Intravenous catheter for the continuous propofol administration: Intravenous catheter 24 G, 0.75 in (Becton Dickinson Medical) is connected to the Tygon R-3603 tube (I.D. of 3/32 in., 7 feet length, and O.D. of 5/32 in., VWR, Mississauga, ON, Canada) by a male luer integral lock ring (3/32 in. ID natural polypropylene tubing, Cole-Parmer Instrument Company, Montreal, QC, Canada). The other end of the tube is connected to the 10 mL syringe by a female luer integral lock ring (3/32 in. ID natural polypropylene tubing, Cole-Parmer Instrument Company).

5. Small animal monitoring system (SA Instruments, Stony Brook, NY, USA).

6. Magnevist (Gd-DTPA), diluted 1:3.5 (1.43 mM) dose :0.24 mg/kg (Berlex Canada Inc, Montreal, QC, Canada).

7. Heparin solution (Hospira).

3. Methods

3.1. Animal Surgery and BBBD Procedure

1. Induce anesthesia with O_2 (2 L/min)/Isoflurane 5% mixture for 1 min; thereafter, isoflurane is reduced to 2–3%.

2. Prepare a caudal i.v. access for the propofol infusion (3.9 mL/kg/h). Ten minute after the initiation of the propofol infusion, isoflurane is discontinued.

3. Place an i.v. catheter in the rat trachea and ventilate with a tidal volume of 2.5 mL and a respiratory rate of 65/min.

4. Expose the right carotid complex of the animal and isolate the right external carotid via a ventral incision. Branches arising from the external carotid are coagulated and separated. Care should be taken not to damage the vagus nerve running close to the bifurcation of the internal and the external carotid arteries.

5. The external carotid is dissected, isolated, and two sutures are placed around the vessel. One of the sutures is used to ligate the external carotid distally.

6. Clamp the proximal part of the external carotid with the vascular clip.

7. Make an incision in the external carotid using Vannas micro-scissors and insert the catheter in a retrograde fashion, so that the tip of the polyethylene tubing lies above the bifurcation (Fig. 5).

8. The second suture is tied around the vessel to secure the catheter in place after its insertion in the external carotid. Tie the suture with a double-knot on the catheter to secure it in place in the external carotid. To improve stability, another double-knot can be applied with the sutures of the distal ligation of the external carotid.

9. Remove from the external carotid the vascular clip, and to make sure that the cannulation is correct, briefly ensure that there is blood backflow into the catheter.

10. Discontinue the propofol infusion 1 min prior to the infusion of mannitol, which is prewarmed to 37°C prior to infusion and infused at a rate of 0.12 mL/s for 30 s via the previously installed catheter using a constant infusion pump (see Note 7).

11. Prior to mannitol infusion, a vascular clip is applied temporarily on the right common carotid artery to decrease the hemo-dynamic effects related to the anesthetic agents (see Note 3). The vascular clip is removed at the end of the mannitol infusion.

12. The propofol infusion is resumed and the therapeutic agent to be tested is administered via the intra-arterial catheter (see Note 8).

13. The accumulation of the therapeutic agent in brain paren-chyma is optimal when administered 3 min after the BBBD procedure (12).

14. Ligate the external carotid and retrieve the catheter.

15. The wound is closed with sutures, the propofol infusion dis-continued, and the animal is allowed to recover.

3.2. Evans Blue Extravasation

1. Perform the procedure described in Subheading 3.1, steps 1–9.

2. The left jugular vein is surgically isolated, and two sutures are placed around the vessel. The first suture is used to ligate the jugular vein distally.

3. Make a nick in the left jugular vein above the ligature and insert the catheter into the vessel lumen.

4. The second suture is tied around the vessel with a double-knot to secure the catheter in place. To improve stability, another double-knot can be applied.

5. The 2% solution of Evans blue is administered via the left jugular vein 5 min prior to the BBBD procedure (6). This

marker binds tightly but reversibly to albumin in vivo resulting in a 68,500 Da molecular weight marker that does not cross the intact BBB.

3.2.1. Interpretation

The Evans blue staining as visualized at the surface of the brain is evaluated against an arbitrary staining scale, giving a qualitative score of the intensity of delivery (4, 6, 13, 14) (Fig. 2 and Table 1). Results obtained are inherently subjective and the reliability of the generated results is therefore questionable (see Note 9).

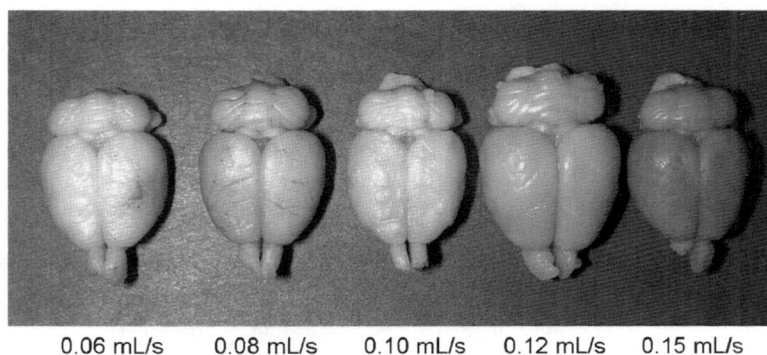

| 0.06 mL/s | 0.08 mL/s | 0.10 mL/s | 0.12 mL/s | 0.15 mL/s |

Fig. 2. Sequential discoloration of the right hemisphere in a series of Fischer rats infused with i.v. Evans blue and subjected to BBBD according to our modified model. The optimal infusion rate was identified to be 0.12 mL/s. Table 1 shows Grades 0–IV corresponding to infusion rates of 0.06, 0.08, 0.10, 0.12, and 0.15 mL/s, respectively.

Table 1
Grading scale for descriptive quantification of blood-brain barrier disruption detected by Evans blue staining

BBBD grade	Description
Grade 0	No blue staining of the cerebral parenchyma
Grade I	Slight blue tint to the cerebral parenchyma in the territory supplied by the parent artery infused with the mannitol
Grade II	Clearly demarcated blue staining of the cerebral parenchyma in the territory supplied by the parent artery infused with the mannitol
Grade III	Blue staining of the cerebral parenchyma which tends to surpass the territory supplied by the parent artery infused with the mannitol via the circle of Willis
Grade IV	Extreme blue staining of the cerebral parenchyma which surpasses the vascular territory infused with the mannitol

3.3. Albumin
Immunohistochemistry

1. Osmotic BBBD with mannitol is performed in adult male rats as described in Subheading 3.1.

2. The animals are perfused 15 min post-BBBD by an intracardiac infusion of 4% paraformaldehyde.

3. The brains are removed and fixed for 48 h in 4% paraformaldehyde.

4. Using a brain matrix and an industrial razor blade, the brains are cut into three coronal slabs, each having a thickness of 4 mm. The olfactory bulb and cerebellum are excluded from sectioning. The brain slabs are processed using standard techniques and embedded in paraffin.

5. 5 μm brain sections are cut using a microtome for the immunohistochemical demonstration of albumin using the Vectastain ABC Kit.

6. The sections are deparaffinized in three changes of Xylene, 5 min each, and rehydrated in two changes of 100% EtOH for 3 min each and one change of 70% EtOH for 3 min.

7. Sections are washed in tap water for 1 min and then in distilled water for 1 min.

8. Quench the endogenous peroxidase by placing sections in $PBS/0.5\%H_2O_2$ for 10 min.

9. Repeat step 7.

10. Preheat citrate buffer in a pressure cooker.

11. Place the brain sections in the preheated citrate buffer and seal the pressure cooker. Heat the pressure cooker until it is pressurized and maintain the pressure for 5 min.

12. Place the pressure cooker in a warm water bath until the pressure is completely released, then remove the lid.

13. Rinse in two changes of PBS, for 5 min each.

14. Two blocking steps are done. First incubate sections for 1 h at RT in PBC (100 μL/slide) and then for 30 min in PBC containing 2% serum from the Vectastain Kit (100 μL/slide).

15. Repeat step 13 and incubate sections in antialbumin antibody in a humidity chamber at 4°C, overnight.

16. Repeat step 13 and place sections in biotinylated secondary antibody in a humidity chamber for 1 h at RT.

17. Repeat step 13 and place slides in the Vectastain conjugation solution for 30 min at RT.

18. Repeat step 13 and place sections in the DAB/metal Concentrate diluted 1/10 in the peroxide Buffer (150 μL/slide).

19. Stop reaction by rinsing in distilled H_2O.

20. Repeat step 13.

21. Counterstain in Harris Hematoxylin for 4–30 s and rinse in circulating water, until the water is clear.

22. Wash slides with 3 Lithium carbonate washes followed by tap water for 5 min.

23. Dehydrate sections by placing in graded ethanol and toluene:

    (a) EtOH 50%, 2 washes
    (b) EtOH 70%, 2 washes
    (c) EtOH 100%, 2 washes
    (d) Toluene, 6 washes

24. The slides are mounted and dried overnight and these mounted brain slices are digitized using QCapture.

25. Using Sigmascan Pro, the two hemispheres are delimited and the brain pixel density is calculated. Stained pixels corresponding to the albumin detection by immunohistochemistry are calculated. The ratios of stained pixel density and total brain pixel density are calculated using Sigmascan Pro. Percentages of delivery are obtained from these ratios (Fig. 3) (see Note 10).

3.3.1. Interpretation

1. These data give a conservative estimate of the extent of delivery, as albumin is a large protein.

2. It is expressed as a percentage of the treated hemisphere in a given coronal slice and can also be used as a composite score

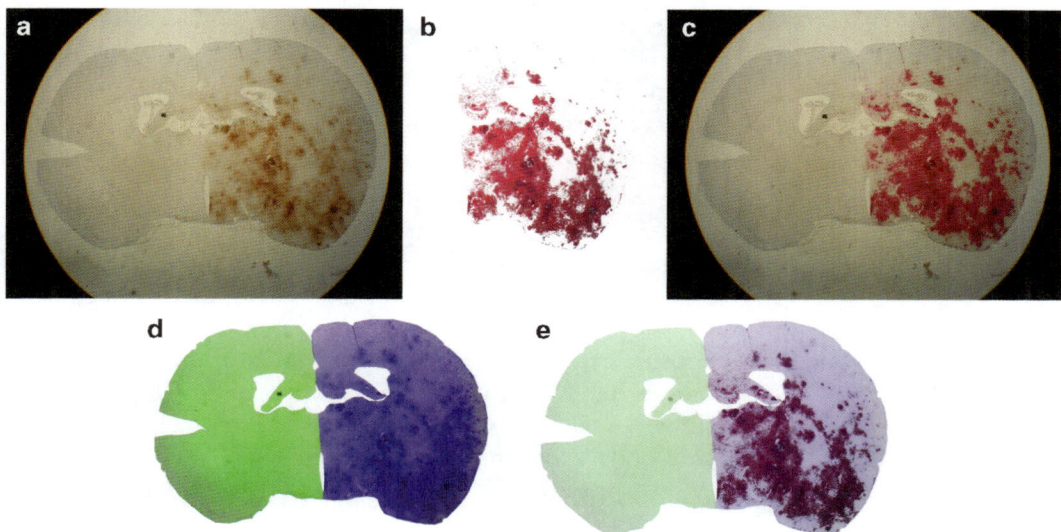

Fig. 3. Coronal slices of nontumor-bearing Fischer rat brain exposed to the BBBD procedure. (**a**) Albumin immunostaining produces a brown color in the treated hemisphere that can be isolated from the hemisphere using an intensity threshold set to retain the pixels above a certain value (**b**, **c**). After the definition of each hemisphere (**d**), a ratio of stained pixels expressed as a fraction of the treated hemisphere is produced (**e**).

translating into global delivery, by simply summating the scores obtained on multiple contiguous slices.

3. As the brain samples are cut in a standardized manner using a brain matrix, the number of slices is always consistent, thus ensuring that the composite score is reproducible.

4. Its objectiveness should help investigators to evaluate the impact of various surrogates on CNS delivery and thus contribute to a better comprehension and characterization of the BBBD process.

3.4. MRI Quantification of BBB Permeability

1. The surgical procedure is the same as described in Subheading 3.1 with a few modifications (see Note 11).

2. After the surgical procedure, the animal is placed in the MRI gantry.

3. A T_1 map is acquired using a multiflip angle approach (10°, 20°, 25°, 35°, 50°).

4. The dynamic acquisition (T_1-weighted images, TR/TE: 100/2.49 ms, FOV: 4×4 cm^2, matrix: $(128)^2$, α: 30°, NA: 4) is started and an image is produced every ~51 s. After three sets of images, mannitol is infused.

5. Maximal accumulation of the contrast agent is observed when the contrast agent is administered 3 min post-BBBD (600 μL with an infusion rate of 600 μL/min).

3.4.1. Interpretation

1. This methodology allows osmotic BBBD to be conducted directly in the animal placed in a MRI gantry, thus permitting the acquisition of images dynamically before, during, and after the procedure. This setup allows the study of the distribution of contrast agents of different molecular weight, to estimate brain exposure against several surrogates following the osmotic BBBD, and to estimate the window of barrier opening.

2. To quantify brain exposure, initial brain tissue MRI signal intensity needs to be recorded and subsequent signals measured during and after the BBBD are correlated with contrast agent concentrations by mathematical calculations (15). Brain exposure can thus be assessed as a function of the amount of contrast agent in a given brain compartment (Fig. 4a). The following calculations are used to determinate the contrast agent concentrations in brain parenchyma.

(a) S0 = I0*(1-cos(alpha*pi/180)*exp(-TR/T1))/ ((1-exp(-TR/T1))*sin(alpha*pi/180))

(b) R1max = 1/TR   *log((S0*sin(alpha*pi/180)-Imax* cos(alpha*pi/180))/(S0*sin(alpha*pi/180)-Imax))

(c) Cmax = (R1max - 1/T1)/r1

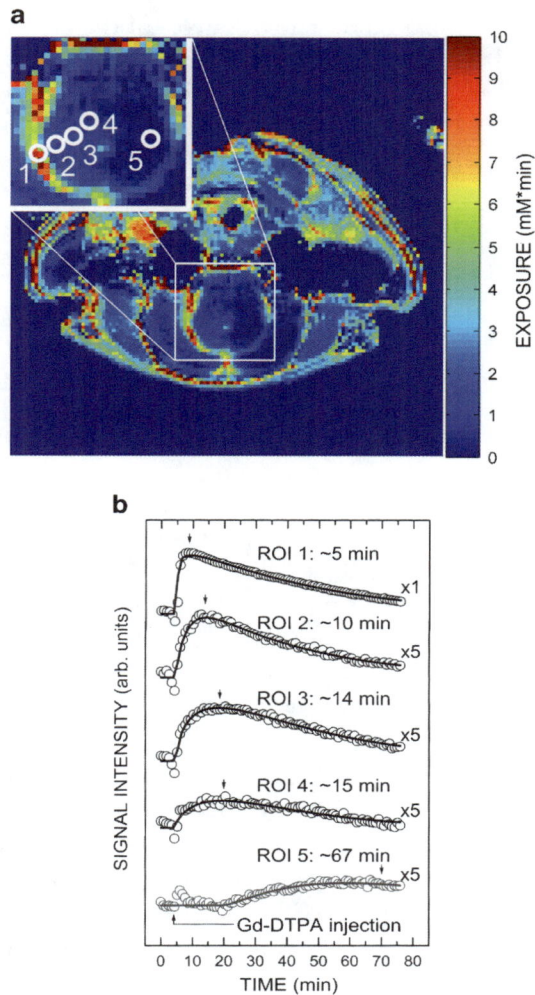

Fig. 4. Exposure and diffusion of Gd-DTPA in brain parenchyma after a BBBD procedure is shown. (**a**) Exposure (in mM × min) to Gd-DTPA for the first 17 min calculated from a series of T_1-weighted axial images. The delay between the infusion of mannitol and the injection of Gd-DTPA was 3 min. The enhancement patterns for each region of interest (ROI) are plotted in (**b**). Solid lines are fitted curves that determine the time to reach the maximum exposure as indicated for each dataset.

(d) TR (repetition time) = 100 ms

(e) T1 (T1 relaxation time of the tissue on study) = determined by the T_1 map of the region of interest (ROI)

(f) I0 = signal intensity of the ROI before treatment

(g) Imax = maximal signal intensity of the ROI obtained after the contrast agent administration

(h) Cmax = maximal concentration obtained of the ROI after the contrast agent administration

(i) r1 = 3.6 (mM*ms)$^{-1}$ (Gd-DTPA relaxivity)

3. The dynamics of the exposure of the brain to Gd-DTPA over time highlight two different mechanisms by which the procedure increases BBB permeability: (a) direct permeabilization of the BBB, and (b) a diffusion of the contrast agent within the brain from the area described above (Fig. 4). We have demonstrated this using Gd-DTPA, a low molecular weight contrast agent (0.5 kDa), The molecular weight of Gd-DTPA is 0.9 kDa administered i.v. The MRI signal intensity was analyzed in several regions of interest scattered across different areas of the brain over time. Initial results highlight the heterogeneity in CNS penetration, and the subsequent diffusion of contrast agent from an area representing the cleft between the brainstem and the cerebral hemisphere (Fig. 4). This area was identified as consistently showing the largest exposure to Gd-DTPA after BBBD.

4. Conclusions

1. In the human, thus far, the clinical application of BBBD has been limited to the use of chemotherapy agents and of boronophenylalanine in the context of brain tumor treatment (16, 17).

2. It is of concern that a number of clinical studies in the field of brain tumors are still utilizing intravenous administration of agents that possess poor penetration of the BBB.

3. The scientific community involved in the field of neuroscience therapeutics needs to realize and acknowledge the role played by the BBB in preventing drug delivery to the CNS. Beside efficacy of a molecule for its target, the molecule must reach the target in sufficient concentration and time.

4. The BBBD procedure represents a concerted effort to produce a rational approach to this delivery impediment. It has shown its safety and its efficacy in increasing delivery (17). Its potency to modify clinical outcome in certain neoplastic diseases has been established (18). Other strategies are also under investigation. With continuous effort from groups dedicating research toward this interesting problem, the neuroscience field will continue to accumulate data that will eventually allow more translational use of these strategies in the clinic.

5. Notes

1. BBBD requires general anesthesia for a number of reasons (3, 4, 6). The procedure generates a significant level of pain. It causes a transient rise in intracranial pressure requiring

cerebral protection. An actual increase in brain water (vasogenic edema) of 1–1.5% has also been documented after the procedure.

2. Propofol has been shown to be the most efficient agent, producing greater than 95% of good to excellent BBBD in animals (19). With increase in both the consistency and the intensity of the BBBD, propofol also produces neurotoxicity with certain chemotherapy agents not previously reported as toxic with other anesthetics (20). Hence, the choice of a specific anesthetic agent may be accompanied by undesirable toxicity eliminating any potential advantages.

3. The choice of the anesthetic agent is important in the quality of BBBD, as well as the potential for toxicity related to the drug treatments (19, 21). If Ketamine is used as an anesthetic agent, the following modification to the model is recommended to improve the effectiveness and consistency of the BBBD procedure and to decrease the risks of neurotoxicity observed with the preclinical use of propofol (16, 32). The modification simply involves the placement of a temporary vascular clip to the common carotid artery approximately 1 cm proximal to its bifurcation prior to the infusion of mannitol (Fig. 5). This simplifies the hemodynamic system by decreasing the number of variables involved. The rate and duration of infusion then become the sole relevant modifiable parameters.

4. Different animal BBBD models have been developed, but the rat, for its ease of use, is the most frequently used (22). Studies in normal rats demonstrated that greater than 95% of the animals attained good or excellent BBBD when infused with 1.4 mol/L mannitol when using propofol. In contrast, in rats with brain tumor xenografts, only approximately 60% of animals had good or excellent disruption (14).

5. It is of note that the parameters considered optimal for the BBBD procedure are different and must be adjusted for the strain of rat, the size of the animals, and the anesthetic used. Reviewing the BBBD approach in seven experimental brain tumor models, Blasberg et al. concluded that vascular permeability varies tremendously among models, and the threshold for BBBD within tumor is directly related to permeability of the tumor before disruption (23).

6. The mannitol infusion induces hemodynamic instability. This is illustrated in the treated animals by a brief (20–30 s) period of apnea, hypotension, and bradycardia (6). The same phenomenon is observed in the humans undergoing the procedure and must be pharmacologically prevented. It is for this reason that all animals are ventilated and when the mannitol infusion is higher than 0.12 mL/s, atropine (0.04 mg/kg) is administered s.c. to the animals.

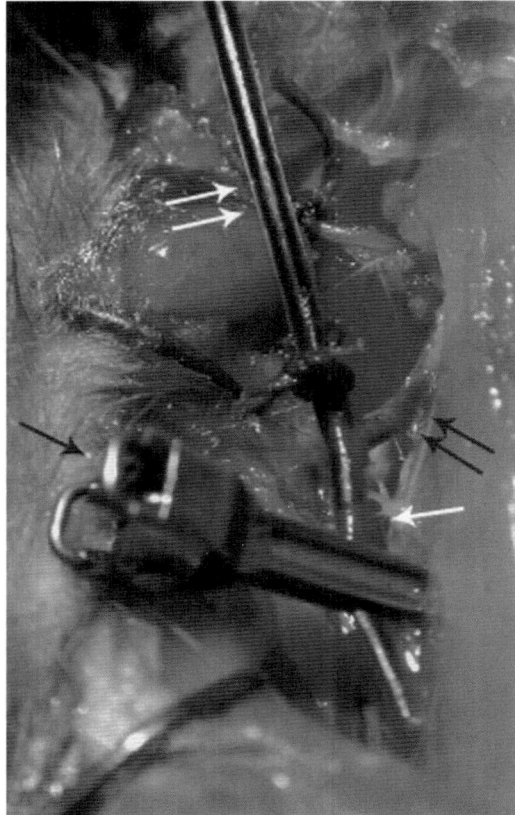

Fig. 5. Dissection of the right carotid artery complex in the rat demonstrates the application of a temporary vascular clip (*black arrow*) to the right common carotid artery (*white arrow*). A PE-50 intramedic catheter (*double white arrow*) is present in the external carotid artery allowing infusion of mannitol in a retrograde fashion into the internal carotid artery (*double black arrow*) to produce osmotic blood-brain barrier disruption.

7. Two parameters are paramount in the ability to mediate a hyperosmolar modification of the barrier: the osmolarity of the solution and the infusion time. Using a solution of 1.6 mOsm/mL arabinose in pentobarbital-anesthetized rats, Rapoport determined an interval of 30 s as the optimal infusion time for BBBD (24). The normal osmolarity of blood/serum is about 0.3 mOsm/mL. The same infusion time was applied to the use of mannitol with similar findings in the same animal model (6). Our laboratory has identified 22 s as a minimum infusion time to produce BBBD in the Long-Evans rat model. However, this was produced at the expense of a higher infusion rate with an increase rate of hemorrhagic complication. Therefore, the 30 s infusion interval is considered standard by most authors.

8. While testing the animal model of BBBD using propofol as the anesthetic agent with the goal of further improving delivery, Fortin et al. demonstrated an increase in toxicity with therapeutic agents that were never identified as toxic prior to the use of this anesthetic agent (13). Thus, agents routinely used without toxicity in animals and humans become toxic in the context of BBBD when the enhanced delivery is further maximized. Neurotoxicity, however, might also be produced by increasing the delivery of molecules that are normally restricted in their CNS penetration. Severe neurotoxicity has been documented when using BBBD in animal models to increase the delivery of chemotherapy agents such as cisplatin and doxorubicin (6). Different molecules have been used in conjunction with the BBBD procedure in animal models. Different chemotherapeutic agents, monoclonal antibodies, superparamagnetic particles, and viruses have been successfully delivered with this approach (25–30).

9. The fact that the Evans blue binds tightly to albumin has recently been questioned. Another weakness of this approach is that only the cortical surface is surveyed, while the deep structures are not visualized and no data can be obtained on the topographic extent of delivery.

10. A previous study documented the extent of delivery following BBBD by ultrastructural quantitative immunohistochemistry using antibody conjugated with protein A-gold. This technique was used to study the dynamics of BBBD by quantification of the density of gold particles in the different compartments (31). Image analysis was performed on electron micrographs; therefore, a global estimation of delivery could not be determined.

11. Due to the high magnetic field, ferromagnetic material needs to be replaced by plastic material or smaller ferromagnetic objects. Due to these limitations, intra-arterial and intravenous catheters are different from those described in Subheading 2.1. Also, during the acquisition, the animal is placed in the core of the MRI scanner and is unreachable, so the vascular clip cannot be used either while the osmotic BBBD is performed in the MRI scanner.

Acknowledgments

This work was supported by a grant from the CIHR (DF), as well as the National Bank of Canada Research Chair on Brain Tumor treatment (DF).

References

1. Fine HA, Dear KB, Loeffler JS, Black PM, Canellos GP (1993) Meta-analysis of radiation therapy with and without adjuvant chemotherapy for malignant gliomas in adults. Cancer 71:2585–2597

2. Holsi P, Sappino A P, de Tribolet N, Dietrich PY (1998) Malignant glioma: should chemotherapy be overthrown by experimental treatments ? Ann Oncol 6:589–600

3. Fortin D, Neuwelt EA (2002) Therapeutic manipulation of the blood-brain barrier. Neurobase-neurosurgery. First edition. Medlink CD-ROM

4. Kroll RA, Neuwelt EA (1998) Outwitting the blood-brain barrier for therapeutic purposes: osmotic opening and other means. Neurosurgery 42:1083–1100

5. Rapoport SI, Hori M, Klatzo I (1972) Testing of a hypothesis for osmotic opening of the blood-brain barrier. Am J Physiol 223: 323–331

6. Neuwelt EA (ed) (1989) Implications of the Blood-Brain Barrier and Its Manipulation. Vol 1 and 2. New York, Plenum

7. Pardridge WM (1991) Advances in cell biology of blood-brain barrier transport. Semin Cell Biol 2:419–426

8. Pardridge WM (2002) Targeting neurotherapeutic agents through the blood-brain barrier. Arch Neurol 59:35–40

9. Pardridge WM (1997) Drug delivery to the brain. J Cereb Blood Flow Metab 17:713–731

10. Bradbury MWB (1986) Appraisal of the role of endothelial cells and glia in barrier breakdown. In Suckling AJ, Rumsby MG, Bradbury MWB (eds): The Blood-Brain Barrier in Health and Disease. Chichester, Ellis Horwood, pp 128–129

11. Selman WR, Lust WD, Ratcheson RA (1996) Cerebral blood flow. In: RH Wilkins, Rengachary SS (eds): Neurosurgery. New York, McGraw-HILL, 1997–2007

12. Blanchette M, Pellerin M, Tremblay L, Lepage M, Fortin D (2009) Real time monitoring of Gd-DTPA during osmotic blood-brain barrier disruption using MRI in normal Wistar rats. Neurosurgery, accepted 30 march 2009

13. Bhattacharjee AK, Nagashima T, Kondoh T, Tamaki N (2001) Quantification of early bloodbrain barrier disruption by in situ brain perfusion technique. Brain Res Prot 8:126–131

14. Rawson RA (1942) The binding of T-1824 and structurally related diazo dyes by the plasma proteins. Am J Physiol 138:708–717

15. Landis CS, Li X, Telang FW, Coderre JA, Micca PL, Rooney WD, Latour LL, Vétek G, Pályka I, Springer CS (2000) Determination of the MRI contrast agent concentration time course in vivo following bolus injection: effect of equilibrium transcytolemmal water exchange. Magn Reson Med 44: 563–574

16. Doolittle ND, Miner ME, Hall WA, Siegal T, Hanson EJ, Osztie E, McAllister LD, Bubalo JS, Kraemer DF, Fortin D, Nixon R, Muldoon LL, Neuwelt EA (2000) Safety and efficacy of a multicenter study using intraarterial chemotherapy in conjunction with osmotic opening of the blood-brain barrier for the treatment of patients with malignant brain tumors. Cancer 88:637–647

17. Yang W, Barth RF, Carpenter DE, Moeschberger ML, Goodman JH (1996) Enhanced delivery of boronophenylalanine by means of intracarotid injection and bloodbrain barrier disruption for neutron capture therapy. Neurosurgery 38:985–992

18. Kraemer DF, Fortin D, Doolittle ND, Neuwelt EA (2001) Association of total dose intensity of chemotherapy in primary central nervous system lymphoma and survival. Neurosurgery 48:1033–1041

19. Remsen LG, Pagel MA, McCormack CI, Fiamengo S, Sexton G, Neuwelt EA (1999) The influence of anesthetic choice, PaCO$_2$, and other factors on osmotic blood brain barrier disruption in rats with brain tumor xenografts. Anesth Analg 88:559–567

20. Fortin D, McCormick CI, Remsen LG, Nixon R, Neuwelt EA (2000) Unexpected neurotoxicity of etoposide phosphate administered in combination with other chemotherapeutic agents after blood-brain barrier modification to enhance delivery, using propofol for general anesthesia, in a rat model. Neurosurgery 47:199–207

21. Gummerlock MK, Neuwelt EA (1990) The effects of anesthesia on osmotic blood-brain barrier disruption. Neurosurgery 26:268–277

22. Vorbrodt AW, Dobrogowska DH, Tarnawski M, Lossinsky AS (1994) A quantitative immunocytochemical study of the osmotic opening of the blood–brain barrier to endogenous albumin. J Neurocytol 23:792–800

23. Blasberg RG, Groothuis D, Molnar P (1990) A review of hyperosmotic blood-brain barrier disruption in seven experimental brain tumor models, in Johansson BB, Owman C, Widner H (eds): Pathophysiology of the Blood-Brain Barrier. Amsterdam, Elsevier, 197–220

24. Rapoport SI, Fredericks WR, Ohno K, Pettigrew KD (1980) Quantitative aspects of

reversible osmotic opening of the blood-brain barrier. Am J Physiol 238:R421–R431

25. Doran SE, Ren XD, Betz AL, Pagel MA, Neuwelt EA, Roessler BJ, Davidson BL (1995) Gene expression from recombinant viral vectors in the CNS following blood-brain barrier disruption. Neurosurgery 36: 965–970

26. Muldoon LL, Nilaver G, Kroll RA, Pagel MA, Breakefield XO, Chiocca EA, Davidson BL, Weissleder R, Neuwelt EA (1995) Comparison of intracerebral inoculation and osmotic blood-brain barrier disruption for delivery of adenovirus, herpes virus and iron oxide particles to normal rat brain. Am J Pathol 147:1840–1851

27. Neuwelt EA, Barnett PA, Hellström I, Hellström KE, Beaumier P, McCormick CI, Weigel RM (1988) Delivery of melanoma-associated immunoglobulin monoclonal antibody and Fab fragments to normal brain utilizing osmotic blood-brain barrier disruption. Cancer Res 48:4725–4729

28. Neuwelt EA, Specht HD, Barnett PA, Dahlborg SA, Miley A, Larson SM, Brown P, Eckerman KF, Hellström KE, Hellström I (1987) Increased delivery of tumor-specific monoclonal antibodies to brain after osmotic blood-brain barrier modification in patients with melanoma metastatic to the central nervous system. Neurosurgery 20:885–895

29. Neuwelt EA, Weissleder R, Nilaver G, Kroll RA, Roman-Goldstein S, Szumowski J, Pagel MA, Jones RS, Remsen LG, McCormick CI, et al. (1994) Delivery of virus-sized iron oxide particles to rodent CNS neurons. Neurosurgery 34:777–784

30. Nilaver G, Muldoon LL, Kroll RA, Pagel MA, Breakefield XO, Davidson BL, Neuwelt EA (1995) Delivery of herpes virus and adenovirus to nude rat intracerebral tumors following osmotic blood-brain barrier disruption. Proc Natl Acad Sci U S A 92:9829–9833

31. Freedman FB, Johnson JA (1969) Equilibrium and kinetic properties of the Evans blue–albumin system. Am J Physiol 216:675–681

32. Fortin D, Adams R, Gallez A (2004) A blood-brain barrier disruption model eliminating the hemodynamic effect of ketamine. Can J Neurol Sci 31:248–53

Chapter 24

Integrated Platform for Brain Imaging and Drug Delivery Across the Blood–Brain Barrier

Umar Iqbal, Abedelnasser Abulrob, and Danica B. Stanimirovic

Abstract

The development of imaging and therapeutic agents against neuronal targets is hampered by the limited access of probes into the central nervous system across the blood–brain barrier (BBB). The evaluation of drug penetration into the brain in experimental models often requires complex procedures, including drug radiolabeling, as well as determinations in multiple animals for each condition or time point. Prospective in vivo imaging of drug biodistribution may provide an alternative to "classical" pharmacokinetics and biodistribution studies in that a contrast-enhanced imaging signal could serve as a surrogate for the amount of drug or biologic delivered to the organ of interest. For the brain-targeting applications, it is necessary to develop formulation strategies that enable a simultaneous drug and contrast agent delivery across the BBB. In this chapter, we describe methods for encapsulating drugs into liposome nanocarriers with surface display of both the imaging contrast agent for one or multiple imaging modalities and the single-domain antibody that undergoes receptor-mediated transcytosis across the BBB. Contrast-enhanced imaging signal detected in the brain after intravenous injection of such formulation(s) is proportional to the amount of drug delivered into the brain parenchyma. This method allows for a prospective, noninvasive estimation of drug delivery, accumulation, and elimination from the brain.

Key words: Liposome, Imaging agent, Doxorubicin, In vivo time-domain optical imaging, Single-domain antibodies, Drug delivery, FC5, Blood–brain barrier

1. Introduction

Transport of drugs across this blood–brain barrier (BBB) continues to be a major pharmaceutical challenge. The development of approaches for transvascular brain delivery of neurotherapeutics, biologics, or imaging agents has relied on strategies that increase drug lipophilicity and/or positive charge, synthesis of mimics of naturally transported compounds, and use of receptor-mediated transcytosis (RMT) triggered by a ligand binding to receptors

Sukriti Nag (ed.), *The Blood-Brain and Other Neural Barriers: Reviews and Protocols*, Methods in Molecular Biology, vol. 686, DOI 10.1007/978-1-60761-938-3_24, © Springer Science+Business Media, LLC 2011

expressed on the luminal side of the brain endothelium, including transferrin receptor (1), insulin receptor (2), diphtheria toxin receptor (3), melanotransferrin P97 receptor (4), and low density lipoprotein receptor-related protein family of receptors (5). Using a phage-display library of llama single-domain antibodies (sdAb), we have recently identified a novel receptor–ligand complex that undergoes RMT across the BBB in vitro and in vivo (6). sdAbs consist solely of the variable region of the heavy chain camelid antibody (7–9) and are the smallest known antibody fragments (13 kD). FC5, a sdAb which selectively binds to a luminal $\alpha(2,3)$-sialoglycoprotein receptor epitope of brain endothelial cells (BEC) and undergoes RMT via a clathrin-coated pits (6), was isolated (10, 11) and engineered to display a free cysteine, enabling its conjugation with polyethylene glycol (PEG) chains, biologics, and nanoparticles (6).

Liposomes are nanocarriers that are composed of a lipid bilayer surrounding an aqueous core. The liposome platform is multifunctional and can be used to encapsulate drugs, biologics such as proteins, peptides, small interfering ribonucleic acid, and imaging agents such as optical, magnetic, and computed tomography. Liposomes offer the benefit of increased stability for administered agents, improved efficacy, and reduced adverse reactions. Liposomes are one of the most promising nanotechnology formats, as they are both biologically inert and highly biocompatible (12). By attaching a targeting antibody to liposomes, the therapeutic agent could be delivered to the specific antigen-expressing sites. Liposomes could be functionalized with the BBB-transmigrating antibody, as previously described for the transferrin receptor antibody (1) and the insulin receptor antibody (13) to target drug into the brain. In animal models, the measurements of the drug amount delivered into the brain using these carrier systems can be performed in brain samples ex vivo. However, for clinical translation and evaluation of efficacy of these carrier systems in patients, it would be important to develop noninvasive imaging approaches capable of quantifying and prospectively monitoring the amount and fate of the drug delivered into the brain, analogous to positron emission tomography-based imaging approaches for brain receptor occupancy (14).

In this chapter, we describe methods and procedures for synthesizing liposome carriers loaded with the brain-targeting therapeutic and functionalized with both the BBB-transmigrating sdAb, FC5, and with the imaging contrast agent to enable both drug delivery across the BBB and noninvasive detection of the formulation in the brain by in vivo imaging. In vivo imaging has the potential to improve preclinical and clinical trials by providing surrogate biomarker of treatment efficacy and/or disease progression. In this example, we utilize near-infrared optical imaging to collect information on the amount of drug delivered into the brain via a nanocarrier.

2. Materials

2.1. Liposome Synthesis

1. Chemicals:

   (a) Lecithin (L-α phosphatidycholine) (Avanti Polar Lipids, Alabaster, AL).

   (b) Cholesterol (Sigma-Aldrich Canada, Oakville, ON).

   (c) Maleimide-PEG-DSPE (1,2-distearoyl-*sn*-glycero-3-phosphoethanolamine-N-[maleimide(polyethylene glycol)2000](Ammonium Salt)) (Avanti Polar Lipids).

   (d) mPEG-DSPE (1,2-distearoyl-sn-glycero-3-phosphoethanolamine-N-[methoxy(polyethylene glycol)-2000] (ammonium salt)) (Avanti Polar Lipids).

   (e) Dodecylamine-PE (1,2-dipalmitoyl-*sn*-glycero-3-phosphoethanolamine-N-(dodecanylamine)) (Avanti Polar Lipids).

   (f) DOGS-NTA (1,2-dioleoyl-*sn*-glycero-3-{[N(5-amino-1 carboxypentyl)iminodiacetic acid]succinyl} (Nickel Salt)) (Avanti Polar Lipids).

2. Chloroform.

3. Eppendorf Vortex Mixer.

4. 1.5 mL Eppendorf tubes.

5. Eppendorf Vaccufuge.

6. PBS, pH 7.4; 3.2 mM Na_2HPO_4, 0.5 mM KH_2PO_4, 1.3 mM KCl, 135 mM NaCl.

7. pH meter (Corning Life Sciences, Lowell, MA).

8. Hot water bath with temperature up to 60°C (Precision Scientific, Mumbai, Maharashtra).

9. Liposofast Extruder (Avestin, Ottawa, Canada).

10. 100 nm polycarbonate membrane (Avestin).

11. Nicomp particle sizer, model 370 (Nicomp, Santa Barbara, CA, USA).

2.2. Liposome Functionalizing Agents and Drugs

1. FC5 sdAb (National Research Council of Canada – Institute for Biological Sciences) (see Note 1).

2. Doxorubicin-HCl (Sigma-Aldrich Canada) (see Note 2).

3. Cy5.5-NHS ester (GE Healthcare, Chalfont St. Giles, UK) (see Note 3).

4. Dimethyl Sulfoxide (DMSO).

5. Bicinchonic acid (BCA) protein assay kit (Thermo Fisher Scientific Inc, Waltham MA).

6. Amicon® Ultra-4 Centrifugal Filter Unit-100k cutoff membrane (Millipore, Billerica MA).

7. Ammonium Sulfate (Sigma-Aldrich Canada).

8. Triton X-100.

9. Isopropanol – HCl (0.75 N HCl).

10. TCEP-HCl (Sigma-Aldrich Canada).

11. 0.5 M EDTA, pH 8.0.

2.3. Animals, Imaging Equipment, and Software

1. CD-1 mouse, 6–8 weeks old (Charles River Laboratories, Wilmington, MA).

2. Isofluorane (Baxter, Mississauga, ON).

3. 30% O_2N_2 balance gas tank (Praxair, Danbury, CT).

4. Time-domain eXplore Optix preclinical imager with 670 nm pulsed laser diode (Advanced Research Technologies, Montreal, QC) (see Note 4).

5. eXplore Optix OptiView software (Advanced Research Technologies).

6. Heat lamp.

7. 0.5 cc Insulin syringe with 28-gauge needle.

2.4. Brain Tissue Analyses

1. Dextran (60–70 k) (Sigma-Aldrich Canada).

2. Spatula.

3. Fluorometric plate reader (BioTek, Winooski, VT).

4. Desktop centrifuge for speeds up to $15,000 \times g$.

5. Pasteur pipette.

6. 22-gauge needle.

7. Surgical instruments: Blunt small scissors, sharp small scissors.

8. Microperfusion pump (Bioptechs, Butler, PA).

9. Heparin solution (1,000 i.u/mL).

3. Methods

The methods described in this chapter will outline: (1) synthesis, production, and characterization of liposome nanocarriers; (2) liposome functionalization with the BBB vector; (3) liposome functionalization with the imaging contrast agent; (4) loading liposomes with the drug; (5) animal imaging using eXplore Optix preclinical imager; and (6) the determination of drug concentration in the brain. The schematic of the composition of the liposome carrier functionalized with the imaging agent and the BBB-transmigrating antibody described in this method is shown in Fig. 1.

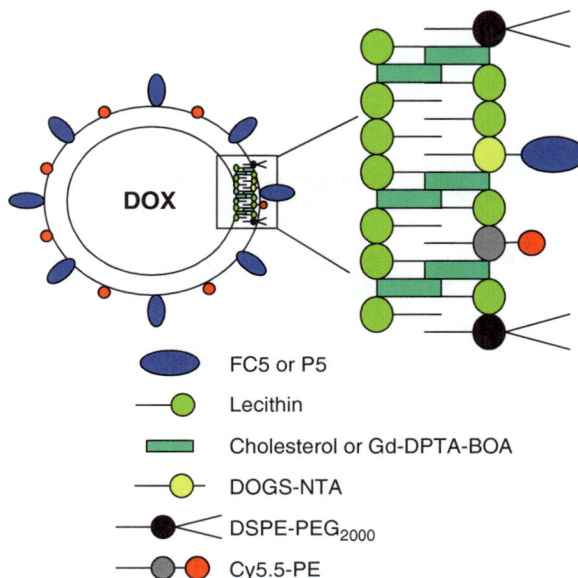

Fig. 1. Schematic drawing of the liposome functionalized with the optical (Cy5.5) and/or MRI imaging agent (Gadolinium, Gd), the blood–brain barrier transmigrating vector (FC5 or P5), and loaded with the brain-targeting drug (doxorubicin). Liposome nanoparticle is also PEGylated to avoid uptake and removal by the reticuloendothelial system (such liposomes are often referred to as "stealth").

3.1. Synthesis, Production, and Characterization of Liposomes

Liposomes were synthesized according to a previously described method (15).

1. Dissolve lecithin (9 mg), cholesterol (2.3 mg), mPEG2000-DSPE (1.33 mg), maleimide-derivatized PEG2000-DSPE (0.35 mg), and dodecylamine-PE (0.05 mg) in 800 µL of chloroform to achieve a molar ratio of 2:1:0.8:0.2:0.1 for the lipid mixture. Mix the lipids on a vortex mixer (see Note 5).

2. Pipette equal volumes (200 µL) of the lipid mixture into four 1.5 mL Eppendorf tubes (see Note 6).

3. Evaporate the chloroform overnight from the Eppendorf tubes under reduced pressure and elevated temperature (>45°C) using a Vacufuge™ to ensure complete removal of the chloroform (see Note 7).

4. Rehydrate the thin lipid film with 1.13 mL of PBS, pH 7.4 at 60°C using a hot water bath to create a 1% w/v liposome solution (see Note 8). If liposomes are to be loaded with the drug doxorubicin, refer to Subheading 3.3.

5. Extrude the multilamellar vesicle dispersion 21 times through two stacked 100 nm polycarbonate membranes using LiposoFast Extruder to create monodisperse unilamellar vesicles with an average size distribution of 100 nm.

6. Dilute the extruded sample 10–100× to decrease the turbidity of the solution and run 5 µL of sample on a Nicomp particle sizer. This will give a number-weighted Gausian size distribution of the average diameter of the particles (see Note 9).

3.2. Liposome Functionalization with the Imaging Contrast Agent

1. Dilute 1 mg of cy5.5-NHS ester in 100 µL of DMSO to create a concentration of 10 mg/mL cy5.5-NHS ester.

2. Add 10× molar excess of cy5.5-NHS ester (0.335 mg) to the number of mols of dodecylamine-PE lipid in the outer lipid membrane (half of the dodecylamine-PE added) of the liposome. Mix well and incubate the reaction at room temperature for 2 h (see Note 10).

3. Remove the unreacted cy5.5-NHS ester dye using an Amicon® Ultra-4 Centrifugal Filter Unit with a 100k cutoff membrane and resuspend in 1.13 mL PBS, pH 7.4 (see Note 11).

3.3. Loading Liposome with the Drug

Doxorubicin is remotely loaded into unilamellar liposomes via an ammonium sulfate gradient as described previously (16) (see Note 12).

1. Create a thin lipid film as described in Subheading 3.1, steps 1–3.

2. Rehydrate lipids with 1.13 mL of an aqueous solution containing 250 µM Ammonium sulfate and 0.15M NaCl at pH 5.5 at 60°C in a water bath.

3. Extrude the liposome solution as described in Subheading 3.1, **step 5**.

4. Incubate the extruded unilamellar liposomes with the ammonium sulfate buffer for 1 h at 60°C (see Note 13).

5. Remove the external buffer using an Amicon® Ultra-4 Centrifugal Filter Unit with a 100k cutoff membrane and resuspend sample in PBS, pH 7.4 (see Note 11).

6. Remotely load 1.13 mg doxorobucin (~80 µg doxorobucin/ µmol of phospholipid) dissolved in 500 µl PBS, pH 7.4 into liposomes by combining the doxorubicin solution and the liposome solution, followed by incubating the mixture for 60 min at 60°C (see Note 14).

7. Remove free doxorubicin using an Amicon® Ultra-4 Centrifugal Filter Unit with a 100k cutoff membrane and resuspend in 1.13 mL PBS, pH 7.4 (see Notes 11 and 15).

3.4. Liposome Functionalization with the BBB Carrier

1. Reduce 0.89 mg of cys-FC5 sdAb dimers by adding 10 mM TCEP and 20 mM EDTA at pH 8.0 in an Eppendorf tube and flush tube with nitrogen gas and seal. React for 30 min at room temperature (see Note 16).

2. Combine the reduced cys-FC5 with the maleimide-PEG liposome solution under nitrogen gas and react overnight at 4°C (see Note 17).

3. Remove unreacted cys-FC5, TCEP, and EDTA using an Amicon® Ultra-4 Centrifugal Filter Unit with a 100k cutoff membrane and resuspend in 1.13 PBS, pH 7.4 (see Notes 11and 18)

3.5. Animal Imaging Using eXplore Optix

Optical imaging using eXplore Optix time-domain small animal imaging system allows for a real-time, prospective, noninvasive in vivo monitoring of the biodistribution of fluorescently labeled molecules, in particular proteins and nanoparticles, such as liposomes. This section describes the steps required for animal imaging using the eXplore Optix preclinical imager. These steps include: basic anesthesia of mice and their placement in the animal imaging bed (see Subheading 3.5.1), scanning parameters for the eXplore Optix acquisition software (see Subheading 3.5.2), methods for intravenous injection in mouse tail vein (see Subheading 3.5.3), and methods for intracardiac perfusion and removal of the mouse brain (see Subheading 3.5.4).

3.5.1. Anesthesia and Animal Placement

1. Deeply anesthetize a CD-1 mouse using isoflurane (3.5 mL/min of 30% O_2 N_2 balance gas and 4% isofluorane).

2. Reduce anesthesia to a maintenance level (0.8 mL/min of 30% O_2 N_2 balance and 1.5% isofluorane) and shave the mouse on both the dorsal and ventral sides.

3. Place the animal in the dorsal position on a mouse bed in the eXplore Optix preclinical imager. Set the bed temperature at 36°C.

4. Set the appropriate elevation of the animal using the side-viewing digital camera inside the machine, such that the animal body is just below the green line limit.

3.5.2. Scanning Parameters for eXplore Optix Acquisition Software

1. Draw a region of interest around the animal using the polygon draw tool in the eXplore optix acquisition software version 2.0 (see Note 4).

2. Set the step size of the scan to 3.0 mm (see Note 19).

3. Select the appropriate laser to be used (cy5.5 laser – 670 nm excitation) (see Note 20).

4. Prescan the mouse to obtain a background fluorescence intensity image.

5. During the scan, the animal will be taken into the scanning chamber. The program will run its power automation sequence (see Note 21).

6. Once the scan is complete, remove the mouse from the chamber and allow it to recover from the anesthesia.

3.5.3. Intravenous Injection of Labeled Liposomes

1. Heat the animal in the cage using a heat lamp for 5 min.

2. Place the mouse in an injection cone, and slowly inject the mouse with cy5.5-labeled doxorubicin-loaded liposomes at a concentration of 6 mg/kg in a volume up to 200 μL via the tail vein using a 28-gauge needle.

3. Scan the mouse at various time points: 30 min, 1, 2, 4, and 24 h as described in Subheading 3.5, steps 1–9.

3.5.4. Intracardiac Perfusion and Removal of Mouse Brain

1. Under isofluorane anesthesia, using blunt scissors, make a midline incision through the skin and muscle to expose the abdomen.

2. Cut the diaphragm on both sides and extend the cut along the rib cage on both sides.

3. Expose and cut the pleura and pericardium covering the lungs and heart, respectively.

4. Snip the right atrium to provide an exit route for the blood.

5. Insert a 22-gauge needle into the left ventricle and perfuse 30 mL of saline containing heparin (1 i.u./mL) via the aorta using a microperfusion pump with a 2 mL/min flow rate (see Note 22).

6. Decapitate the mouse using small sharp scissors. Make a midline incision through the skin overlying the skull and expose the skull bones. Carefully cut the skull bones and expose the brain. Use a spatula to scoop out the brain. Keep the brain on ice.

3.6. Determination of Drug Concentration in the Brain

This section describes the extraction of doxorubicin from the brain (see Subheading 3.6.1) (see Note 23) and the capillary depletion method (see Subheading 3.6.2) to separate capillaries from brain parenchyma (17). Doxorubicin tissue content was quantified fluorometrically in a plate reader (excitation 470 nm, Emission 590 nm) using a method described previously (18).

3.6.1. Doxorubicin Extraction from Brain

1. Weigh dissected brain on an electronic scale.

2. Gently homogenize brain in PBS, pH 7.4 (20% w/v) by hand using a pestle in a 10 mL glass tube.

3. Place homogenate (100 μL) in a 1.5 mL Eppendorf tube and add 50 μl of 10% (v/v) Triton X-100, 100 μL of distilled water, 1,000 μL of acidified isopropanol (containing 0.75 N HCl).

4. Place the tube on a vortex mixer for 10 min and extract doxorubicin overnight at –25°C in a freezer.

5. The next day, warm the tube up to room temperature.

6. Vortex for 10 min.

7. Centrifuge at $15,000 \times g$ for 20 min.

8. Quantify Doxorubicin in supernatant fluorimetrically using a fluorescence plate reader (excitation 470 nm and emission 590 nm).

9. Determine Doxorubicin concentration by interpolating concentration values from a fluorescent standard curve of different dilutions of doxorubicin spiked in brain homogenates, followed by the Doxorubicin extraction procedure as described in Subheading 3.6, steps 3–8 (see Note 24).

10. The data can be expressed as nanogram doxorubicin per gram of brain.

3.6.2. Capillary Depletion Method

1. Gently homogenize brain by hand in PBS, 20% (w/v).

2. Add an equal volume of 30% dextran to yield a final concentration of 15% dextran.

3. Centrifuge the homogenate at $5,800 \times g$ (10,000 rpm) for 20 min at 4°C (see Note 25).

4. Use a 1 mL pipette and withdraw the upper layer with minimal dextran contamination.

5. Withdraw dextran and wipe out walls for residual brain parenchyma, if any.

6. Centrifuge brain parenchyma at $5,800 \times g$ for 10 min at 4°C.

7. Insert a Pasteur pipette to the bottom of the tube and slowly withdraw the dextran layer.

8. Dissolve vessels in PBS, using sonication for a few seconds.

9. Dilute vessels and parenchyma to the same degree, to make them comparable for fluorescence measurements.

10. Continue with doxorubicin extraction and fluorometric quantification as described in Subheading 3.6, steps 3–10.

3.7. Data analyses, Interpretation, and Conclusions

The methods and procedures in this chapter describe a unified platform that uses imaging as a surrogate marker to noninvasively monitor drug delivery to the brain. To "validate" the described platform, the user is encouraged to perform initial studies that correlate obtained imaging signal with the measured drug concentration in the brain parenchyma, as shown in Fig. 2. In the described example, the fluorescence intensity (FI) signals obtained in the brain region of interest (ROI) in vivo using the eXplore Optix imaging system were correlated with the measured concentrations of doxorubicin in the brain parenchyma after capillary depletion in separate groups of animals using the same liposome formulations. The results show a direct correlation between fluorescence intensity signal ($r = 0.99$) and parenchymal concentration of doxorubicin delivered to the brain using nontargeted liposomes or liposomes functionalized with the BBB vector carrier, FC5. Alternatively, using this multifunctional

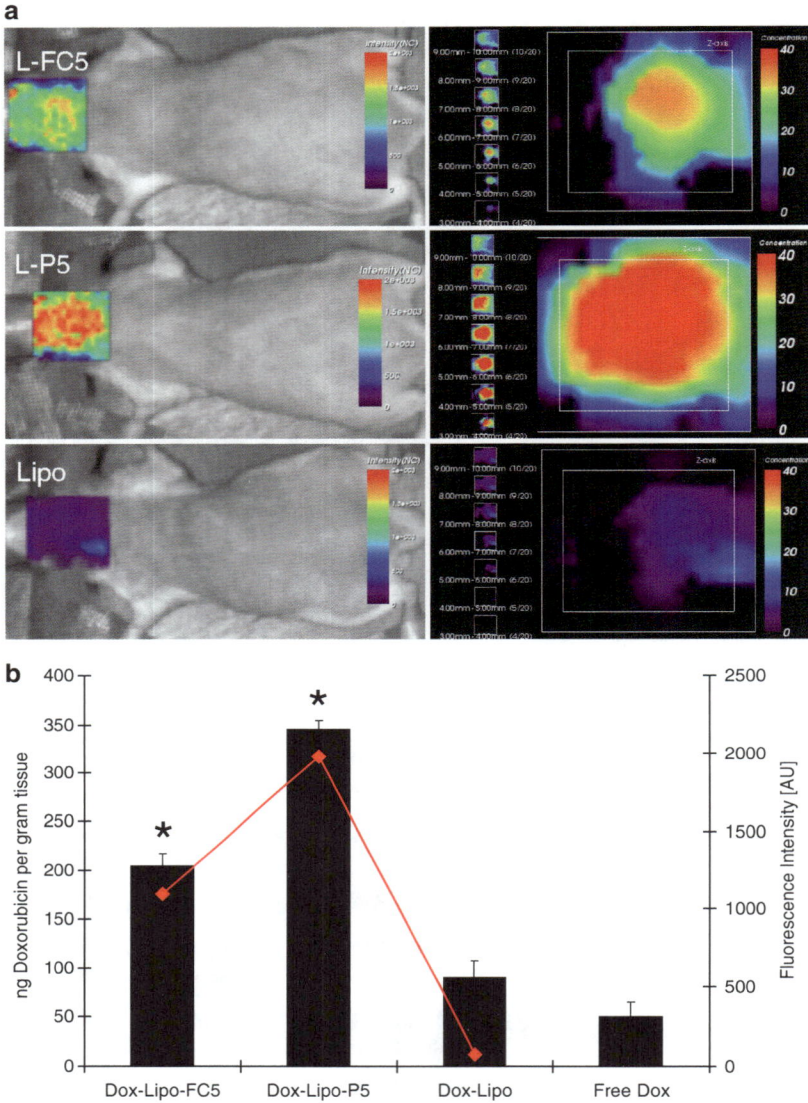

Fig. 2. Demonstration of correlative relationship of optical imaging signal in vivo and the amount of drug (doxorubicin) delivered into the brain using liposomes functionalized with the BBB-transmigrating sdAbs (FC5 and P5) and labeled with the near-infrared fluoroprobe, Cy5.5. (**a**) *Left panels* are representative near-infrared fluorescence intensity images obtained using eXplore Optix (670/700 nm laser) of a head region of mice, 24 h after injection of Cy5.5-FC5-, Cy5.5-P5-functionalized, or Cy5.5-labeled unfunctionalized liposomes. *Right panels* are depth-concentration optical tomography z-axis sections (10, 1-mm sections) through the thickness of the head. (**b**) Doxorubicin (Dox) concentration in the brain parenchyma (*closed bars*) 24 h after intravenous injection of 6 mg/kg. Free doxorubicin or doxorubicin loaded into either nontargeted liposomes or liposomes functionalized with FC5 or P5 was determined fluorometrically after animal perfusion and capillary depletion. Superimposed line is fluorescence intensity (FI) measured in a separate group of animals injected with same formulations labeled with Cy5.5. *Asterisks* indicate $p < 0.01$ as compared with free doxorubicin. ($n = 5$/group).

(containing both imaging agent and therapeutic drug) nanocarrier, correlative assessments can be done in the same group of animals. Similar principles would apply to a variety of chosen BBB vector carriers and encapsulated drugs. Once the correlative nature of the imaging signal and the amount of delivered drug is experimentally demonstrated, the platform could be used to noninvasively monitor drug delivery into as well as drug elimination from the brain using prospective imaging protocols. This can be further correlated with the drug-mediated pharmacological effects in the same animal over prolonged time periods. It is important to note that the individual components of this platform can be modified to suit a wide array of biological applications and imaging modalities.

Imaging surrogate biomarkers are considered a key area of development in the FDA Critical Path Initiative (19) and are becoming essential in the drug approval process. The described integrated drug delivery/imaging platform could therefore serve as a model for developing imaging biomarkers for noninvasively monitoring and quantifying drug delivery into the brain in relation to pharmacological drug efficacy.

4. Notes

1. FC5 sdAb is a BBB-transmigrating antibody isolated from the llama sdAb phage-display library (10). FC5, similar to other sdAbs, is ten times smaller (15 kDa) than IgG, is highly stable and resistant to proteases, pH, and temperature. These characteristics of sdAbs make them ideal targeting moieties for liposomal carriers, since many sdAbs can attach to a single liposome, without significantly changing the overall diameter of the formulation (20, 21). We have shown previously that increased avidity (i.e., higher number of antigen-binding molecules) of FC5 achieved by pentamerization results in increased efficiency of RMT across BEC (22). FC5 is further engineered to contain a free cysteine (FC5-cys) (22), which can be reacted with the maleimide group, enabling chemical attachment of various compounds to FC5. The free cysteine moiety in FC5 is remote from the antigen-binding site (22) to prevent stearic hindrance by attached molecules. Another formulation used in this example is P5, a pentamerized version of FC5 developed as described previously (22).

2. Doxorubicin is a widely used chemotherapeutic drug that is prevented from entering the brain by the efflux pumps, such as P-glycoprotein which is highly expressed by BEC (23).

3. Cy5.5 is an optical probe that emits in the near-infrared spectrum (Excitation 675 nm/Emission 694 nm) and is suitable for use with the eXplore Optix system which is equipped with a 670 nm laser. NHS ester form of Cy5.5 is used in this application to enable Cy5.5 conjugation to maleimide-PEG-DSPE component of liposomes.

4. eXplore Optix is an in vivo small animal optical imaging equipment that uses Time-Domain (TD) imaging technology. Optical imaging technologies analyze the propagation of light particles (photons) through a medium such as tissue. For in vivo optical imaging, the observation of photon behavior in the near-infrared (NIR) region is favored because of tissue's low absorption properties in this spectral band (between 650 and 1,100 nm), thus allowing light to penetrate several centimeters of tissue (24). NIR photons traveling through tissue are highly scattered before either being totally absorbed by the tissue or emerging at the surface where they are detected. Traditional in vivo optical molecular imaging systems grossly measure all of the photons that propagate through tissue without any temporal discrimination – this is known as the Continuous Wave (CW) technique. This method cannot discriminate absorption events from scattering events, which impedes its capability to uncouple location (depth) from concentration in the image. Bioluminescence measurements by definition are CW since they are not generated in response to a light stimulus. Since scattered photons have no preferential direction or orientation, it is possible to statistically differentiate one from another by observing the time at which they emerge from the scattering medium; this is known as TD imaging. In TD optical imaging, short pulses of light are sent to illuminate the specimen under study. The system then detects the photons according to their time-of-flight within the tissue; this time-of-flight distribution (generally called a TPSF or Temporal Point Spread Function) is used to recover the optical characteristics of the specimen, discriminating absorption from scattering properties. Whereas CW intensity measurement includes all photons, TD temporally discriminates photons, which have probed different depths, resulting in greater depth sensitivity. Depth information leads to an accurate recovery of fluorophore concentration when the data are reconstructed in postprocessing. Due to the temporal dimension in TD measurements, the signal already contains volumetric information about the tissue and enables tomographic (3D) image reconstruction. For the application described in this chapter, other optical imaging scanners can be used instead of eXplore Optix, for example, Kodak Optical Imaging System FX Pro (Carestream Health,

Rochester, NY) and the IVIS optical imager (Caliper Life Sciences, Hopkinton, MA). However, since all other currently commercially available optical imaging scanners use CW technique, a direct correlation between the optical signal, representing concentration of the fluorophore, and the concentration of delivered drug may not be possible or may require a very complex image processing algorithms.

5. PEG is a polymer produced by the interaction of ethylene oxide and water with the general formula $H(OCH_2CH_2)_nOH$, where n is the average number of repeating oxyethylene groups. PEGylation is the chemical process of covalently attaching PEG chains to drugs, biologics, or nanoparticles to achieve their longer circulating half-life, to reduce their uptake by the reticuloendothelial system, and to evade the immune system (25). The nanoparticle PEGylation is achieved by a covalent attachment of PEG chains to the nanoparticle surface. Typically, such PEG chains have MW of 1–5 kDa and may be either branched or linear (26).

6. The use of multiple Eppendorf tubes provides a sufficient surface area for the creation of a thin lipid film. Alternatively, a small round bottom flask can be used.

7. This process will create a thin lipid film in the bottom of each Eppendorf tube.

8. Heating at 60°C is above the membrane transition temperature of lecithin, leading to increased lipid fluidity. Hydration of the lipids with an aqueous buffer results in spontaneous formation of multilamellar vesicles of varying sizes.

9. To calculate the number of liposomes in the solution: (a) calculate the surface area (SA_{total}) of half of the lipids added (assuming half of the lipids distribute to the outer layer of the liposome and half distribute to the inner layer), where 0.55 nm^2 is the approximate area of a single phospholipid; (b) calculate the surface area ($SA_{liposome}$) of a single liposome using the formula $4\pi r^2$, where the radius is derived from the average size of the liposomes after extrusion; and then (c) divide the $SA_{total}/SA_{liposome}$ to obtain an estimate of the number of liposomes in the solution. In this example there are 9.74×10^{13} vesicles in the solution.

10. Alternatively, or in addition to Cy5.5 (or other optical probes), imaging agents for other imaging modalities can be incorporated within or on the surface of the liposome to achieve multimodality. For example, the MRI contrast agent Gadolinium-DPTA (Gd-DTPA) linked to a phospholipid can be added to the initial mixture of lipids; Gd-DTPA will be incorporated into the liposomes after rehydration of the sample (27) (Fig. 1).

11. It is important to wash sample at least 3 times with 3–4 mL of PBS, pH 7.4 during the ultrafiltration process to ensure a complete removal of the free, unreacted cy5.5-NHS ester dye. Alternatively, dialysis tubing with an 8000 MWCO against >1,000 volumes of buffer can be used.

12. Agents other than doxorubicin can be encapsulated inside the liposomes, including peptides, DNA plasmids, and hydrophobic drugs. In addition to loading water-soluble drugs into the aqueous compartment of the liposome, water-insoluble drugs can be loaded into the hydrophobic compartment of the liposome.

13. This will create a pH gradient between the intraliposomal compartment and the external solution.

14. The red color of the doxorubicin solution will rapidly change to a darker purple color due to the fluorescence quenching that occurs when doxorubicin is loaded into liposomes.

15. To determine the efficiency of doxorubicin loading into the liposomes, take a 100 µL sample and add 8 mL of PBS, 1 mL of isopropanol-HCL, and 1 mL of 10% Triton X-100 solution. Heat the sample at 60°C for 30 min in a water bath to release all of the encapsulated doxorubicin. Use a fluorescent plate reader (excitation 470 nm, emission 590 nm) to relate the total release of doxorubicin to the amount of doxorubicin added initially (1.13 mg). To determine the concentration from the fluorescence reading, create a standard curve of known amounts of doxorubicin plotted against fluorescence intensity. Calculate % loading efficiency by dividing total encapsulated doxorubicin by total doxorubicin added ×100% (28, 29). The expected loading efficiency is greater than 90% for this method. If lower efficiency is recorded, increasing the rigidity of the liposome to prevent drug leakage (i.e., adding more cholesterol) or allowing more incubation time in the creation of the ammonium sulfate pH gradient is recommended.

16. Approximately 100–150 sdAbs are expected to attach to each liposome. This number is calculated by measuring the protein concentration of the targeted and nontargeted liposome formulation using a standard Bicinchonic acid (BCA) protein assay kit (Thermo Fisher Scientific Inc, Waltham MA). Subtracting the estimated protein value for the targeted liposome formulation from the nontargeted liposome formulation yields an estimate of the sdAb concentration in the targeted solution. The sdAb concentration is converted to the number of sdAbs in the solution by using a molar mass of 15,000 $g/$ mol for sdAbs. Dividing the number of sdAb molecules by the number of liposome in the targeted liposome solution will yield an estimate of the number of antibodies per liposome.

17. Alternatively, FC5 engineered with a polyhistidine tag can be coupled noncovalently to liposomes that contain a metal ion-chelating lipid such as DOGS-NTA (30).

18. Antibody attachment to liposomes can add up to 1–10 nm to the overall diameter of the liposomes (100±15 nm) (31). Dynamic light scattering using a NiComp particle sizer is used to determine the liposome size (see Subheading 3.1, step 6).

19. The step size can be reduced to a minimum of 0.5 mm for higher resolution images, but the scanning time increases.

20. Other lasers are also available, including a 780 nm laser for the near-infrared imaging, which can be used with alternative dyes emitting in appropriate spectrum range, such as Li-Cor IR800CW-NHSester (Li-Cor Biosciences, Lincoln, NB).

21. Laser power and integration time per pixel are automatically optimized by the acquisition software. The data are recorded as temporal point spread functions (TPSF) and the images can be reconstructed as fluorescence intensity, fluorescence lifetime, concentration-depth maps, and 3-D volumetric representation (optical tomography) using Optiview analysis software version 2.0.

22. Make sure organs are completely cleared of blood.

23. Doxorubicin is a fluorescent molecule and can be quantified easily using fluorometric methods. Alternative methods can be used to quantify other encapsulated drugs, including high performance liquid chromatography, ELISA, or liquid chromatography-mass spectrometry.

24. Create the standard curve of brain tissue homogenate to correct for nonspecific background fluorescence of the brain tissue.

25. Upper fraction is the brain parenchyma, and the bottom fraction is capillaries.

Acknowledgments

The work described in this chapter was in part supported by the Canadian Institutes of Health Research Emerging Team grant #79031.

References

1. Pardridge, W. M., Buciak, J. L., and Friden, P.M. (1991) Selective transport of an anti-transferrin receptor antibody through the blood-brain barrier in vivo *J Pharmacol Exp Ther* 259, 66–70.

2. Coloma, M. J., Lee, H. J., Kurihara, A., Landaw, E. M., Boado, R.J., Morrison, S. L., and Pardridge, W. M. (2000) Transport across the primate blood-brain barrier of a genetically engineered chimeric monoclonal antibody

to the human insulin receptor *Pharm Res* **17**, 266–74.

3. Gaillard, P. J., Visser, C.C., and de Boer, A. G. (2005) Targeted delivery across the blood-brain barrier *Expert Opin Drug Deliv* **2**, 299–309.

4. Demeule, M., Regina, A., Jodoin, J., Laplante, A., Dagenais, C., Berthelet, F., Moghrabi, A., and Beliveau, R. (2002) Drug transport to the brain: key roles for the efflux pump P-glycoprotein in the blood-brain barrier *Vascul Pharmacol* **38**, 339–48.

5. Lillis, A. P., Van Duyn, L. B., Murphy-Ullrich, J. E., and Strickland, D. K. (2008) LDL receptor-related protein 1: unique tissue-specific functions revealed by selective gene knockout studies *Physiol Rev* **88**, 887–918.

6. Abulrob, A., Sprong, H., Van Bergen en Henegouwen, P., and Stanimirovic, D. (2005) The blood-brain barrier transmigrating single domain antibodies: mechanisms of transport and antigenic epitopes in human brain endothelial cells *J Neurochem* **95**, 1201–14.

7. Hamers-Casterman, C., Atarhouch, T., Muyldermans, S., Robinson, G., Hamers, C., Songa, E. B., Bendahman, N., and Hamers, R. (1993) Naturally occurring antibodies devoid of light chains *Nature* **363**, 446–8.

8. Holliger, P. and Hudson, P. J. (2005) Engineered antibody fragments and the rise of single domains *Nature Biotechnology* **23**, 1126–36.

9. Muyldermans, S. (2001) Single domain camel antibodies: current status *J Biotechnol* **74**, 277–302.

10. Muruganandam, A., Tanha, J., Narang, S., and Stanimirovic, D. (2002) Selection of phage-displayed llama single-domain antibodies that transmigrate across human blood-brain barrier endothelium *FASEB J* **16**, 240– 2.

11. Tanha, J., Muruganandam, A., and Stanimirovic, D. (2003) Phage display technology for identifying specific antigens on brain endothelial cells *Methods Mol Med* **89**, 435–49.

12. Torchilin, V. P. (2005) Recent advances with liposomes as pharmaceutical carriers *Nat Rev Drug Discov* **4**, 145–60.

13. Boado, R. J. (2007) Blood-brain barrier transport of non-viral gene and RNAi therapeutics *Pharm Res* **24**, 1772–87.

14. Summerfield, S. G., Lucas, A.J., Porter, R. A., Jeffrey, P., Gunn, R. N., Read, K. R., Stevens, A. J., Metcalf, A. C., Osuna, M. C., Kilford, P. J., Passchier, J., and Ruffo, A. D. (2008) Toward an improved prediction of human in vivo brain penetration *Xenobiotica* **38**, 1518–35.

15. Hansen, C. B., Kao, G. Y., Moase, E. H., Zalipsky, S., and Allen, T. M. (1995) Attachment of antibodies to sterically stabilized liposomes: evaluation, comparison and optimization of coupling procedures *Biochim Biophys Acta* **1239**, 133–44.

16. Drummond, D. C., Meyer, O., Hong, K., Kirpotin, D. B., and Papahadjopoulos, D. (1999) Optimizing liposomes for delivery of chemotherapeutic agents to solid tumors *Pharmacol Rev* **51**, 691–743.

17. Bowman, P. D., Ennis, S. R., Rarey, K. E., Betz, A. L., and Goldstein, G. W. (1983) Brain microvessel endothelial cells in tissue culture: a model for study of blood-brain barrier permeability *Am Neurol* **14**, 396–402.

18. Charrois, G. J. and Allen, T. M. (2003) Multiple injections of pegylated liposomal doxorubicin: pharmacokinetics and therapeutic activity *J Pharmacol Exp Ther* **306**, 1058–67.

19. Richter, W. S. (2006) Imaging biomarkers as surrogate endpoints for drug development *Eur J Nucl Med Mol Imaging* **33** Suppl 1, 6–10.

20. Cheng, W. W. and Allen, T. M. (2008) Targeted delivery of anti-CD19 liposomal doxorubicin in B-cell lymphoma: a comparison of whole monoclonal antibody, Fab' fragments and single chain Fv *J Control Release* **126**, 50–8.

21. Zhou, Y., Drummond, D. C., Zou, H., Hayes, M. E., Adams, G. P., Kirpotin, D. B., and Marks, J. D. (2007) Impact of single-chain Fv antibody fragment affinity on nanoparticle targeting of epidermal growth factor receptor-expressing tumor cells *J Mol Biol* **371**, 934–47.

22. Abulrob, A., Zhang, J., Tanha, J., Mackenzie, R. and Stanimirovic, D. (2005b) Single domain antibodies: blood brain barrier delivery vectors. Theme *"Drug Transport(ers) & Diseased Brain" International Congress Series* **1277**, 212–23.

23. Takamiya, Y., Abe, Y., Tanaka, Y., Tsugu, A., and Kazuno, M. (1997) Murine P-glycoprotein on stromal vessels mediates multidrug resistance in intracerebral human glioma xenografts *Br J Cancer* **76**, 445–50.

24. Frangioni, J. V. (2003) In vivo near-infrared fluorescence imaging *Curr Opin Chem Biol* **7**, 626–34.

25. Gabizon, A., Shmeeda, H., and Barenholz, Y. (2003) Pharmacokinetics of pegylated liposomal Doxorubicin: review of animal and human studies *Clin Pharmacokinet* **42**, 419–36.

26. Ryan, S. M., Mantovani, G., Wang, X., Haddleton, D. M., and Brayden, D. J. (2008) Advances in PEGylation of important biotech molecules: delivery aspects *Expert Opin Drug Deliv* **5**, 371–83.

27. Caruthers, S. D., Winter, P. M., Wickline, S. A., and Lanza, G. M. (2006) Targeted magnetic resonance imaging contrast agents *Methods Mol Med* **124**, 387–400.

28. Kulkarni, S. B., Betageri, G. V., and Singh, M. (1995) Factors affecting microencapsulation of drugs in liposomes *J Microencapsul* **12**, 229–46.

29. Kepczyński, M., Nawalany, K., Kumorek, M., Kobierska, A., Jachimska, B., and Nowakowska, M. (2008) Which physical and structural factors of liposome carriers control their drug-loading efficiency? *Chem Phys Lipids* **155**, 7–15.

30. Chikh, G. G., Li, W. M., Schutze-Redelmeier, M. P., Meunier, J. C., and Bally, M. B. (2002) Attaching histidine-tagged peptides and proteins to lipid-based carriers through use of metal ion chelating lipids *Biochem Biophys Acta* **1567**, 204–12.

31. Emanuel, N., Eli, K., Bolotin, E. M., Smorodinsky, N. I., and Barenholz, Y. (1996) Preparation and characterization of doxorubicin-loaded sterically stabilized immunoliposomes *Pharmaceutical Res* **3**, 352–9.

Chapter 25

Targeting the Choroid Plexus-CSF-Brain Nexus Using Peptides Identified by Phage Display

Andrew Baird, Brian P. Eliceiri, Ana Maria Gonzalez, Conrad E. Johanson, Wendy Leadbeater, and Edward G. Stopa

Abstract

Drug delivery to the central nervous system requires the use of specific portals to enable drug entry into the brain and, as such, there is a growing need to identify processes that can enable drug transfer across both blood-brain and blood–cerebrospinal fluid barriers. Phage display is a powerful combinatorial technique that identifies specific peptides that can confer new activities to inactive particles. Identification of these peptides is directly dependent on the specific screening strategies used for their selection and retrieval. This chapter describes three selection strategies, which can be used to identify peptides that target the choroid plexus (CP) directly or for drug translocation across the CP and into cerebrospinal fluid.

Key words: Blood-brain barrier, Choroid plexus, Choroid epithelial cells, CNS targeting, Library screening, Combinatorial biology, Blood-CSF barrier, Drug translocation, *Ex vivo* biopanning, Ligand internalization, Phage display, Screening

1. Introduction

Filamentous bacteriophage are a group of viruses that contain circular, single stranded DNA encased in a long (1 μm) protein capsid. By electron microscopy, they appear as cylinders and, at the very tip of the Ff class of phage (e.g., M13), there are five copies of the plll protein that is encoded by gene lll. Normally, this protein is used by the virus to target and enter *Escherichia coli* that contains the F plasmid and expresses a receptor for this phage. Once inside the bacteria, these phage replicate and are released by the bacteria at a very high titer (10^{11} pfu/mL).

In 1985, Smith (1) conceived of a method whereby short nucleic acid sequences of DNA inserted into the glll gene generate

Sukriti Nag (ed.), *The Blood-Brain and Other Neural Barriers: Reviews and Protocols*, Methods in Molecular Biology, vol. 686, DOI 10.1007/978-1-60761-938-3_25, © Springer Science+Business Media, LLC 2011

phage that display a peptide-pIII fusion protein but retain infectivity for *E. coli*. Reasoning that the displayed peptides could confer phage with new intrinsic activity, he proposed that it should be possible to introduce random sequences of DNA into the gIII gene and create complex libraries of peptides. Because each peptide is displayed in five copies on an individual phage, mixtures of these phage, called libraries, can be screened and individual phage particles harvested to select novel peptides that confer novel activities to these phage. They are then enriched by a process called "biopanning" and any phage carrying the screened activity is recovered and replicated in bacteria. Sequencing the gene III of these phage reveals the identity of the peptide that confers the new biological activity to the phage. In a library containing 10^9 different peptide sequences, there is a statistically significant probability that there are 100 different candidate peptides that could be recovered from the library even if the probability of its existence is so low (e.g., $1:10^7$) as to be unrealistic.

Over the past several years, we and other investigators have been adapting the original phage display technique to identify various peptides with different specificities and activities (2, 3). For example, peptides have been identified that induce physical stability of particles in organic solvents like chloroform, decrease complement activation of macromolecules in blood, modify immunogenicity, alter viral tropism *in vitro* and *in vivo*, internalize particles into cells, transduce cells, promote transcytosis *in vitro* and *in vivo*, and even promote transmigration of particles across cell barriers *in vitro* and *in vivo* (3–23). Specifically relevant to the methods described here are biopanning approaches used to characterize organ and cell homing peptides that can target the vasculature and parenchyma (3, 14–17, 24–27), peptides that can mediate transcytosis across epithelial cells *in vitro* (11), and antibodies that transmigrate phage across the blood-brain barrier (13). Our own laboratories' focus has been to identify and exploit ligands that internalize into target cells. To this end, we have re-engineered phage vectors for binding to mammalian cells (9) and monitored their entry into cells by immunostaining for internalization (28), transfection for drug delivery (7), PCR for DNA delivery (29) and transduction for gene expression (30).

Recently, we have been using these methods to evaluate whether it is possible to target the CNS and, specifically, whether the unique features of the choroid plexus (CP) at the interface between blood and cerebrospinal fluid (CSF) (31–36) would allow exploitation of molecular translocation into the CP, CSF, and brain parenchyma. To this end, we have used EGF-targeted phage (30) as a test substance to explore phage targeted to the CP nexus (Fig. 1). We reasoned that if EGF-phage could be targeted to the epithelium, then targeted-phage could be identified to assist in drug translocation directly to the CP. Alternatively,

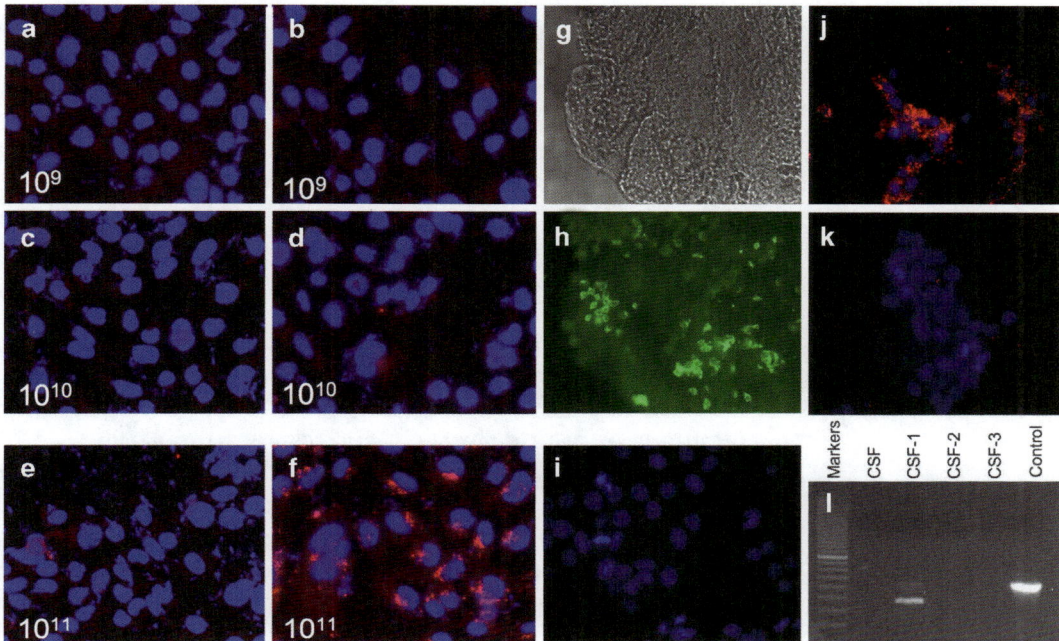

Fig. 1. Targeting cultured CP epithelial cells and explants with EGF phage is shown. Using the methods described here, control (untargeted phage) and EGF-targeted phage were incubated with either cultured CP epithelial cells (**a–f**), explants of mouse CP (**k**), or injected i.v. (**i**) to demonstrate the feasibility of CP targeting with phage. Controls (**a, c, e** and **i**) used untargeted phage, whereas EGF-targeted phage was evaluated in **b, d, f, h** and **j**. When the particles are added to cells after adding exogenous EGF (1 μg/mL), then specificity can be demonstrated by eliminating the internalization (**k**). (**l**) In another approach, PCR can be used to assess recoveries from CSF in control untreated animals (CSF) and in three treatment groups (CSF-1, CSF-2, and CSF-3) as long as there is a positive control.

genes could be delivered to the CP to restore its function or to secrete therapeutic proteins into CSF (31–33, 37–42). If successful, the CP could be re-engineered to modulate its production of CSF and deliver biological agents to treat CNS disease (32, 36, 37, 43–49).

With cultured CP cells (Fig. 1), it is possible to demonstrate that particle internalization is dose dependant. In the example shown here, EGF-targeted phage were added in concentrations of 10^9–10^{11} to cells in culture and incubated as described below for 2 h before internalization analysis. Under these conditions, internalization of untargeted particles is nearly undetectable (Fig. 1a, c, e), but the EGF-targeted particle is readily detected inside the cultured CP cells (Fig. 1b, d, f). These data establish that CP cells in culture can be used to test targeting ligands and that the methods are applicable for biopanning libraries for CP targeting. In other studies, we used the methods described below to mine a phage display library to characterize novel targeting peptides that also target these cells (not shown).

Using explants of CP cells, it is also possible to show specific epithelial cell targeting. The example shown in Fig. 1 illustrates how it is possible to demonstrate the specificity of internalization. When explants of CP (Fig. 1g) are incubated with phage, there is internalization if the particles are targeted with EGF (Fig. 1h). Very little targeting is observed when untargeted phage is added to the media. When cells are preincubated in the presence of EGF prior to the addition of EGF-targeted phage, the EGF-dependant internalization that is normally detected (Fig. 1j) is not observed (Fig. 1k). In other studies, we used the methods described below to mine a phage display library to characterize novel targeting peptides that can also target these cells (not shown).

To identify CP-CSF targeting peptides for these different applications of drug delivery, it is necessary to mine libraries of peptide sequences. With the establishment of target validation by EGF-targeted phage (Fig. 1), it is possible that libraries can be explored for various classes of CP-targeting peptides in several different screening assays (Fig. 2). They include (a) targeting the CP by *in vitro* screening for targeting peptides after injecting peptide libraries into mice, (b) targeting CP epithelium by biopanning CP explants ex vitro and even (c) evaluating particle translocation across the CP and into CSF. To explore the possibility of translocation across the CP and into CSF, PCR rather than biopanning can be used to detect particles in CSF after an intra-arterial injection of libraries into rat (Fig. 1l). However, because the concentration

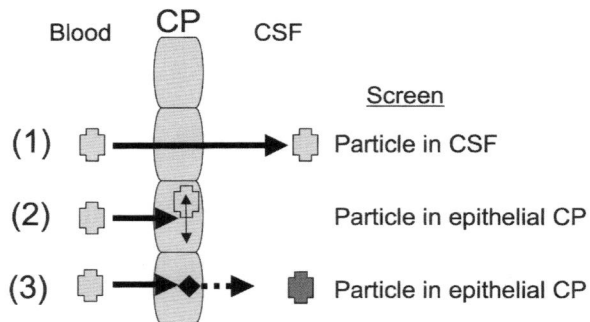

Fig. 2. Strategies for drug targeting to the CP are shown. There are three ways to target drugs to the brain via the CP. Firstly, drugs could be translocated directly into CSF from the apical (ventricular) side of the epithelium to find their targets via CSF bulk flow. In the second mechanism, the CP itself is the drug target, and the epithelium could be the therapeutic target of the drug and modulate its natural functions. The third mechanism targets the CP, for example with a gene, with the goal of exploiting the CPs natural ability to produce CSF and secretes biotherapeutic factors. Phage display could be used to identify each of the three categories of CP targeting agents depending on the different biopanning screens deployed. In the first, the CP is targeted "transchoroidally" from blood to CSF, in the second, it is targeted to the basolateral (blood-facing) epithelium, and in the third to the apical (CSF-facing) epithelium.

of recovered phage is very low and the sensitivity of the PCR is so high, the possibility of cross-contamination is a major concern. Still, these data point to possibly using combinatorial biology to identify transmigrating peptides *in vivo*.

Herein we describe how targeting can be screened for CP-targeting peptides using cultured CP cells, explants of intact CP, brain intracerebroventricular injections (i.c.v.), and CSF collection. Screening methodologies can be applied toward identifying CP-targeting peptides using these same approaches by analyzing particle internalization with phage display.

2. Materials

2.1. Peptide Libraries

1. Use either the PhD-7 display (#E8102L), the PhD-12 display (#E8111L), or the PhD-C7C display (#E8120S), (New England Biolabs, Ipswich, MA, USA) (see Note 1).

2. Alternatively, phage can be engineered with known targeting ligands, e.g., epidermal growth factor (EGF), fibroblast growth factor (FGF), anti EGF receptor, ciliary neurotrophic factor (CNTF), and interleukin-2 IL2 (8, 20, 30, 50, 51).

2.2. Culture of Primary Choroid Plexus Epithelial Cells

1. Phosphate buffered saline (PBS) without Ca^{2+} and Mg^{2+}: 2.69 mM KCl, 1.47 mM KH_2PO_4, 8.1 mM Na_2HPO_4, pH 7.6.

2. 0.1 mg/mL Pronase (Sigma-Aldrich, St Louis, MO, USA).

3. 0.025%Trypsin (Invitrogen Inc, Carlsbad, CA, USA).

4. 12.5 μg/mL DNAseI (Roche Diagnostics, Indianapolis, IN, USA).

5. Ham's F-12 and DMEM (1:1) (Invitrogen Inc).

6. 10% Fetal bovine serum (Invitrogen Inc).

7. 2 mM Glutamine (Invitrogen Inc).

8. 50 μg/mL Gentamycin (Sigma-Aldrich).

9. 1 μg/mL Insulin (Sigma-Aldrich).

10. 5 μg/mL Transferrin (Sigma-Aldrich).

11. 5 ng/mL Sodium selenite (Sigma-Aldrich).

12. 10 ng/mL Epidermal growth factor (Sigma-Aldrich).

13. 2 μg/mL Hydrocortisone (Sigma-Aldrich).

14. 5 ng/mL Basic fibroblast growth factor (Invitrogen Inc).

15. Plastic tissue culture dishes (Falcon Laboratories, Colorado Springs, CO, USA).

16. Laminin (Boehringer-Ingelheim GmbH, Ingelheim, Germany).

17. Transwells (Corning, Lowell, MA, USA).

2.3. Dissection of CP

1. Donor Balb/c mice (4–6 week old).
2. 3–5% Isofluorane or suitable IACUC approved terminal anesthesia.
3. Surgical instruments for dissection: scalpel, scissors, and forceps.
4. Dissecting microscope.
5. Razor blade.

2.4. Explants of CP

1. Tissue culture 10 cm plastic dishes (Falcon).
2. Tissue culture 12, 48, and 96 well plastic dishes (Falcon).
3. Glass Pasteur pipette, modified for tissue harvesting (see Note 2).
4. Ligand-targeted phagemid (10^{10}–10^{11} pfu/mL).
5. Dissected choroid plexus from Subheading 2.3 (1 CP/animal/well).
6. RPMI medium (Invitrogen Inc) containing 5% normal horse serum (NHS, Invitrogen Inc) and 10% fetal calf serum (FCS, Sigma-Aldrich).

2.5. Biopanning Peptide Libraries for CP Targeting

1. Peptide library (10^{10}–10^{11} particles, New England Biolabs) in culture medium incubation buffer.

2.6. Visualization of Phage Internalization

1. Rabbit anti-fd bacteriophage antibody 1:700 (Sigma-Aldrich). Control M13KE phage (New England Biolabs, N0316S) and ligand-targeted M13KE phage that are identified by biopanning.
2. Blocking solution: 5% NGS and 1% BSA in PBS.
3. Alexa 594-labeled goat anti-rabbit antibody (Invitrogen Inc): 1:1000 dilution in PBS/BSA and 1.5% NGS.
4. Mounting media containing DAPI (Vector Laboratories, Burlingame, USA).
5. 0.3% PBS-Tween 20.
6. Absolute methanol.
7. Normal goat serum (Vector Laboratories): 5% in PBS.
8. Bovine serum albumin (Jackson Immunoresearch Laboratories Inc, Westgrove, USA): 1% in PBS.
9. Fixative solution: 2% paraformaldehyde containing 2% glucose in PBS, pH 7.4.
10. Fluorescence Microscope.

2.7. Establishing Targeting Specificity

1. Prepare synthetic peptide (10–100 µg) that corresponds to the ligand sequence.

2. 50 mM HCl, pH 2.

3. 50 mM Glycine buffer pH 2.8 and 500 mM NaCl.

4. Anti-receptor antibody, 1:1000 (see Note 3).

2.8. Evaluating Intracerebroventricular Targeting In Vivo

1. Stereotaxic frame.

2. 50 uL Hamilton syringe.

3. Tissue-Tek OCT Solution (Sakura-America, Torrance, CA, USA).

4. Cryostat.

3. Methods

3.1. Peptide Libraries for Biopanning

1. Peptide libraries are handled as described by the manufacturer (see Note 1).

2. Aliquots containing 10^{11}–10^{12} pfu/mL are added to target CP cells in culture or CP explants obtained as described in Subheading 3.3.

3.2. In Vitro Biopanning Using Cultured Rat CP Cells

1. Dissect the CP under conventional light microscopy (see Note 4).

2. Rinse the tissue twice in Ca^{2+} and Mg^{2+} free PBS.

3. Digest 25 min with 0.1 mg/mL pronase at 37°C.

4. Recover predigested tissue by sedimentation.

5. Shake briefly in 0.025% trypsin containing 12.5 μg/mL DNAse-I.

6. Remove supernatant and keep on ice in 10% fetal bovine serum.

7. Repeat digestion five times.

8. Pellet cells by centrifugation.

9. Resuspend in full Ham's culture medium.

10. Incubate resuspended cells on plastic dishes for 2 h at 37°C.

11. Collect supernatant containing unattached cells.

12. Place medium for seeding on laminin-coated transwells.

13. Perform experiments after 7 days on confluent cell monolayers (see Note 5).

3.3. Ex Vivo Biopanning Using Explants of Mouse or Rat CP

1. Terminal anesthesia for 4–6 week mice (or rats) is performed using methods approved by the Institutional Animal Care and Use Committee of the University of California (IACUC) (see Note 6).

2. Dissect brains from the skull (see Note 4).

3. Harvest CP from the fourth ventricle (see Note 7) and lateral ventricles (see Note 8).

4. Place fragments of CP in 1 mL of RPMI media containing 5% NHS and 10% FCS.

3.4. Collection of CSF

1. Immobilize a deeply anaesthetized rat in a stereotaxic frame.

2. Raise the head above the body so that the head is elevated but parallel to the table top.

3. Make a central incision overlying the top of the skull to the base of the neck.

4. Expose the posterior aspect of the skull using blunt ended scissors.

5. Retract muscle and fascia.

6. Observe landmarks of bone (white and hard), separation of bony area (white line), and the presence of a 2–5 mm circular yellow/white membrane at their intersection.

7. Perforate membrane with a 25–100 uL Hamilton syringe.

8. Withdraw CSF slowly and place in a cryovial and freeze.

3.5. CP Dissection After In Vivo Biopanning

1. Perform terminal anesthesia on mice using methods approved by the IACUC (see Note 6).

2. Dissect brain from the skull (see Note 4).

3. Harvest choroid plexuses from the fourth ventricle (see Note 7) and lateral ventricles (see Note 8).

3.6. Biopanning, Recovery, and Amplification of Targeted Phage

1. Dilute phage to desired concentration (10^9–10^{12} particles/mL) in culture media.

2. Add particles to target cells or explants (see Note 9).

3. Incubate cells (or explants) for 2 h at 37°C in a CO_2 tissue culture incubator.

4. Wash cells (or explants) 3–5 times with PBS containing Ca^{+2} and Mg^{+2}.

5. Incubate cells (or explants) with 50 mM glycine buffer pH 2.8 containing 0.5 M NaCl for 5 min at room temperature and process for immunostaining.

3.7. Detection of Internalized Phage (see Note 10)

1. Transfer CP fragments to individual wells of a 96 well plate containing 100 uL RPMI with 10% FCS and 5% NHS.

2. Add phage (purified or library) in 10 μL to each well.

3. Incubate for 2 h in 37°C incubator under 5% CO_2.

4. Add 100–200 μL PBS with Ca^{+2} and Mg^{+2} to each well (see Note 11).

5. Using a clean glass scoop (see Note 2), transfer CP to the PBS wells of a previously prepared 48 well plate containing 1 mL of PBS (see Note 12).

6. Move one piece at a time and when all tissues are moved, transfer them to the next wash step (<1 min/step).

7. Repeat cycle through PBS two changes, PBS/Tween three changes, PBS three changes, and fix tissue in the fixative solution at room temperature for 20 min.

8. Wash once with PBS and transfer to methanol for permeabilization for 10 min.

9. Wash twice with PBS.

10. Transfer to a blocking solution of 5% NGS/1% BSA in PBS and incubate for 20 min at room temperature.

11. Transfer to rabbit anti-M13 (fd) for 1 h at room temperature.

12. Wash twice with PBS and transfer to Alexa 594 labeled goat anti-rabbit antibody for 45 min at room temperature.

13. Wash twice with PBS, and PBS-DAPI prepared by adding one drop of DAPI to mounting medium in 100 uL PBS.

14. Mount tissue on slide with a fine brush and place a coverslip over the tissue.

15. Evaluate immunostaining by fluorescence microscopy.

3.8. Evaluation of Targeting Specificity

1. Prepare synthetic peptide (10–100 μg) that corresponds to the sequence identified by phage display in 50 mM HCl, pH 2.

2. Incubate with cells (or target).

3. Add phage (10^{10}) to target.

4. Incubate as given in Subheading 3.6, step 3.

5. Process for internalization.

3.9. Evaluating Intracerebro- ventricular Targeting In Vivo

1. Immobilize the head of the deeply anesthetized rat (200–250 gm) in stereotaxic apparatus.

2. Make a midline incision in the skin overlying the skull and identify the bregma on the skull surface.

3. Using a microdrill, perform a craniotomy at the following coordinates: 1 mm posterior to the bregma and 1.5 mm lateral to the midline.

4. Insert the needle attached to a glass Hamilton syringe through the craniotomy into the brain to a depth of 4 mm and inject 20 μL of the phage solution into the lateral ventricle (see Note 9).

5. Leave the needle in place for 1–2 min to avoid reflux of the injection fluid.

6. At specific times after injection (24–72 h), kill deeply anesthetized the animals using techniques approved by local Institutional Animal Care Committee such as overdose or CO_2 inhalation.

7. Perfuse the animals with 4% paraformaldehyde in PBS, pH 7.4 via a cannula in the left ventricle.

8. Dissect the brain out and postfix overnight at 4°C in 4% paraformaldehyde in PBS.

9. Place brain in 30% sucrose overnight.

10. Embed in OCT compound and store at –80°C.

11. Mount 12 μm-thick cryostat sections onto positively charged slides.

12. Perform immunostaining.

3.10. Evaluating Targeting In Vivo

1. Inject 100 μL of phage into a tail vein or carotid artery (see Note 13).

2. Harvest CP as described above in Subheading 2.3.

3. Perform immunostaining.

4. Notes

1. Although an EGF-targeted phage made in our laboratory (8, 30) is described in this chapter, the availability of commercial peptide display libraries from companies such as New England Biolabs puts phage display into the hands of any laboratory with basic equipment for molecular biology. Vectors obtained from these suppliers can be used to construct personalized libraries with oligonucleotide sequences of different length, antibodies or even cDNA libraries. These methods are described extensively in several reviews (2). New England Biolabs offers three premade random peptide libraries, as well as the cloning vector M13KE for construction of custom libraries. The premade libraries consist of linear heptapeptide (Ph.D.-7) and dodecapeptide (Ph.D.-12) libraries, as well as a disulfide-constrained heptapeptide (Ph.D.-C7C) library. The randomized segment of the Ph.D.-C7C library is flanked by a pair of cysteine residues that are oxidized during phage assembly to a disulfide linkage, resulting in the displayed peptides being presented as loops. All of the libraries have complexities in excess of two billion independent clones. The randomized peptide sequences in all three libraries are

expressed at the N-terminus of the minor coat protein pIII, resulting in a valency of five copies of the displayed peptide per virion. In both the Ph.D.-7 and the Ph.D.-12 libraries, the first residue of the peptide-pIII fusion is the first randomized position, while the first randomized position in the Ph.D.-C7C library is preceded by Ala-Cys. All of the libraries contain a short linker sequence (Gly-Gly-Gly-Ser) between the displayed peptide and the pIII protein sequence.

2. Glass Pasteur pipettes are modified for tissue harvesting by heating the end to form a loop-like structure that can be used to pull explanted tissue out of cell culture plates without damaging the tissue. Any similar device (e.g., small brush) is acceptable, but the use of forceps is discouraged because of tissue damage that leads to false positive staining. To avoid cross contamination, it is imperative that a different collection device be used for each sample. This is why the modified Pasteur pipette is the cheapest, most reliable alternative that does not cause damage to tissue.

3. Peptide specificity can be evaluated in many ways. The most effective is to preincubate the explants or cells in culture for 10–20 min at 4°C; however, ligand can also be added at the same time as the ligand-phage and incubated together with the cell targets at 37°C for 2 h. For the ex vivo studies, for example, the explants can also be incubated with the ligand-peptides for 40 min at 37°C under 5% CO_2 before beginning the incubation with the ligand-targeted phage. The approach is ligand dependant and should be changed according to the best approximation of the ligand–receptor interaction. If the targeting ligand's receptor is known (e.g., EGF targeting through EGF receptor), then antibodies to the receptor can be preincubated with cells. In either case, the signal detected by immunostaining will be decreased in co-incubated samples indicating that internalization is specific and receptor mediated.

4. It is critical that the cerebellum be removed from the skull intact if the choroid plexus needs to be recovered from the fourth ventricle. To this end, the animal is killed with an overdose of anesthetic and the brain is carefully dissected out from the skull by making a midline incision in the skull, with care not to damage the cerebellum. A second incision is made in a coronal plane at the level of the orbit. With small forceps, the brain is separated from the skull, the optic and trigeminal nerves are cut and the brain placed in a petri dish containing cold PBS.

5. The protocol for cell culture of rat CP derives from published methods (48) optimized for neonatal CP dissected from 3–6-day-old pups. However, cell lines (52) are also compatible with the experimental approach of ligand targeting. Once established, the upper chamber of cultured epithelial cells

represents the fluid in contact with the apical side of epithelial cells (the CSF side), while the lower chamber represents the basolateral side (the blood side).

6. Isofluorane or halothane overdose appears suitable for CP dissection and yields choroidal tissues compatible for analysis. Nontraumatic sacrifice prevents hemorrhage in the brain upon decapitation and dissection although the effects of anesthesia on CP function deserve consideration. Direct decapitation is discouraged because of the hemorrhage in the brain and blood in the microvasculature of the CP.

7. The cerebellum is carefully separated from the rest of the brain. The anterior portion of the cerebellum is placed facing down on a Petri dish (Fig. 3a). Under a stereomicroscope and with the aid of two pairs of tweezers, the cerebellum is carefully lifted and separated from the brain stem to visualize the fourth ventricle. The CP, resembling a "Y" attached to the roof of the cerebellum, can then be easily dislodged with tweezers. The CP is placed in RPMI media containing 10% FCS and 5% NHS.

8. The lateral ventricle CP is dissected after removal of the fourth ventricle CP. The brain is immersed in PBS, and two parallel sagittal incisions are made with a scalpel 10 mm from the midline along the length of the brain and to a depth of 4 mm in such a way that it cuts through the corpus callosum

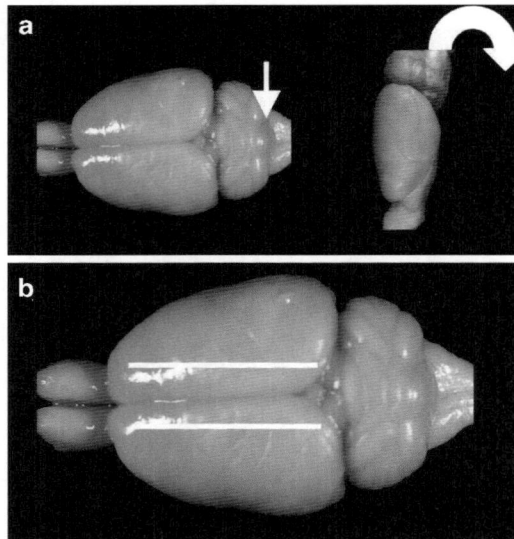

Fig. 3. (**a**, **b**) Dissection of brain for choroid plexus sampling: Three cuts with a flat edged razor blade enable dissection of the CP from the brain. While it is possible to accomplish this on the bench top, a dissecting microscope is highly recommended.

(Fig. 3b). The cortex is then pulled away to the side exposing the lateral ventricles and choroid plexuses. With a pair of tweezers, each end of the CP is gently pulled away. The CP is then placed in RPMI media containing 10% FCS and 5% NHS.

9. In order to evaluate whether the peptides identified by bio-panning epithelial cells in culture or explants are active (and specific) *in vivo*, it is important that they be tested by inject-ing peptide-targeted phage i.c.v. and subsequently examining whether they can enter the CP epithelium, ependyma, and even brain parenchyma. Immunohistochemical staining for M13 coat protein is used to detect the distribution of parti-cles throughout the brain.

10. There are several ways to monitor internalization of phage. The technique described here is immunofluorescence. Because the display peptide ligand enables internalization of the par-ticles into endosomes of the target cells, the major coat pro-tein (pVIII) can be visualized in permeabilized cells for up to 24 h after entry into cell. Later, the coat protein is found to be degraded so that immunostaining is not possible. In con-trast, the internalized phage DNA appears to have lived much longer than the coat protein and it can be used for recovery and identification of targeting agents. We have found phage DNA in cells 3 months after internalization *in vitro*.

11. The PBS is added to wells to make it easier to collect the dis-sected tissue from the well and transfer it to the next treatment.

12. It is recommended to pre-prepare a template for washes depending on the size of the experiment (Fig. 4). In the example given here, a 48 well plate is prepared with the washes and incubations that are necessary for two treatments.

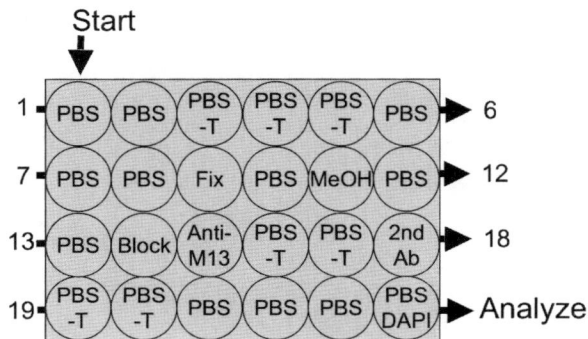

Fig. 4. Template for immunostaining. A 24 well plate is prepared prior to immunostaining using a template like the one shown here. Tissues are carefully transferred from one well to the next rather than washing in a single well and risking losing tissue.

Identical treatments can be combined for each of the steps without compromising data quality. It is most critical not to damage tissue during transfer.

13. Because the biopanning protocols described here identify epithelial targeting *in vitro* and ex vivo, they will be suitable for systemic drug delivery if they can translocate across the permeable choroidal endothelium (likely) and enter epithelial cells from the basolateral (blood) side of the barrier. Particles (10^{12} in 100 µL PBS) are injected intra-arterially and CSF collected for PCR analyses of phage DNA.

Acknowledgments

This work was supported in part by the National Institutes of Health (USA) and the Biochemistry and Biotechnology Research Council (UK). The authors would like to thank Drs. Paul Kassner and David Larocca who first engineered the initial EGF-displayed phage that helped establish the feasibility of CP targeting, Dr. Michael Burg who helped identify CP-targeting peptides in peptide libraries, and Ms. Emelie Amburn and Dr. Karen Sims who assisted in their characterization.

References

1. Smith GP (1985) Filamentous fusion phage: novel expression vectors that display cloned antigens on the virion surface. Science 228:1315–1317

2. Barbas CF (2001) Phage display: a laboratory manual. Cold Spring Harbor: Cold Spring Harbor Laboratory

3. Pasqualini R, Arap W, McDonald DM (2002) Probing the structural and molecular diversity of tumor vasculature. Trends Mol Med 8:563–571

4. Larocca D, Baird A (2001) Receptor-mediated gene transfer by phage-display vectors: applications in functional genomics and gene therapy. Drug Discov Today 6:793–801

5. Larocca D, Burg MA, Jensen-Pergakes K, Ravey EP, Gonzalez AM, Baird A (2002) Evolving phage vectors for cell targeted gene delivery. Curr Pharm Biotechnol 3:45–57

6. Larocca D, Jensen-Pergakes K, Burg MA, Baird A (2001) Receptor-targeted gene delivery using multivalent phagemid particles. Mol Ther 3:476–484

7. Larocca D, Jensen-Pergakes K, Burg MA, Baird A (2002) Gene transfer using targeted filamentous bacteriophage. Methods Mol Biol 185:393–401

8. Larocca D, Kassner PD, Witte A, Ladner RC, Pierce GF, Baird A (1999) Gene transfer to mammalian cells using genetically targeted filamentous bacteriophage. Faseb J 13:727–734

9. Larocca D, Witte A, Johnson W, Pierce GF, Baird A (1998) Targeting bacteriophage to mammalian cell surface receptors for gene delivery. Hum Gene Ther 9:2393–2399

10. Hart SL, Knight AM, Harbottle RP, Mistry A, Hunger HD, Cutler DF, Williamson R, Coutelle C (1994) Cell binding and internalization by filamentous phage displaying a cyclic Arg-Gly-Asp-containing peptide. J Biol Chem 269:12468–12474

11. Ivanenkov VV, Menon AG (2000) Peptide-mediated transcytosis of phage display vectors in MDCK cells. Biochem Biophys Res Commun 276:251–257

12. Merril CR, Biswas B, Carlton R, Jensen NC, Creed GJ, Zullo S, Adhya S (1996) Long-circulating bacteriophage as antibacterial agents. Proc Natl Acad Sci U S A 93:3188–3192

13. Muruganandam A, Tanha J, Narang S, Stanimirovic D (2002) Selection of phage-displayed llama single-domain antibodies that transmigrate across human blood-brain barrier endothelium. Faseb J 16:240–242

14. Pasqualini R (1999) Vascular targeting with phage peptide libraries. Q J Nucl Med 43:159–162

15. Pasqualini R, Ruoslahti E (1996) Organ targeting in vivo using phage display peptide libraries. Nature 380:364–366

16. Pasqualini R, Ruoslahti E (1996) Searching for a molecular address in the brain. Mol Psychiatry 1:421–422

17. Pasqualini R, Ruoslahti E (1996) Tissue targeting with phage peptide libraries. Mol Psychiatry 1:423

18. Petrenko VA, Smith GP (2000) Phage from landscape libraries as substitute antibodies. Protein Eng 13:589–592

19. Petrenko VA, Smith GP, Gong X, Quinn T (1996) A library of organic landscapes on filamentous phage. Protein Eng 9:797–801

20. Poul MA, Marks JD (1999) Targeted gene delivery to mammalian cells by filamentous bacteriophage. J Mol Biol 288:203–211

21. Rajotte D, Arap W, Hagedorn M, Koivunen E, Pasqualini R, Ruoslahti E (1998) Molecular heterogeneity of the vascular endothelium revealed by in vivo phage display. J Clin Invest 102:430–437

22. Sokoloff AV, Bock I, Zhang G, Sebestyen MG, Wolff JA (2000) The interactions of peptides with the innate immune system studied with use of T7 phage peptide display. Mol Ther 2:131–139

23. Yip YL, Hawkins NJ, Smith G, Ward RL (1999) Biodistribution of filamentous phage-Fab in nude mice. J Immunol Methods 225:171–178

24. Koivunen E, Arap W, Valtanen H, Rainisalo A, Medina OP, Heikkila P, Kantor C, Gahmberg CG, Salo T, Konttinen YT, Sorsa T, Ruoslahti E, Pasqualini R (1999) Tumor targeting with a selective gelatinase inhibitor. Nat Biotechnol 17:768–774

25. Koivunen E, Restel BH, Rajotte D, Lahdenranta J, Hagedorn M, Arap W, Pasqualini R (1999) Integrin-binding peptides derived from phage display libraries. Methods Mol Biol 129:3–17

26. Trepel M, Arap W, Pasqualini R (2000) Exploring vascular heterogeneity for gene therapy targeting. Gene Ther 7:2059–2060

27. Trepel M, Grifman M, Weitzman MD, Pasqualini R (2000) Molecular adaptors for vascular-targeted adenoviral gene delivery. Hum Gene Ther 11:1971–1981

28. Burg MA, Jensen-Pergakes K, Gonzalez AM, Ravey P, Baird A, Larocca D (2002) Enhanced phagemid particle gene transfer in camptothecin-treated carcinoma cells. Cancer Res 62:977–981

29. Burg M, Ravey EP, Gonzales M, Amburn E, Faix PH, Baird A, Larocca D (2004) Selection of internalizing ligand-display phage using rolling circle amplification for phage recovery. DNA Cell Biol 23:457–462

30. Kassner PD, Burg MA, Baird A, Larocca D (1999) Genetic selection of phage engineered for receptor-mediated gene transfer to mammalian cells. Biochem Biophys Res Commun 264:921–928

31. Chodobski A, Szmydynger-Chodobska J (2001) Choroid plexus: target for polypeptides and site of their synthesis. Microsc Res Tech 52:65–82

32. Emerich DF, Vasconcellos AV, Elliott RB, Skinner SJ, Borlongan CV (2004) The choroid plexus: function, pathology and therapeutic potential of its transplantation. Expert Opin Biol Ther 4:1191–1201

33. Ghersi-Egea JF, Strazielle N (2002) Choroid plexus transporters for drugs and other xenobiotics. J Drug Target 10:353–357

34. Johanson CE, Duncan JA, Stopa EG, Baird A (2005) Enhanced prospects for drug delivery and brain targeting by the choroid plexus-CSF route. Pharm Res 22:1011–1037

35. Silverberg GD, Huhn S, Jaffe RA, Chang SD, Saul T, Heit G, Von Essen A, Rubenstein E (2002) Downregulation of cerebrospinal fluid production in patients with chronic hydrocephalus. J Neurosurg 97:1271–1275

36. Spector R, Johanson C (2006) Micronutrient and urate transport in choroid plexus and kidney: implications for drug therapy. Pharm Res 23:2515–2524

37. Carro E, Spuch C, Trejo JL, Antequera D, Torres-Aleman I (2005) Choroid plexus megalin is involved in neuroprotection by serum insulin-like growth factor I. J Neurosci 25:10884–10893

38. Hakvoort A, Haselbach M, Galla HJ (1998) Active transport properties of porcine choroid plexus cells in culture. Brain Res 795:247–256

39. Johanson CE, Gonzalez AM, Stopa EG (2001) Water-imbalance-induced expression of FGF-2 in fluid-regulatory centers: choroid plexus and neurohypophysis. Eur J Pediatr Surg 11 Suppl 1:S37–38

40. Liao CW, Fan CK, Kao TC, Ji DD, Su KE, Lin YH, Cho WL (2008) Brain injury-associated biomarkers of TGF-beta1, S100B, GFAP, NF-L, tTG, AbetaPP, and tau were concomitantly enhanced and the UPS was impaired during acute brain injury caused by Toxocara canis in mice. BMC Infect Dis 8:84

41. Parandoosh Z, Johanson CE (1982) Ontogeny of blood-brain barrier permeability to, and cerebrospinal fluid sink action on, [14C] urea. Am J Physiol 243:R400–407

42. Schreiber G (2002) The evolution of transthyretin synthesis in the choroid plexus. Clin Chem Lab Med 40:1200–1210

43. Borlongan CV, Thanos CG, Skinner SJ, Geaney M, Emerich DF (2008) Transplants of encapsulated rat choroid plexus cells exert neuroprotection in a rodent model of Huntington's disease. Cell Transplant 16:987–992

44. Emerich DF, Skinner SJ, Borlongan CV, Thanos CG (2005) A role of the choroid plexus in transplantation therapy. Cell Transplant 14:715–725

45. Johanson CE, Szmydynger-Chodobska J, Chodobski A, Baird A, McMillan P, Stopa EG (1999) Altered formation and bulk absorption of cerebrospinal fluid in FGF-2-induced hydrocephalus. Am J Physiol 277:R263–271

46. Logan A, Frautschy SA, Gonzalez AM, Sporn MB, Baird A (1992) Enhanced expression of transforming growth factor beta 1 in the rat brain after a localized cerebral injury. Brain Res 587:216–225

47. Stopa EG, Berzin TM, Kim S, Song P, Kuo-LeBlanc V, Rodriguez-Wolf M, Baird A, Johanson CE (2001) Human choroid plexus growth factors: What are the implications for CSF dynamics in Alzheimer's disease? Exp Neurol 167:40–47

48. Strazielle N, Ghersi-Egea JF (1999) Demonstration of a coupled metabolism-efflux process at the choroid plexus as a mechanism of brain protection toward xenobiotics. J Neurosci 19:6275–6289

49. Vercellino M, Votta B, Condello C, Piacentino C, Romagnolo A, Merola A, Capello E, Mancardi GL, Mutani R, Giordana MT, Cavalla P (2008) Involvement of the choroid plexus in multiple sclerosis autoimmune inflammation: a neuropathological study. J Neuroimmunol 199:133–141

50. Saggio I, Gloaguen I, Laufer R (1995) Functional phage display of ciliary neurotrophic factor. Gene 152:35–39

51. Buchli PJ, Wu Z, Ciardelli TL (1997) The functional display of interleukin-2 on filamentous phage. Arch Biochem Biophys 339:79–84

52. Hosoya K, Hori S, Ohtsuki S, Terasaki T (2004) A new in vitro model for blood-cerebrospinal fluid barrier transport studies: an immortalized choroid plexus epithelial cell line derived from the tsA58 SV40 large T-antigen gene transgenic rat. Adv Drug Deliv Rev 56:1875–1885

Sukriti Nag (ed.), *The Blood-Brain and Other Neural Barriers: Reviews and Protocols*, Methods in Molecular Biology, vol. 686,
DOI 10.1007/978-1-60761-938-3, © Springer Science+Business Media, LLC 2011